S0-AQC-468

Introductory Algebra: A Worktext

Introductory Algebra: A Worktext

Charles D. Miller / Margaret L. Lial

American River College Sacramento, California

Scott, Foresman and Company Glenview, Illinois

Dallas, Tex. Oakland, N.J. Palo Alto, Cal. Tucker, Ga. London

Library of Congress Cataloging in Publication Data

Miller, Charles David, 1942–
Introductory algebra.

Includes index.
1. Algebra. I. Lial, Margaret L., joint author.
II. Title.
QA152.2.M54 512.9 78-27210
ISBN 0-673-15217-0
AMS 1970 Subject Classification 98A10

Printed in the United States of America.

1 2 3 4 5 6-STR-86 85 84 83 82 81 80 79 78

Preface

Introductory Algebra: A Worktext is for a course in basic algebra for college students. The elements of arithmetic are covered in the first two sections, which review the key ideas of fractions and decimals, with other ideas reviewed later as needed. In our experience, students often start this course worried about their background in fractions. Because of this, we have been careful to include all the material on fractions that is needed in algebra.

This book has been extensively class-tested with a large variety of remedial algebra students. Based on this class testing and the comments of reviewers, the book was designed with the following features:

- **Objectives** Each section begins with a list of skills the students will learn in that section.
- **Quick Checks** A student never has to read more than a few lines before coming to an instruction to work a Quick Check problem in the margin of the page. Thus, students do not have to go far before they get an opportunity to test their understanding of a given topic. Any difficulties can be pinpointed immediately with the use of these Quick Checks.
- **Clear explanations** We spend several hours a day in our mathematics learning center. Therefore, any explanation that is unclear in the text itself will cause a good number of students to come to us for help. This practical experience enabled us to locate spots that needed improvement.
- **Word problems** Word problems cause more difficulties than almost any other aspect of algebra. To remedy this, we have introduced word problems very carefully. The student sees word problems very early in the text, and gradually gets used to them. We try to give extensive practice in translating from words to the symbols of algebra.
- **Testing program** In the book, chapter tests, a diagnostic pretest, and a sample final examination help students review for class tests. The Instructor's Guide contains five tests (one multiple choice) for each chapter and two sample final examinations. A diagnostic pretest in the Guide can be used to identify students who need the course.
- **Audio tapes** A complete set of tapes to accompany this book has been developed; information regarding the tapes is available from the publisher. We have placed sets of tapes in our school library, as well as in the mathematics learning center. Students needing help with a particular topic, or students who have missed class, have benefited greatly from the use of these tapes. Students say that the tapes are "believable"—they sound like an instructor and not a professional announcer.

This book is suitable for a traditional lecture class, a self-paced class, or in small-group situations. The package of supplementary items, including audio tapes and a complete Instructor's Guide, makes the book very flexible. In addition to the tests described above, the Instructor's Guide has the answers not included in the book, a complete explanation of how to teach a self-paced class, and solutions for the Quick Check problems. These solutions may be reproduced and given to students if desired.

Many people helped us prepare this book. In particular, Vern Heeren of American River College helped with the class testing. Ronald Rose and Michael Karelius of American River College helped insure that the book is indeed suitable for self-paced classes. Other help came from Richard Spangler,

Tacoma Community College; Michelle Harrell, Miami-Dade Community College; Judith Hall, West Virginia University; Carla Hager, Metropolitan State College; Ray Lewin, College of South Idaho; Jerry Hattaway, Bainbridge Junior College; Max Gerling, Tallahassee Community College; Adele LeGere, Oakton Community College; and Lynn Hensel, Henry Ford Community College. Matthew Pittner of American River College helped prepare the index. At Scott, Foresman we received valuable assistance from editors Nancy Liskar, Pamela Carlson, and Robert Runck. To all these people we offer our thanks.

Charles D. Miller
Margaret L. Lial

Contents

name date hour

DIAGNOSTIC PRETEST

The questions in this pretest are given later in the text as worked out examples. For help on any problem, see the indicated page.

1. Find the product of 3/8 and 4/9 and write the answer in lowest terms. (page 2)
2. Divide 1852.882 by 73.82. (page 13)
3. Find the value of $\frac{4(5+3)+3}{2(3)-1}$. (page 21)
4. Simplify $-|-8|$. (page 38)
5. Add $-4.6 + 8.1$. (page 43)
6. Subtract $8 - (-5)$. (page 49)
7. Name the property which tells why $-8 + 5 = 5 + (-8)$. (page 69)
8. Name the property which tells why $4(x + 5) = 4x + 20$. (page 72)
9. Simplify the equation $(5 + x) + (-2) = (-3) + [11 + (-1)]$. (page 79)
10. Solve $4r + 5r - 3 + 8 - 3r - 5r = 12 + 8$. (page 86)
11. Solve $\frac{3}{4}h = 6$. (page 90)
12. Solve $4(k - 3) - k = k - 6$. (page 96)
13. Solve $A = lw$ for w. (page 114)
14. Solve $6 + 3y \geqslant 4y - 5$. (page 122)
15. Solve $-4t \geqslant 8$. (page 128)
16. Evaluate 5^4. (page 141)
17. Simplify $5^8/5^3$. (page 143)
18. Subtract the polynomial $6x^3 - 4x^2 + 2$ from the polynomial $11x^3 + 2x^2 - 8$. (page 158)
19. Find $(3b + 5r)^2$. (page 169)
20. Divide $4x^3 - 4x^2 + 5x - 8$ by $2x - 1$. (page 178)
21. Evaluate 5^{-3}. (page 183)
22. Write 6.2×10^3 without exponents. (page 187)
23. Factor out the greatest common factor: $48y^{12} - 36y^{10} + 12y^7$. (page 195)
24. Factor $m^2 + 9m + 14$. (page 200)
25. Factor $3y^2 + 14y + 8$. (page 205)
26. Factor $49z^2 - 64$. (page 211)
27. Solve the equation $x^2 - 5x = -6$. (page 215)
28. Reduce $14k^2/2k^3$ to lowest terms. (page 238)

29. Multiply $\frac{6}{x} \cdot \frac{x^2}{12}$. (page 243)

30. Add $\frac{2}{3y} + \frac{1}{4y}$. (page 249)

31. Solve $\frac{x}{3} + \frac{x}{4} = 10 + x$. (page 255)

32. Complete the following ordered pairs for the equation $y = 4x + 5$: (7,); (−9,). (pages 275–76)

33. Graph the following ordered pairs: (1, 5); (−2, 3); (7, −2). (page 281)

34. Graph $3x + 2y = 6$. (page 288)

35. Graph $x = 3$. (page 290)

36. Graph $x - y > 5$. (page 297)

37. Solve the system $\begin{aligned} 2x + 3y &= 4 \\ 3x - y &= -5 \end{aligned}$ by graphing. (page 308)

38. Solve the system $\begin{aligned} x + y &= 5 \\ x - y &= 3 \end{aligned}$ by addition. (page 315)

39. Solve the system $\begin{aligned} 2x + 4y &= 5. \\ 4x + 8y &= -9 \end{aligned}$ (page 321)

40. Solve the system $\begin{aligned} 3x + 5y &= 26 \\ y &= 2x \end{aligned}$ by substitution. (page 325)

41. The sum of two numbers is 63. Their difference is 19. Find the numbers. (page 331)

42. Simplify $\sqrt{20}$. (page 353)

43. Simplify $3\sqrt{2} + \sqrt{8}$. (page 361)

44. Rationalize the denominator of $9/\sqrt{6}$. (page 365)

45. Simplify the product $(4 - \sqrt{3})(4 + \sqrt{3})$. (page 370)

46. Solve the equation $\sqrt{x + 1} = 3$. (page 375)

47. Solve the equation $(x - 1)^2 = 6$. (page 384)

48. Solve the equation $2x^2 - 7x - 9 = 0$ using the quadratic formula. (page 395)

49. Solve $x^2 = 2x + 1$. (page 395)

50. Graph $y = -x^2$. (page 402)

Introductory Algebra: A Worktext

1 Number Systems

1.1 FRACTIONS

OBJECTIVES

- Multiply fractions
- Reduce fractions to lowest terms
- Divide fractions
- Write a fraction as an equal fraction having a given denominator
- Add fractions
- Subtract fractions

In many ways, algebra is a sort of advanced arithmetic. We spend a lot of time in algebra adding, subtracting, multiplying, and dividing, but we do these things with *letters,* such as x or y, instead of numbers, as in arithmetic.

To prepare for our study of algebra, we begin this chapter with two sections offering a quick review of the fundamentals of arithmetic.

In everyday life, the numbers we most often see are the whole numbers, 0, 1, 2, 3, 4, 5, . . . , and the fractions, 1/2, 2/3, 11/12, and so on. $\left(1/2 \text{ is another way of saying } \frac{1}{2}.\right)$ In a fraction, the number on top is called the **numerator** and the number on the bottom is called the **denominator.**

A fraction is a **proper fraction** if the numerator is smaller than the denominator; otherwise it is an **improper fraction.** For example,

$$\frac{3}{4}, \quad \frac{7}{8}, \quad \frac{9}{10}, \quad \frac{125}{126} \quad \text{are proper fractions, and}$$

$$\frac{5}{4}, \quad \frac{17}{15}, \quad \frac{28}{3} \quad \text{are improper fractions.}$$

An improper fraction can also be expressed as a **mixed number,** which is a combination of a whole number and a proper fraction. For instance, 4/3, an improper fraction, can be written as 1 1/3, since 1 1/3 equals $3/3 + 1/3 = 4/3$.

A fraction is said to be **reduced to lowest terms** when both the numerator and the denominator cannot be divided by any number other than 1. To reduce a fraction to lowest terms, we find the largest number that will divide evenly into both the numerator and denominator. This number is called the **greatest common factor.**

Example 1 Reduce to lowest terms.

(a) $\dfrac{10}{15} = \dfrac{10 \div 5}{15 \div 5} = \dfrac{2}{3}$

Since the largest number that can be divided evenly into both 10 and 15 is 5, 5 is the greatest common factor of 10 and 15.

QUICK CHECKS

1. Reduce to lowest terms.

(a) $\frac{8}{14}$

(b) $\frac{35}{42}$

(c) $\frac{9}{18}$

(d) $\frac{10}{50}$

(e) $\frac{12}{20}$

2. Find each product and reduce to lowest terms.

(a) $\frac{5}{8} \cdot \frac{2}{10}$

(b) $\frac{3}{4} \cdot \frac{2}{3}$

(c) $\frac{1}{10} \cdot \frac{12}{5}$.

(d) $\frac{7}{9} \cdot \frac{12}{14}$

Quick Check Answers

1. (a) 4/7 (b) 5/6 (c) 1/2 (d) 1/5 (e) 3/5

2. (a) 1/8 (b) 1/2 (c) 6/25 (d) 2/3

(b) $\frac{9}{12} = \frac{9 \div 3}{12 \div 3} = \frac{3}{4}$

3 is the largest number that can be divided evenly into both 9 and 12. Therefore, 9/12 reduced to lowest terms equals 3/4.

(c) $\frac{15}{45} = \frac{15 \div 15}{45 \div 15} = \frac{1}{3}$

Both 15 and 45 can be divided by 3, 5, and 15. The largest of these three numbers, 15, is the greatest common factor, so we use it to divide both the numerator and denominator.

Work Quick Check 1 at the side.

In this section, we study the basic operations on fractions, namely addition, subtraction, multiplication, and division. The easiest operation for fractions is multiplication, so we begin with it.

To **multiply** two fractions, first multiply their numerators and then multiply their denominators. In symbols, if a/b and c/d are fractions, then

$$\frac{a}{b} \cdot \frac{c}{d} = \frac{a \cdot c}{b \cdot d}.$$

The dot (·) indicates multiplication. For example,

$$3 \cdot 5 = 15 \quad \text{and} \quad 7 \cdot 8 = 56.$$

When two numbers are multiplied, the answer is called the **product** of the two numbers.

Example 2 Find the product of 3/8 and 4/9 and express the answer in lowest terms.

First, multiply 3/8 and 4/9.

$$\frac{3}{8} \cdot \frac{4}{9} = \frac{3 \cdot 4}{8 \cdot 9} = \frac{12}{72}$$

Reduce the answer to lowest terms.

$$\frac{12}{72} = \frac{12 \div 12}{72 \div 12} = \frac{1}{6}$$

Work Quick Check 2 at the side.

To **divide** two fractions, we invert the second fraction and replace the division sign with a multiplication sign, as shown in the following definition.

$$\frac{a}{b} \div \frac{c}{d} = \frac{a}{b} \cdot \frac{d}{c}$$

The reason this works will be explained in a later chapter. The answer to a division problem is called a **quotient**. For example, the quotient of 20 and 10 is 2, since $20 \div 10 = 2$.

Example 3 Find the following quotients and reduce to lowest terms.

(a) $\frac{3}{4} \div \frac{8}{5} = \frac{3}{4} \cdot \frac{5}{8} = \frac{15}{32}$

(b) $\frac{3}{4} \div \frac{5}{8} = \frac{3}{4} \cdot \frac{8}{5} = \frac{3 \cdot 8}{4 \cdot 5} = \frac{24}{20} = \frac{6}{5}$

(c) $\frac{2}{30} \div \frac{6}{15} = \frac{2}{30} \cdot \frac{15}{6} = \frac{2 \cdot 15}{30 \cdot 6} = \frac{30}{180} = \frac{1}{6}$

Work Quick Check 3 at the side.

3. Find each quotient and reduce to lowest terms.

(a) $\frac{9}{10} \div \frac{3}{5}$

(b) $\frac{3}{4} \div \frac{9}{16}$

(c) $\frac{1}{2} \div \frac{1}{8}$

(d) $\frac{2}{3} \div 6$

To work addition and subtraction problems, we need to make sure that all the fractions in a problem have the same denominator. We can do this by rewriting the fraction with a new denominator. For example, suppose we need to rewrite 3/4 as a fraction with denominator 32.

$$\frac{3}{4} = \frac{?}{32}$$

In this case, we need to ask ourselves what number times 4 will equal 32? Since $4 \cdot 8 = 32$, 8 is the number we use to rewrite 3/4. We want the value of the original fraction, 3/4, to stay the same, so we must multiply both the numerator and the denominator by the *same* number, 8.

$$\frac{3}{4} = \frac{3 \cdot 8}{4 \cdot 8} = \frac{24}{32}$$

Example 4 Write each of the following as fractions having the indicated denominators.

(a) $\frac{5}{8} = \frac{?}{72}$

Since 8 divided into 72 is 9, we multiply the numerator and denominator by 9.

$$\frac{5}{8} = \frac{5 \cdot 9}{8 \cdot 9} = \frac{45}{72}$$

(b) $\frac{2}{3} = \frac{?}{18}$

Since 3 divided into 18 is 6, we can multiply the numerator and denominator by 6.

$$\frac{2}{3} = \frac{2 \cdot 6}{3 \cdot 6} = \frac{12}{18}$$

Work Quick Check 4 at the side.

4. Write as fractions having the indicated denominators.

(a) $\frac{9}{10} = \frac{}{40}$

(b) $\frac{4}{5} = \frac{}{60}$

(c) $\frac{1}{7} = \frac{}{49}$

(d) $\frac{3}{2} = \frac{}{8}$

Quick Check Answers

3. (a) 3/2 or 1 1/2 (b) 4/3 or 1 1/3 (c) 4 (d) 1/9

4. (a) 36 (b) 48 (c) 7 (d) 12

The **sum** of two fractions having the same denominator is found by adding the numerators. If a/b and c/b are fractions, then

$$\frac{a}{b} + \frac{c}{b} = \frac{a+c}{b}.$$

For example,

$$\frac{3}{7} + \frac{2}{7} = \frac{3+2}{7} = \frac{5}{7} \quad \text{and} \quad \frac{2}{10} + \frac{5}{10} = \frac{2+5}{10} = \frac{7}{10}.$$

Usually, the two fractions to be added do not have the same denominators. In this case, work the problem as shown in the next example.

Example 5 Add the pairs of fractions.

(a) $\frac{1}{2} + \frac{1}{3}$

We cannot add until the fractions have the same denominator. One number that we can use as a common denominator is 6, since both 2 and 3 divide into 6. Write both $1/2$ and $1/3$ as fractions with denominator 6.

$$\frac{1}{2} = \frac{1 \cdot 3}{2 \cdot 3} = \frac{3}{6} \quad \text{and} \quad \frac{1}{3} = \frac{1 \cdot 2}{3 \cdot 2} = \frac{2}{6}$$

Now we can add.

$$\frac{1}{2} + \frac{1}{3} = \frac{3}{6} + \frac{2}{6} = \frac{3+2}{6} = \frac{5}{6}$$

(b) $\frac{3}{10} + \frac{5}{12}$

A common denominator here is 60, since both 10 and 12 divide into 60.

$$\frac{3}{10} = \frac{3 \cdot 6}{10 \cdot 6} = \frac{18}{60} \quad \text{and} \quad \frac{5}{12} = \frac{5 \cdot 5}{12 \cdot 5} = \frac{25}{60}$$

Finally,

$$\frac{3}{10} + \frac{5}{12} = \frac{18}{60} + \frac{25}{60} = \frac{18+25}{60} = \frac{43}{60}.$$

(c) $3\frac{1}{2} + 2\frac{3}{4}$

Change both mixed numbers to improper fractions as follows.

$$3\frac{1}{2} = 3 + \frac{1}{2} = \frac{3}{1} + \frac{1}{2} = \frac{6}{2} + \frac{1}{2} = \frac{7}{2}$$

Also,

$$2\frac{3}{4} = 2 + \frac{3}{4} = \frac{8}{4} + \frac{3}{4} = \frac{8+3}{4} = \frac{11}{4}.$$

Then

$$3\frac{1}{2} + 2\frac{3}{4} = \frac{7}{2} + \frac{11}{4} = \frac{14}{4} + \frac{11}{4} = \frac{25}{4} \text{ or } 6\frac{1}{4}.$$

Work Quick Check 5 at the side.

Subtraction of fractions is very similar to addition. Just subtract the numerators instead of adding them, according to the following definition.

$$\frac{a}{b} - \frac{c}{b} = \frac{a-c}{b}$$

Example 6 Subtract the following pairs of fractions.

(a) $\frac{5}{8} - \frac{3}{8} = \frac{5-3}{8} = \frac{2}{8} = \frac{1}{4}$

(b) $\frac{3}{4} - \frac{1}{3}$

A common denominator is 12.

$$\frac{3}{4} - \frac{1}{3} = \frac{9}{12} - \frac{4}{12} = \frac{9-4}{12} = \frac{5}{12}$$

(c) $\frac{7}{9} - \frac{1}{6} = \frac{14}{18} - \frac{3}{18} = \frac{14-3}{18} = \frac{11}{18}$

In this case, 18 is a common denominator of both 9 and 6.

(d) $2\frac{1}{2} - 1\frac{3}{4}$

First, change the mixed numbers $2\frac{1}{2}$ and $1\frac{3}{4}$ into improper fractions.

$$2\frac{1}{2} = 2 + \frac{1}{2} = \frac{4}{2} + \frac{1}{2} = \frac{5}{2},$$

$$1\frac{3}{4} = 1 + \frac{3}{4} = \frac{4}{4} + \frac{3}{4} = \frac{7}{4}$$

Then

$$2\frac{1}{2} - 1\frac{3}{4} = \frac{5}{2} - \frac{7}{4}.$$

Multiply $\frac{5}{2}$ by $\frac{2}{2}$ to get a denominator of 4.

$$\frac{5}{2} \cdot \frac{2}{2} = \frac{10}{4}$$

Finally,

$$\frac{10}{4} - \frac{7}{4} = \frac{3}{4}.$$

Work Quick Check 6 at the side.

5. Add.

(a) $\frac{1}{2} + \frac{1}{4}$

(b) $\frac{2}{3} + \frac{1}{12}$

(c) $\frac{3}{4} + \frac{1}{6}$

(d) $\frac{7}{30} + \frac{2}{45}$

(e) $\frac{7}{10} + \frac{2}{15}$

(f) $3\frac{1}{5} + 2\frac{3}{10}$

6. Subtract.

(a) $\frac{9}{11} - \frac{3}{11}$

(b) $\frac{2}{3} - \frac{1}{2}$

(c) $\frac{3}{10} - \frac{1}{4}$

(d) $2\frac{3}{8} - 1\frac{1}{2}$

Quick Check Answers

5. (a) 3/4 (b) 3/4 (c) 11/12 (d) 5/18 (e) 5/6 (f) 5 1/2

6. (a) 6/11 (b) 1/6 (c) 1/20 (d) 7/8

WORK SPACE

name date hour

1.1 EXERCISES

Reduce each fraction to lowest terms. See Example 1.

1. $\frac{7}{14} =$ ________
2. $\frac{3}{9} =$ ________
3. $\frac{10}{12} =$ ________
4. $\frac{8}{10} =$ ________
5. $\frac{16}{18} =$ ________
6. $\frac{14}{20} =$ ________
7. $\frac{50}{75} =$ ________
8. $\frac{32}{48} =$ ________
9. $\frac{72}{108} =$ ________
10. $\frac{96}{120} =$ ________

Find the products or quotients. Reduce answers to lowest terms. See Examples 1, 2, and 3.

11. $\frac{3}{4} \cdot \frac{9}{5} =$ ________
12. $\frac{3}{8} \cdot \frac{2}{7} =$ ________
13. $\frac{1}{10} \cdot \frac{6}{5} =$ ________
14. $2 \cdot \frac{1}{3} =$ ________
15. $\frac{9}{4} \cdot \frac{8}{15} =$ ________
16. $\frac{3}{5} \cdot \frac{5}{3} =$ ________
17. $\frac{3}{8} \div \frac{5}{4} =$ ________
18. $\frac{9}{16} \div \frac{3}{8} =$ ________
19. $\frac{5}{12} \div \frac{15}{4} =$ ________
20. $\frac{15}{16} \div \frac{30}{8} =$ ________
21. $\frac{15}{32} \cdot \frac{8}{25} =$ ________
22. $\frac{24}{25} \cdot \frac{50}{3} =$ ________
23. $\frac{2}{3} \cdot \frac{5}{8} =$ ________
24. $\frac{2}{4} \cdot \frac{3}{9} =$ ________
25. $\frac{13}{2} \cdot \frac{2}{3} =$ ________
26. $\frac{9}{4} \cdot \frac{8}{7} =$ ________
27. $\frac{28}{3} \cdot \frac{6}{2} =$ ________
28. $\frac{121}{9} \cdot \frac{18}{11} =$ ________

Add or subtract. Reduce answers to lowest terms. See Examples 5 and 6.

29. $\frac{1}{12} + \frac{3}{12} =$ ________
30. $\frac{2}{3} + \frac{2}{3} =$ ________
31. $\frac{1}{10} + \frac{6}{10} =$ ________
32. $\frac{3}{4} + \frac{8}{4} =$ ________
33. $\frac{4}{9} + \frac{2}{3} =$ ________
34. $\frac{3}{5} + \frac{2}{15} =$ ________
35. $\frac{8}{11} + \frac{3}{22} =$ ________
36. $\frac{9}{10} - \frac{3}{5} =$ ________
37. $\frac{2}{3} - \frac{3}{5} =$ ________
38. $\frac{8}{12} - \frac{5}{9} =$ ________
39. $\frac{5}{6} - \frac{3}{10} =$ ________
40. $\frac{11}{4} - \frac{11}{8} =$ ________

41. $3\frac{1}{4} + 6\frac{1}{8} =$ ____________ 42. $5\frac{2}{3} + \frac{1}{4} =$ ____________ 43. $4\frac{1}{2} + \frac{2}{3} =$ ____________

44. $7\frac{5}{8} + 3\frac{3}{4} =$ ____________ 45. $6\frac{2}{3} - 5\frac{1}{4} =$ ____________ 46. $8\frac{8}{9} - 7\frac{4}{5} =$ ____________

47. $\frac{2}{5} + \frac{1}{3} + \frac{9}{10} =$ ____________ 48. $\frac{3}{8} + \frac{5}{6} + \frac{2}{3} =$ ____________ 49. $\frac{5}{7} + \frac{1}{4} - \frac{1}{2} =$ ____________

50. $\frac{2}{3} + \frac{1}{6} - \frac{1}{2} =$ ____________

Work each of the following word problems.

51. John Rizzo paid 1/8 of a debt in January, 1/3 in February, and 1/4 in March. What portion of the debt was paid in these three months? ____________

52. A rectangle is 5/16 yard on each of two sides, and 7/12 yard on each of the other two sides. Find the total distance around the rectangle. ____________

5/16

7/12

53. The Eastside Wholesale Market sold 3 1/4 tons of broccoli last month, 2 3/8 tons of spinach, 7 1/2 tons of corn, and 1 5/16 tons of turnips. Find the total number of tons of vegetables sold by the firm during the month. ____________

54. Sharkey's Casino decided to expand by buying a piece of property next to the casino. The property has an irregular shape, with five sides. The lengths of the five sides are 146 1/2 feet, 98 3/4 feet, 196 feet, 76 5/8 feet, and 100 7/8 feet. Find the total distance around the piece of property. ____________

55. Joann Kaufmann worked 40 hours during a certain week. She worked 8 1/4 hours on Monday, 6 3/8 hours on Tuesday, 7 2/3 hours on Wednesday, and 8 3/4 hours on Thursday. How many hours did she work on Friday? ____________

56. A concrete truck is loaded with 9 7/8 cubic yards of concrete. The driver gives out 1 1/2 cubic yards at the first stop and 2 3/4 cubic yards at the second stop. At a third stop, the customer needs 3 5/12 cubic yards. How much concrete is left in the truck? ____________

57. Rosario wants to make 16 dresses to sell at the company bazaar. Each dress needs 2 1/4 yards of material. How many yards should be bought? ____________

58. Lindsay allows 1 3/5 bottles of beverage for each guest at a party. If he expects 35 guests, how many bottles of beverage will he need? ____________

OBJECTIVES

- Write a number in expanded form
- Convert decimals to fractions and fractions to decimals
- Add and subtract decimals
- Multiply and divide decimals
- Convert percents to decimals and decimals to percents
- Find percentages by multiplication

A **decimal** is a number written with a decimal point, such as 4.2. Each digit in a decimal number has a place value, as shown below.

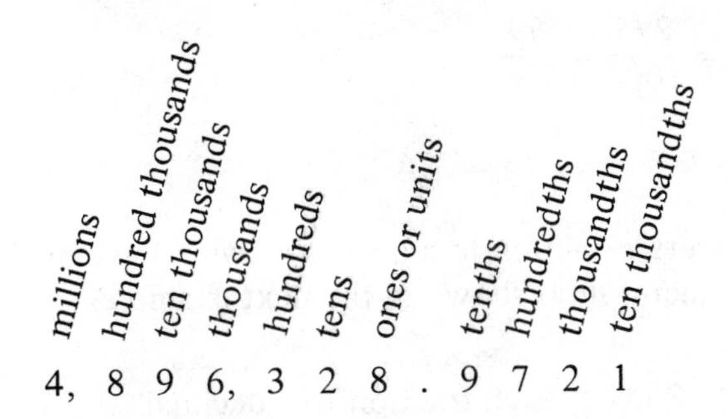

Notice that each successive place value is ten times larger than the place value to its right; each successive place value is one tenth as large as the place value to its left.

To find the place value for a number, write it in **expanded form**. For example,

$$1729 = 1000 + 700 + 20 + 9.$$

From this form of the number, we see that the digit 1 is in the thousands position, 7 is in the hundreds position, 2 is in the tens position, and 9 is in the ones or units position.

Example 1 Write the following numbers in expanded form.

(a) $98.42 = 90 + 8 + .4 + .02$

(b) $.618 = .6 + .01 + .008$

Work Quick Check 1 at the side.

QUICK CHECKS

1. Write in expanded form.

(a) 478

(b) 15,813

(c) 12.46

(d) .438

Using our example of an expanded number, we can see, for example, that 3.64 has a 3 in the ones position, a 6 in the tenths position (written as 6/10), and a 4 in the hundredths position (written as 4/100). With this information, we can then express the decimal number in the form of a fraction. That is,

$$3.64 = 3 + 6/10 + 4/100.$$

We multiply the numerator and denominator of 6/10 by 10 in order to get a common denominator of 100.

$$\begin{aligned} 3.64 &= 3 + (6/10 \cdot 10/10) + 4/100 \\ &= 3 + 60/100 + 4/100 \\ &= 3\ 64/100 \end{aligned}$$

Example 2 Convert each decimal to a fraction.

(a) .9

This decimal is the same as 9/10.

Quick Check Answers

1. (a) 400 + 70 + 8 (b) 10,000 + 5000 + 800 + 10 + 3 (c) 10 + 2 + .4 + .06 (d) .4 + .03 + .008

2. Convert to fractions.

(a) .5

(b) .80

(c) .15

(d) .305

(b) $.6 = \frac{6}{10} = \frac{3}{5}$

(c) $.25 = \frac{25}{100} = \frac{1}{4}$

(d) $.295 = \frac{295}{1000} = \frac{59}{200}$

Work Quick Check 2 at the side.

To convert a fraction to a decimal, divide the denominator into the numerator, as shown in the next example.

Example 3 Convert each fraction to a decimal.

(a) $\frac{1}{2}$

Divide 2 into 1. Annex 0's following the decimal point as needed.

$$2\overline{)1.0} = .5 \qquad \text{Decimal point comes straight up}$$

$$\begin{array}{r} .5 \\ 2\overline{)1.0} \\ 1\,0 \\ \hline 0 \end{array}$$

$$\frac{1}{2} = .5$$

(b) $\frac{4}{5}$

Divide 5 into 4.

$$\begin{array}{r} .8 \\ 5\overline{)4.0} \\ 4\,0 \\ \hline 0 \end{array}$$

$$\frac{4}{5} = .8$$

(c) $\frac{3}{8}$

$$\begin{array}{r} .375 \\ 8\overline{)3.000} \\ 2\,4 \\ \hline 60 \\ 56 \\ \hline 40 \\ 40 \\ \hline 0 \end{array}$$

$$\frac{3}{8} = .375$$

Quick Check Answers

2. (a) 1/2 (b) 4/5 (c) 3/20 (d) 61/200

(d) $\frac{2}{3}$

$$\begin{array}{r} .6666 \\ 3\overline{)2.000\ldots} \\ \underline{1\,8} \\ 20 \\ \underline{18} \\ 20 \\ \underline{18} \\ 20 \end{array}$$

The division here does not terminate. An answer like this is called a **repeating decimal.** When a solution like this occurs, we round the answer to the nearest thousandth, such as $2/3 = .667$. If the number that repeats is 5 or more, round to the next higher number. If it is 4 or less, round to the next lower number. Thus,

$$.5555 = .556 \quad \text{and} \quad .3333 = .333.$$

Work Quick Check 3 at the side.

3. Convert to decimals. If a division does not come out even, round to the nearest thousandth.

(a) $\frac{1}{5}$

(b) $\frac{17}{20}$

(c) $\frac{5}{8}$

(d) $\frac{2}{9}$

(e) $\frac{3}{7}$

The operations on decimals—addition, subtraction, multiplication, and division—are explained in the next example.

Example 4 Add or subtract as indicated.

(a) $6.92 + 14.8 + 3.217$

Place the numbers in a column, with decimal points lined up. Then add as you normally would.

$$\begin{array}{r} 6.92 \\ 14.8 \\ +\ \ 3.217 \\ \hline 24.937 \end{array} \quad \text{Decimal points lined up}$$

If you want, annex 0's to make all the numbers the same length; this is a good way to avoid errors.

$$\begin{array}{r} 6.92 \\ 14.8 \\ +\ \ 3.217 \\ \hline \end{array} \quad \text{becomes} \quad \begin{array}{r} 6.920 \\ 14.800 \\ +\ \ 3.217 \\ \hline 24.937 \end{array}$$

(b) $47.6 - 32.509$

Write the numbers in a column, annexing 0's to 47.6.

$$\begin{array}{r} 47.6 \\ -\,32.509 \\ \hline \end{array} \quad \text{becomes} \quad \begin{array}{r} 47.600 \\ -\,32.509 \\ \hline 15.091 \end{array}$$

(c) $3 - .253$

$$\begin{array}{r} 3.000 \\ -\ \ .253 \\ \hline 2.747 \end{array}$$

Quick Check Answers

3. (a) .2 (b) .85 (c) .625 (d) .222 (e) .429

4. Add or subtract.

(a)
```
   68.9
   42.72
+  8.973
```

(b)
```
  32.5
− 21.72
```

(c) 42.83 + 71.629 + 3.074

(d) 351.8 − 2.706

5. Multiply.

(a) 2.13 × .05

(b) 69.32 × 1.04

(c) 397.12 × .06

(d) 42,980 × .012

Quick Check Answers
4. (a) 120.593 (b) 10.78
(c) 117.533 (d) 349.094
5. (a) .1065 (b) 72.0928
(c) 23.8272 (d) 515.76

Work Quick Check 4 at the side.

To multiply decimals, ignore the decimal points and multiply normally. Then add together the number of decimal places in each number being multiplied and place the decimal point that many places from the right in the answer.

Example 5 Multiply.

(a) 29.3 × 4.52
Multiply normally:

```
    29.3    ← 1 decimal in top number
  × 4.52    ← 2 decimals in second number
    5 86
   14 6 5
  117 2
  132.4 36  ← 3 decimals in answer
```

(b) 7.003 × 55.8

```
    7.003   ← 3 decimals
  ×  55.8   ← 1 decimal
   5 602 4
  35 015
 350 15
 390.767 4  ← 4 decimals in answer
```

Work Quick Check 5 at the side.

Division of decimal numbers uses a slightly different process. Change the divisor (the number you are dividing by) into a whole number by moving the decimal point as many places as necessary to the right. Move the decimal point in the dividend (the number being divided) to the right by the same number of places. Then divide as you normally would.

Example 6 Divide.

(a) 279.45 ÷ 24.3
Step 1 Write the problem.

24.3)279.45

Step 2 Move the decimal point in 24.3 one place to the right, to get 243. Do the same thing with 279.45. By doing this, we make 24.3 into the whole number 243.

24.3.)279.4.5

Step 3 Bring the decimal point straight up and divide normally.

```
        11.5
243)2794.5
     243
     364
     243
     121 5
     121 5
         0
```

(b) $73.82\overline{)1852.882}$

Move the decimal point two places to the right in 73.82, to get 7382. Do the same thing with 1852.882.

$$73.82.\overline{)1852.88.2}$$

Bring the decimal point straight up and divide normally.

```
          25.1
7382)185288.2
     14764
      37648
      36910
       7382
       7382
          0
```

Work Quick Check 6 at the side.

One of the main uses of decimals comes in working problems with percent. The word **percent** means "per one hundred" and is written with the sign %. One percent means "one per one hundred" and is written

$$1\% = .01.$$

Example 7 Convert.

(a) 75% to a decimal
Since $1\% = .01$, we have

$$75\% = 75 \cdot 1\% = 75 \cdot (.01) = .75.$$

(b) 125% to a decimal

$$125\% = 125 \cdot 1\% = 125 \cdot (.01) = 1.25$$

(c) .32 to a percent

$$.32 = 32 \cdot (.01) = 32 \cdot 1\% = 32\%$$

(d) 2.63 to a percent

$$2.63 = 263 \cdot (.01) = 263 \cdot 1\% = 263\%$$

Work Quick Check 7 at the side.

6. Divide.

(a) $32.3\overline{)481.27}$

(b) $.37\overline{)5.476}$

(c) $1.06\overline{)61.692}$

(d) $375.125 \div 3.001$

7. Convert.

(a) 50% to a decimal

(b) 310% to a decimal

(c) .71 to a percent

(d) 1.32 to a percent

Quick Check Answers
6. (a) 14.9 (b) 14.8 (c) 58.2 (d) 125
7. (a) .50 (b) 3.10 (c) 71% (d) 132%

8. Find the percentages.

(a) 20% of 70

(b) 36% of 500

(c) The amount of discount on a television set with a list price of \$270 if the set is on sale at 25% off.

Quick Check Answers

8. (a) 14 (b) 180 (c) \$67.50

The next example shows how to find percentages by using multiplication.

Example 8 Find the percentages.

(a) 15% of 600

The word *of* indicates multiplication. For this reason, we can find 15% of 600 by multiplying.

$$15\% \cdot 600 = (.15) \cdot 600 = 90$$

(b) 125% of 80

$$125\% \cdot 80 = (1.25) \cdot 80 = 100$$

(c) A camera with a regular price of \$18 is on sale this week at 22% off. Find the amount of the discount.

Multiply 22% and \$18.

$$22\% \cdot \$18 = (.22) \cdot \$18 = \$3.96$$

The discount is \$3.96.

Work Quick Check 8 at the side.

name date hour

1.2 EXERCISES

Write in expanded form. See Example 1.

1. 86 = ______
2. 15 = ______
3. 694 = ______
4. 856 = ______
5. 5237 = ______
6. 4761 = ______
7. 36.81 = ______
8. 78.92 = ______
9. .567 = ______
10. .146 = ______

Convert the decimals to fractions. Reduce to lowest terms. See Example 2.

11. .2 = ______
12. .8 = ______
13. .36 = ______
14. .72 = ______
15. .336 = ______
16. .215 = ______
17. .805 = ______
18. .625 = ______

Convert the fractions to decimals. Round to the nearest thousandth, if necessary. See Example 3.

19. $\frac{5}{8}$ = ______
20. $\frac{3}{8}$ = ______
21. $\frac{3}{4}$ = ______
22. $\frac{7}{16}$ = ______
23. $\frac{9}{16}$ = ______
24. $\frac{15}{16}$ = ______
25. $\frac{2}{3}$ = ______
26. $\frac{5}{6}$ = ______
27. $\frac{7}{9}$ = ______
28. $\frac{9}{11}$ = ______

Perform the indicated operations. See Examples 4, 5, and 6.

29. 14.23 + 9.81 + 74.63 + 18.715 = ______
30. 89.416 + 21.32 + 478.91 + 298.213 = ______
31. 19.74 − 6.53 = ______

32. 27.96 − 8.39 = ______

33.
$$\begin{array}{r} 48.96 \\ 37.421 \\ +\ 9.72 \\ \hline \end{array}$$

34.
$$\begin{array}{r} 9.71 \\ 4.8 \\ 3.6 \\ 5.2 \\ +\ 8.17 \\ \hline \end{array}$$

35.
$$\begin{array}{r} 8.6 \\ -\ 3.751 \\ \hline \end{array}$$

36.
$$\begin{array}{r} 27.8 \\ -\ 13.582 \\ \hline \end{array}$$

37. 39.6 · (4.2) = ______

38. 18.7 · (2.3) = ______

39. 42.1 · (3.9) = ______

40. 19.63 · (4.08) = ______

41. .042 · (3.2) = ______

42. .571 · (2.9) = ______

43. 24.84 ÷ 6 = ______

44. 32.84 ÷ 4 = ______

45. 7.6266 ÷ 3.42 = ______

46. 14.9202 ÷ 2.43 = ______

47. 2.496 ÷ .52 = ______

48. 3.57 ÷ .034 = ______

Convert to percent. See Examples 7(c) and 7(d).

49. .80 = ______
50. .75 = ______
51. .007 = ______
52. 1.4 = ______

53. .67 = ______
54. .003 = ______
55. .125 = ______
56. .983 = ______

Convert the following percents to decimals. See Examples 7(a) and 7(b).

57. 53% = ______
58. 38% = ______
59. 129% = ______
60. 174% = ______

61. 96% = ______
62. 11% = ______
63. .9% = ______
64. .1% = ______

Find each of the following. Round your answer to the nearest hundredth if necessary. See Example 8.

65. 14% of 780 = ______

66. 12% of 350 = ______

67. 22% of 1086 = ______

68. 20% of 1500 = ______

69. 5% of 80 = ______

70. 125% of 2000 = ______

name date hour

71. 175% of 5820 = ____________________

72. 15% of 75 = ____________________

73. 25% of 484 = ____________________

74. 400% of 3200 = ____________________

75. 118% of 125.8 = ____________________

76. 3% of 128 = ____________________

77. 90% of 5930 = ____________________

78. 12.4% of 56 = ____________________

79. 7% of 150 = ____________________

80. 18.5% of 780 = ____________________

Solve the word problems.

81. A retailer has $23,000 invested in her business. She finds that she is earning 12% per year on this investment. How much is she earning per year? ____________________

82. A family of four with a monthly income of $1500 spends 90% of its earnings and saves the rest. What is the *annual* savings of this family? ____________________

83. Harley Dabler recently bought a duplex for $72,000. He expects to earn 8% per year on the purchase price. How many dollars per year will he earn? ____________________

84. For a recent tour of the eastern United States, a travel agent figured that the trip totaled 2300 miles, with 35% of the trip by air. How many miles of the trip were by air? ____________________

85. Capitol Savings Bank pays 5.5% interest per year. What is the annual interest on an account of $3000? ____________________

86. Beth's Bargain Basement is having a sale this week–15% off on any purchase. What would the discount be on a purchase of $250? ____________________

87. When installing carpet, Delta Carpet Layers wastes 6% of the carpet. Of a total of 4150 yards laid in April, how much was wasted? ____________________

88. Scott Samuels must pay 6.5% sales tax on a new car. The cost of the car is $4600. Find the amount of the tax. ____________

89. An ad for steel-belted radial tires promises 15% better mileage. Alexandria's Toyota now goes 420 miles on a tank of gas. If she switched to the new tires, how many extra miles could she go on a tank of gas? ____________

90. A home worth $42,000 is located in an area where home prices are increasing at a rate of 12% per year. By how much would the value of this home increase in one year? ____________

1.3 SYMBOLS

So far in this chapter we have seen some of the common symbols of arithmetic, such as $+$, $-$, $\times$, $\div$, and $=$. We also saw the dot $(\cdot)$, used for multiplication. In this section, we learn some new symbols.

The symbol $=$ tells us that two numbers are equal. If we place a slash through this symbol, we get $\neq$, a symbol for "does *not* equal." For example,

$$7 \neq 8$$

says that 7 is not equal to 8.

If two numbers are not equal, then one of the numbers must be smaller than the other. The symbol $<$ represents "is less than," so that "7 is less than 8" is written

$$7 < 8.$$

Also, "6 is less than 9" is written $6 < 9$.

The symbol $>$ means "is greater than." Since 8 is greater than 2, we have

$$8 > 2.$$

The statement "17 is greater than 11" is written

$$17 > 11.$$

To keep the symbols $<$ and $>$ straight, remember that the symbol always points to the smaller number. To write "8 is less than 15," point the symbol toward the 8:

$$8 < 15.$$

Work Quick Check 1 at the side.

Just as we write "is not equal to" as $\neq$, we write "is not less than" as $\nless$. The statement "5 is not less than 4" is written $5 \nless 4$. "Is not greater than" is written $\ngtr$. For example, $6 \ngtr 9$ is read as "6 is not greater than 9."

Example 1 Write each word statement in symbols.

(a) Twelve equals ten plus two.
In symbols: $12 = 10 + 2$

(b) Nine is less than ten.

$$9 < 10$$

(c) Fifteen is not equal to eighteen.

$$15 \neq 18$$

(d) Seven is not greater than ten.

$$7 \ngtr 10$$

Work Quick Check 2 at the side.

OBJECTIVES

- Know the meaning of $<$, $>$, $\leqslant$, and $\geqslant$
- Translate word phrases to symbols
- Understand the order of operations

QUICK CHECKS

1. Write *true* or *false*.

(a) $7 < 5$

(b) $12 > 6$

(c) $4 \neq 10$

(d) $10 < 21$

2. Write in symbols.

(a) Nine equals eleven minus two.

(b) Seventeen is less than thirty.

(c) Eight is not equal to ten.

(d) Fourteen is not greater than fifteen.

(e) Six is not less than four.

Quick Check Answers
1. (a) False (b) True
(c) True (d) True
2. (a) $9 = 11 - 2$ (b) $17 < 30$
(c) $8 \neq 10$ (d) $14 \ngtr 15$
(e) $6 \nless 4$

3. Write *true* or *false*.

(a) $30 \leqslant 40$

(b) $25 \geqslant 10$

(c) $40 \leqslant 10$

(d) $21 \leqslant 21$

(e) $3 \geqslant 3$

Quick Check Answers
3. (a) True (b) True (c) False (d) True (e) True

Two other symbols, $\leqslant$ and $\geqslant$, also represent the idea of inequality. The symbol $\leqslant$ means "is less than or equal to," so that

$$5 \leqslant 9$$

means "5 is less than or equal to 9." This statement is true, since $5 < 9$ is true. If either the $<$ part or the $=$ part is true, then the inequality $\leqslant$ is true.

The symbol $\geqslant$ means "is greater than or equal to." Again,

$$9 \geqslant 5$$

is true because $9 > 5$ is true. Also, $8 \leqslant 8$ is true since $8 = 8$ is true. But it is not true that $13 \leqslant 9$ because neither $13 < 9$ nor $13 = 9$ is true.

Example 2 Write *true* or *false*.

(a) $15 \leqslant 20$
Since $15 < 20$, the statement $15 \leqslant 20$ is true.

(b) $25 \geqslant 30$
Both parts of this statement, $25 > 30$ and $25 = 30$, are false. Because of this, $25 \geqslant 30$ is false.

(c) $12 \geqslant 12$
Since $12 = 12$, this statement is true.

Work Quick Check 3 at the side.

Many of the problems that we will work will involve more than one symbol of arithmetic. For example, let us find the value of

$$5 + 2 \cdot 3.$$

At first glance it appears that we can first multiply 2 and 3, or we can first add 5 and 2. If we first multiply 2 and 3, and then add 5, the result is

$$5 + 2 \cdot 3 = 5 + 6 = 11.$$

If we first add 5 and 2, and then multiply by 3, the answer is

$$5 + 2 \cdot 3 = 7 \cdot 3 = 21.$$

To make sure that everyone who works a problem like this always gets the same answer, the following **order of operations** has been agreed upon.

1. Do any operations inside parentheses.
2. Do any multiplications or divisions, working from left to right.
3. Do any additions or subtractions, working from left to right.
4. If the problem involves a fraction bar, do all work above the bar and below the bar separately, and then simplify, if possible, to get the final answer.

Example 3 Find the value of $5 + 2 \cdot 3$

Using the order of operations given above, we first multiply.

$$5 + 2 \cdot 3 = 5 + 6$$

Then add.

$$5 + 6 = 11$$

Therefore, $5 + 2 \cdot 3 = 11$. According to the order of operations given above, there is only one possible answer for this problem.

Work Quick Check 4 at the side.

Example 4 We have used a dot to show multiplication; another way to show multiplication is with parentheses. For example, 3(7) means $3 \cdot 7$. Find the simplest answer for the following.

(a) $9(6 + 11)$

This expression can be simplified by first working inside the parentheses.

$$\begin{aligned} 9(6 + 11) &= 9(17) && \text{Work inside parentheses} \\ &= 153 && \text{Multiply} \end{aligned}$$

(b)
$$\begin{aligned} 2(5 + 6) + 7 \cdot 3 &= 2(11) + 7 \cdot 3 && \text{Work inside parentheses} \\ &= 22 + 21 && \text{Multiply} \\ &= 43 && \text{Add} \end{aligned}$$

(c) $\dfrac{4(5 + 3) + 3}{2(3) - 1}$

Simplify the top and bottom of the fraction separately.

$$\begin{aligned} \frac{4(5 + 3) + 3}{2(3) - 1} &= \frac{4(8) + 3}{2(3) - 1} && \text{Work first inside parentheses} \\ &= \frac{32 + 3}{6 - 1} && \text{Multiply} \\ &= \frac{35}{5} && \text{Add or subtract} \\ &= 7 \end{aligned}$$

Work Quick Check 5 at the side.

An expression with double parentheses, such as $2(8 + 3(6 + 5))$, can be confusing. To eliminate this, square brackets, [ ], are used instead of one pair of parentheses. For example, $2[8 + 3(6 + 5)]$ is simplified by first working with the innermost parentheses, until a single number is found for the terms inside the brackets.

$$\begin{aligned} 2[8 + 3(6 + 5)] &= 2[8 + 3(11)] \\ &= 2[8 + 33] \\ &= 2[41] \\ &= 82 \end{aligned}$$

Work Quick Check 6 at the side.

4. Solve.

(a) $3 \cdot 8 + 7$

(b) $12 \cdot 6 + 9$

(c) $6 \cdot 11 + 4$

5. Find the simplest possible answer.

(a) $2 \cdot 9 + 7 \cdot 3$

(b) $5 \cdot 8 + 6 \cdot 7$

(c) $4(3 + 7)$

(d) $2(9 + 8) - 3 \cdot 5$

(e) $7 \cdot 6 - 3(8 + 1)$

(f) $\dfrac{2(7 + 8) + 2}{3 \cdot 5 + 1}$

6. Find the simplest possible answer.

(a) $4[7 + 3(6 + 1)]$

(b) $3[6 + (4 + 2)]$

(c) $9[(4 + 8) - 3]$

Quick Check Answers

4. (a) 31 (b) 81 (c) 70
5. (a) 39 (b) 82 (c) 40 (d) 19 (e) 15 (f) 2
6. (a) 112 (b) 36 (c) 81

WORK SPACE

name ______ date ______ hour ______

1.3 EXERCISES

Answer true *or* false *for each statement. See Example 2.*

1. $8 + 2 = 10$ ______
2. $9 + 12 = 21$ ______
3. $6 \neq 5 + 1$ ______
4. $8 \neq 9 - 1$ ______
5. $9 < 12$ ______
6. $15 < 21$ ______
7. $12 \geqslant 10$ ______
8. $25 \geqslant 19$ ______
9. $11 < 11$ ______
10. $45 < 45$ ______
11. $18 \nless 20$ ______
12. $29 \nless 30$ ______
13. $0 < 15$ ______
14. $9 < 0$ ______
15. $8 \ngtr 0$ ______
16. $0 \ngtr 4$ ______
17. $15 \leqslant 32$ ______
18. $26 \geqslant 50$ ______

In Exercises 19–44, work the problems, then decide whether the given statement is true *or* false. *See Examples 3 and 4.*

19. $3 \cdot 4 + 7 < 10$ ______
20. $8 \cdot 2 - 5 > 12$ ______
21. $9 \cdot 3 - 11 \leqslant 16$ ______
22. $6 \cdot 5 - 12 \leqslant 18$ ______
23. $3 \cdot 8 - 4 \cdot 6 \leqslant 0$ ______
24. $2 \cdot 20 - 8 \cdot 5 \geqslant 0$ ______
25. $5 \cdot 11 + 2 \cdot 3 \leqslant 60$ ______
26. $9 \cdot 3 + 4 \cdot 5 \geqslant 48$ ______
27. $9 \cdot 2 - 6 \cdot 3 \geqslant 2$ ______
28. $8 \cdot 3 - 4 \cdot 6 < 1$ ______
29. $12 \cdot 3 - 6 \cdot 6 \leqslant 0$ ______
30. $13 \cdot 2 - 15 \cdot 1 \geqslant 10$ ______
31. $6 \cdot 5 + 3 \cdot 10 \leqslant 0$ ______
32. $5 \cdot 8 + 10 \cdot 4 \geqslant 0$ ______

33. $4[2 + 3(4)] \geqslant 50$ ______

34. $3[5(2) - 3] < 20$ ______

35. $60 < 5[8 + (2 + 3)]$ ______

36. $80 < 9[(14 + 5) - 10]$ ______

37. $2[2 + 3(2 + 5)] \leqslant 45$ ______

38. $3[4 + 3(4 + 1)] \leqslant 55$ ______

39. $\dfrac{5(4 - 1) + 3}{2 \cdot 4 + 1} \geqslant 3$ ______

40. $\dfrac{7(3 + 1) - 2}{5 \cdot 2 + 3} \leqslant 2$ ______

41. $\dfrac{2(5 + 3) + 2 \cdot 2}{2(4 - 1)} > 4$ ______

42. $\dfrac{9(7 - 1) - 8 \cdot 2}{4(6 - 1)} > 2$ ______

43. $\dfrac{2(5 + 1) - 3(1 + 1)}{5(8 - 6) - 4 \cdot 2} \leqslant 3$ ______

44. $\dfrac{3(8 - 3) + 2(4 - 1)}{9(6 - 2) - 11(5 - 2)} \geqslant 7$ ______

Insert parentheses in each expression (if needed) so that the resulting statement is true.

Example The statement $9 - 3 - 2 = 8$ will be true if parentheses are inserted around $3 - 2$. Thus

$$9 - (3 - 2) = 8.$$

It is not true that $(9 - 3) - 2 = 8$, because $6 - 2 \neq 8$.

45. $10 - 7 - 3 = 6$ ______

46. $3 \cdot 5 + 7 = 36$ ______

47. $3 \cdot 5 + 7 = 22$ ______

48. $3 \cdot 5 - 4 = 3$ ______

49. $3 \cdot 5 - 4 = 11$ ______

50. $3 \cdot 5 + 2 \cdot 4 = 23$ ______

51. $3 \cdot 5 + 2 \cdot 4 = 84$ ______

52. $3 \cdot 5 + 2 \cdot 4 = 68$ ______

53. $3 \cdot 5 - 2 \cdot 4 = 36$ ______

54. $3 \cdot 5 - 2 \cdot 4 = 7$ ______

55. $100 \div 20 \div 5 = 1$ ______

56. $360 \div 18 \div 4 = 5$ ______

OBJECTIVES

- Know the meaning of the word *variable*
- Find the value of an expression when given a value of the variable
- Know the meaning of *equation* and *domain*
- Identify solutions of an equation from a given domain
- Solve inequalities from a given domain and graph the solutions on a number line

A **variable** is a letter, such as x, y, or z, that represents an unknown number. For example, x is the variable in the **algebraic expression** $x + 5$. The expression $x + 5$ takes on different numerical values as the variable x itself takes on different numerical values.

For example, if x represents 12, we can find the value of $x + 5$ by substituting 12 for x. Doing this, we get

$$\begin{aligned} x + 5 &= 12 + 5 \\ &= 17. \end{aligned}$$

On the other hand, if $x = 3$, then $x + 5$ becomes $3 + 5$, or 8. For $x = 7$, the value of $x + 5$ is $7 + 5 = 12$.

Example 1 Find the value of $x + 9 + x + 5$ when:

(a) $x = 1$

Replace x by 1.

$$\begin{aligned} x + 9 + x + 5 &= 1 + 9 + 1 + 5 \\ &= 16 \end{aligned}$$

(b) $x = 6$

$$\begin{aligned} x + 9 + x + 5 &= 6 + 9 + 6 + 5 \\ &= 26 \end{aligned}$$

Work Quick Check 1 at the side.

QUICK CHECKS

1. Find the value of each expression when $x = 2$ and when $x = 8$.

(a) $x + 9$

(b) $x - 1$

(c) $x + 6 + x - 2$

As we have seen, multiplication can be shown with a dot or with parentheses. For example, both $9 \cdot 2$ and $9(2)$ mean 9×2, or 18. The product of the number 8 and the variable x can also be written as $8 \cdot x$ or $8(x)$. However, with a variable, a product can be written without any special symbols at all. The product of 8 and x is written $8x$, and the product of 12 and z is written $12z$. The product of x and z is written xz.

Example 2 Find the value of each expression when $x = 5$ and $y = 3$.

(a) $2x + 5y$

Replace x with 5 and y with 3. Do the multiplications first, then add.

$$\begin{aligned} 2x + 5y &= 2 \cdot 5 + 5 \cdot 3 && \text{Let } x = 5 \text{ and } y = 3 \\ &= 10 + 15 && \text{Multiply} \\ &= 25 && \text{Add} \end{aligned}$$

(b) $\dfrac{9x - 8y}{2x - y}$

Replace x with 5 and y with 3.

Quick Check Answers

1. (a) 11; 17 (b) 1; 7 (c) 8; 20

2. Find the value of each expression when $x = 6$ and $y = 9$.

(a) $5y$

(b) $12x$

(c) xy

(d) $4x + 7y$

(e) $\dfrac{x+2}{y-5}$

(f) $\dfrac{4x-2y}{x+1}$

Quick Check Answers

2. (a) 45 (b) 72 (c) 54 (d) 87 (e) 2 (f) 6/7

$$\frac{9x-8y}{2x-y} = \frac{9 \cdot 5 - 8 \cdot 3}{2 \cdot 5 - 3} \quad \text{Let } x = 5 \text{ and } y = 3$$

$$= \frac{45-24}{10-3} \quad \text{Multiply}$$

$$= \frac{21}{7} \quad \text{Subtract}$$

$$= 3 \quad \text{Simplify}$$

Work Quick Check 2 at the side.

An **equation** is a mathematical sentence which says that two algebraic expressions are equal. Examples of equations include

$$x + 4 = 11, \quad 2y = 16, \quad \text{and} \quad 4p + 1 = 25 - p.$$

To solve an equation, it is necessary to find all values of the variable that make the equation true. The values of the variable that make the equation true are called the **solutions** of the equation.

In this chapter, whenever you are given an equation to solve, you will also be given a set of numbers which might be solutions. (A **set** of numbers is just a collection of numbers—we use the symbols { } to keep the numbers together. See Appendix A.) You must then decide which if any of these numbers are really solutions of the equations.

Example 3 Which numbers in the set {6, 7, 8} are solutions of the equation $x + 4 = 11$?

To find out, use each number from the set {6, 7, 8} in turn and substitute it for x.

If $x = 6$,	If $x = 7$,	If $x = 8$,
$x + 4 = 11$	$x + 4 = 11$	$x + 4 = 11$
$6 + 4 = 11$	$7 + 4 = 11$	$8 + 4 = 11$
$10 = 11$	$11 = 11$	$12 = 11$
False	*True*	*False*

Only $x = 7$ leads to a true result. Therefore, the only number from the set {6, 7, 8} which is a solution for the equation $x + 4 = 11$ is 7.

The set of all possible replacements ({6, 7, 8} in Example 3) for the variable is called the **domain** of the variable.

Example 4 Solve the equation $3x - 6 = 24$. The domain is {9, 10, 11}.

To find the solution, try each number from the domain.

If $x = 9$,	If $x = 10$,	If $x = 11$
$3x - 6 = 24$	$3x - 6 = 24$	$3x - 6 = 24$
$3 \cdot 9 - 6 = 24$	$3 \cdot 10 - 6 = 24$	$3 \cdot 11 - 6 = 24$
$27 - 6 = 24$	$30 - 6 = 24$	$33 - 6 = 24$
$21 = 24$	$24 = 24$	$27 = 24$
False	*True*	*False*

A true statement is obtained only when the number 10 replaces the variable. Therefore, the solution of $3x - 6 = 24$ is 10.

Work Quick Check 3 at the side.

3. Solve each equation. The domain is $\{2, 3, 4, 5, 6\}$.

(a) $2x = 10$

(b) $x - 1 = 3$

(c) $2k + 3 = 15$

(d) $8p - 11 = 5$

(e) $9k - 8 = 19$

Sometimes we want to say that one expression is less than or greater than another expression. Remember that "less than" is written using the symbol $<$. "Greater than" is written using the symbol $>$.

If we want to say that an expression is either less than or equal to a second expression, we use the symbol $\leqslant$. If we want to say that the first expression is greater than or equal to the second expression, we use the symbol $\geqslant$.

Such statements are called **inequalities.** Examples of inequalities include

$$x + 2 \leqslant 6, \quad 3y \geqslant 18, \quad 2x + 1 < 9, \text{ and } \quad 5x - 2 > 13.$$

To solve an inequality, we must find all numbers from the given domain that make the inequality true.

Example 5 Solve the inequality $2x + 1 < 9$. The domain is $\{1, 2, 3, 4, 5\}$.

Replace the variable x, in turn, with the numbers 1, 2, 3, 4, and 5. You should find that $2x + 1 < 9$ is true for 1, 2, and 3, but not for 4 or 5. Therefore, the solutions of $2x + 1 < 9$ from the given domain are 1, 2, and 3.

Work Quick Check 4 at the side.

4. Solve each inequality. The domain is $\{2, 3, 4, 5, 6, 7\}$.

(a) $x + 2 > 6$

(b) $2p - 1 > 8$

(c) $11r - 9 \leqslant 24$

(d) $3x - 1 \geqslant 11$

It is often helpful to draw a diagram which shows the solutions of an equation or inequality. To do so, first draw a straight line. Choose any point on the line and label it 0. Then choose another point to the right of 0 and label it 1. The distance between 0 and 1 sets up a unit measure that can be used to locate more points, labeled 2, 3, 4, 5, 6, and so on. We call this diagram a **number line.** (See Figure 1.)

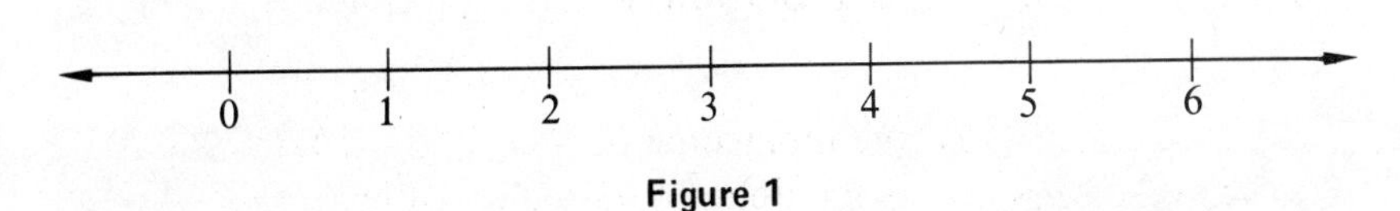

Figure 1

Other points could be found by dividing the units into halves, thirds, and so on. Each number is called the **coordinate** of the point that it labels. For example, 2 is the coordinate of the point labeled A in Figure 2.

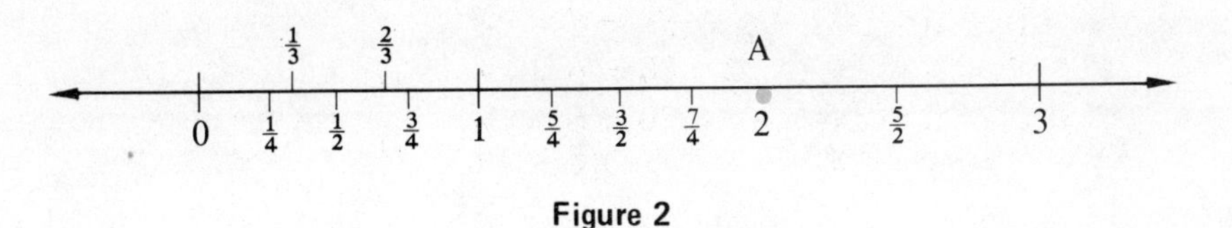

Figure 2

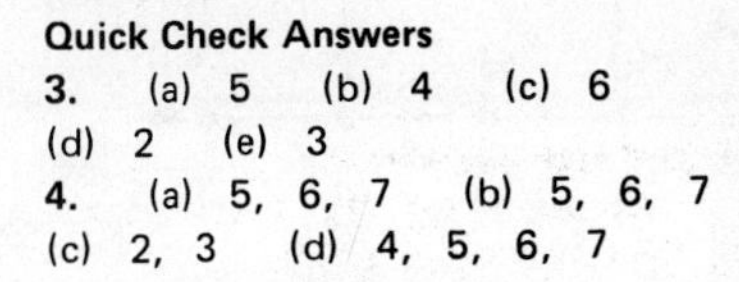
Quick Check Answers
3. (a) 5 (b) 4 (c) 6 (d) 2 (e) 3
4. (a) 5, 6, 7 (b) 5, 6, 7 (c) 2, 3 (d) 4, 5, 6, 7

Example 6 Find and graph the solutions of $5r - 2 \geq 18$. The domain is {2, 3, 4, 5, 6, 7}.

By trying each of the numbers from the domain, you should find that the solutions of the given inequality are the numbers 4, 5, 6, and 7. These numbers have been located on the number line in Figure 3. The number line and the points on it make up the **graph** of the solutions.

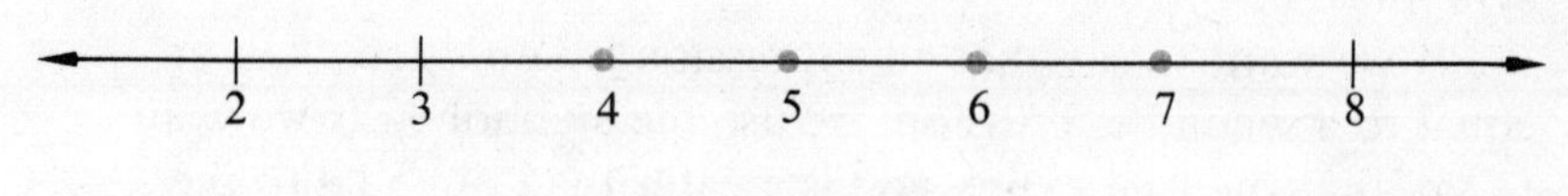

Figure 3

Example 7 Find and graph the solutions of $x + 1 = 10$. The domain is {0, 2, 4, 6, 8, 10}.

None of the numbers in the domain makes the given equation true. (The only solution to the given equation is 9, which is not in the domain.) Thus, we must say that there are no solutions from the given domain, or that the solution set contains no numbers. Any set containing no numbers or other elements is called the **empty set** or **null set**. The symbol for the empty set is $\emptyset$. The graph of the empty set is a number line with no points marked on it, as in Figure 4.

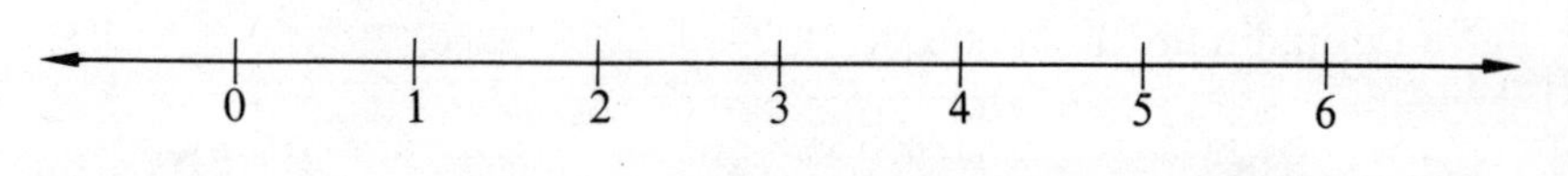

Figure 4

Work Quick Check 5 at the side.

5. Find and graph the solutions. The domain is {0, 1, 2, 3, 4, 5}.

(a) $5m = 10$

(b) $2k + 3 < 10$

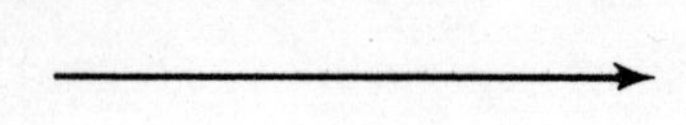

(c) $5 + 9r \geq 23$

(d) $5 - m > 10$

Quick Check Answers

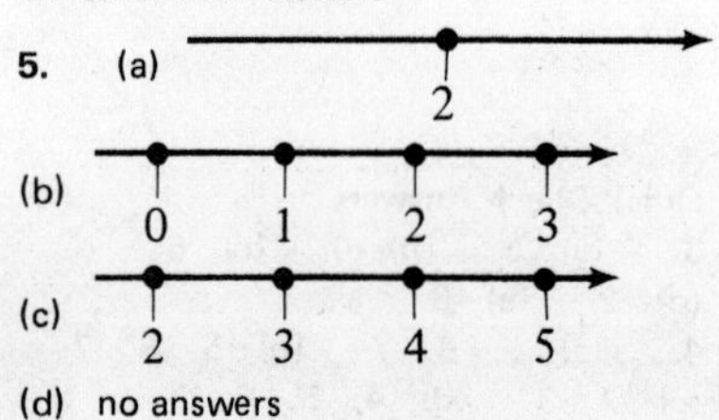

name date hour

1.4 EXERCISES

Find the numerical value of the given expression when x = 3. *See Example 1.*

1. $x + 9 =$ ______
2. $x - 2 =$ ______
3. $5x =$ ______
4. $7x =$ ______
5. $2x + 8 =$ ______
6. $9x - 5 =$ ______

Find the numerical value of the given expression when x = 4 *and* y = 2. *See Example 2.*

7. $x + y =$ ______
8. $x - y =$ ______
9. $8x + 3y + 5 =$ ______
10. $4x + 2y + 2 =$ ______
11. $3(x + 2y) =$ ______
12. $2(2x - y) =$ ______
13. $5(4x - 7y) =$ ______
14. $8(5x - 9y) =$ ______
15. $\dfrac{2x + 3y}{x + y + 1} =$ ______
16. $\dfrac{5x - 3y + 1}{2x} =$ ______
17. $\dfrac{2x + 4y - 6}{5y + 2} =$ ______
18. $\dfrac{4x + 3y - 1}{x} =$ ______

Decide whether or not the number following each equation is a solution for the given equation. See Example 3.

19. $x + 6 = 15; 9$ ______
20. $p - 5 = 12; 17$ ______
21. $3r + 5 = 8; 2$ ______
22. $5m + 2 = 7; 2$ ______
23. $6a + 2(a + 3) = 14; 1$ ______
24. $2y + 3(y - 2) = 14; 4$ ______
25. $2x + 3x + 8 = 38; 6$ ______
26. $6x + 4x - 9 = 11; 2$ ______
27. $2 + 3y + 4y = 20; 3$ ______
28. $6 + 8r - 3r = 11; 5$ ______

Find and graph the solution of each equation from the given domain. See Example 7.

29. $2 + x = 4$; $\{2, 4, 6\}$

30. $3 + p = 6$; $\{1, 2, 3\}$

31. $y + 8 = 12$; $\{1, 3, 4, 5\}$

32. $r + 2 = 10$; $\{7, 8, 9\}$

33. $2k = 12$; $\{5, 6, 7\}$

34. $3m = 15$; $\{4, 5, 6\}$

35. $4a + 3 = 19$; $\{3, 4, 5\}$

36. $9z + 1 = 19$; $\{2, 3, 4\}$

37. $11k - 5 = 17$; $\{4, 5, 6\}$

38. $7s - 2 = 12$; $\{2, 3, 4\}$

Find and graph the solution of each inequality from the given domain. See Example 6.

39. $y + 1 \leq 6$; $\{3, 4, 5, 6\}$

40. $k - 3 \geq 5$; $\{6, 7, 8, 9, 10\}$

41. $3r - 2 \leq 7$; $\{1, 2, 3, 4, 5\}$

42. $5m - 1 \leq 9$; $\{1, 2, 3, 4\}$

43. $2z + 3 < 12$; $\{2, 3, 4, 5\}$

44. $4k - 5 > 8$; $\{2, 3, 4, 5, 6\}$

45. $4r - 3 \geq 13$; $\{0, 1, 2, 3\}$

name date hour

46. $5p + 1 \leqslant 14$; $\{3, 4, 5, 6\}$

47. $2x + 1 \neq 9$; $\{2, 3, 4, 5, 6\}$

48. $5x - 2 \neq 13$; $\{1, 2, 3, 4, 5, 6\}$

Write each word phrase using mathematical symbols. Use x *to represent the variable.*

Example "A number plus 18" would be written $x + 18$. "Four times a number, subtracted from 19" is $19 - 4x$.

49. Twice a number ______

50. 5 times a number ______

51. 6 added to a number ______

52. 4 added to a number ______

53. A number subtracted from 8 ______

54. 9 subtracted from a number ______

55. 8 added to three times a number ______

56. 6 subtracted from 5 times a number ______

57. Twice a number, subtracted from 15 ______

58. 8 times a number, added to 52 ______

Write the given word statement in mathematical symbols. Use x *as the variable. Find the solution for the statement if the domain is* $\{0, 2, 4, 6, 8, 10\}$.

Example The sum of a number and four is six.

First write this sentence in symbols as $x + 4 = 6$. Then find the solution from the domain. The solution is the number 2.

	Statement	Solution
59. The sum of a number and 8 is 12.	______	______
60. A number minus 3 equals 7.	______	______
61. The sum of a number and 2 is less than 11.	______	______

	Statement	Solution
62. The sum of twice a number and 6 is less than 10.	__________	__________
63. 5 more than twice a number is less than 10.	__________	__________
64. The product of a number and three is greater than eight.	__________	__________
65. The sum of a number and 2, divided by 4, is 3.	__________	__________
66. Three times a number is equal to two more than twice the number.	__________	__________
67. Twelve divided by a number equals three times that number.	__________	__________
68. Six times a number, minus three, is greater than twenty-four.	__________	__________

name date hour

CHAPTER 1 TEST

Reduce to lowest terms.

1. $\frac{15}{40} =$ ________ 2. $\frac{84}{132} =$ ________ 3. $\frac{48}{72} =$ ________

Work the problems.

4. $\frac{5}{8} + \frac{9}{10} =$ ________ 5. $\frac{3}{8} + \frac{7}{12} + \frac{11}{15} =$ ________

6. $21\frac{3}{4} - 7\frac{3}{8} =$ ________ 7. $\frac{3}{2} \cdot \frac{4}{9} =$ ________ 8. $\frac{6}{5} \div \frac{19}{15} =$ ________

9. Johnson Forest Products needed to raise cash quickly. To do so, the company sold 46 1/3 acres out of a 104 7/9 acre piece of land which it owned. How many acres of land were left? ________

10. Convert .625 to a fraction. ________

11. Convert 9/16 to a decimal. ________

Perform the indicated operations.

12. $9.6 + 8.42 + 3.75 =$ ________

13. $123.4 - 98.7 =$ ________

14. $21.98 \cdot (.72) =$ ________

15. $252.008 \div 21.8 =$ ________

16. Convert .19 to a percent. ________

17. Convert 76.2% to a decimal. ________

Find each of the following.

18. 8% of 170 = ________

19. 14.2% of 4600 = ________

20. The number of employees of Thompson Construction Company is only 85% of last year's number of 240. Find the number of employees this year. ________

21. A clock radio with a regular price of $48 is on sale this week at 20% off. Find the amount of the discount. ________

Write true *or* false.

22. $5 \cdot 9 + 6 \geqslant 51$ ______

23. $4[5(1) - 3] < 8$ ______

24. $\dfrac{9(4) + 3(2)}{5 \cdot 4 + 1} < 3$ ______

25. $6 \cdot 5 + 4 < 8 \cdot 5 - 9$ ______

Find and graph the solution from the given domain.

26. $x > 5$; {2, 4, 6, 8, 10}

27. $3k \leqslant 9$; {0, 1, 2, 3, 4, 5}

28. $4y + 1 > 5$; {0, 1, 2, 3, 4}

2 Operations with the Real Numbers

2.1 THE REAL NUMBERS AND THE NUMBER LINE

OBJECTIVES

- Identify negative numbers
- Identify whole numbers
- Find the additive inverse of a number
- Tell which of two real numbers is smaller
- Find the absolute value of a number
- Solve equations with absolute value

Set up a number line and let the distance between 0 and 1 be the unit measure. Now go to the *left* of 0 and mark off points one unit to the left, two units to the left, three units to the left, and so on. (See Figure 1.) The points to the left of 0 are labeled with minus signs:

$$\ldots, -4, -3, -2, -1.$$

The three dots show that the numbers continue without end to the left.

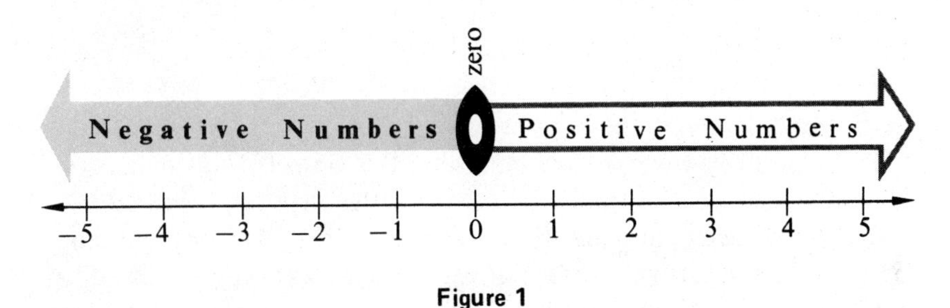

Figure 1

The numbers to the left of 0 on the number line are called **negative numbers.** To the right of 0, the numbers are called **positive numbers.** The number 0 itself is neither positive nor negative.

There are many examples of negative numbers used in practical applications. For example, the altitude of Badwater in Death Valley, California, is −282 feet (282 feet below sea level, which is taken to be 0). Some land in certain river deltas is at an elevation of −12 feet when compared to the river–the land is 12 feet lower than the river, which is held in levees. A temperature on a cold January day can be written as −10°, or 10 degrees below zero. A business which spends more than it takes in has a negative profit.

The set of numbers which corresponds to all the points on the number line is the set of **real numbers.** The set of real numbers is made up of negative numbers, positive numbers, and zero.

Several separate sets of numbers (called **subsets**) are contained within the total set of real numbers. These subsets are used so often in mathematics that they have been given special names. The subsets of the set of real numbers that we use in this text are as follows.

Whole Numbers	$\{0, 1, 2, 3, 4, \ldots\}$
Integers	$\{\ldots, -2, -1, 0, 1, 2, \ldots\}$
Rational Numbers	{all quotients of two integers, with nonzero denominators}
Irrational Numbers	{all real numbers that are not rational}

One property of the real numbers tells us that if we select any real number x (except 0), we can find exactly one number on the number line the same distance from 0 as x, but on the opposite side of 0.

For example, Figure 2 shows that the numbers 3 and -3 are each the same distance from 0 but are on opposite sides of 0. The numbers 3 and -3 are called **additive inverses** or **negatives** of each other.

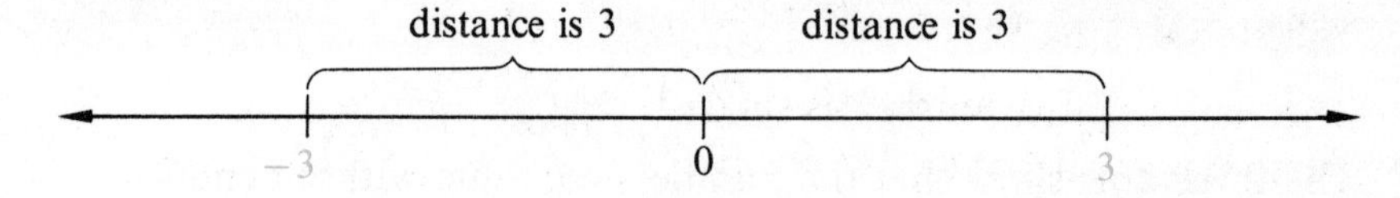

Figure 2

In general, the additive inverse of a number x is that number which is the same distance from 0 on the number line as x, but on the opposite side of 0.

The additive inverse of the number 0 is 0 itself. This makes 0 the only real number that is its own additive inverse. Other additive inverses occur in pairs. For example, 4 and -4, 3 and -3, and 5 and -5 are additive inverses of each other. Several pairs of additive inverses are shown in Figure 3.

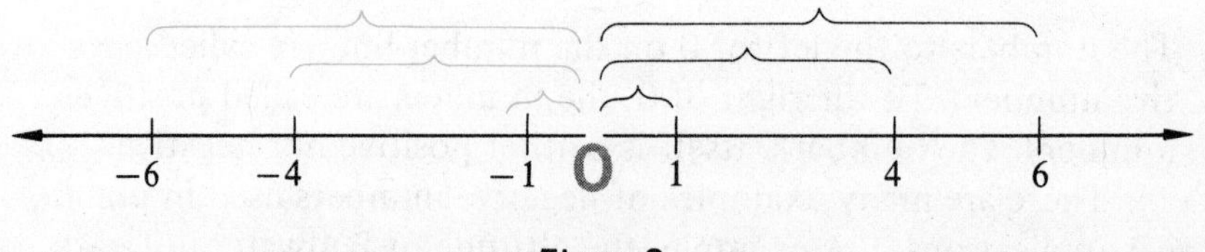

Figure 3

The additive inverse of a number can be indicated by writing the symbol $-$ in front of the number. With this symbol, the additive inverse of 7 is written -7. The additive inverse of -3 can be written $-(-3)$. We know that 3 is an additive inverse of -3. Since a number can have only one additive inverse, the symbols 3 and $-(-3)$ must really represent the same number, which means that

$$-(-3) = 3.$$

In general, for any real number x, it is true that $-(-x) = x$.

Example 1 Find the additive inverse of each number.

Number	Additive inverse
-3	3
-4	$-(-4)$, or 4
0	-0, or 0
-2	$-(-2)$, or 2
5	-5
19	-19

Work Quick Check 1 at the side.

If you are given any two whole numbers, you probably can tell which number is the smaller of the two. What happens when we look at negative numbers, such as those in the set of integers? Positive numbers increase as the corresponding points on the number line go to the right. For example, $8 < 12$, and 8 is to the left of 12 on the number line. We can extend this idea to all real numbers.

> The smaller of any two different real numbers is the one farther to the left on the number line.

Then any negative number is smaller than 0, and any negative number is smaller than any positive number. Also, 0 is smaller than any positive number.

Example 2 Is it true that $-3 < -1$?

To decide whether the statement $-3 < -1$ is true, locate both numbers, -3 and -1, on a number line, as in Figure 4. Since -3 is to the left of -1 on the number line, -3 is smaller than -1. The statement $-3 < -1$ is true.

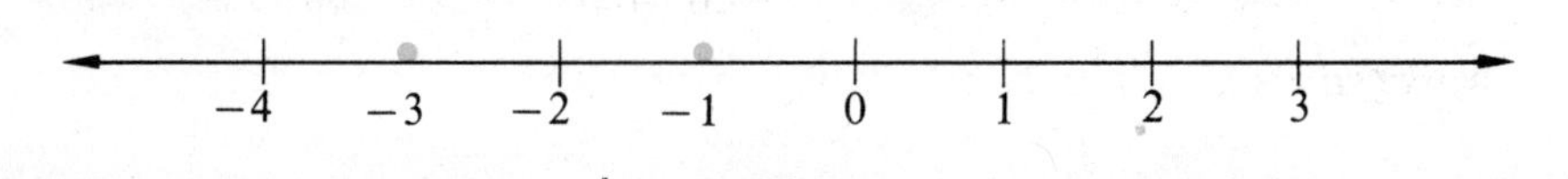

Figure 4

Work Quick Check 2 at the side.

Two numbers which are additive inverses of each other are the same distance from 0 on the number line. For example, 2 and -2 are additive inverses of each other, and 2 and -2 are each the same distance from 0 on the number line. To express this, we say that 2 and -2 have the same absolute value. The **absolute value** of a number is defined as its distance from 0 on the number line.

The symbol for the absolute value of the number x is $|x|$, read "the absolute value of x." For example, the distance from 2 to 0 on the number line is 2 units, so that

$$|2| = 2.$$

QUICK CHECKS

1. Find the additive inverses.

(a) 6

(b) 15

(c) -9

(d) -12

(e) 0

2. Answer *true* or *false*.

(a) $-2 < 4$

(b) $6 > -3$

(c) $-9 < -12$

(d) $-4 \geqslant -1$

(e) $-6 \leqslant 0$

Quick Check Answers
1. (a) -6 (b) -15 (c) 9 (d) 12 (e) 0
2. (a) True (b) True (c) False (d) False (e) True

Also, the distance from -2 to 0 on the number line is 2 units, so that

$$|-2| = 2.$$

Since distance is a physical measurement which is never negative, the absolute value of a number can never be negative. For example, $|12| = 12$ and $|-12| = 12$, since both 12 and -12 lie 12 units from 0 on the number line. Since 0 is 0 units from 0, we have $|0| = 0$.

Example 3 Simplify by removing absolute value symbols.

(a) $|5| = 5$

(b) $|-5| = 5$

(c) $-|5| = -(5) = -5$

(d) $-|-14| = -(14) = -14$

(e) $-|-8| = -(8) = -8$

Work Quick Check 3 at the side.

Example 4 Find the solutions. The domain of the variable is $\{-4, -3, -2, -1, 0, 1, 2, 3, 4\}$.

(a) $|x| = 2$

The equation $|x| = 2$ is true if x is replaced with 2, since $|2| = 2$. The equation is also true if x is replaced with -2, since $|-2| = 2$. Therefore, the numbers 2 and -2 are the solutions of the equation $|x| = 2$.

(b) $|x| \leq 2$

By substitution from the domain above, check that $-2, -1, 0, 1$, and 2 are the solutions of this inequality.

Work Quick Check 4 at the side.

3. Simplify by removing absolute value symbols.

(a) $|-6| =$

(b) $|9| =$

(c) $|-36| =$

(d) $-|15| =$

(e) $-|-9| =$

(f) $-|-30| =$

4. Find the solutions. The domain is $\{-4, -3, -2, -1, 0, 1, 2, 3, 4\}$.

(a) $|x| = 3$

(b) $|x| \leq 3$

(c) $|x| < 1$

(d) $|x| > 1$

Quick Check Answers

3. (a) 6 (b) 9 (c) 36 (d) -15 (e) -9 (f) -30

4. (a) 3, -3 (b) $-3, -2, -1, 0, 1, 2, 3$ (c) 0 (d) $-4, -3, -2, 2, 3, 4$

name date hour

2.1 EXERCISES

Give the additive inverse of each number. For the exercises with absolute value, simplify first before deciding on the additive inverse. See Examples 1 and 3.

1. 8 ______ 2. 12 ______ 3. -9 ______

4. -11 ______ 5. -2 ______ 6. -3 ______

7. $|15|$ ______ 8. $|5|$ ______ 9. $|8|$ ______

Circle the smaller of the two given numbers. See Examples 2 and 3.

10. $-5, 5$ 11. $9, -3$ 12. $-12, -4$

13. $-9, -14$ 14. $-8, -1$ 15. $-15, -16$

16. $3, |-4|$ 17. $5, |-2|$ 18. $|-3|, |-4|$

19. $|-8|, |-9|$ 20. $-|-6|, -|-4|$ 21. $-|-2|, -|-3|$

Write true *or* false *for each statement. See Examples 2 and 3.*

22. $-2 < -1$ ______ 23. $-8 < -4$ ______ 24. $-3 \geqslant -7$ ______

25. $-9 \geqslant -12$ ______ 26. $-15 \leqslant -20$ ______ 27. $-21 \leqslant -27$ ______

28. $6 > -(-2)$ ______ 29. $-8 > -(-2)$ ______ 30. $-4 < -(-5)$ ______

31. $|-6| < |-9|$ ______ 32. $|-12| < |-20|$ ______ 33. $-|8| > |-9|$ ______

34. $-|12| > |-15|$ ______ 35. $-|-5| \geqslant -|-9|$ ______

Graph each group of numbers on the indicated number line. Simplify the expressions having absolute value bars before graphing them.

36. $0, 3, -5, -6$

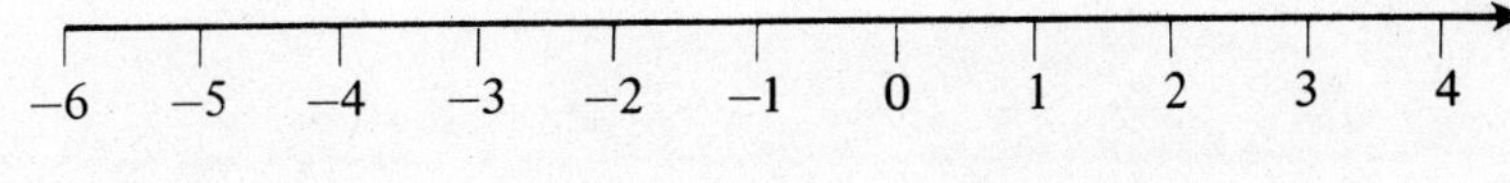

37. $2, 6, -2, -1$

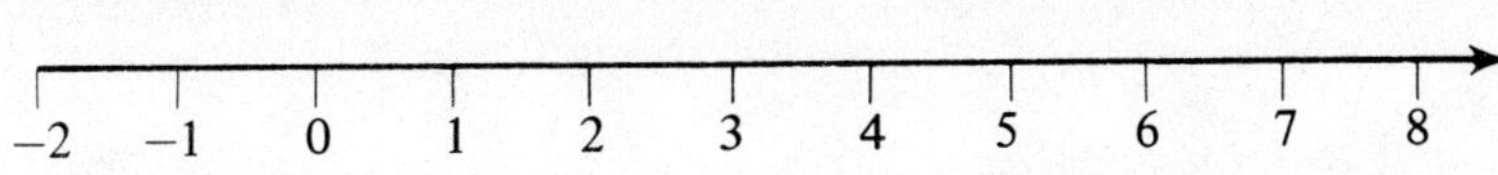

38. $-2, -6, |-4|, 3, -|4|$

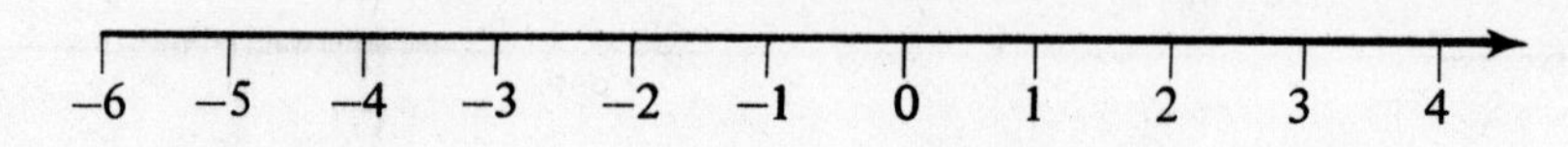

39. $-5, -3, -|-2|, -0, |-4|$

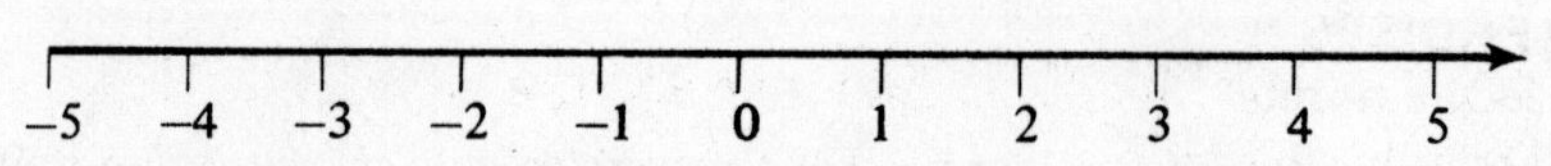

40. $|3|, -|3|, -|-4|, -|-2|$

–5 –4 –3 –2 –1 0 1 2 3 4 5

Solve each equation or inequality. See Example 4. Assume that the domain of the variable is $\{-4, -3, -2, -1, 0, 1, 2, 3, 4\}$.

41. $|x| = 3$ ______________________ 42. $|x| = 4$ ______________________

43. $|x| < 2$ ______________________ 44. $|x| \leqslant 4$ ______________________

45. $|x| \leqslant 1$ ______________________ 46. $|x| < 3$ ______________________

47. $|x| > 1$ ______________________ 48. $|x| > 2$ ______________________

49. $|x| < 1$ ______________________ 50. $|x| \geqslant 4$ ______________________

51. $|x| \geqslant 0$ ______________________ 52. $|x| \leqslant 0$ ______________________

OBJECTIVES

- Add real numbers on the number line
- Add in your head
- Use the order of operations with real numbers

We can use the number line to illustrate the addition of real numbers, as shown by the following examples.

Example 1 Use the number line to find the sum $2 + 3$.

To add the positive numbers 2 and 3 on the number line, start at 0 and draw an arrow two units to the right as shown in Figure 5. This arrow represents the number 2 in the sum $2 + 3$. Then, from the right end of this arrow, draw another arrow 3 units to the right, as shown in Figure 5. The coordinate at the end of this second arrow shows that $2 + 3 = 5$.

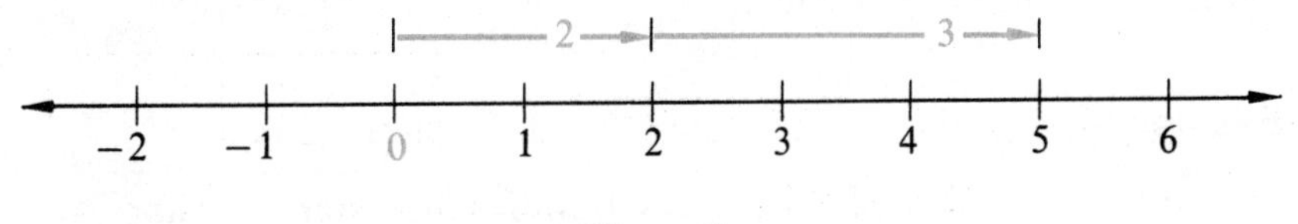

Figure 5

Example 2 Use the number line to find the sum $-2 + (-4)$. (We placed parentheses around the -4 to avoid the confusing use of $+$ and $-$ next to each other.)

To add the negative numbers -2 and -4 on the number line, again start at 0 and draw an arrow two units to the *left*. We draw the arrow to the left to represent the addition of a *negative* number. From the left end of this first arrow, draw a second arrow four units to the left, as shown in Figure 6. The coordinate at the end of this second arrow shows that $-2 + (-4) = -6$.

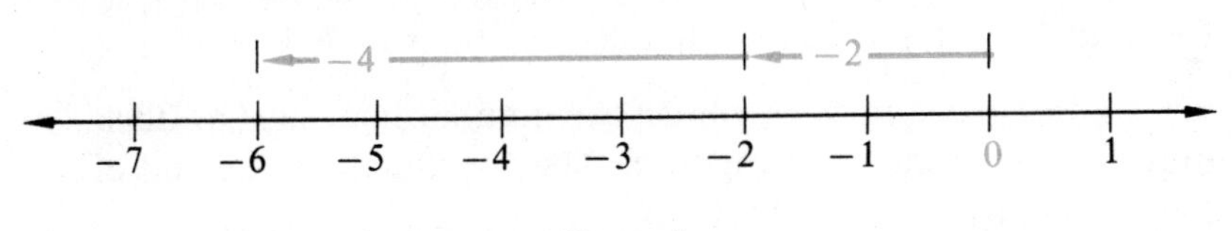

Figure 6

Work Quick Check 1 at the side.

QUICK CHECKS

1. Use the number line to find the sums.

(a) $1 + 4$

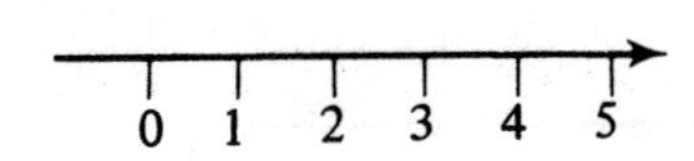

(b) $-2 + (-5)$

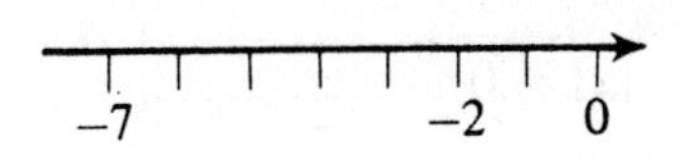

(c) $-3 + (-1)$

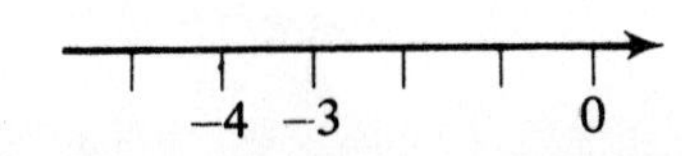

Refer back to Example 2 to see that the sum of the two negative numbers -2 and -4 is a negative number whose distance from 0 is the sum of the distance of -2 from 0 and the distance of -4 from 0. That is, *the sum of two negative numbers is the negative of the sum of their absolute values.*

$$-2 + (-4) = -(|-2| + |-4|) = -6$$

If both x and y are negative numbers, then

$$x + y = -(|x| + |y|).$$

Quick Check Answers

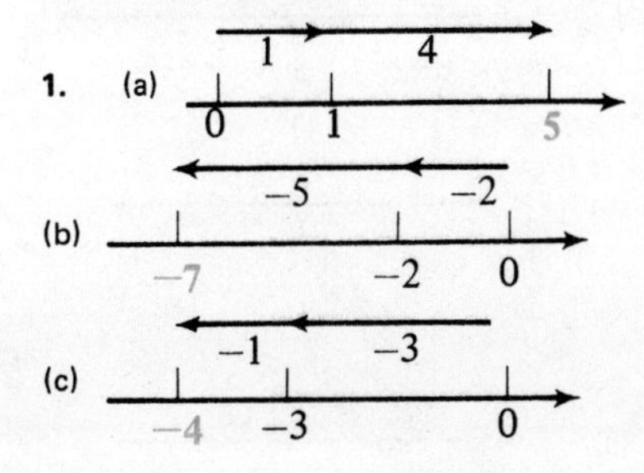

2. Find the sums. Use the formula $x+y=-(|x|+|y|)$.

(a) $-3+(-9)$

(b) $-8+(-11)$

(c) $-5+(-12)$

3. Use the number line to find the sums.

(a) $6+(-3)$

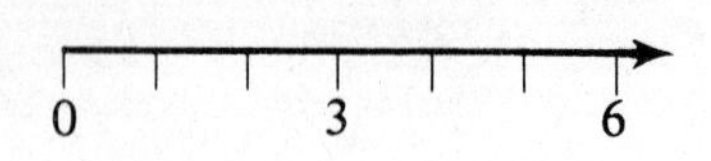

(b) $-5+1$

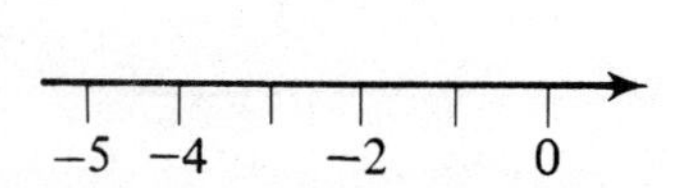

(c) $-7+2$

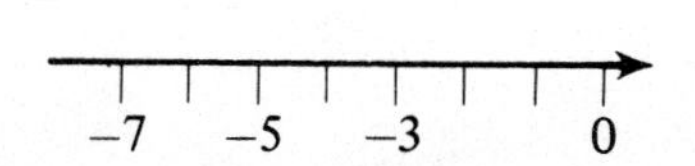

Quick Check Answers

2. (a) -12 (b) -19 (c) -17

3. (a)

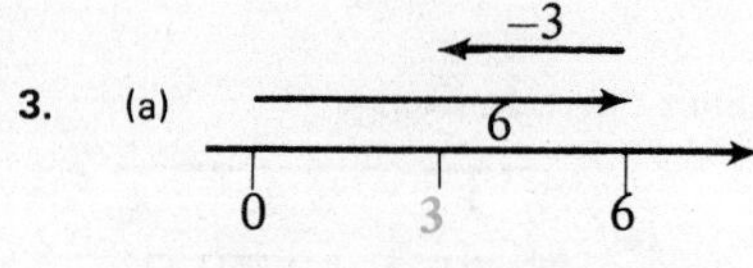

(b)

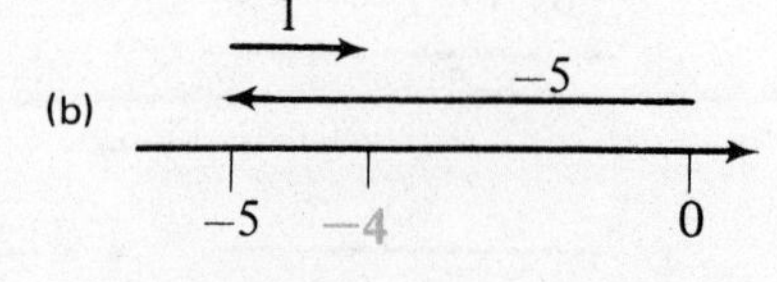

(c)

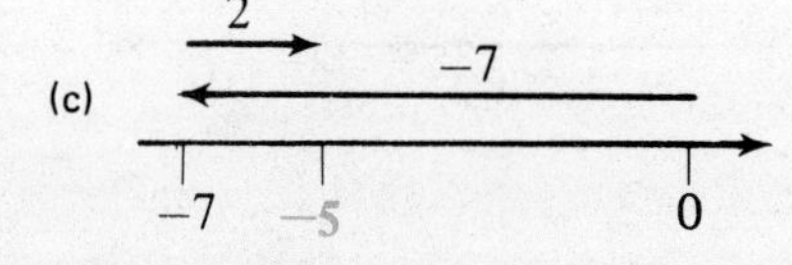

Example 3 Find the sums using the formula

$$x+y=-(|x|+|y|).$$

(a) $-2+(-9)=-11$
$=-(|-2|+|-9|)=-(2+9)=-11$

(b) $-8+(-12)=-20$
$=-(|-8|+|-12|)=-(8+12)=-20$

(c) $-15+(-3)=-18$
$=-(|-15|+|-3|)=-(15+3)=-18$

Work Quick Check 2 at the side.

To find the sum of a positive number and a negative number, we again use the number line.

Example 4 Use the number line to find the sum $-2+5$.

To find the sum $-2+5$ on the number line, start at 0 and draw an arrow two units to the left. From the left end of this arrow, draw a second arrow 5 units to the right, as shown in Figure 7. The coordinate at the end of the second arrow shows that $-2+5=3$.

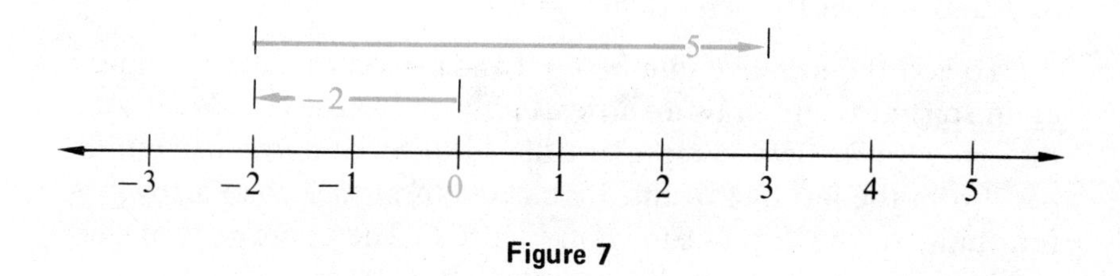

Figure 7

Example 5 Use the number line to find the sum $4+(-6)$.

To find the sum $4+(-6)$ on the number line, start at 0 and draw arrows as shown in Figure 8. The coordinate at the end of the second arrow shows that $4+(-6)=-2$.

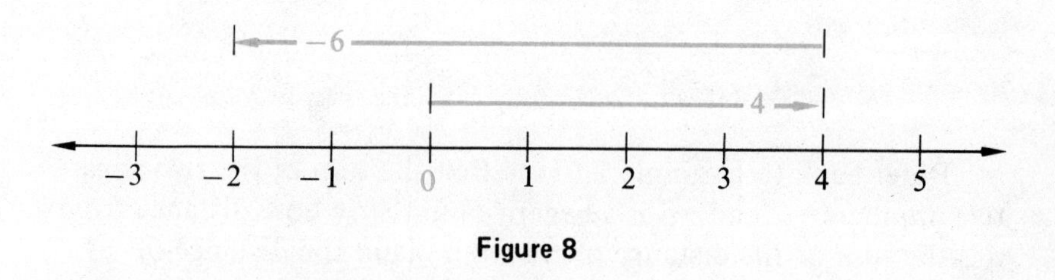

Figure 8

Work Quick Check 3 at the side.

Example 6 Check each answer, trying to work the addition in your head. If you get stuck, use a number line.

(a) $7+(-4)=3$

(b) $-8+12=4$

(c) $-11 + 15 = 4$

(d) $-\frac{1}{2} + \frac{1}{8} = -\frac{3}{8}$ (Remember to obtain a common denominator first.)

(e) $\frac{5}{6} + \left(-\frac{4}{3}\right) = -\frac{1}{2}$

(f) $-4.6 + 8.1 = 3.5$

Work Quick Check 4 at the side.

Sometimes a problem involves square brackets, [ ]. As mentioned in Chapter 1, brackets are treated just like parentheses. Do the calculations inside the brackets until a single number is obtained. Remember to use the order of operations given in Section 1.3 when adding more than two numbers.

Example 7 Find the sums.

(a) $-3 + [4 + (-8)]$

First work inside the brackets. Follow the rules for the order of operations given in Section 1.3.

$$-3 + [4 + (-8)] = -3 + (-4) = -7$$

(b) $$8 + [(-2 + 6) + (-3)] = 8 + [4 + (-3)] = 8 + 1 = 9$$

Work Quick Check 5 at the side.

4. Check each answer, trying to work the addition in your head. If you get stuck, use a number line.

(a) $-8 + 2 = -6$

(b) $-15 + 4 = -11$

(c) $17 + (-10) = 7$

(d) $\frac{3}{4} + \left(-\frac{11}{8}\right) = -\frac{5}{8}$

(e) $-9.5 + 3.8 = -5.7$

5. Find the sums.

(a) $2 + [7 + (-3)]$

(b) $6 + [(-2 + 5) + 7]$

(c) $-9 + [-4 + (-8 + 6)]$

Quick Check Answers

4. All are correct.

5. (a) 6 (b) 16 (c) −15

WORK SPACE

name date hour

2.2 EXERCISES

Find the sums. See Examples 1-6.

1. $5 + (-3) =$ ______

2. $11 + (-8) =$ ______

3. $6 + (-8) =$ ______

4. $3 + (-7) =$ ______

5. $-6 + (-2) =$ ______

6. $-8 + (-3) =$ ______

7. $-9 + (-2) =$ ______

8. $-15 + (-6) =$ ______

9. $-3 + (-9) =$ ______

10. $-11 + (-5) =$ ______

11. $12 + (-8) =$ ______

12. $10 + (-2) =$ ______

13. $4 + [13 + (-5)] =$ ______

14. $6 + [2 + (-13)] =$ ______

15. $8 + [-2 + (-1)] =$ ______

16. $12 + [-3 + (-4)] =$ ______

17. $-2 + [5 + (-1)] =$ ______

18. $-8 + [9 + (-2)] =$ ______

19. $-6 + [6 + (-9)] =$ ______

20. $-3 + [4 + (-8)] =$ ______

21. $[9 + (-2)] + 6 =$ ______

22. $[8 + (-14)] + 10 =$ ______

23. $[(-9) + (-14)] + 12 =$ ______

24. $[(-8) + (-6)] + 10 =$ ______

25. $-\frac{1}{6} + \frac{2}{3} =$ ______

26. $\frac{9}{10} + \left(-\frac{3}{5}\right) =$ ______

27. $\frac{5}{8} + \left(-\frac{17}{12}\right) =$ ______

28. $-\frac{6}{25} + \frac{19}{20} =$ ______

29. $2\frac{1}{2} + \left(-3\frac{1}{4}\right) =$ ______

30. $-4\frac{3}{8} + 6\frac{1}{2} =$ ______

31. $7.9 + (-8.4) =$ ______

32. $11.6 + (-15.4) =$ ______

33. $-6.1 + [3.2 + (-4.8)] =$ ______

34. $-9.4 + [-5.8 + (-1.4)] =$ ______

35. $[-3 + (-4)] + [5 + (-6)] =$ ______

36. $[-8 + (-3)] + [-7 + (-6)] =$ ______

37. $[-4 + (-3)] + [8 + (-1)] =$ ______

38. $[-5 + (-9)] + [16 + (-21)] =$ ______

39. $[-4 + (-6)] + [(-3) + (-8)] + [12 + (-11)] =$ ______

40. $[-2 + (-11)] + [12 + (-2)] + [18 + (-6)] =$ ______

Write true *or* false.

41. $-4 + 0 = -4$ ______

42. $-6 + 5 = -1$ ______

43. $-8 + 12 = 8 + (-12)$ ______

44. $15 + (-8) = 8 + (-15)$ ______

45. $-9 + 5 + 6 = -2$ ______

46. $-6 + 3 = 3$ ______

47. $-3 + 5 = 5 + (-3)$ ______

48. $11 + (-6) = -6 + 11$ ______

49. $[4 + (-6)] + 6 = 4 + (-6 + 6)$ ______

50. $[(-2) + (-3)] + (-6) = 12 + (-1)$ ______

51. $-7 + [-5 + (-3)] = [(-7) + (-5)] + 3$ ______

52. $6 + [-2 + (-5)] = [(-4) + (-2)] + 5$ ______

53. $-5 + (-|-5|) = -10$ ______

54. $|-3| + (-5) = -2$ ______

name date hour

Find all solutions for each equation. Let the domain of x *be* $\{-3, -2, -1, 0, 1, 2, 3\}$.

55. $x + 2 = 0$ ________

56. $x + 3 = 0$ ________

57. $x + 1 = -2$ ________

58. $x + 2 = -1$ ________

59. $14 + x = 12$ ________

60. $x + 8 = 7$ ________

61. $x + (-4) = -6$ ________

62. $x + (-2) = -5$ ________

63. $-8 + x = -6$ ________

64. $-2 + x = -1$ ________

Solve the problems.

65. Joann has \$15. She then spends \$6. What is her balance at that time? ________

66. An airplane is at an altitude of 6000 feet. It then descends 4000 feet. What is the final altitude? ________

67. Chuck is standing 15 feet below sea level in Death Valley. He then goes down another 120 feet. Find his final altitude. ________

68. Hiram has \$11 and spends \$19. What is his final balance? ________

69. One measure of Nancy's blood pressure is 120. It then changes by −30. Find the new blood pressure. ________

70. The temperature is −14°. It then goes down 12°. Find the new temperature. ________

WORK SPACE

OBJECTIVES

- Find a difference on the number line
- Use the definition of subtraction
- Work subtraction problems involving brackets

We already know how to subtract a positive number from a larger positive number (for example, $7-4$). The answer to such a subtraction problem is called a **difference**, and can also be found by using a number line. Since addition of a positive number on the number line is shown by an arrow to the right, it is reasonable to represent subtraction of a positive number by an arrow going to the left.

Example 1 Use the number line to find the difference of $7-4$.

To find the difference of $7-4$ on the number line, begin at 0 and draw an arrow 7 units to the right. From the right end of this arrow, draw an arrow 4 units to the left, as shown in Figure 9. The coordinate at the end of the second arrow shows that $7-4=3$.

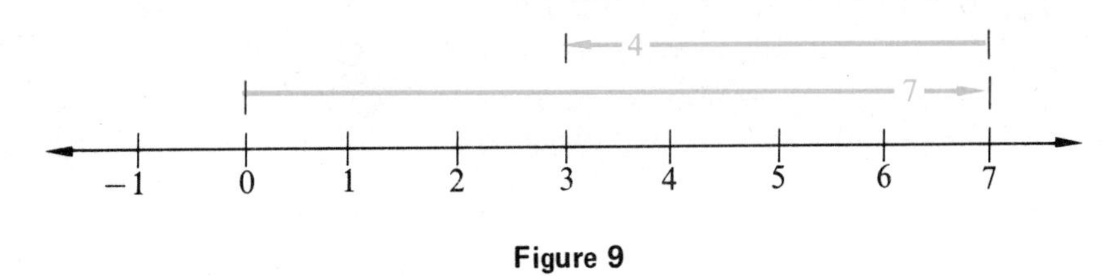

Figure 9

Work Quick Check 1 at the side.

The procedure used above to find $7-4$ is exactly the same procedure that would be used to find $7+(-4)$ so that

$$7-4=7+(-4).$$

Based on this, it seems that subtraction of a positive number from a larger positive number is the same as adding the additive inverse of the smaller number to the larger. We extend this definition of subtraction to all real numbers.

For any two real numbers x and y,

$$x-y=x+(-y).$$

In other words, to subtract y from x, add the additive inverse of y to x.

Example 2 Find the differences using the formula $x-y=x+(-y)$.

(a) $12-3=12+(-3)=9$

(b) $5-7=5+(-7)=-2$

(c) $-3-(-5)=-3+(5)=2$

(d) $8-(-5)=8+(5)=13$

Work Quick Check 2 at the side.

QUICK CHECKS

1. Use the number line to find the differences.

(a) $5-1$

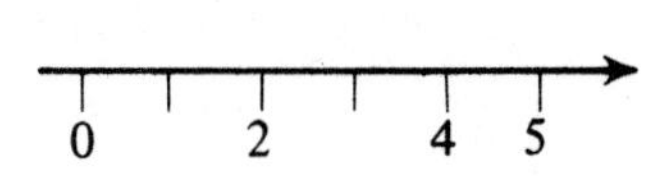

(b) $6-2$

0 2 4 6

2. Find the differences using the formula $x-y=x+(-y)$.

(a) $6-10$

(b) $-2-4$

(c) $3-(-5)$

(d) $-8-(-12)$

Quick Check Answers

1. (a)

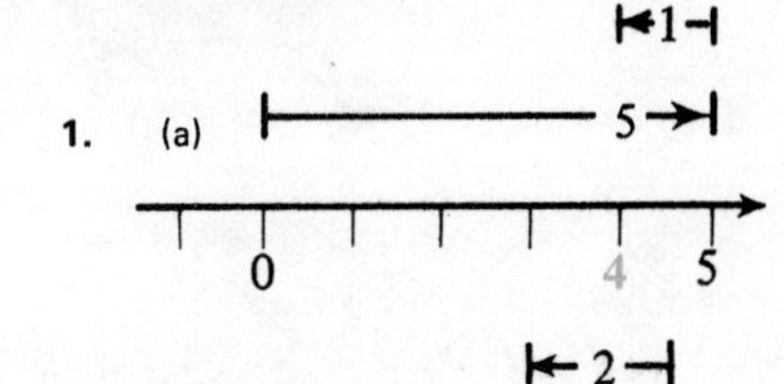

(b)

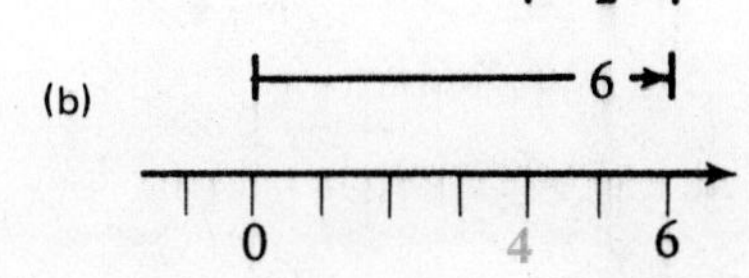

2. (a) -4 (b) -6 (c) 8 (d) 4

3. Perform the indicated operations.

(a) $2 - [(-3) - (4 + 6)]$

(b) $[(5 - 7) + 3] - 8$

The operation of subtraction is the inverse of the operation of addition since subtraction can be used to reverse the result of an addition problem. For example, if 4 is added to a number, and then subtracted from the sum, the original number is the result.

We have now used the symbol $-$ for a variety of purposes. One use is to represent subtraction, as in $9 - 5$. The symbol $-$ is also used to represent negative numbers, such as $-10, -2, -3,$ and so on. Finally, the symbol $-$ is used to represent the additive inverse of a number. More than one use may appear in the same problem, such as $-6 - (-9)$, where -9 is subtracted from -6. The meaning of the symbol is usually clear from the context of the problem.

Example 3 When working problems involving both parentheses and brackets, first do any operations inside the parentheses or brackets from the inside out.

(a)
$$\begin{aligned} -6 - [2 - (8 + 3)] &= -6 - [2 - 11] \\ &= -6 - (-9) \\ &= -6 + (9) \\ &= 3 \end{aligned}$$

(b)
$$\begin{aligned} 5 - [(-3 - 2) - (4 - 1)] &= 5 - [(-3 + (-2)) - 3] \\ &= 5 - [(-5) - 3] \\ &= 5 - [(-5) + (-3)] \\ &= 5 - (-8) \\ &= 5 + 8 = 13 \end{aligned}$$

Work Quick Check 3 at the side.

Quick Check Answers

3. (a) 15 (b) −7

name ________ date ________ hour ________

2.3 EXERCISES

Subtract. See Examples 1 and 2.

1. $3 - 6 =$ ________

2. $7 - 12 =$ ________

3. $5 - 9 =$ ________

4. $8 - 13 =$ ________

5. $-6 - 2 =$ ________

6. $-11 - 4 =$ ________

7. $-9 - 5 =$ ________

8. $-12 - 15 =$ ________

9. $6 - (-3) =$ ________

10. $8 - (-5) =$ ________

11. $5 - (-12) =$ ________

12. $12 - (-2) =$ ________

13. $-6 - (-2) =$ ________

14. $-7 - (-5) =$ ________

15. $2 - (3 - 5) =$ ________

16. $5 - (6 - 13) =$ ________

17. $-2 - (5 - 8) =$ ________

18. $-3 - (4 - 11) =$ ________

19. $\frac{1}{2} - \left(-\frac{1}{4}\right) =$ ________

20. $\frac{1}{3} - \left(-\frac{4}{3}\right) =$ ________

21. $-\frac{3}{4} - \frac{5}{8} =$ ________

22. $-\frac{5}{6} - \frac{1}{2} =$ ________

23. $\frac{5}{8} - \left(-\frac{1}{2} - \frac{3}{4}\right) =$ ________

24. $\frac{9}{10} - \left(\frac{1}{8} - \frac{3}{10}\right) =$ ________

25. $3.4 - (-8.2) =$ ________

26. $5.7 - (-11.6) =$ ________

27. $-6.4 - 3.5 =$ ________

28. $-4.4 - 8.6 =$ ________

Perform the indicated operations. See Example 3.

29. $(4-6)+12=$ ______

30. $(3-7)+4=$ ______

31. $(8-1)-12=$ ______

32. $(9-3)-15=$ ______

33. $6-(-8+3)=$ ______

34. $8-(-9+5)=$ ______

35. $2+(-4-8)=$ ______

36. $6+(-9-2)=$ ______

37. $(-5-6)-(9-2)=$ ______

38. $(-4-8)-(6-1)=$ ______

39. $(-8-2)-(-9-3)=$ ______

40. $(-4-2)-(-8-1)=$ ______

41. $-9-[(3-2)-(-4-2)]=$ ______

42. $-8-[(-4-1)-(9-2)]=$ ______

43. $-3+[(-5-8)-(-6+2)]=$ ______

44. $-4+[(-12+1)-(-1-9)]=$ ______

Write the given problem in symbols (no variables are needed). Then solve.

	Problem in symbols	Solution
45. Subtract −6 from 12.	______	______
46. Subtract −8 from 15.	______	______
47. From −25, subtract −4.	______	______
48. What number is 6 less than −9?	______	______
49. −24 is how much greater than −27?	______	______
50. How much greater is 8 than −5?	______	______

name date hour

Work the word problems.

51. The temperature dropped 10° below the previous temperature of −5°. Find the new temperature. ______

52. Bill owed his brother $10. He repaid $6 and later borrowed $7. What positive or negative number represents his present financial status? ______

53. The bottom of Death Valley is 282 feet below sea level. The top of Mt. Whitney, visible from Death Valley, has an altitude of 14,494 feet above sea level. Using zero as sea level, find the difference between these two elevations. ______

54. Harriet has $15, while Thomasina is $12 in debt. Find the difference in their financial positions. ______

OBJECTIVES

- Multiply a positive and a negative number
- Multiply two negative numbers
- Simplify numerical expressions with positive and negative numbers
- Substitute numerical values for variables

Any rules we develop for multiplication of real numbers ought to be consistent with the rules for multiplication of positive numbers. For example, we would want the product of 0 and any real number (positive or negative) to be 0.

$$x \cdot 0 = 0$$

In order to define the product of a positive and a negative number so that the result is consistent with our definition of multiplying two positive numbers, we look at the following pattern.

$$3 \cdot \mathbf{5} = 15$$
$$3 \cdot \mathbf{4} = 12$$
$$3 \cdot \mathbf{3} = 9$$
$$3 \cdot \mathbf{2} = 6$$
$$3 \cdot \mathbf{1} = 3$$
$$3 \cdot \mathbf{0} = 0$$
$$3 \cdot (-1) = ?$$

What number should we assign to the product $3(-1)$ so that the pattern is maintained? The numbers on the left of the equals sign (in heavy type) decrease by 1 as we go down the list. Also, the products on the right decrease by 3 as we go down the list. To maintain this pattern, the number on the right in the bottom row must be 3 less than 0, which is -3. Therefore, we must have

$$3(-1) = -3.$$

The pattern continues with

$$3(-2) = -6$$
$$3(-3) = -9$$
$$3(-4) = -12,$$

and so on.

In general, if both x and y represent positive numbers, then

$$\mathbf{x(-y) = -(xy) \quad \text{or}}$$
$$\mathbf{(-x)y = -(xy).}$$

In other words, the product of a positive number and a negative number is negative.

Example 1 Find the products using the multiplication definitions given above.

(a) $8(-5) = -(8 \cdot 5) = -40$

(b) $5(-4) = -(5 \cdot 4) = -20$

(c) $(-7)(2) = -(7 \cdot 2) = -14$

(d) $(-9)(3) = -(9 \cdot 3) = -27$

Work Quick Check 1 at the side.

QUICK CHECKS

1. Find the products.

(a) $2(-6)$

(b) $(7)(-8)$

(c) $(-9)(2)$

(d) $(-10)(3)$

Quick Check Answers

1. (a) -12 (b) -56 (c) -18 (d) -30

2. Find the products.

(a) $(-5)(-6)$

(b) $(-7)(-3)$

(c) $(-8)(-5)$

(d) $(-11)(-2)$

We know that the product of two positive numbers is positive, and we have seen that the product of a positive and a negative number is negative. What about the product of two negative numbers? Consider another pattern.

$$(-5)(4) = -20$$
$$(-5)(3) = -15$$
$$(-5)(2) = -10$$
$$(-5)(1) = -5$$
$$(-5)(0) = 0$$
$$(-5)(-1) = ?$$

The numbers on the left of the equals sign decrease by 1 as we go down the list. Also, the products on the right increase by 5 as we go down the list. To maintain this pattern, we will have to agree that $(-5)(-1)$ is 5 more than $(-5)(0)$. Therefore, we must have

$$(-5)(-1) = 5.$$

The pattern continues with

$$(-5)(-2) = 10$$
$$(-5)(-3) = 15$$
$$(-5)(-4) = 20$$
$$(-5)(-5) = 25,$$

and so on.

In general, if x and y both represent positive numbers, then

$$(-x)(-y) = xy.$$

In other words, the product of two negative numbers is positive.

Example 2 Find the products using the multiplication formula given above.

(a) $(-9)(-2) = 9 \cdot 2 = 18$

(b) $(-6)(-12) = 6 \cdot 12 = 72$

(c) $(-8)(-1) = 8 \cdot 1 = 8$

(d) $(-15)(-2) = 15 \cdot 2 = 30$

Work Quick Check 2 at the side.

Example 3 Simplify according to the multiplication definitions given in this chapter and the rules for order of operations given in Chapter 1.

(a) $(-9)(2) - (-3)(2)$

To simplify, first find all the products working from left to right.

$$(-9)(2) - (-3)(2) = -18 - (-6)$$

Now perform the subtraction.

$$= -18 + 6$$
$$= -12$$

Quick Check Answers

2. (a) 30 (b) 21 (c) 40 (d) 22

(b) $(-6)(-2) - (3)(-4) = 12 - (-12)$
$= 12 + 12$
$= 24$

(c) $-5(-2 - 3) = -5(-5) = 25$

Work Quick Check 3 at the side.

3. Simplify.

(a) $(-3)(4) - (2)(6)$

(b) $(-5)(-6) - (8)(-3)$

(c) $-7(-2 - 5)$

Example 4 Evaluate $(3x + 4y)(-2m)$ given $x = -1$, $y = -2$, and $m = -3$.

First substitute the given values. Then find the value of the expression.

$$\begin{aligned}(3x + 4y)(-2m) &= [3(-1) + 4(-2)]\,[-2(-3)]\\ &= [-3 + (-8)]\,[6]\\ &= (-11)(6)\\ &= -66\end{aligned}$$

Work Quick Check 4 at the side.

4. Evaluate.

(a) $(-3x)(4x - 2y)$ if $x = 2$ and $y = -1$.

(b) $(5x + 3y)(-2x)$ if $x = 1$ and $y = -3$.

Quick Check Answers

3. (a) -24 (b) 54 (c) 49
4. (a) -60 (b) 8

WORK SPACE

name date hour

2.4 EXERCISES

Find the products. See Examples 1 and 2.

1. $(-3)(-4) =$ ______
2. $(-3)(4) =$ ______
3. $3(-4) =$ ______
4. $-2(-8) =$ ______
5. $(-1)(-5) =$ ______
6. $(-9)(-5) =$ ______
7. $(-4)(-11) =$ ______
8. $(-5)(7) =$ ______
9. $(-10)(-12) =$ ______
10. $9(-5) =$ ______
11. $(8)(-6) =$ ______
12. $(13)(-2) =$ ______
13. $(-6)(5) =$ ______
14. $(-9)(0) =$ ______
15. $(13)(-5) =$ ______
16. $(12)(5) =$ ______
17. $0(-11) =$ ______
18. $3(-15) =$ ______
19. $(15)(-11) =$ ______
20. $(-9)(-4) =$ ______
21. $\left(-\frac{3}{8}\right)\left(-\frac{10}{9}\right) =$ ______
22. $\left(-\frac{5}{4}\right)\left(\frac{6}{15}\right) =$ ______
23. $(-5.1)(.02) =$ ______
24. $(-3.7)(-2.1) =$ ______

Perform the indicated operations. See Example 3.

25. $9(6 - 10) =$ ______
26. $5(12 - 15) =$ ______
27. $-6(2 - 4) =$ ______
28. $-9(5 - 8) =$ ______
29. $(4 - 9)(2 - 3) =$ ______
30. $(6 - 11)(3 - 6) =$ ______
31. $(2 - 5)(3 - 7) =$ ______
32. $(5 - 12)(2 - 6) =$ ______
33. $(-4 - 3)(-2) + 4 =$ ______
34. $(-5 - 2)(-3) + 6 =$ ______
35. $5(-2) - 4 =$ ______
36. $9(-6) - 8 =$ ______
37. $3(-4) - (-2) =$ ______
38. $5(-2) - (-9) =$ ______

39. $(-8-2)(-4)-(-5) =$ ______

40. $(-9-1)(-2)-(-6) =$ ______

41. $|-4(-2)| + |-4| =$ ______

42. $|8(-5)| + |-2| =$ ______

43. $|2|(-4) + |6| \cdot |-4| =$ ______

44. $|-3|(-2) + |-8| \cdot |5| =$ ______

Evaluate the expressions for $x = -2$, $y = 3$, and $a = -4$. See Example 4.

45. $2x + 7y =$ ______

46. $3x + 5y =$ ______

47. $5x - 2y + 3a =$ ______

48. $6x - 5y + 4a =$ ______

49. $(2x + y)(3a) =$ ______

50. $(5x - 2y)(-2a) =$ ______

51. $(3x - 4y)(-5a) =$ ______

52. $(6x + 2y)(-3a) =$ ______

Find the solution for each equation. The domain of the variable is $\{-3, -2, -1, 0, 1, 2, 3\}$.

53. $2x = -4$ ______

54. $3x = -6$ ______

55. $-4x = 0$ ______

56. $-9x = 0$ ______

57. $-8x = 16$ ______

58. $-9x = 27$ ______

2.5 DIVISION OF REAL NUMBERS

OBJECTIVES

- Find the reciprocal or multiplicative inverse of a number
- Divide positive and negative numbers
- Learn the meaning of *factor*

We have seen that a subtraction problem can be worked by adding the additive inverse of the second number to the first. Division problems are related to multiplication in a similar way. To subtract, we use the additive inverse–to divide, we use the multiplicative inverse. The multiplicative inverse of 8 is 1/8, and of 5/4 is 4/5. Note that

$$8 \cdot \frac{1}{8} = \frac{8}{8} = 1 \quad \text{and} \quad \frac{5}{4} \cdot \frac{4}{5} = \frac{20}{20} = 1.$$

(Remember that any nonzero number divided by itself is 1.)

Pairs of numbers whose product is 1 are **multiplicative inverses** or **reciprocals** of each other.

Example 1 Find the multiplicative inverse of each number.

Number	Multiplicative Inverse (Reciprocal)
4	$\frac{1}{4}$
-5	$-\frac{1}{5}$ or $\frac{1}{-5}$
$\frac{3}{4}$	$\frac{4}{3}$
$-\frac{5}{8}$	$-\frac{8}{5}$
0	None

Why is there no multiplicative inverse for the number 0? Suppose k is to be the multiplicative inverse of 0. Then $k \cdot 0$ should equal 1. But $k \cdot 0 = 0$ for any number k. So there is no value of k which is a solution of the equation

$$k \cdot 0 = 1.$$

Thus, the number 0 has no multiplicative inverse.

Work Quick Check 1 at the side.

1. Complete the chart.

	Number	Multiplicative Inverse
(a)	6	______
(b)	-2	______
(c)	$\frac{2}{3}$	______
(d)	$-\frac{1}{4}$	______
(e)	0	______

QUICK CHECKS

Division of real numbers x and y (that is, the division of x by y) is defined as the multiplication of x by the multiplicative inverse of y. That is, for any real numbers x and y, where $y \neq 0$, the quotient of x and y is

$$\frac{x}{y} = x \cdot \frac{1}{y}.$$

In the definition above, we said that y, the number we divide by, cannot be 0. The reason is that 0 has no multiplicative inverse,

Quick Check Answers
1. (a) 1/6 (b) −1/2 (c) 3/2 (d) −4 (e) none

so that 1/0 is not a number. For this reason, division by 0 is meaningless and is never permitted. If a division problem turns out to involve division by 0, write "no such number."

Example 2 Find the quotients using the definition of division.

(a) $\frac{12}{3} = 12 \cdot \frac{1}{3} = 4$

(b) $\frac{-10}{2} = -10 \cdot \frac{1}{2} = -5$

(c) $\frac{8}{-4} = 8 \cdot \left(\frac{1}{-4}\right) = -2$

(d) $\frac{-14}{-7} = -14\left(\frac{1}{-7}\right) = 2$

(e) $\frac{-100}{-20} = -100\left(\frac{1}{-20}\right) = 5$

(f) $\frac{-10}{0}$ no such number

Work Quick Check 2 at the side.

In practice, when we divide with positive or negative numbers, we disregard all signs and divide normally. The answer will be positive if both numbers (the divisor and the dividend) have like signs. The answer is negative if the numbers have unlike signs.

Example 3 Find the quotients.

(a) $\frac{8}{-2} = -4$

(b) $\frac{-45}{-9} = 5$

Work Quick Check 3 at the side.

From the definitions of multiplication and division of real numbers, we see that

$$\frac{-40}{8} = -40 \cdot \frac{1}{8} = -5,$$

and

$$\frac{40}{-8} = 40\left(\frac{1}{-8}\right) = -5,$$

so that

$$\frac{-40}{8} = \frac{40}{-8}.$$

In general, if $y \neq 0$, then we can express the quotient of a positive and a negative number in three forms.

$$\frac{-x}{y} = \frac{x}{-y} = -\frac{x}{y}$$

2. Use the formula for division to find the quotients.

(a) $\frac{42}{7}$

(b) $\frac{-36}{6}$

(c) $\frac{-12}{-4}$

(d) $\frac{18}{-9}$

(e) $\frac{-3}{0}$

3. Find the quotients.

(a) $\frac{-8}{-2}$

(b) $\frac{-16}{2}$

(c) $\frac{15}{-5}$

Quick Check Answers

2. (a) 6 (b) −6 (c) 3 (d) −2 (e) no such number

3. (a) 4 (b) −8 (c) −3

We usually use the forms

$$\frac{-x}{y} \quad \text{or} \quad -\frac{x}{y}.$$

The form $x/-y$ is seldom used.

In general,

$$\frac{-x}{-y} = \frac{x}{y}.$$

Because x/y is simpler than $-x/-y$, the form $-x/-y$ is seldom used.

Example 4 Simplify $\frac{5(-2)-(3)(4)}{2(1-6)}$.

Simplify the numerator and denominator separately. Then divide.

$$\frac{5(-2)-(3)(4)}{2(1-6)} = \frac{-10-12}{2(-5)} = \frac{-22}{-10} = \frac{11}{5}$$

Work Quick Check 4 at the side.

If one integer is exactly divisible by another (that is, if the quotient is an integer, not a fraction), the second integer is a **factor** of the first. For example, 12 is exactly divisible by 3, so that 3 is a factor of 12. The factors of 12 are the numbers −12, −6, −4, −3, −2, −1, 1, 2, 3, 4, 6, and 12.

Example 5 List all of the factors for the numbers 18, 20, 15, 7, and 1.

Number	Factors
18	−18, −9, −6, −3, −2, −1, 1, 2, 3, 6, 9, 18
20	−20, −10, −5, −4, −2, −1, 1, 2, 4, 5, 10, 20
15	−15, −5, −3, −1, 1, 3, 5, 15
7	−7, −1, 1, 7
1	−1, 1

Work Quick Check 5 at the side.

4. Simplify.

(a) $\frac{5(-4)}{-2-8}$

(b) $\frac{-6(-8)+(-3)9}{(-2)[4-(-3)]}$

5. Give all factors of each number.

(a) 24

(b) 30

Quick Check Answers

4. (a) 2 (b) −3/2
5. (a) −24, −12, −8, −6, −4, −3, −2, −1, 1, 2, 3, 4, 6, 8, 12, 24 (b) −30, −15, −10, −6, −5, −3, −2, −1, 1, 2, 3, 5, 6, 10, 15, 30

WORK SPACE

name date hour

2.5 EXERCISES

Give the reciprocal where one exists. See Example 1.

1. 9 ____________ 2. 8 ____________ 3. -4 ____________

4. -10 ____________ 5. $2/3$ ____________ 6. $3/4$ ____________

7. $-9/10$ ____________ 8. $-4/5$ ____________ 9. 0 ____________

Find the quotients. See Examples 2 and 3.

10. $\frac{-10}{5} =$ ____________ 11. $\frac{-12}{3} =$ ____________ 12. $\frac{-15}{5} =$ ____________

13. $\frac{-20}{2} =$ ____________ 14. $\frac{18}{-3} =$ ____________ 15. $\frac{24}{-6} =$ ____________

16. $\frac{100}{-20} =$ ____________ 17. $\frac{250}{-25} =$ ____________ 18. $\frac{-12}{-6} =$ ____________

19. $\frac{-25}{-5} =$ ____________ 20. $\frac{-150}{-10} =$ ____________ 21. $\frac{-280}{-20} =$ ____________

22. $\frac{-180}{-5} =$ ____________ 23. $\frac{-350}{-7} =$ ____________ 24. $\frac{0}{-2} =$ ____________

25. $\frac{0}{12} =$ ____________ 26. $\left(-\frac{1}{2}\right) \div \left(-\frac{3}{4}\right) =$ ____________ 27. $\left(-\frac{5}{8}\right) \div \left(-\frac{3}{16}\right) =$ ____________

28. $(-4.2) \div (-2) =$ ____________ 29. $(-9.8) \div (-7) =$ ____________

30. $4 \div (-.8) =$ ____________ 31. $-6 \div (.3) =$ ____________

32. $\frac{12}{2-5} =$ ____________ 33. $\frac{15}{3-8} =$ ____________ 34. $\frac{50}{2-7} =$ ____________

35. $\frac{30}{5-8} =$ ____________ 36. $\frac{-30}{2-8} =$ ____________

37. $\frac{-50}{6-11} =$ ______ 38. $\frac{-40}{8-(-2)} =$ ______ 39. $\frac{-72}{6-(-2)} =$ ______

40. $\frac{-120}{-3-(-5)} =$ ______ 41. $\frac{-200}{-6-(-4)} =$ ______

42. $\frac{-15-3}{3} =$ ______ 43. $\frac{16-(-2)}{-6} =$ ______

44. $\frac{-30-(-8)}{-11} =$ ______ 45. $\frac{-17-(-12)}{5} =$ ______

Simplify the numerator and denominator separately. Then find the quotient. See Example 4.

46. $\frac{-8(-2)}{3-(-1)} =$ ______ 47. $\frac{-12(-3)}{-15-(-3)} =$ ______

48. $\frac{-15(2)}{-7-3} =$ ______ 49. $\frac{-20(6)}{-5-1} =$ ______

50. $\frac{-2(6)+3}{2-(-1)} =$ ______ 51. $\frac{3(-8)+4}{-6+1} =$ ______

52. $\frac{-5(2)+3(-2)}{-3-(-1)} =$ ______ 53. $\frac{4(-1)+3(-2)}{-2-3} =$ ______

54. $\frac{2-4(2)}{4-1} =$ ______ 55. $\frac{-4-3(-2)}{5-3} =$ ______

56. $\frac{-9(-2)-(-4)(-2)}{-2(3)-2(2)} =$ ______ 57. $\frac{5(-2)-3(4)}{-2[3-(-2)]-1} =$ ______

58. $\frac{4(-2)-5(-3)}{2[-1+(-3)]-(-8)} =$ ______ 59. $\frac{5(-3)-(-2)(-4)}{5[-4+(-2)]+3(10)} =$ ______

Find all integer factors of each number.

60. 36 ______ 61. 32 ______

62. 25 ______ 63. 14 ______

64. 40 ______ 65. 50 ______

name date hour

66. 17 ____________ 67. 13 ____________

68. 29 ____________ 69. 37 ____________

Find the solution of each equation. The domain of the variable is {−8, −6, −4, −2, 0, 2, 4, 6, 8}.

70. $\frac{x}{4} = -2$ ____________ 71. $\frac{x}{2} = -1$ ____________

72. $\frac{x}{-2} = 3$ ____________ 73. $\frac{x}{-2} = -2$ ____________

74. $\frac{x}{-3} = 0$ ____________ 75. $\frac{x}{5} = 0$ ____________

76. $\frac{x}{-2} = -4$ ____________ 77. $\frac{x}{-1} = 2$ ____________

78. $\frac{x}{4} = -1$ ____________ 79. $\frac{x}{3} = -2$ ____________

Write each sentence in symbols and find the solution. The domain is the set of all real numbers.

	Symbols	Solution
80. Six times a number is −42.	____________	____________
81. Four times a number is −32.	____________	____________
82. When a number is divided by 5, the answer is 15.	____________	____________
83. When a number is divided by 6, the result is −3.	____________	____________
84. When a number is divided by 3, the result is −9.	____________	____________
85. When a number is divided by −3, the answer is −4.	____________	____________

	Symbols	Solution
86. The quotient of a number and 2 is −6. (Write the quotient as $x/2$.)	______	______
87. The quotient of a number and −9 is 2.	______	______
88. The quotient of 6 and one more than a number is 3.	______	______

OBJECTIVES

Identify the use of the following properties

- Closure
- Commutative
- Associative
- Identity
- Inverse
- Distributive

In this section we list some of the properties of addition and multiplication of real numbers. In the following statements, x, y, and z are all real numbers.

1. **Closure properties** The closure properties tell us that the sum of two real numbers is a real number and the product of two real numbers is a real number.

$x + y$ is a real number

xy is a real number

Although the closure property may seem obvious, it is not true for all operations or for all sets of numbers.

Example 1 Does the set of positive numbers have a closure property for subtraction?

For the answer to be *yes*, then we must be able to subtract any two positive numbers and get a positive answer. The numbers 4 and 6 are both positive. However, their difference, $4 - 6$, is not positive. Because of this, subtraction with positive numbers does not have the closure property.

Work Quick Check 1 at the side.

QUICK CHECKS

1. Does the set of positive numbers have a closure property for division?

2. **Commutative properties** The commutative properties tell us that two numbers can be added or multiplied in any order.

$$x + y = y + x$$

$$xy = yx$$

Example 2 Use a commutative property to complete each statement.

(a) $-8 + 5 = 5 +$ ______

By the commutative property for addition, the missing number is -8, since $-8 + 5 = 5 + (-8)$.

(b) $(-2)(7) =$ ______ (-2)

By the commutative property for multiplication, the missing number is 7, since $(-2)(7) = (7)(-2)$.

Work Quick Check 2 at the side.

2. Complete each statement. Use a commutative property.

(a) $x + 9 = 9 +$ ______

(b) $(-12)(4) =$ ______ (-12)

3. **Associative properties** The associative properties tell us that when we are adding or multiplying three numbers, the first two may be grouped together, or the last two may be grouped together, without affecting the answer.

$$(x + y) + z = x + (y + z)$$

$$(xy)z = x(yz)$$

Quick Check Answers

1. Yes, since any two positive numbers may be divided, giving a positive answer.

2. (a) x (b) 4

3. Complete each statement. Use an associative property.

(a) $(9 + 10) + (-3) =$
$9 + [\_\_\_\_\_ + (-3)]$

(b) $-5 + (2 + 8) = (\_\_\_\_\_) + 8$

4. Decide whether each statement is an example of the commutative property, the associative property, or both.

(a) $2(4 \cdot 6) = (2 \cdot 4)6$

(b) $(2 \cdot 4)6 = (4 \cdot 2)6$

(c) $(2 + 4) + 6 = 4 + (2 + 6)$

Quick Check Answers
3. (a) 10 (b) $-5 + 2$
4. (a) associative (b) commutative (c) both

Example 3 Use the associative properties to complete each statement.

(a) $8 + (-1 + 4) = (8 + \_\_\_\_\_) + 4$

The missing number is -1.

(b) $[2 + (-7)] + 6 = 2 + \_\_\_\_\_$

The completed expression on the right should be $2 + (-7 + 6)$.

Work Quick Check 3 at the side.

By the associative property of addition, the sum of three numbers will be the same however we "associate" the numbers in groups. For this reason, parentheses can be left out in many addition or multiplication problems. For example, we can write

$$-1 + 2 + 3$$

instead of

$$(-1 + 2) + 3 \quad \text{or} \quad -1 + (2 + 3).$$

Example 4 **(a)** Is $(2 + 4) + 5 = 2 + (4 + 5)$ an example of the associative property?

The order of the three numbers is the same on both sides of the equals sign. The only change is in the grouping of the numbers. Therefore, this is an example of the associative property.

(b) Is $6(3 \cdot 10) = 6(10 \cdot 3)$ an example of the associative property or the commutative property?

Here, the same numbers, 3 and 10, are grouped. However, on the left, the 3 appears first in $(3 \cdot 10)$. On the right, the 10 appears first. Since the only change involves the order of the numbers, this is an example of the commutative property.

(c) Is $(8 + 1) + 7 = 8 + (7 + 1)$ an example of the associative property or the commutative property?

In the statement, both the order and the grouping are changed. On the left the order of the three numbers is 8, 1, and 7. On the right it is 8, 7, and 1. On the left the 8 and 1 are grouped, while on the right the 7 and 1 are grouped. Therefore, in this example both the associative and the commutative properties are used.

Work Quick Check 4 at the side.

4. Identity properties The identity properties tell us that we can add 0 to any number and get the same number, and we can multiply 1 by any number and get the same number.

$$x + 0 = x \quad \text{and} \quad 0 + x = x$$

$$x \cdot 1 = x \quad \text{and} \quad 1 \cdot x = x$$

The number 0 leaves the identity, or value, of any real number unchanged by addition. For this reason, 0 is called the **identity**

element for addition. In a similar way, multiplication by 1 leaves any real number unchanged, so 1 is the **identity element for multiplication.**

Example 5 These statements are examples of the identity properties.

(a) $-3 + 0 = -3$

(b) $0 + \frac{1}{2} = \frac{1}{2}$

(c) $-\frac{3}{4} \cdot 1 = -\frac{3}{4}$

(d) $1 \cdot 25 = 25$

5. **Inverse properties** The inverse properties tell us that to any number x we can add the number $-x$ to get 0, and for any nonzero number x we can multiply by $1/x$ to get 1.

$$x + (-x) = 0 \quad \text{and} \quad -x + x = 0$$

$$x \cdot \frac{1}{x} = 1 \quad \text{and} \quad \frac{1}{x} \cdot x = 1 \quad (x \neq 0)$$

We call $-x$ the **additive inverse** of x and we call $1/x$ the **multiplicative inverse** of x.

Example 6 These statements are examples of the inverse properties.

(a) $\frac{2}{3} \cdot \frac{3}{2} = 1$

(b) $(-5)\left(\frac{-1}{5}\right) = 1$

(c) $-\frac{1}{2} + \frac{1}{2} = 0$

(d) $4 + (-4) = 0$

Work Quick Check 5 at the side.

5. Complete the statements so that they are examples of either an identity property or an inverse property.

(a) $-6 +$ ______ $= 0$

(b) $\frac{4}{3} \cdot$ ______ $= 1$

(c) $\frac{-1}{9} \cdot$ ______ $= 1$

(d) $275 +$ ______ $= 275$

The final property of real numbers is one which relates addition and multiplication. Using the distributive property, we change a product to a sum.

6. **Distributive property** One form of the distributive property tells us that we can multiply a number x by a sum of numbers $y+z$ by multiplying x by each number in the sum and then adding the resulting products.

$$x(y + z) = xy + xz \quad \text{and} \quad (y + z)x = yx + zx$$

Another form of the distributive property is valid for subtraction.

$$x(y - z) = xy - xz \quad \text{and} \quad (y - z)x = yx - zx$$

Quick Check Answers

5. (a) 6 (b) 3/4 (c) −9 (d) 0

6. Simplify using the distributive property.

(a) $2(p + 5)$

(b) $9(x + 2)$

(c) $-4(y + 7)$

(d) $5(m - 4)$

(e) $9 \cdot k + 9 \cdot 5$

(f) $3a - 3b$

The distributive property can also be extended to more than two numbers.

$$a(b + c + d) = ab + ac + ad, \quad \text{and so on.}$$

Example 7 Simplify using the distributive property.

(a) $5(9 + 6) = 5 \cdot 9 + 5 \cdot 6 = 45 + 30 = 75$

(b) $4(x + 5) = 4x + 4 \cdot 5 = 4x + 20$

(c) $-2(x + 3) = -2x + (-2)(3) = -2x - 6$

(d) $3(k - 9) = 3k - 3 \cdot 9 = 3k - 27$

(e) $6 \cdot 8 + 6 \cdot 2 = 6(8 + 2) = 6(10) = 60$

(f) $4x - 4m = 4(x - m)$

(g) $8(3r + 5z) = 8(3r) + 8(5z)$
$= (8 \cdot 3)r + (8 \cdot 5)z$ Associative property
$= 24r + 40z$

Work Quick Check 6 at the side.

In summary, the set of real numbers has the following properties.

For all real numbers x, y, and z:

	Addition	*Multiplication*
Closure	$x + y$ is a real number	xy is a real number
Commutative	$x + y = y + x$	$xy = yx$
Associative	$(x + y) + z = x + (y + z)$	$(xy)z = x(yz)$
Identity	$x + 0 = x$ and $0 + x = x$	$x \cdot 1 = x$ and $1 \cdot x = x$
Inverse	$x + (-x) = 0$ and $-x + x = 0$	$x \cdot \frac{1}{x} = 1$ and $\frac{1}{x} \cdot x = 1$, $(x \neq 0)$
Distributive	$x(y + z) = xy + xz$ and $(y + z)x = yx + zx$	

Quick Check Answers
6. (a) $2p + 10$ (b) $9x + 18$
(c) $-4y - 28$ (d) $5m - 20$
(e) $9(k + 5)$ (f) $3(a - b)$

name date hour

2.6 EXERCISES

Label each statement as an example of the commutative, associative, closure, identity, inverse, or distributive property. See Example 4.

1. $6 + 15 = 15 + 6$ ____________

2. $9 + (11 + 4) = (9 + 11) + 4$ ____________

3. $5(15 \cdot 8) = (5 \cdot 15)8$ ____________

4. $(23)(9) = (9)(23)$ ____________

5. $12(-8) = (-8)(12)$ ____________

6. $(-9)[6(-2)] = [-9(6)](-2)$ ____________

7. $2 + (p + r) = (p + r) + 2$ ____________

8. $(m + n) + 4 = (n + m) + 4$ ____________

9. $-6 + 12$ is a real number ____________

10. $(-9)(-11)$ is a real number ____________

11. $6 + (-6) = 0$ ____________

12. $-8 + 8 = 0$ ____________

13. $-4 + 0 = -4$ ____________

14. $0 + (-9) = -9$ ____________

15. $3\left(\frac{1}{3}\right) = 1$ ____________

16. $-7\left(-\frac{1}{7}\right) = 1$ ____________

17. $\frac{2}{3} \cdot 1 = \frac{2}{3}$ ____________

18. $-\frac{9}{4} \cdot 1 = -\frac{9}{4}$ ____________

19. $6(5 - 2x) = 6 \cdot 5 - 6(2x)$ ____________

20. $5(2m) + 5(7n) = 5(2m + 7n)$ ____________

Use the indicated property to write a new expression which is equal to the given expression. Simplify the new expression if possible. See Examples 2, 3, 5, and 6.

21. $9 + k$; commutative ____________

22. $z + 5$; commutative ____________

23. $m + 0$; identity ____________

24. $(-9) + 0$; identity ______

25. $3(r + m)$; distributive ______

26. $11(k + z)$; distributive ______

27. $8 \cdot \frac{1}{8}$; inverse ______

28. $\frac{1}{6} \cdot 6$; inverse ______

29. $12 + (-12)$; inverse ______

30. $-8 + 8$; inverse ______

31. $5 + (-5)$; commutative ______

32. $-9 + 9$; commutative ______

33. $-3(r + 2)$; distributive ______

34. $4(k - 5)$; distributive ______

35. $9 \cdot 1$; identity ______

36. $1(-4)$; identity ______

37. $(k + 5) + (-6)$; associative ______

38. $(m + 4) + (-2)$; associative ______

39. $(4z + 2r) + 3k$; associative ______

40. $(6m + 2n) + 5r$; associative ______

name date hour

Use the distributive property to rewrite each expression. Simplify if possible. See Example 5.

41. $5(m + 2)$ ______

42. $6(k + 5)$ ______

43. $-4(r + 2)$ ______

44. $-3(m + 5)$ ______

45. $-8(k - 2)$ ______

46. $-4(z - 5)$ ______

47. $-9(a + 3)$ ______

48. $-3(p + 5)$ ______

49. $(r + 8)4$ ______

50. $(m + 12)6$ ______

51. $(8 - k)(-2)$ ______

52. $(9 - r)(-3)$ ______

53. $2(5r + 6m)$ ______

54. $5(2a + 4b)$ ______

55. $-4(3x - 4y)$ ______

56. $-9(5k - 12m)$ ______

57. $5 \cdot 8 + 5 \cdot 9$ ______

58. $4 \cdot 3 + 4 \cdot 9$ ______

59. $7 \cdot 2 + 7 \cdot 8$ ______

60. $6x + 6m$ ______

61. $9p + 9q$ ______

62. $8(2x) + 8(3y)$ ______

63. $5(7z) + 5(8w)$ ______

64. $11(2r) + 11(3s)$ ______

Decide whether or not each set of numbers has the closure property for (a) addition, (b) subtraction, and (c) multiplication.

Example $\{0, 2, 4, 6, 8, \ldots\}$

This set contains all the even numbers that are not negative. The sum of two even numbers is even, so the answer for part (a) is *yes.* When two even numbers are subtracted, the answer might well be negative ($8 - 12$, for example) and there are no negative numbers in the set. Thus, the answer for (b) is *no.* For (c), the answer is *yes.*

65. $\{0, 1, 2, 3, 4, 5, 6, 7, \ldots\}$ (a) ______ (b) ______ (c) ______

66. {0, −1, −2, −3, −4, −5, . . .} (a) ______ (b) ______ (c) ______

67. {. . . , −3, −2, −1, 0, 1, 2, 3, . . .} (a) ______ (b) ______ (c) ______

68. {. . . , −4, −2, 0, 2, 4, 6, . . .} (a) ______ (b) ______ (c) ______

69. {0, 5, 10, 15, 20, 25, . . .} (a) ______ (b) ______ (c) ______

70. {0, 3, 6, 9, 12, 15, 18, . . .} (a) ______ (b) ______ (c) ______

name date hour

CHAPTER 2 TEST

Perform the indicated operations (whenever possible).

1. $-6 + 10 =$ ________
2. $-2 + (-7) =$ ________
3. $12 + (-13) =$ ________
4. $27 - (-5) =$ ________
5. $10 - 18 =$ ________
6. $-1 - 3 =$ ________
7. $-5 + 15 =$ ________
8. $-2(-3) =$ ________
9. $12(-6) =$ ________
10. $-1(3) =$ ________
11. $\frac{27}{-3} =$ ________
12. $\frac{-9}{9} =$ ________
13. $\frac{-8}{-4} =$ ________
14. $\frac{5}{0} =$ ________
15. $\frac{0}{-3} =$ ________
16. $(-5)0 =$ ________
17. $[2 + (-3)] - 4 =$ ________
18. $6(-5) - 3 =$ ________
19. $(3 - 8)(-1 + 4) =$ ________
20. $(-6 + 3) - (-1 - 4) =$ ________
21. $(-3 - 2)(-5) + (-7) =$ ________
22. $8 - (3)(-6) + (-1) =$ ________
23. $-9 - (-5)(-2) =$ ________
24. $\frac{-8 - (-6)}{-1 - 1} =$ ________
25. $\frac{-7 - (-5 + 1)}{-4 - (-3)} =$ ________
26. $\frac{-6[5 - (-1 + 4)]}{-9(6 - 3) - 6(-4)} =$ ________
27. $\frac{15(-4 - 2)}{16(-2) + (-7 - 1)(-3 - 1)} =$ ________

Find the solution for each equation. The domain of the variable is $\{-9, -5, -4, -2, -1, 3, 9\}$.

28. $x + 5 = 8$ ________
29. $3 + x = 2$ ________
30. $-2x = -18$ ________
31. $5x = -25$ ________
32. $2x + 1 = -7$ ________
33. $-4x - 2 = 6$ ________

Select the smaller number from each pair.

34. $-3, -5$ __________
35. $6, -|-8|$ __________
36. $|-4|, 0$ __________

37. $3, |-5|$ __________
38. $|4|, -6$ __________

Match the property in Column I with all examples of it from Column II.

Column I	*Column II*
39. Commutative __________	A. $-2 + 2 = 0$
40. Associative __________	B. $3 + (7 + x) = (3 + 7) + x$
41. Closure __________	C. $8 + 0 = 8$
42. Identity __________	D. $17 \cdot 1 = 17$
43. Inverse __________	E. $3(x + y) = 3x + 3y$
44. Distributive __________	F. $-7 + \sqrt{2}$ is a real number
	G. $3(-2)$ is a real number
	H. $8 + m = m + 8$
	I. $-5\left(\frac{1}{-5}\right) = 1$
	J. $mn = nm$

45. Is the set of numbers $\{\ldots, -6, -3, 0, 3, 6, 9, \ldots\}$ closed for addition? __________

3 Solving Equations and Inequalities

In Section 1.4, we saw how to solve equations and inequalities when the domain of the variable is given. From now on, the domain will be the set of all real numbers. This means that when you have an equation or an inequality to solve, you must find all real number solutions for it.

3.1 SIMPLIFYING EQUATIONS

OBJECTIVES

- Simplify equations
- Solve equations after simplifying

In solving an equation, the first step is to simplify the expressions on the left-hand side and on the right-hand side as much as possible. You can do this by either adding or multiplying, or both.

Example 1 Simplify the equation $x + 8 + 9 = 3 + 20$.

Add the 8 and 9 on the left side of the equals sign. Then add the 3 and 20 on the right side. The result is a simpler equation.

$$x + 17 = 23$$

Work Quick Check 1 at the side.

Example 2 Simplify the equation

$$(5 + x) + (-2) = (-3) + [11 + (-1)].$$

Here, you need to rearrange both numbers and variables. On the left side, use the commutative property and the associative property to reverse the order of x and -2. On the right side, add together the numbers in the brackets first.

$$(5 + x) + (-2) = (-3) + [11 + (-1)]$$
$$[5 + (-2)] + x = (-3) + 10$$
$$3 + x = 7$$

You should be able to see by inspection that the solution of this equation is 4.

QUICK CHECKS

1. Simplify each equation.

(a) $x + 11 - 4 = 9 + 1$

(b) $3x + 8 + 9 = 25 - 2$

Quick Check Answers

1. (a) $x + 7 = 10$
(b) $3x + 17 = 23$

To check that the solution is correct, substitute 4 for the variable in the given equation. Does this result in a true statement?

$$(5 + x) + (-2) = (-3) + [11 + (-1)]$$
$$(5 + 4) + (-2) = (-3) + [11 + (-1)] \qquad \text{Let } x = 4$$
$$9 + (-2) = (-3) + 10$$
$$7 = 7$$

Since this result is true, then 4 is the solution.

Work Quick Check 2 at the side.

To simplify an expression containing a sum such as $3x + 5x$, use the distributive property.

$$3x + 5x = (3 + 5)x = 8x.$$

Example 3 Simplify the following expressions.

(a) $9m + 5m = (9 + 5)m = 14m$

(b) $6r + 3r + 2r = (6 + 3 + 2)r = 11r$

(c) $4x + x = 4x + 1x = 5x$ (Note: $1x = x$)

(d) $16y - 9y = (16 - 9)y = 7y$

Work Quick Check 3 at the side.

Example 4 Simplify the equation $12m - 3m = 27$. Solve by inspection.

Use the distributive property on the left-hand side.

$$12m - 3m = (12 - 3)m = 9m$$

The equation $12m - 3m = 27$ is simplified as

$$9m = 27.$$

By inspection, the solution is 3.

Example 5 Simplify the equation $4x + 2 + 3x - 6x = 9$. Solve by inspection.

Use the associative, commutative, and distributive properties.

$$4x + 2 + 3x - 6x = 9$$
$$4x + 3x - 6x + 2 = 9$$
$$(4 + 3 - 6)x + 2 = 9$$
$$1x + 2 = 9$$
$$x + 2 = 9.$$

By inspection, the solution is 7.

Work Quick Check 4 at the side.

A **term** in an equation is either a single number or the product of a number and a variable.

2. Simplify each equation, then solve by inspection. Check the result.

(a) $x + 2 + 5 + (-4) = 6 + 3$

(b) $(4 + x) + (-8) = 6 + (-2)$

3. Simplify.

(a) $4k + 7k$

(b) $8y + 2y + 3y$

(c) $4r - r$

(d) $8z + 2z - 5z$

4. Simplify each equation. Solve by inspection.

(a) $8r - 4r = 12$

(b) $2y + 7y = 18$

(c) $z + 3z + 4z = 40$

(d) $7m - 6m + 2 = 9$

(e) $6 + 3k + 5k + 2 - 7k = 10$

Quick Check Answers

2. (a) $x + 3 = 9$; 6 (b) $x - 4 = 4$; 8
3. (a) $11k$ (b) $13y$ (c) $3r$ (d) $5z$
4. (a) $4r = 12$; 3 (b) $9y = 18$; 2 (c) $8z = 40$; 5 (d) $m + 2 = 9$; 7 (e) $k + 8 = 10$; 2

Example 6 The terms in the equation $4m + 2 + 3m - 6m = 9$ are as follows.

$$\left.\begin{array}{r} 2 \\ 9 \end{array}\right\} \quad \text{Single numbers (in this case, integers)}$$

$$\left.\begin{array}{r} 4m \\ 3m \\ 6m \end{array}\right\} \quad \text{Products of a number and a variable}$$

Rearranging numbers within each side of an equation is called **collecting terms.** Carrying out the operations of addition or subtraction is called **combining terms.**

Example 7 Simplify the equation $3p + 7 - 2p - 5 + 1 = 2$.

First, rearrange the terms so that terms with variables are next to each other.

$$3p - 2p + 7 - 5 + 1 = 2 \qquad \text{Collect terms}$$

Now use the distributive property to combine the terms having a variable. Add and subtract the numbers as indicated.

$$(3p - 2p) + (7 - 5 + 1) = 2$$
$$p + 3 = 2 \qquad \text{Combine terms}$$

By inspection, the solution is -1.

Example 8 Solve the equation $-1 + 2a + 16 - a = 12$.

Collect and combine terms.

$$(2a - a) + (-1 + 16) = 12$$
$$(2a - 1a) + 15 = 12$$
$$1a + 15 = 12$$
$$a + 15 = 12$$

By inspection, the solution is -3.

Work Quick Check 5 at the side.

All the equations of this section are examples of linear equations. A **linear equation** follows the pattern

$$ax + b = c,$$

where a, b, and c represent any real numbers (except that a cannot be 0).

5. Simplify each equation. Solve by inspection.

(a) $-4 + 5x + 2 - 4x = 5$

(b) $2m + 8 + 3m + 1 - 4m = 14$

(c) $2 = 3r + 8r + 9 - 1 + (-10r)$

(d) $2k - 7 + 5k - 1 - 6k = 4$

Quick Check Answers

5. (a) $x - 2 = 5; 7$
(b) $m + 9 = 14; 5$ (c) $2 = r + 8; -6$
(d) $k - 8 = 4; 12$

WORK SPACE

name date hour

3.1 EXERCISES

In Exercises 1-10, an equation is given, along with a proposed solution. Decide whether or not the proposed solution is correct.

Example $5r - 3r + 1 = 13; 6$. Substitute the proposed solution, 6, for r in the equation: $5 \cdot 6 - 3 \cdot 6 + 1 = 13$. Is this true? From $30 - 18 + 1 = 13$, we get $13 = 13$, a true statement. Therefore, 6 is the correct solution.

1. $2x + 3 = 13; 5$ ______

2. $4x - 1 = 11; 3$ ______

3. $5k + 1 = 20; 4$ ______

4. $3z + 2 = 21; 6$ ______

5. $3h - 2h + 1 + 4 = 9; 4$ ______

6. $5m - 2m + 4 - 3 = 13; 4$ ______

7. $4y + 1 - 5y + 3y - 5 = 6; 5$ ______

8. $3s - 5s + 6s + 5 - 2 = 17; 4$ ______

9. $4x + 1 = 4x + 6 - 3x + 2x - 5; -2$ ______

10. $8j - 4j + 3 - 5 + 6j = 11j + 1; -3$ ______

Simplify each expression by collecting and combining terms.

11. $2k + 9 + 5k + 6 =$ ______

12. $2 + 17z + 1 + 2z =$ ______

13. $m + 1 - m + 2 + m - 4 =$ ______

14. $12 - 13x - 27 + 2x - x =$ ______

15. $-5y + 3 - 1 + 5 + y - 7 =$ ______

16. $2k - 7 - 5k + 7k - 3 - k =$ ______

17. $-2x + 3 + 4x - 17 + 20 =$ ______

18. $r - 6 - 12r - 4 + 6r =$ ______

19. $16 - 5m - 4m - 2 + 2m + 6 =$ ______

20. $6 - 3z - 2z - 5 + z - 3z - 3 =$ ______

Simplify the given equations. Then solve by inspection. See Examples 4, 5, 7, and 8.

	Equation	Solution
21. $3x + 4x = 14$	______	______
22. $x + 5 + 6 = 18 + 2$	______	______
23. $6k - 5k + 5 = 10 + 3$	______	______
24. $5m - 4m - 7 = 2 + 3$	______	______
25. $-3z + 7 - 3z + 7z = 31 - 18$	______	______
26. $15 - 2 + 4 = -4 + 1 - 5k + 7k - k$	______	______
27. $-10 + x + 4x - 7 - 4x = 21 - 19$	______	______
28. $-x + 10x - 3x - 4 - 5x = 2 + 10$	______	______

In Exercises 29-31, write an equation using the given information. Then solve the equation by inspection.

29. The sum of 5 times a number, 3 times the number, and 1 is 49. Find the number. ______

30. The difference of 4 times a number and 3 times a number, added to 8, is 11. Find the number. ______

31. Joann is three times as old as Marilyn. The sum of their ages is 24. Find Joann's age. ______

3.2 THE ADDITION PROPERTY OF EQUALITY

OBJECTIVES

- Use the addition property of equality to solve equations
- Use the property after an equation is simplified

Our goal in solving an equation is to go through a series of steps, ending up with an equation where only the variable x is on one side of the equals sign and some number is on the other side. That is, we want to reduce the given equation to the form

$$x = \text{a number.}$$

Think about the equation

$$x + 5 = 12.$$

We know that $x + 5$ and 12 represent the same number, since this is the meaning of the equals sign. We want to change the left-hand side from $x + 5$ to x. We could do this by adding -5 to $x + 5$.

$$x + 5 + (-5) = x + 0 = x$$

To keep the equality between $x + 5$ and 12, we must also add -5 to 12.

$$12 + (-5) = 7$$

By adding -5 to both sides of the equation $x + 5 = 12$, we get the simpler equation

$$x = 7.$$

The solution is 7.

The **addition property of equality** is the property of the real numbers that lets us add the same number to both sides of an equation.

> If A, B, and C represent algebraic expressions, then the equations
>
> $$\mathbf{A = B}$$
>
> and
>
> $$\mathbf{A + C = B + C}$$
>
> have the same solution.

Example 1 Solve the equation $x - 16 = 7$.

If x were alone on the left, we would have the solution. So use the addition property of equality to add 16 to both sides.

$$x - 16 = 7$$
$$x - 16 + 16 = 7 + 16$$
$$x = 23$$

To check, substitute 23 for x in the original equation.

$$x - 16 = 7$$
$$23 - 16 = 7$$
$$7 = 7$$

Since the check results in a true statement, 23 is the correct solution.

QUICK CHECKS

1. Use the addition property of equality to solve the following equations.

(a) $m - 2 = 6$

(b) $y + 7 = 11$

(c) $a + 2 = -3$

(d) $p + 6 = 2$

2. Solve each equation.

(a) $5m = 4m + 6$

(b) $3y = 2y - 9$

(c) $2k - 8 = 3k$

(d) $7m + 1 = 8m$

3. Solve each equation.

(a) $7p + 2p - 8p + 5 = 9 + 1$

(b) $11k - 6 - 4 - 10k = -5 + 5$

(c) $-4 + 3 - 2m + 3m = 10 - 5 - 7$

(d) $3p - 5p + 3p - 8 = -6 - 2$

Quick Check Answers
1. (a) 8 (b) 4 (c) -5 (d) -4
2. (a) 6 (b) -9 (c) -8 (d) 1
3. (a) 5 (b) 10 (c) -1 (d) 0

In this example, how did we know to add 16 to both sides of the equation $x - 16 = 7$? We want one side of the equation to contain only the variable term and the other side to contain only a number. We know that $x + 0 = x$, so we need to get $x + 0$. What number must be added to $x - 16$ to get $x + 0$? Since the sum of any number and its additive inverse is 0, we must add the additive inverse of -16, which is 16, to both sides of the equation.

Work Quick Check 1 at the side.

Example 2 Solve the equation $3k + 17 = 4k$.

In order to find the solution, we need to get all terms that contain variables on the same side of the equation. One way to do this is to use the addition property of equality, and add $-3k$ to both sides.

$$3k + 17 = 4k$$
$$3k + 17 + (-3k) = 4k + (-3k)$$
$$17 = k$$

The solution is 17.

The equation $3k + 17 = 4k$ could also be solved by first adding $-4k$ to both sides as follows.

$$3k + 17 = 4k$$
$$3k + 17 + (-4k) = 4k + (-4k)$$
$$17 - k = 0$$

Now add -17 to both sides.

$$17 - k + (-17) = 0 + (-17)$$
$$-k = -17$$

This result gives the value of $-k$, but not of k itself. However, we know that the additive inverse of k is -17. Then k must be 17.

$$-k = -17$$
$$k = 17$$

This answer agrees with the first one.

Work Quick Check 2 at the side.

Example 3 Solve the equation $4r + 5r - 3 + 8 - 3r - 5r = 12 + 8$.

First, simplify the equation by combining terms.

$$4r + 5r - 3r - 5r - 3 + 8 = 12 + 8$$
$$r + 5 = 20$$

Add -5 to both sides of this equation.

$$r + 5 + (-5) = 20 + (-5)$$
$$r = 15$$

The solution of the given equation is 15.

Work Quick Check 3 at the side.

name date hour

3.2 EXERCISES

Solve each equation by using the addition property of equality. See Examples 1 and 2.

1. $x - 3 = 7$ ________

2. $x + 5 = 13$ ________

3. $7 + k = 5$ ________

4. $9 + m = 4$ ________

5. $3r - 10 = 2r$ ________

6. $2p = p + 3$ ________

7. $7z = -8 + 6z$ ________

8. $4y = 3y - 5$ ________

9. $m + 5 = 0$ ________

10. $k - 7 = 0$ ________

11. $2 + 3x = 2x$ ________

12. $10 + r = 2r$ ________

13. $2p + 6 = 10 + p$ ________

14. $5r + 2 = -1 + 4r$ ________

15. $2k + 2 = -3 + k$ ________

16. $6 + 7x = 6x + 3$ ________

17. $x - 5 = 2x + 6$ ________

18. $-3r + 7 = -4r - 19$ ________

19. $6z + 3 = 5z - 3$ ________

20. $6t + 5 = 5t + 7$ ________

Solve each equation. First simplify both sides of the equation as much as possible. See Example 3.

21. $4x + 3 + 2x - 5x = 2 + 8$ ________

22. $3x + 2x - 6 + x - 5x = 9 + 4$ ________

23. $9r + 4r + 6 - 8 = 10r + 6 + 2r$ ________

24. $-3t + 5t - 6t + 4 - 3 = -3t + 2$ ________

25. $11z + 2 + 4z - 3z = 5z - 8 + 6z$ ________

26. $2k + 8k + 6k - 4k - 8 + 2 = 2k + 2 + 3k$ ______

27. $4m + 8m - 9m + 2 - 5 = 4m + 6$ ______

28. $15y - 4y + 8 - 2 + 7 - 4 = 4y + 2 + 8y$ ______

29. $-9p + 4p - 3p + 2p - 6 = -5p - 6$ ______

30. $5x - 2x + 3x - 4x + 8 - 2 + 4 = 5x + 10 - 4x$ ______

In Exercises 31-34, write an equation using the given information, then solve it.

31. Three times a number is 17 more than twice the number. Find the number. ______

32. If six times a number is subtracted from seven times a number, the result is -9. Find the number. ______

33. If five times a number is added to three times the number, the result is the sum of seven times the number and nine. Find the number. ______

34. The sum of two consecutive integers is 13 less than three times the smaller integer. Find the integers. (Hint: Let x represent the first of the integers. Then $x + 1$ represents the second.) ______

OBJECTIVES

- Use the multiplication property of equality
- Use the property after an equation is simplified
- Solve equations such as $-r = 4$
- Solve equations with decimals

The addition property of equality by itself is not enough to solve an equation like $3x + 2 = 17$.

$$3x + 2 = 17$$
$$3x + 2 + (-2) = 17 + (-2) \quad \text{Add } -2 \text{ to both sides}$$
$$3x = 15 \quad \text{Simplify}$$

We do not have the variable x alone on one side of the equation; we have $3x$ instead. To go from $3x = 15$ to x = a number, we need another property similar to the addition property.

If $3x = 15$, then $3x$ and 15 both represent the same number. Multiplying both $3x$ and 15 by the same number will also result in an equality. The **multiplication property of equality** states that both sides of an equation can be multiplied by the same term.

> If A, B, and C represent algebraic expressions, the equations
>
> $$A = B$$
>
> and
>
> $$AC = BC$$
>
> have exactly the same solution. (Assume that $C \neq 0$.)

Now we go back and solve $3x = 15$. On the left, we have $3x$. We need $1x$, or x, instead of $3x$. To get x, multiply both sides of the equation by 1/3. This works because $3 \cdot 1/3 = \frac{3}{3} = 1$.

$$3x = 15$$
$$\frac{1}{3}(3x) = \frac{1}{3} \cdot 15$$
$$\left(\frac{1}{3} \cdot 3\right)x = \frac{1}{3} \cdot 15$$
$$1x = 5$$
$$x = 5$$

The solution of the equation is 5.

Example 1 Solve the equation $-5p = -30$.

To get p alone on the left, use the multiplication property of equality and multiply both sides of the equation by $-1/5$.

$$-5p = -30$$
$$-\frac{1}{5} \cdot -5p = -\frac{1}{5} \cdot -30$$
$$1p = 6$$
$$p = 6$$

The solution is 6.

Work Quick Check 1 at the side.

QUICK CHECKS

1. Solve each equation.

(a) $7m = 56$

(b) $3r = -12$

(c) $8y = 108$

(d) $-2m = 16$

Quick Check Answers

1. (a) 8 (b) -4 (c) 27/2 or 13 1/2 (d) -8

2. Solve each equation.

(a) $\frac{m}{2} = 6$

(b) $\frac{y}{5} = 5$

(c) $\frac{p}{4} = -6$

(d) $\frac{a}{-2} = 8$

3. Solve each equation.

(a) $\frac{2}{3}m = 8$

(b) $\frac{3}{5}k = 12$

(c) $\frac{7}{8}m = 28$

(d) $\frac{3}{4}k = -21$

(e) $\frac{1}{2}p = \frac{7}{4}$

Quick Check Answers

2. (a) 12 (b) 25 (c) −24 (d) −16

3. (a) 12 (b) 20 (c) 32 (d) −28 (e) 7/2

Example 2 Solve the equation $\frac{a}{4} = 3$.

We can replace $a/4$ by $\frac{1}{4}a$, since division by 4 is the same as multiplication by its reciprocal 1/4. To get a alone on the left, multiply both sides by 4.

$$\frac{a}{4} = 3$$

$$\frac{1}{4}a = 3$$

$$4 \cdot \frac{1}{4}a = 4 \cdot 3$$

$$1a = 12$$

$$a = 12$$

We multiplied by 4 since $4 \cdot 1/4 = 1$, and we want $1a$ on the left, and not $(1/4)a$.

Check the answer:

$$\frac{a}{4} = 3 \quad \text{Original equation}$$

$$\frac{12}{4} = 3 \quad \text{Let } a = 12$$

$$3 = 3$$

The solution 12 is correct.

Work Quick Check 2 at the side.

Example 3 Solve the equation $\frac{3}{4}h = 6$.

To get h alone on the left, multiply both sides of the equation by 4/3. We use 4/3 because $(4/3) \cdot (3/4)h = 1 \cdot h = h$.

$$\frac{3}{4}h = 6$$

$$\frac{4}{3}\left(\frac{3}{4}h\right) = \frac{4}{3} \cdot 6$$

$$1 \cdot h = \frac{4}{3} \cdot \frac{6}{1}$$

$$h = 8$$

The solution is 8.

Work Quick Check 3 at the side.

Example 4 Solve the equation $5m + 6m = 33$.

First, use the distributive property to combine terms.

$$5m + 6m = 33$$

$$11m = 33$$

Now multiply both sides by the reciprocal of 11, which is $\frac{1}{11}$.

$$\frac{1}{11} \cdot 11m = \frac{1}{11} \cdot 33$$
$$m = 3$$

The solution is 3.

Example 5 Solve the equation $-r = 4$.

To find the solution for this equation, we need to get r itself, and not $-r$, on one side of the equals sign. To do this, first write $-r$ as $-1 \cdot r$. (Remember that multiplying any positive number by -1 will give you the additive inverse of that number.)

$$-r = 4$$
$$-1 \cdot r = 4$$

Multiplying both sides of this last equation by -1 will remove the minus sign from r, since the product of two negative numbers is positive.

$$-1(-1 \cdot r) = -1 \cdot 4$$
$$(-1)(-1) \cdot r = -4$$
$$1 \cdot r = -4$$
$$r = -4$$

The solution of the equation $-r = 4$ is thus -4.

Work Quick Check 4 at the side.

Example 6 Solve the equation $2.3x = 6.9$.

Multiply both sides by $1/2.3$.

$$\frac{1}{2.3}(2.3x) = \frac{1}{2.3}(6.9)$$
$$1x = \frac{6.9}{2.3}$$
$$x = 3$$

Work Quick Check 5 at the side.

4. Solve each equation.

(a) $5p + 2p = 28$

(b) $9k - k = -56$

(c) $7m - 5m = -12$

(d) $4r - 9r = 20$

(e) $-m = 2$

(f) $-p = -7$

5. Solve each equation.

(a) $-1.5p = 4.5$

(b) $12.5k = -62.5$

(c) $-.7m = -4.9$

Quick Check Answers

4. (a) 4 (b) −7 (c) −6 (d) −4 (e) −2 (f) 7
5. (a) −3 (b) −5 (c) 7

WORK SPACE

name date hour

3.3 EXERCISES

Solve each equation. See Examples 1–6.

1. $5x = 25$ ______
2. $7x = 28$ ______
3. $2m = 50$ ______
4. $6y = 72$ ______
5. $3a = -24$ ______
6. $5k = -60$ ______
7. $8s = -56$ ______
8. $9t = -36$ ______
9. $-4x = 16$ ______
10. $-6x = 24$ ______
11. $-12z = 108$ ______
12. $-11p = 77$ ______
13. $5r = 0$ ______
14. $2x = 0$ ______
15. $-y = 6$ ______
16. $-m = 2$ ______
17. $-n = -4$ ______
18. $-p = -8$ ______
19. $2x + 3x = 20$ ______
20. $3k + 4k = 14$ ______
21. $5m + 6m - 2m = 72$ ______
22. $11r - 5r + 6r = 84$ ______
23. $k + k + 2k = 80$ ______
24. $4z + z + 2z = 28$ ______
25. $3r - 5r = 6$ ______
26. $9p - 13p = 12$ ______
27. $7r - 13r = -24$ ______
28. $12a - 18a = -36$ ______
29. $6y + 8y - 17y = 9$ ______
30. $14a - 19a + 2a = 15$ ______
31. $-7y + 8y - 9y = -56$ ______
32. $-11b + 7b + 2b = -100$ ______
33. $\frac{m}{2} = 16$ ______
34. $\frac{p}{5} = 3$ ______
35. $\frac{x}{7} = 7$ ______
36. $\frac{k}{8} = 2$ ______
37. $\frac{2}{3}t = 6$ ______
38. $\frac{3}{4}m = 18$ ______

39. $\frac{5}{2}z = 20$ __________

40. $\frac{9}{5}r = 18$ __________

41. $\frac{3}{4}p = -60$ __________

42. $\frac{5}{8}z = -40$ __________

43. $\frac{2}{3}k = 5$ __________

44. $\frac{5}{3}m = 6$ __________

45. $\frac{-2}{7}p = -7$ __________

46. $-\frac{3}{11}y = -2$ __________

47. $1.7p = 5.1$ __________

48. $2.3k = 11.5$ __________

49. $-4.2m = 25.2$ __________

50. $-3.9a = -15.6$ __________

Write an equation for each problem. Then solve the equation.

51. When a number is divided by 4, the result is 6. Find the number. __________

52. The quotient of a number and -5 is 2. Find the number. __________

53. Chuck decided to divide a sum of money equally among four relatives, Dennis, Mike, Ed, and Joyce. Each relative received $62. Find the sum that was originally divided. __________

54. If twice a number is divided by 5, the result is 4. Find the number. __________

OBJECTIVES

- Learn the four steps in solving a linear equation and how to use them

We now combine the methods we have used to help us solve more complicated equations.

Step 1 Combine terms to simplify. Use the commutative, associative, and distributive properties as needed.

Step 2 If necessary, use the addition property of equality to simplify further, so that the variable term is on one side of the equals sign, and the number term is on the other.

Step 3 If necessary, use the multiplication property of equality to simplify further. This gives an equation of the form $x = a$ number.

Step 4 Check the solution by substituting into the original equation.

Example 1 Solve the equation $2x + 3x + 3 = 38$. Follow the four steps of the summary above.

Step 1 Combine terms.

$$2x + 3x + 3 = 38$$
$$5x + 3 = 38$$

Step 2 Use the addition property of equality. Add -3 to both sides.

$$5x + 3 + (-3) = 38 + (-3)$$
$$5x = 35$$

Step 3 Use the multiplication property of equality. Multiply both sides by $1/5$.

$$\frac{1}{5} \cdot 5x = \frac{1}{5} \cdot 35$$
$$x = 7$$

Step 4 Check the solution. Substitute 7 for x in the original equation.

$$2x + 3x + 3 = 38$$
$$2(7) + 3(7) + 3 = 38 \qquad \text{Let } x = 7$$
$$14 + 21 + 3 = 38$$
$$38 = 38$$

Since the final statement is true, 7 is the correct solution.

Work Quick Check 1 at the side.

Example 2 Solve the equation $3r + 4 - 2r - 7 = 4r + 3$.

Step 1 $3r + 4 - 2r - 7 = 4r + 3$

$$r - 3 = 4r + 3 \qquad \text{Combine terms}$$

QUICK CHECKS

1. Solve each equation.

(a) $3k + 2k + 7 = 17$

(b) $7m + 9m - 5 = 43$

(c) $9p - 5p + p - 8 = -18$

(d) $-8y + 7y + 3y - 6 = 7$

Quick Check Answers

1. (a) 2 (b) 3 (c) -2 (d) $13/2$

Step 2 $\quad r - 3 + 3 = 4r + 3 + 3 \quad$ Add 3

$$r = 4r + 6$$

$$r + (-4r) = 4r + 6 + (-4r) \quad \text{Add } -4r$$

$$-3r = 6$$

Step 3 $\quad \left(-\frac{1}{3}\right)(-3r) = \left(-\frac{1}{3}\right)6 \quad$ Multiply by $-\frac{1}{3}$

$$r = -2$$

Step 4 Substitute -2 for r in the original equation.

$$3r + 4 - 2r - 7 = 4r + 3$$

$$3(-2) + 4 - 2(-2) - 7 = 4(-2) + 3$$

$$-6 + 4 + 4 - 7 = -8 + 3$$

$$-5 = -5$$

The correct solution for the equation is -2.

Work Quick Check 2 at the side.

Example 3 Solve the equation $4(k - 3) - k = k - 6$.

Step 1 Before combining terms, use the distributive property to simplify $4(k - 3)$.

$$4(k - 3) = 4k - 4 \cdot 3 = 4k - 12$$

Now combine terms.

$$4k - 12 - k = k - 6$$

$$3k - 12 = k - 6$$

Step 2 $\quad 3k - 12 + 12 = k - 6 + 12 \quad$ Add 12

$$3k = k + 6$$

$$3k + (-k) = k + 6 + (-k) \quad \text{Add } -k$$

$$2k = 6$$

Step 3 $\quad \frac{1}{2}(2k) = \frac{1}{2} \cdot 6 \quad$ Multiply by $\frac{1}{2}$

$$k = 3$$

Step 4 Check your answer by substituting 3 for k in the original equation. Remember to do all the work inside the parentheses first.

$$4(k - 3) - k = k - 6$$

$$4(3 - 3) - 3 = 3 - 6$$

$$4(0) - 3 = 3 - 6$$

$$0 - 3 = 3 - 6$$

$$-3 = -3$$

The correct solution of the equation is 3.

Work Quick Check 3 at the side.

2. Solve each equation.

(a) $7 + 4p - 3p + 8 = 9p + 7$

(b) $5y - 7y + 6y - 9 = 3 + 2y$

(c) $-3k - 5k - 6 + 11 = 2k - 5$

(d) $2y + 5y - 6 + 8 = 9y - 1$

3. Solve each equation.

(a) $7(p - 2) + p = 2p + 4$

(b) $11 + 3(a + 1) = 5a + 16$

(c) $3(m + 5) - 5 + 2m = 2(m - 10)$

(d) $4(8 - 3t) = -8(t + 2) + 32$

Quick Check Answers

2. (a) 1 (b) 6 (c) 1 (d) 3/2

3. (a) 3 (b) -1 (c) -10 (d) 4

Example 4 Solve the equation $6a - (3 + 2a) = 3a + 1$.

On the left of the equals sign, we have $6a - (3 + 2a)$, which tells us to subtract $(3 + 2a)$ from $6a$. We subtract by changing the signs of all of the terms in parentheses and adding:

$$\begin{aligned} 6a - (3 + 2a) &= 6a + (-3 - 2a) \\ &= 6a - 3 - 2a. \end{aligned}$$

The original equation now becomes

$$6a - 3 - 2a = 3a + 1.$$

Step 1 Simplify.

$$4a - 3 = 3a + 1$$

Step 2 First, add 3 to both sides, then do the same with $-3a$.

$$\begin{aligned} 4a - 3 + 3 &= 3a + 1 + 3 \\ 4a &= 3a + 4 \\ 4a + (-3a) &= 3a + 4 + (-3a) \\ a &= 4 \end{aligned}$$

There is no reason to go further; check that the solution is 4.

Work Quick Check 4 at the side.

4. Solve each equation.

(a) $2y - (y + 7) = 9$

(b) $7m - (2m - 9) = 39$

(c) $5(m - 2) - (3m - 6) = 3m - 1$

(d) $-(3k + 2) - (k + 1) = -5$

Quick Check Answers

4. (a) 16 (b) 6 (c) -3 (d) 1/2

WORK SPACE

name date hour

3.4 EXERCISES

Simplify the expressions. Use the distributive property and combine terms wherever possible.

1. $3(k - 6) =$ ____________
2. $5(m + 4) =$ ____________
3. $6(5t + 11) =$ ____________
4. $3(2x + 4) =$ ____________
5. $-3(n + 5) =$ ____________
6. $-4(v - 8) =$ ____________
7. $-2(3x - 4) =$ ____________
8. $-5(4t + 6) =$ ____________
9. $7(r + 2) - 3r =$ ____________
10. $3(m - 6) + 5m =$ ____________
11. $-5(2r - 3) + 2(5r + 3) =$ ____________
12. $-4(5y - 7) + 3(2y - 5) =$ ____________
13. $8(2k - 1) - (4k + 5) =$ ____________
14. $6(3p - 2) - (5p + 1) =$ ____________
15. $-2(-3k + 2) + 2(5k - 6) - 3k - 5 =$ ____________
16. $-2(3r - 4) - (6 - r) + 2r - 5 =$ ____________

In Exercises 17-28, solve the equation, then check the solution. See Examples 1-5.

17. $4h + 8 = 16$ ____________
18. $3x - 15 = 9$ ____________
19. $6k + 12 = -12$ ____________
20. $2m - 6 = 6$ ____________
21. $12p + 18 = 14p$ ____________
22. $10m - 15 = 7m$ ____________
23. $3x + 9 = -3(2x + 3)$ ____________
24. $4z + 2 = -2(z + 2)$ ____________

25. $2(2r-1)=-3(r+3)$ ________________

26. $3(3k+5)=2(5k+5)$ ________________

27. $2(3x+4)=8(2+x)$ ________________

28. $4(3p+3)=3(3p-1)$ ________________

Combine terms as necessary. Then solve the equation. See Examples 1-5.

29. $-4-3(2x+1)=11$ ________________

30. $8-2(3x-4)=2x$ ________________

31. $3k-5=2(k+6)+1$ ________________

32. $4a-7=3(2a+5)-2$ ________________

33. $5(2m-1)=4(2m+1)+7$ ________________

34. $3(3k-5)=4(2k-5)+7$ ________________

35. $5(4t+3)=6(3t+2)-1$ ________________

36. $7(2y+6)=9(y+3)+5$ ________________

37. $5(x-3)+2=5(2x-8)-3$ ________________

38. $6(2v-1)-5=7(3v-2)-24$ ________________

39. $-2(3s+9)-6=-3(3s+11)-6$ ________________

40. $-3(5z+24)+2=2(3-2z)-10$ ________________

41. $6(2p-8)+24=3(5p-6)-6$ ________________

name date hour

42. $2(5x + 3) - 3 = 6(2x - 3) + 15$ ______

43. $3(m - 4) - (2m - 3) = -4$ ______

44. $4(2a + 6) - (7a - 5) = 2$ ______

45. $-(4m + 2) - (-3m - 5) = 3$ ______

46. $-(6k - 5) - (-5k + 8) = -4$ ______

47. $2x + 6x - 9 + 4 = 3x - 9 + 3$ ______

48. $3(z - 2) + 4z = 8 + z + 1 - z$ ______

49. $2(r - 3) + 5(r + 4) = 9$ ______

50. $-4(m - 8) + 3(2m + 1) = 6$ ______

In Exercises 51-54, use the given information to write an equation. Then solve it.

51. If 17 is subtracted from a number, and the result is multiplied by 3, the product is 102. What is the number? ______

52. A teacher says, "If I had three times as many students in my class, I would have 46 more than I have now." How many students are presently in the class? ______

53. If three times the sum of a number and 4 is subtracted from 8, the result is 2. Find the number. ______

54. If five times the sum of a number and 3 is added to 5, the result is -5. Find the number. ______

WORK SPACE

OBJECTIVES

- Write verbal phrases as mathematical phrases
- Translate phrases into equations and solve the equations

In this section we see how to solve many common types of word problems. To get the answer to a word problem, you must first read it and determine what facts are given and what you are asked to find. Then go through the following four steps.

Step 1 Choose a variable to represent the numerical value that you are asked to find—the unknown number.

Step 2 Translate the problem into an equation.

Step 3 Solve the equation.

Step 4 Check your solution by using the original words of the problem.

Step 2 is often the hardest. To translate the problem into an equation, we must write facts stated in words as mathematical expressions.

You are likely to see some of the same words again and again in word problems. Some of these are explained in the next few examples.

Example 1 Write each of the following as a mathematical expression. Use x to represent the unknown. (We could use other letters to represent this unknown quantity.)

(a) 5 plus a number.

The word *plus* indicates addition. If x represents the unknown number, then "5 plus a number" can be written as either

$$5 + x \quad \text{or} \quad x + 5.$$

(b) Add 20 to a number.

If x represents the unknown number, then "add 20 to a number" becomes

$$20 + x \quad \text{or} \quad x + 20.$$

(c) "The sum of a number and 12" is $x + 12$ or $12 + x$.

(d) "7 more than a number" is $7 + x$ or $x + 7$.

Work Quick Check 1 at the side.

QUICK CHECKS

1. Write each of the following as a mathematical expression.

(a) 8 more than a number

(b) A number added to -6

(c) The sum of a number and 9

Example 2 Write each of the following as a mathematical expression. Use x as the variable.

(a) 3 less than a number

"Less than" indicates subtraction. "3 less than a number" is

$$x - 3.$$

(Note: $3 - x$ *would not* be correct.)

Quick Check Answers

1. (a) $8 + x$ or $x + 8$
(b) $x + (-6)$ or $-6 + x$
(c) $9 + x$ or $x + 9$

2. Write each of the following as a mathematical expression. Use x for any variables.

(a) 2 less than a number

(b) A number decreased by 4

(c) 15 minus a number

3. Write each of the following as a mathematical expression. Use x for any variables.

(a) The product of a number and 5

(b) -3 times a number

(c) Double the amount

(d) Five eighths of a number

(e) The quotient of a number and 10

4. Write each of the following as a mathematical expression. Use x as the variable.

(a) 10 added to twice a number

(b) The product of 5, and 2 less than a number

(c) 8 added to a number, divided by three times the number

(d) Twice a number, added to the reciprocal of 5

Quick Check Answers

2. (a) $x-2$ (b) $x-10$ (c) $15-x$
3. (a) $5x$ (b) $-3x$ (c) $2x$ (d) $(5/8)x$ (e) $x/10$
4. (a) $2x+10$ (b) $5(x-2)$ (c) $(8+x)/3x$ (d) $2x+1/5$

(b) "A number decreased by 14" is $x - 14$.

(c) "Ten fewer than x" is $x - 10$.

Work Quick Check 2 at the side.

Example 3 Write each of the following as a mathematical expression. Use x as the variable.

(a) "The product of a number and 3" is written $3 \cdot x$ or just $3x$, since *product* indicates multiplication.

(b) "Three times a number" is also $3x$.

(c) "Two thirds of a number" is $\frac{2}{3}x$.

(d) "The quotient of a number and 2" is $\frac{x}{2}$. [The word *quotient* indicates division—use a fraction bar instead of ÷.]

(e) "The reciprocal of a number" is $\frac{1}{x}$.

Work Quick Check 3 at the side.

Some word problems must be translated with a combination of symbols, as the next example shows.

Example 4 Write each of the following as a mathematical expression. Use x as the variable.

(a) "7 subtracted from 4 times a number" is written $4x - 7$.

(b) "A number plus its reciprocal" is $x + \frac{1}{x}$.

(c) "The sum of a number and 2, multiplied by 5" is $(x + 2) \cdot 5$, or the preferred form, $5(x + 2)$.

(d) "The quotient of a number, and 4 plus the number" is $\frac{x}{4 + x}$.

Work Quick Check 4 at the side.

Since equal mathematical expressions are names for the same number, any words that mean *equals* or *same* translate as =. The = sign gives us an equation which we can then solve.

Example 5 Translate "the product of 4, and a number decreased by 7 is 100" into an equation. Use x as the variable. Solve the equation.

Translate as follows:

The product of 4 and a number decreased by 7 is 100.

$$4 \cdot (x - 7) = 100$$

Simplify:

$$4 \cdot (x - 7) = 100$$
$$4x - 28 = 100$$
$$4x = 128 \qquad \text{Add 28 to both sides}$$
$$x = 32 \qquad \text{Multiply by } 1/4$$

Now check the answer by substituting $x = 32$ into the words of the problem.

$$4(32 - 7) = 100$$
$$4(25) = 100$$
$$100 = 100$$

Since the last statement is true, $x = 32$ is the correct answer.

Work Quick Check 5 at the side.

5. Write equations for each of the following and then solve.

(a) When 3 times a number is added to 9, the answer is 12.

(b) If you add 10 to a number, the result is 20.

(c) Twice a number, increased by 3, equals 17.

Example 6 If three times the sum of a number and 4 is decreased by twice the number, the result is -6. Find the number.

Let x represent the unknown number. "Three times the sum of a number and 4" translates into symbols as $3(x + 4)$. "Twice the number" is $2x$. Now write an equation using the information of the problem.

three times the sum of a number and 4	decreased by	twice the number	the result is	-6
↓	↓	↓	↓	↓
$3(x + 4)$	$-$	$2x$	$=$	-6

We can now solve the equation.

$$3(x + 4) - 2x = -6$$
$$3x + 12 - 2x = -6$$
$$x + 12 = -6$$
$$x = -18$$

Check that $x = -18$ is the correct answer by substituting this value into the words of the problem.

$$3(-18 + 4) - 2(-18) = -6$$
$$3(-14) - (-36) = -6$$
$$-42 + 36 = -6$$
$$-6 = -6$$

Example 7 In a given amount of time, Alice drove 40 miles more than Fred. The total distance that both of them traveled was 204 miles. Find the number of miles driven by Fred.

Let x represent the number of miles driven by Fred. Since Alice drove 40 miles more than Fred, the number of miles she drove is $x + 40$. The problem gives you the total number of miles, 204.

Quick Check Answers

5. (a) $3x + 9 = 12$; 1
(b) $10 + x = 20$; 10
(c) $2x + 3 = 17$; 7

6. Work the following word problems.

(a) If 2 is added to the product of 7 and a number, the result is 8 more than the number. Find the number.

(b) In a given amount of time, Larry drove 30 miles more than Rick. Altogether, they drove 90 miles. Find the total number of miles driven by each.

miles (Fred)		miles (Alice)	is	total miles
x	$+$	$x + 40$	$=$	204

Solve the equation.

$$x + (x + 40) = 204$$
$$2x + 40 = 204$$
$$2x = 164$$
$$x = 82$$

Fred drove 82 miles. Alice drove 40 miles more, or $82 + 40 = 122$ miles. The sum of their miles is $82 + 122 = 204$ miles. This checks with the information in the problem.

Work Quick Check 6 at the side.

Quick Check Answers

6. (a) 1 (b) 60 for Larry, 30 for Rick

name date hour

3.5 EXERCISES

Write each of the following as a mathematical expression. Use x *as the variable. See Examples 1–4.*

1. 8 plus a number ____________________

2. A number added to −6 ____________________

3. −1 added to a number ____________________

4. The sum of a number and 12 ____________________

5. A quantity is increased by −18 ____________________

6. The total of x and 12 ____________________

7. 5 less than a number ____________________

8. A number decreased by 6 ____________________

9. Subtract 9 from a number ____________________

10. 16 fewer than a number ____________________

11. The product of a number and 9 ____________________

12. Double a number ____________________

13. Triple a number ____________________

14. Three fifths of a number ____________________

15. The quotient of a number and 6 ____________________

16. The quotient of −9 and a number ____________________

17. A number divided by -4 ______

18. 7 divided by a number ______

19. The product of 8 and the sum of a number and 3 ______

20. A number is added to twice the number ______

21. A number is subtracted from its reciprocal ______

22. Three times the quotient of a number and 2 ______

23. 8 times the difference of a number and 8 ______

24. The difference of a number and 2, multiplied by -7 ______

For each word problem, follow steps (a)-(d). See Examples 5-7.

(a) Choose a variable to represent the unknown quantity.
(b) Translate the problem into an equation.
(c) Solve the equation.
(d) Check your solution by using the original words of the problem.

25. If three times a number is decreased by 2, the result is 22. Find the number. ______

26. When 6 is added to four times a number, the result is 42. Find the number. ______

27. The sum of a number and 3 is multiplied by 4, giving 36 as a result. Find the number. ______

28. The sum of a number and 8 is multiplied by 5, giving 60 as the answer. Find the number. ______

29. Twice a number is added to the number, giving 90. Find the number. ______

name date hour

30. If the sum of a number and 8 is multiplied by -2, the result is -8. Find the number. ______

31. When 6 is subtracted from a number, the result is 7 times the number. Find the number. ______

32. If 4 is subtracted from twice a number, the result is 4 less than the number. Find the number. ______

33. When five times a number is added to twice the number, the result is 10. Find the number. ______

34. If seven times a number is subtracted from eleven times a number, the result is 9. Find the number. ______

35. Tony has a board 44 inches long. He wishes to cut it into two pieces so that one piece will be six inches longer than the other. How long should the shorter piece be? ______

36. Nevarez and Smith were opposing candidates in the school board election. Nevarez received 30 more votes than did Smith, with 516 total votes cast. How many votes did Smith receive? ______

37. On an algebra test, the highest grade was 42 points more than the lowest grade. The sum of the two grades was 138. Find the lowest grade. ______

38. In a physical fitness test, Alfonso did 25 more pushups than Chuck did. The total number of pushups for both men was 173. Find the number of pushups that Chuck did. ______

39. A pharmacist found that at the end of the day she had 12 more prescriptions for antibiotics than she had for tranquilizers. She had 84 prescriptions altogether for these two types of drugs. How many did she have for tranquilizers? ______

40. Mark White gives glass-bottom boat rides in the Bahama Islands. One day he noticed that the boat contained 17 more men (counting himself) than women, with a total of 165 people on the boat. How many women were on the boat? ______

41. Joann McKillip runs a dairy farm. Last year, her cow Bessie gave 238 more gallons of milk than one of her other cows, Bossie. Between them, the two cows gave 1464 gallons of milk. How many gallons of milk did Bossie give? ______

42. Kevin is three times as old as Bob. Three years ago the sum of their ages was 22 years. How old is each now? (Hint: First write an expression for the age of each now, then for the age of each three years ago.) ______

The following two problems are real "head-scratchers."

43. A store has 39 quarts of milk, some in pint cartons and some in quart cartons. There are six times as many quart cartons as pint cartons. How many quart cartons are there? (Hint: 1 quart = 2 pints.) ______

44. A table is three times as long as it is wide. If it were three feet shorter and three feet wider, it would be square (with all sides equal). How long and how wide is it? ______

OBJECTIVES

- Identify the formulas needed for a problem
- Use a formula to give an equation
- Solve a formula for a specified variable

Many word problems can be solved if you know a formula giving the relationships between certain dimensions, amounts, or quantities. Formulas exist for geometric figures such as squares and circles, for distance, for money earned on bank savings, or for converting English measurements to metric measurements. A list of the formulas you will need in this book is given in the back of the book.

Suppose a word problem talks about putting a fence around a rectangular piece of land. You have to find out how much fencing is needed to enclose the field.

There is a formula for the distance around a rectangle. This distance is called the **perimeter** of the rectangle, and the formula is

$$P = 2l + 2w.$$

In the formula, P stands for perimeter (of the rectangle), l stands for the long side (length), and w stands for the short side (width). The perimeter of a rectangle equals the sum of twice the length and twice the width.

Example 1 The perimeter of a rectangle is 80 meters, and the length is 25 meters.* (See Figure 1.) Find the width of the rectangle.

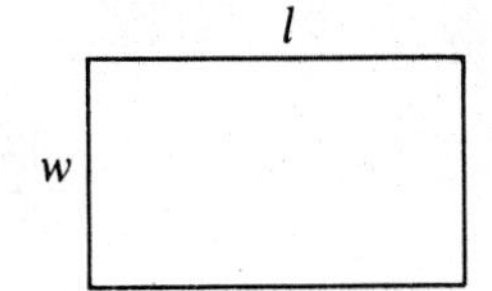

$P = 80$ meters
$l = 25$ meters

Figure 1

To find the width, substitute 80 for P and 25 for l in the formula $P = 2l + 2w$.

$$P = 2l + 2w$$
$$80 = 2(25) + 2w$$

Solve this equation to find w. First simplify it.

$$80 = 50 + 2w$$

Add -50 to both sides.

$$80 + (-50) = 50 + 2w + (-50)$$
$$30 = 2w$$

Multiply both sides by $1/2$.

$$15 = w$$

The width of the rectangle is 15 meters.

Work Quick Check 1 at the side.

QUICK CHECKS

1. Work the following word problems.

(a) The width of a swimming pool is 50 feet, and the length is 75 feet. Find the perimeter of the pool.

(b) A farmer has 1000 meters of fencing material to enclose a rectangular field. The width of the field will be 200 meters. Find the length.

*A meter is a unit of length in the metric system, which is explained in Appendix B.

Quick Check Answers
1. (a) 250 feet (b) 300 meters

2. Work the following word problems.

(a) The perimeter of a triangle is 96 meters. One side is 21 meters long, and a second side is 37 meters long. Find the length of the third side.

(b) The base of a triangle is 35 centimeters. The height is 40 centimeters. Find the area of the triangle.

(c) The area of a triangle is 54 square meters. The height is 18 meters. Find the length of the base of the triangle.

Example 2 The perimeter of a square is 96 inches. Find the length of a side.

You need to know the formula for the perimeter of a square. From the list in the back of the book, you can get the formula $P = 4s$, where s is the length of a side of a square.

The perimeter is given as 96 inches, so that $P = 96$. Substitute 96 for P in the formula.

$$P = 4s$$

$$96 = 4s \qquad P = 96$$

$$\frac{1}{4}(96) = \frac{1}{4}(4s) \qquad \text{Multiply by } 1/4$$

$$24 = s$$

Each side of the square is 24 inches long.

Example 3 The area of a triangle is 126 square meters. The base of the triangle is 21 meters. Find the height.

The formula for the area of a triangle is $A = \frac{1}{2}bh$, where A is area, b is the base, and h is the height. Substitute 126 for A and 21 for b in the formula.

$$A = \frac{1}{2}bh$$

$$126 = \frac{1}{2}(21)h \qquad A = 126, b = 21$$

Simplify the problem by eliminating the fraction 1/2. Multiply both sides of the equation by 2.

$$2(126) = 2\left(\frac{1}{2}\right)(21)h$$

$$252 = 21h$$

Now multiply both sides by 1/21.

$$\frac{1}{21}(252) = \frac{1}{21}(21h)$$

$$12 = h$$

The height of the triangle is 12 meters.

Work Quick Check 2 at the side.

Example 4 The length of a rectangle is 2 meters more than the width. The perimeter is 40 meters. (See Figure 2.) Find the width and the length.

$l = x + 2$

$w = x$ $\qquad$ $P = 40$

Figure 2

Quick Check Answers
2. (a) 38 meters (b) 700 sq. cm
(c) 6 meters

Let x represent the width of the rectangle. Then $x + 2$ represents the length. The formula for the perimeter of a rectangle is $P = 2l + 2w$. Substitute x for w, $x + 2$ for l, and 40 for P.

$$P = 2l + 2w$$
$$40 = 2(x + 2) + 2x \qquad P = 40, l = x + 2, w = x$$
$$40 = 2x + 4 + 2x \qquad \text{Simplify}$$
$$40 = 4x + 4$$
$$36 = 4x \qquad \text{Add } -4 \text{ to both sides}$$
$$9 = x \qquad \text{Multiply both sides by } 1/4$$

The width of the rectangle is 9, and the length is $x + 2 = 9 + 2 =$ 11 meters.

Work Quick Check 3 at the side.

3. Work the following word problems.

(a) The width of a rectangle is 5 meters less than the length. The perimeter is 50 meters. Find the length and width of the rectangle.

(b) The longest side of a triangle is 7 inches longer than the medium side and the shortest side is 3 inches shorter than the medium side. The perimeter of the triangle is 40 inches. Find the length of each side of the triangle.

Example 5 How much simple interest will be earned on a deposit of \$600 for 4 years at an interest rate of 6% per year?

The formula for simple interest is $I = prt$, where I represents interest, p represents principal (the amount of money either borrowed or deposited), r is rate, and t is time in years. Using the given information, we have

$$I = prt$$
$$I = (600)(.06)(4) \qquad 6\% = .06$$
$$I = 144.$$

The deposit will earn \$144 interest.

Work Quick Check 4 at the side.

4. How much simple interest will be earned on \$1200 for 6 years at an interest rate of 5% per year?

Example 6 Two cars start from the same point at the same time and travel in the same direction at constant speeds of 34 and 45 miles per hour, respectively. In how many hours will they be 33 miles apart? (See Figure 3.)

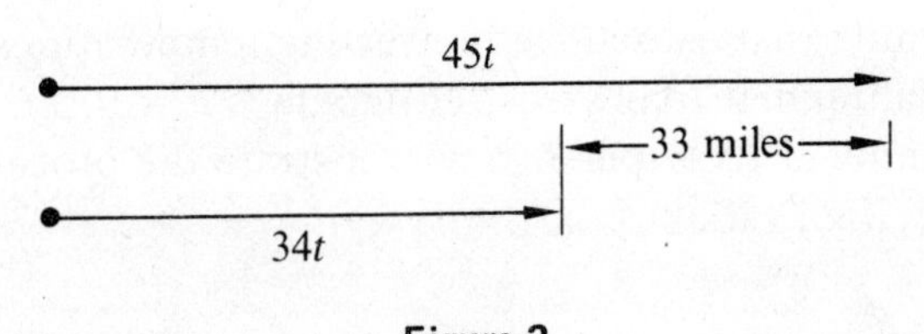

Figure 3

To work this problem, we need to know the formula for distance:

$$d = rt,$$

where d represents distance, r is rate (or speed), and t is time.

Let t represent the unknown number of hours. The distance traveled by the slower car is its rate multiplied by its time, or $34t$. The distance traveled by the faster car is $45t$. The numbers $34t$

Quick Check Answers
3. (a) 15 meters by 10 meters
(b) 12 inches, 19 inches, 9 inches
4. \$360

5. Two cars start from the same point at the same time and travel in the same direction at constant speeds of 42 and 55 miles per hour, respectively. In how many hours will they be 78 miles apart?

and $45t$ represent different distances. From the information in the problem we know that these distances differ by 33 miles, or $45t - 34t = 33$. Now solve the equation to find t.

$$45t - 34t = 33$$
$$11t = 33$$
$$t = 3$$

In 3 hours the two cars will be 33 miles apart.

Work Quick Check 5 at the side.

Sometimes it is necessary to solve a large number of problems which use the same formula. For example, we might be taking a surveying class and need to solve several problems which involve the formula for the area of a rectangle, $A = lw$. Suppose that in each problem we are given the area (A), and the length (l) of a rectangle and want to find the width (w). We save time in the long run by rewriting the formula $A = lw$ so that the value we do not know, w, is alone on one side of the equals sign. The next example shows how to do this. The process of isolating one letter from a formula is called **solving for a specified variable.**

6. Solve each formula for the specified variable.

(a) $d = rt$, for t

(b) $i = prt$, for t

(c) $p = a + b + c$, for a

Example 7 Solve $A = lw$ for w.

We want to get w alone on one side of the equals sign. To do this, multiply both sides of $A = lw$ by $1/l$.

$$A = lw$$
$$\frac{1}{l}(A) = \frac{1}{l}(lw)$$
$$\frac{A}{l} = w \qquad \frac{1}{l}(A) = \frac{A}{l}$$

The formula is now solved for w.

Work Quick Check 6 at the side.

The formula that is used for converting temperatures expressed in degrees Fahrenheit to degrees centigrade is $F = 9/5\ C + 32$. We use this formula in Example 8 to demonstrate the process of solving for a specified variable.

Example 8 Solve $F = \frac{9}{5}C + 32$ for C.

First, get $\frac{9}{5}C$ alone on one side of the equals sign. Do this by adding -32 on both sides.

$$F = \frac{9}{5}C + 32$$
$$F + (-32) = \frac{9}{5}C + 32 + (-32)$$

Quick Check Answers
5. 6 hours
6. (a) $t = d/r$ (b) $t = i/pr$
(c) $a = p - b - c$ or $a = p - (b + c)$

$$F - 32 = \frac{9}{5}C$$

Now, multiply both sides by $\frac{5}{9}$.

$$\frac{5}{9}(F - 32) = \left(\frac{5}{9}\right)\frac{9}{5}C$$

$$\frac{5}{9}(F - 32) = C$$

WORK SPACE

name date hour

3.6 EXERCISES

In Exercises 1-14, a formula is given, along with the values of all but one of the variables in the formula. Find the value of the variable that is not given. See Examples 1 and 2.

1. $P = 4s$; $s = 32$ ____________

2. $P = 2l + 2w$; $l = 5$; $w = 3$ ____________

3. $A = \frac{1}{2}bh$; $b = 6$, $h = 12$ ____________

4. $A = \frac{1}{2}bh$; $b = 6$, $h = 24$ ____________

5. $A = \frac{1}{2}bh$; $A = 20$, $b = 5$ ____________

6. $A = \frac{1}{2}bh$; $A = 30$, $b = 6$ ____________

7. $P = 2l + 2w$; $P = 40$; $w = 6$ ____________

8. $V = \frac{1}{3}Bh$; $V = 80$; $B = 24$ ____________

9. $A = \frac{1}{2}(b + B)h$; $b = 6$, $B = 8$, $h = 3$ ____________

10. $A = \frac{1}{2}(b + B)h$; $b = 10$, $B = 12$, $h = 3$ ____________

11. $d = rt$; $d = 8$, $r = 2$ ____________

12. $d = rt$; $d = 100$, $t = 5$ ____________

13. $C = 2\pi r$; $C = 9.42$, $\pi = 3.14$* ____________

14. $C = 2\pi r$; $C = 25.12$, $\pi = 3.14$ ____________

Solve the given formula for the specified variable. See Examples 7 and 8.

15. $A = lw$; $l =$ ____________

16. $d = rt$; $r =$ ____________

17. $d = rt$; $t =$ ____________

18. $V = lwh$; $w =$ ____________

---

*Actually π is approximately equal, not exactly equal, to 3.14.

19. $V = lwh$; $h =$ ______

20. $I = prt$; $p =$ ______

21. $I = prt$; $t =$ ______

22. $C = 2\pi r$; $r =$ ______

23. $A = \frac{1}{2}bh$; $b =$ ______

24. $A = \frac{1}{2}bh$; $h =$ ______

25. $P = 2l + 2w$; $w =$ ______

26. $a + b + c = p$; $b =$ ______

27. $A = \frac{1}{2}(b + B)h$; $b =$ ______

28. $C = \frac{5}{9}(F - 32)$; $F =$ ______

In Exercises 29–51, write an equation and then solve it. Check your solution in the original statement of the problem. Formulas are listed in the back of the book. See Examples 3–6.

29. The area of a rectangle is 60 meters and the width is 6 meters. Find the length. ______

30. The perimeter of a square is 80 centimeters. Find the length of a side. ______

31. The radius of a circle is 6 feet. Find the circumference. (Let π be approximated by 3.14.) ______

32. The length of a rectangle is 15 inches, and the perimeter is 50 inches. Find the width. ______

33. The perimeter of a triangle is 72 kilometers. One side is 16 kilometers, and another side is 32 kilometers. Find the third side. ______

34. Dorothy Raymond drove 480 miles in 12 hours on a mountain road. Find her average speed. ______

35. A train goes 55 miles per hour for 7 hours. Find the distance traveled. ______

36. The shorter base of a trapezoid is 16 and the longer base is 20. The height is 6. Find the area. ______

name date hour

37. The perimeter of a certain square is seven times the length of a side, decreased by 12. Find the length of a side. __________

38. The circumference of a certain circle is five times the radius, increased by 2.56 meters. Find the radius of the circle. (Use 3.14 as an approximation for π.) __________

39. The perimeter of a certain rectangle is 16 times the width. The length is 12 centimeters more than the width. Find the dimensions of the rectangle. __________

40. The numerical value of the area of a certain triangle is five times the length of the base. The height is fifty times the reciprocal of the base. Find the length of the base. __________

41. The width of a certain rectangle is one less than the length. The perimeter is five times the length, decreased by 5. Find the length of the rectangle. __________

42. From a point on a straight road, John and Fred ride ten-speed bicycles in opposite directions. John rides 10 miles per hour and Fred rides 12 miles per hour. In how many hours will they be 55 miles apart? (Hint: First write expressions for John's distance and Fred's distance using a variable for the time.) __________

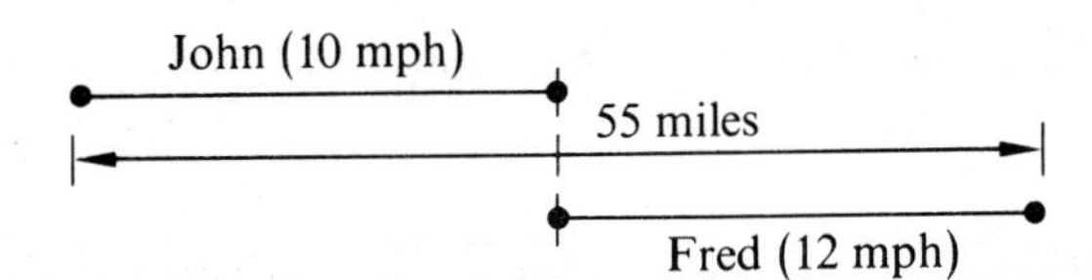

43. Two trains leave Los Angeles at the same time. One travels north at 60 miles per hour and the other south at 80 miles per hour. In how many hours will they be 280 miles apart? __________

44. Two cars are 400 miles apart. Both start at the same time and travel toward one another. They meet four hours later. If the average speed of one car is 20 miles per hour faster than the other, what is the average speed of each car? __________

Slower car

400 miles

Faster car

45. Ann has saved \$163 for a trip to Disneyland. Transportation will cost \$28, tickets for the park entrance and rides will cost \$15 per day, and lodging and meals will cost \$30 per day. How many days can she spend there? ________________

46. At Irv's Burgerville, hamburgers cost 90 cents each, and a bag of french fries costs 40 cents. How many hamburgers and how many bags of french fries can Ted buy with \$8.80 if he wants twice as many hamburgers as bags of french fries? ________________

OBJECTIVES

- Graph intervals on a number line
- Learn the addition property of inequality
- Solve inequalities
- Write verbal phrases as inequalities

Inequalities are statements in which algebraic expressions are related by

$<$ "is less than"
$\leqslant$ "is less than or equal to"
$>$ "is greater than"
$\geqslant$ "is greater than or equal to."

We assume that the domain of any inequality we study is the set of all real numbers. So, when you solve an inequality, you must find all real number solutions for it. For example, $x \leqslant 2$ represents all real numbers that are less than or equal to 2, and not just the integers less than or equal to 2.

A good way to show the solution of an inequality is by graphing. To graph all real numbers satisfying $x \leqslant 2$, place a dot at 2 on a number line and draw an arrow extending from the dot to the left (to represent the fact that all numbers less than 2 are also part of the graph). The graph is shown in Figure 4.

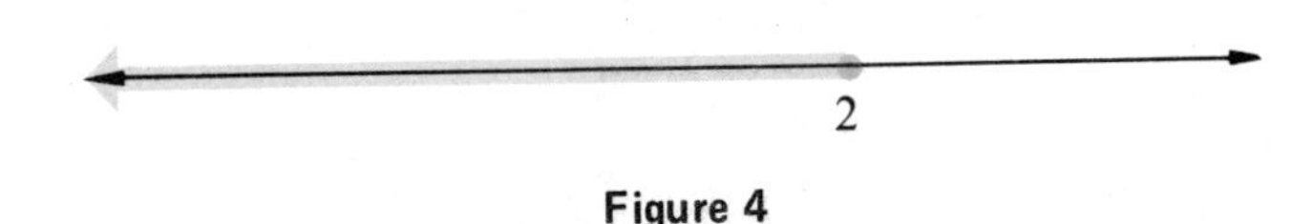

Figure 4

Example 1 Graph $x > -5$.

The statement $x > -5$ says that x can represent numbers greater than -5, but x cannot equal -5 itself. To show this on a graph, place an open circle at -5 and draw an arrow to the right, as in the graph shown in Figure 5. The open circle at -5 shows that -5 is not part of the graph.

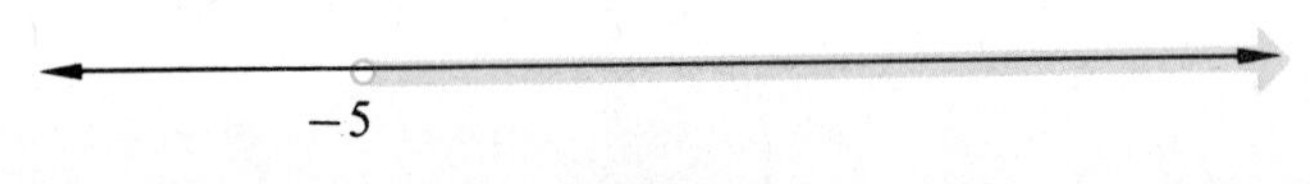

Figure 5

Example 2 Graph $-3 \leqslant x < 2$.

The statement $-3 \leqslant x < 2$ is read "-3 is less than or equal to x and x is less than 2." To graph this inequality, place a heavy dot at -3 (because -3 is part of the graph), and an open circle at 2 (because 2 is not part of the graph). Then draw a line segment between the two circles, as in Figure 6.

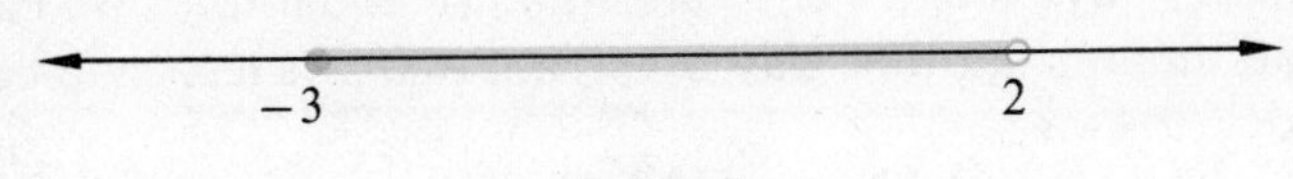

Figure 6

QUICK CHECKS

1. Graph each of the following.

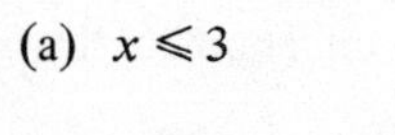

(a) $x \leq 3$

(b) $x > 4$

(c) $x \geq -1$

(d) $4 < x \leq 8$

(e) $-7 < x < -2$

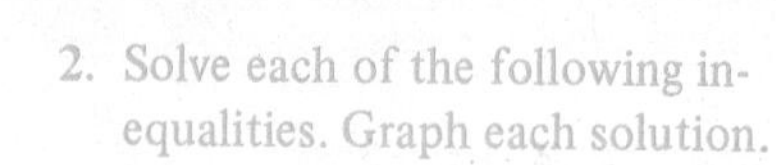

2. Solve each of the following inequalities. Graph each solution.

(a) $x - 1 < 6$

(b) $2m \geq m - 4$

(c) $-1 + 8r < 7r + 2$

(d) $-3r + 2 > -4r - 8$

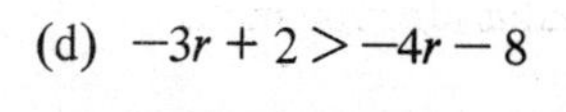

Quick Check Answers

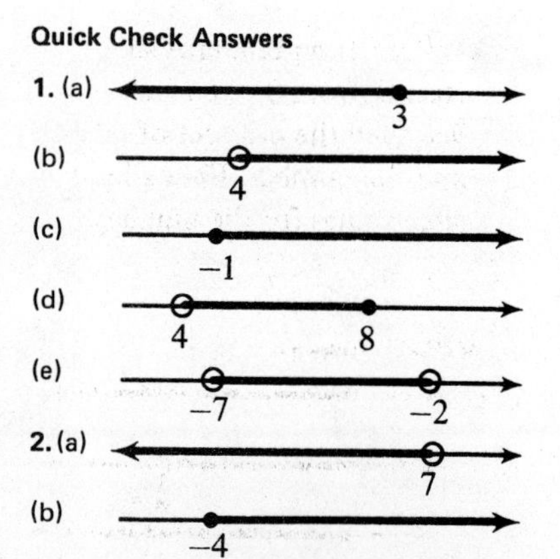

Work Quick Check 1 at the side.

Inequalities such as $x + 4 \leq 9$ can be solved in much the same way that we solved equations. We first use the **addition property of inequality**, which states that the same term can be added to both sides of an inequality.

> For any expressions A, B, and C, the inequalities
>
> $$A < C$$
>
> and
>
> $$A + C < B + C$$
>
> have exactly the same solutions.

The addition property of inequality also works with $>$, $\leq$, or $\geq$.

Example 3 Solve the inequality $7 + 3k > 2k - 5$.

Use the addition property of inequality twice—once to get the terms containing k on one side of the inequality, and a second time to get the integers together on the other side.

$$7 + 3k > 2k - 5$$
$$7 + 3k + (-2k) > 2k - 5 + (-2k)$$
$$7 + k > -5$$
$$7 + k + (-7) > -5 + (-7)$$
$$k > -12$$

The graph of the solution $k > -12$ is shown in Figure 7.

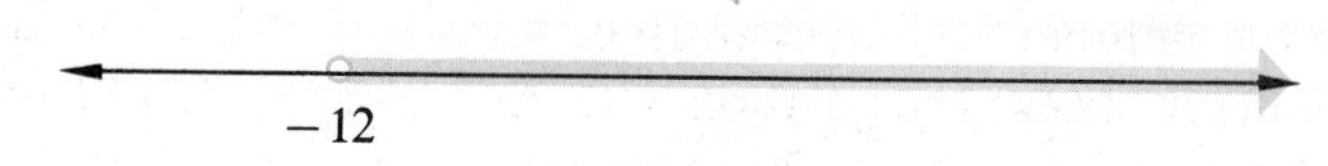

Figure 7

Work Quick Check 2 at the side.

Example 4 Solve $6 + 3y \geq 4y - 5$.

First add $-3y$ to both sides.

$$6 + 3y + (-3y) \geq 4y - 5 + (-3y)$$
$$6 \geq y - 5$$

Then add 5 to both sides.

$$6 + 5 \geq y - 5 + 5$$
$$11 \geq y$$

This solution is perfectly correct, but it is customary to write the solution to an inequality with the variable on the left. The statement $11 \geq y$ says that 11 is greater than or equal to y. We can say the same thing in another way by saying that y is less than or equal to 11, or

$$y \leq 11.$$

The graph of the solution $y \leqslant 11$ is shown in Figure 8.

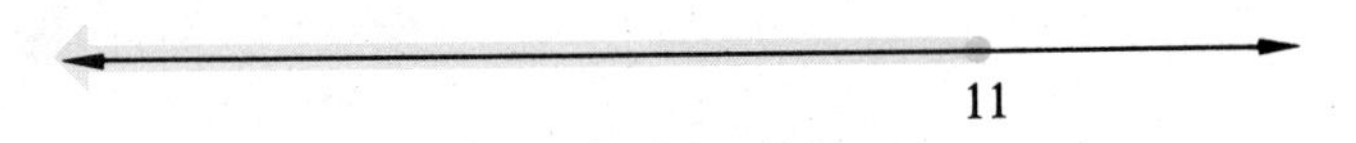

Figure 8

If the inequality $6 + 3y \geqslant 4y - 5$ in Example 4 were solved by first adding the term $-4y$ to both sides, then

$$\begin{aligned} 6 + 3y + (-4y) &\geqslant 4y - 5 + (-4y) \\ 6 - y &\geqslant -5 \\ 6 - y + (-6) &\geqslant -5 + (-6) \\ -y &\geqslant -11. \end{aligned}$$

With an answer like $-y \geqslant -11$, we cannot go any further with the methods discussed so far. To finish this solution, we need the multiplication property of inequality, which we postpone until the next section. In the exercises that follow, if you get an answer like $-y \geqslant -11$, then start over and add terms to both sides in a different way.

Work Quick Check 3 at the side.

Example 5 If 2 is added to five times a number, the result is greater than or equal to 5 more than four times the number. Find the number.

We first translate this word problem into an inequality. Let x represent the number you want to find. Then "2 is added to five times a number" is expressed as $5x + 2$. And "5 more than four times the number" is $4x + 5$. The two expressions are related by "greater than or equal to."

$$5x + 2 \geqslant 4x + 5$$

To solve the inequality $5x + 2 \geqslant 4x + 5$, first add $-4x$ to both sides.

$$\begin{aligned} 5x + 2 + (-4x) &\geqslant 4x + 5 + (-4x) \\ x + 2 &\geqslant 5 \end{aligned}$$

Then add -2 to both sides.

$$\begin{aligned} x + 2 + (-2) &\geqslant 5 + (-2) \\ x &\geqslant 3 \end{aligned}$$

The number we want is greater than or equal to 3.

Work Quick Check 4 at the side.

3. Solve each of the following inequalities. Graph each solution.

(a) $3m \geqslant 2m + 1$

(b) $12y > 13y - 2$

(c) $1 - 6y \leqslant -5y$

(d) $2 + 9a \geqslant 10a - 8$

4. Solve the following word problems.

(a) Twice a number is less than or equal to the sum of the number and 2. Find all possible values for the number.

(b) If 9 times a number is subtracted from 8, the result is less than the product of -10 and the number. Find all possible values for the number.

Quick Check Answers

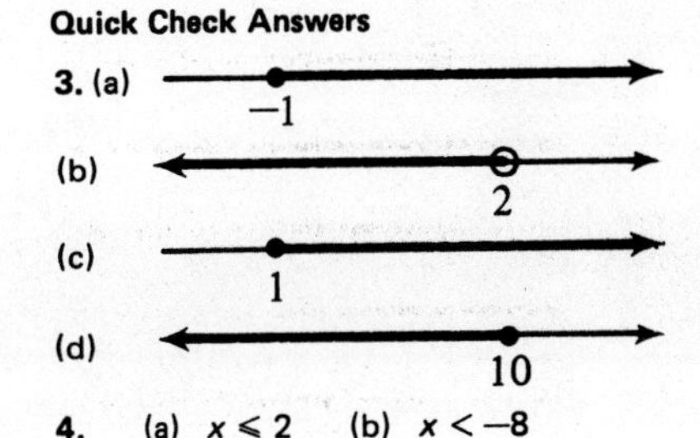

4. (a) $x \leqslant 2$ (b) $x < -8$

WORK SPACE

name date hour

3.7 EXERCISES

Graph each inequality on the given number line. See Examples 1 and 2.

1. $x \leq 4$

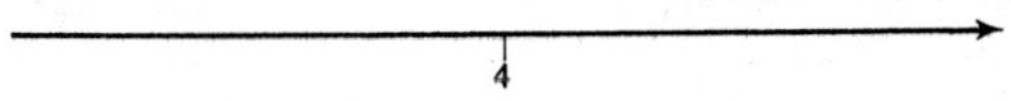

2. $x \leq -3$

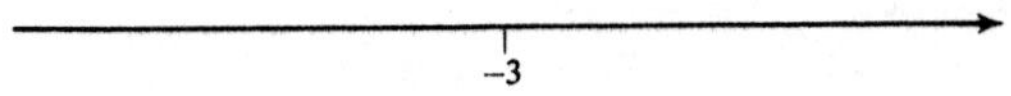

3. $k \geq -5$

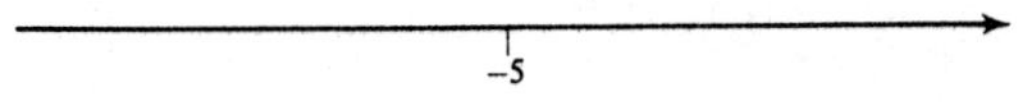

4. $m \geq 6$

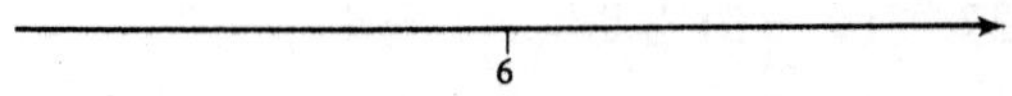

5. $a < 3$

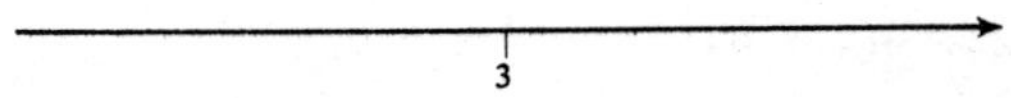

6. $p > 4$

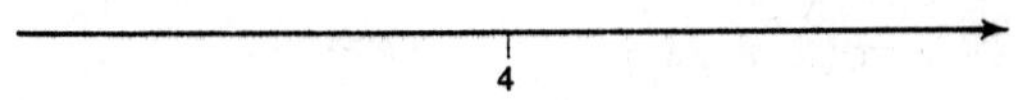

7. $-2 \leq x \leq 5$

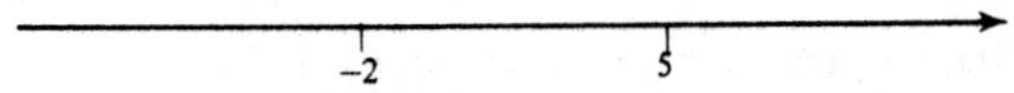

8. $8 \leq m \leq 10$

9. $3 \leq y < 5$

10. $0 < y \leq 10$

Solve each inequality. See Examples 3 and 4.

11. $a + 6 < 8$ ______

12. $k - 4 < 2$ ______

13. $z - 3 \geq -2$ ______

14. $p + 2 \geq -6$ ______

15. $p - 8 \leq 4$ ______

16. $2 + m \geq 5$ ______

17. $-3 + k \geq 2$ ______

18. $-8 + y < -10$ ______

19. $x + 6 \leq 6$ ______

20. $x + 5 > 5$ ______

21. $4x < 3x + 6$ ______

22. $5x \leq 4x - 8$ ______

Solve each inequality and then graph your solution. See Examples 3 and 4.

23. $3n + 5 \leq 2n - 6$

24. $5x - 2 < 4x - 5$

25. $2z - 8 > z - 3$

26. $4x + 6 \leq 3x - 5$

27. $3(y - 5) + 2 < 2(y - 4)$

28. $4(x + 6) - 5 > 3(x + 1)$

29. $-6(k + 2) + 3 \geq -7(k - 5)$

30. $-3(m - 5) + 8 < -4(m + 2)$

In Exercises 31-34, write an inequality using the information given in the problem and then solve it. See Example 5.

31. If four times a number is added to 8, the result is less than three times the number added to 5. Find all possible values of the number. ______

32. The product of 7 and a number is added to 4, giving a result which is greater than or equal to six times the number. Find all possible values of the number. ______

33. If the length of a rectangle is to be twice the width, and the difference between the two dimensions is to be less than or equal to 7 meters, what is the largest possible value for the width? ______

34. The perimeter of a triangle must be no more than 55 centimeters. One side of the triangle is 18 centimeters long. A second side is 13 centimeters. Find the longest possible value for the third side. ______

3.8 THE MULTIPLICATION PROPERTY OF INEQUALITY

OBJECTIVES

- Learn the multiplication property of inequality
- Solve inequalities

Using only the addition property of inequality, we cannot solve inequalities such as $-y \geqslant -11$. We need the *multiplication property of inequality*. To see how this property works, let's look at some examples.

First take the inequality $3 < 7$ and multiply both sides by the positive number 2.

$$3 < 7$$
$$2(3) < 2(7)$$
$$6 < 14 \qquad \text{True}$$

Multiply both sides of $3 < 7$ by the negative number -5.

$$3 < 7$$
$$-5(3) < -5(7)$$
$$-15 < -35 \qquad \text{False}$$

To get a true statement when we multiply both sides by -5, we would have to reverse the direction of the inequality symbol.

$$3 < 7$$
$$-5(3) > -5(7)$$
$$-15 > -35 \qquad \text{True}$$

Take the inequality $-6 < 2$ as another example. Multiply both sides by the positive number 4.

$$-6 < 2$$
$$4(-6) < 4(2)$$
$$-24 < 8 \qquad \text{True}$$

If we multiply both sides of $-6 < 2$ by -5, *and at the same time reverse the direction of the inequality symbol,* we get

$$-6 < 2$$
$$(-5)(-6) > (-5)(2)$$
$$30 > -10. \qquad \text{True}$$

In summary, we have the two parts of the **multiplication property of inequality.**

For any expressions A, B, and C,

(1) if C is positive, then the inequalities

$$A < B$$

and

$$AC < BC$$

have exactly the same solutions;

(2) if C is negative, then the inequalities

$$A < B$$

and

$$AC > BC$$

have exactly the same solutions.

The multiplication property of inequality also works with $>$, $\leq$, or $\geq$.

Important: (1) When you multiply both sides of an inequality by a positive number, the direction of the inequality symbol does not change. Also, adding terms to both sides does not change the symbol.

(2) When you multiply both sides of an inequality by a negative number, the direction of the symbol does change. *You reverse the symbol of inequality only when multiplying by a negative number.*

Example 1 Solve the inequality $3r < 18$.

To simplify this inequality, use the multiplication property of inequality and multiply both sides by 1/3. Since 1/3 is a positive number, the direction of the inequality symbol does not change.

$$3r < 18$$

$$\frac{1}{3}(3r) < \frac{1}{3}(18)$$

$$r < 6$$

The graph of this solution is shown in Figure 9.

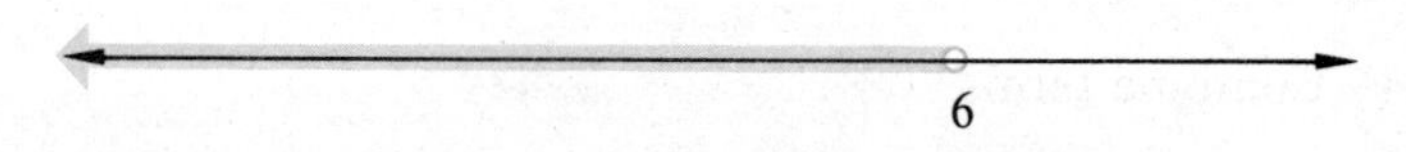

Figure 9

Example 2 Solve the inequality $-4t \geq 8$.

Here we need to multiply both sides of the inequality by $-1/4$, a negative number. This does change the direction of the inequality symbol.

$$-4t \geq 8$$

$$\left(-\frac{1}{4}\right)(-4t) \leq \left(-\frac{1}{4}\right)(8)$$

$$t \leq -2$$

The solution is graphed in Figure 10.

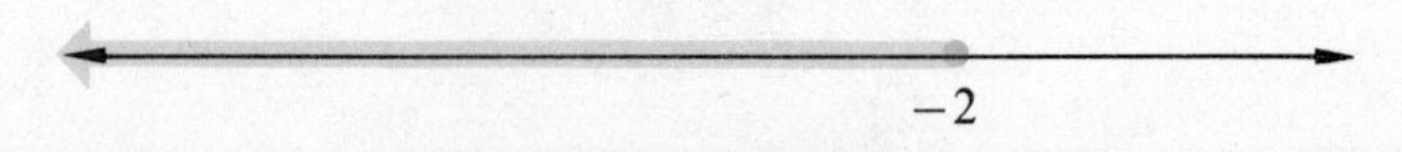

Figure 10

Example 3 Solve the inequality $-x \leq -11$.

Use the multiplication property of inequality and multiply both sides by -1. Since -1 is negative, change the direction of the inequality symbol.

$$-x \leqslant -11$$
$$(-1)(-x) \geqslant (-1)(-11)$$
$$x \geqslant 11$$

The solution is graphed in Figure 11.

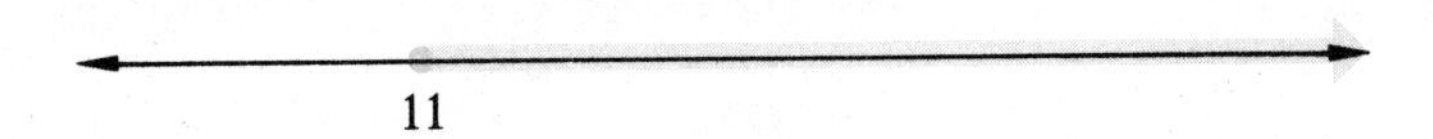

Figure 11

Work Quick Check 1 at the side.

We can now summarize the steps involved in solving an inequality. (Remember that $<$ can be replaced with $>$, $\leqslant$, or $\geqslant$ in this summary.)

Step 1 Use the associative, commutative, and distributive properties to combine terms on both sides of the inequality.

Step 2 Use the addition property of inequality to simplify the inequality to one of the form $ax < b$, where a and b are real numbers.

Step 3 Use the multiplication property of inequality to simplify further to an inequality of the form $x < c$ or $x > c$, where c is a real number.

Example 4 Solve $5(k-3) - 7k \geqslant 4(k-3) + 9$.

Step 1 Combine terms.

$$5(k-3) - 7k \geqslant 4(k-3) + 9$$
$$5k - 15 - 7k \geqslant 4k - 12 + 9$$
$$-2k - 15 \geqslant 4k - 3$$

Step 2 Use the addition property.

$$-2k - 15 + (-4k) \geqslant 4k - 3 + (-4k)$$
$$-6k - 15 \geqslant -3$$
$$-6k - 15 + 15 \geqslant -3 + 15$$
$$-6k \geqslant 12$$

Step 3 Multiply both sides by $-1/6$, a negative number. Change the direction of the inequality symbol.

$$\left(-\frac{1}{6}\right)(-6k) \leqslant \left(-\frac{1}{6}\right)(12)$$
$$k \leqslant -2$$

A graph of the solution is shown in Figure 12.

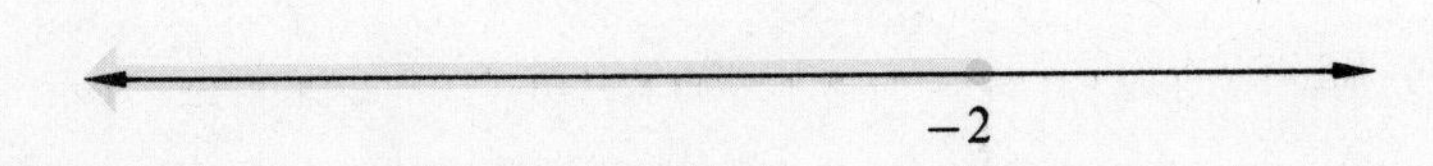

Figure 12

Work Quick Check 2 at the side.

QUICK CHECKS

1. Solve each inequality. Graph each solution.

(a) $5x \geqslant 30$

(b) $9y < -18$

(c) $-4p \leqslant 32$

(d) $-2r > -12$

(e) $-5p \leqslant 0$

2. Solve each of the following inequalities. Graph each solution.

(a) $4(y-1) - 3y > -15 - (2y+1)$

(b) $2(a+6) + 3(a-1) \geqslant 7a - 3$

(c) $9 - 3(2m-1) \leqslant -4(m-3)$

Quick Check Answers

1. (a) 6 (b) −2 (c) −8 (d) 6 (e) 0
2. (a) −4 (b) 6 (c) 0

WORK SPACE

name date hour

3.8 EXERCISES

Solve each inequality and graph the solution. See Examples 1–4.

1. $3x < 27$
2. $5h \geqslant 20$
3. $4r \geqslant -12$
4. $6a < -18$
5. $-2k \leqslant 12$
6. $-3v > 6$
7. $-8y > 72$
8. $-5z \leqslant 40$
9. $-5m > -35$
10. $-8x \leqslant -16$
11. $-6r < -16$
12. $-9a \geqslant -63$
13. $4k + 1 \geqslant 2k - 9$
14. $5y + 3 < 2y + 12$
15. $3 + 2r > 5r - 27$
16. $8 + 6t \leqslant 8t + 12$
17. $4q + 1 - 5 < 8q + 4$
18. $5x - 2 \leqslant 2x + 6 - x$
19. $10p + 20 - p > p + 3 - 23$
20. $-3v + 6 + 3 - 2 > -5v - 19$
21. $-k + 4 + 5k \leqslant -1 + 3k + 5$
22. $6y - 2y - 4 + 7y > 3y - 4 + 7y$

23. $2(x-5)+3x<4(x-6)+3$

24. $5(t+3)-6t\leq 3(2t+1)-4t$

25. $5-(2-r)\leq 3r+5$

26. $-9+(8+y)>7y-4$

27. $3(p+1)-2(p-4)\geq 5(2p-3)+2$

28. $-5(m-3)+4(m+6)<2(m-3)+4$

In Exercises 29-32, write an inequality using the information given in the problem. Then solve it.

29. A student has test grades of 75 and 82. What must he score on a third test to have an average of 80 or higher?

30. In Exercise 29, if 100 is the highest score possible on the third test, how high an average (to the nearest tenth) can the student make? What is the lowest average possible for the three tests?

31. Twice a number added to three times the sum of the number and 2 is more than 17. Find the numbers that satisfy this condition.

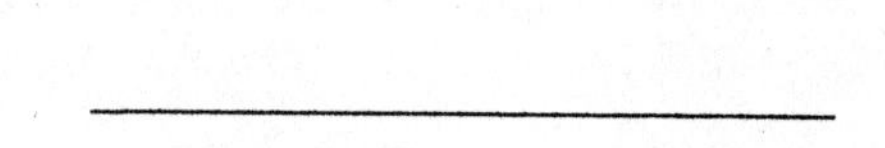

32. Mr. Odjakjian earned \$200 at odd jobs during July, \$300 during August, and \$225 during September. If his average salary for the four months from July through October is to be at least \$250, how much must he earn during October?

OBJECTIVES

- Find percents of given numbers
- Work word problems involving percents

We studied the basics of percent in Section 1.2. Now that we can solve equations, we can work more useful percent problems. First, recall that *percent* means *per one hundred.*

We define 1% as

$$1\% = .01.$$

For example, $32\% = 32 \cdot 1\% = 32 \cdot (.01) = .32$. Also, $125\% = 125 \cdot 1\% = 125 \cdot (.01) = 1.25$.

The next examples show the various types of percent problems.

Example 1 Find 25% of 820.

The word *of* translates as *times.* If we use x to represent the number we need to find, we have

$$x = 25\% \cdot 820$$
$$x = (.25) \cdot 820$$
$$x = 205.$$

Example 2 The sales tax in one state is 6%. Find the amount of sales tax on a new car costing $7100.

Let x represent the amount of the tax. Then

$$x = 6\% \cdot 7100 = (.06) \cdot 7100 = 426.00.$$

The tax is $426.

Work Quick Check 1 at the side.

Example 3 16 is 80% of what number?

Let x be the number we need. Then 80% of x is 16, or

$$80\% \cdot x = 16$$
$$(.80) \cdot x = 16.$$

To solve this equation, multiply both sides by 1/.80.

$$\left(\frac{1}{.80}\right) \cdot (.80) \cdot x = \left(\frac{1}{.80}\right) \cdot 16$$
$$x = \frac{16}{.80}$$
$$x = 20$$

Example 4 75 is 20% of what number?

QUICK CHECKS

1. Find each of the following.

(a) 70% of 110

(b) 41% of 23

(c) 130% of 60

(d) A fight promoter must pay 12% of all money received to rent a hall. She receives $42,000. How much goes to the hall?

(e) Find the amount of a 5% sales tax on a purchase of $172.

Quick Check Answers

1. (a) 77 (b) 9.43 (c) 78 (d) $5040 (e) $8.60

2. (a) 10 is 25% of what number?

(b) 65 is 50% of what number?

(c) 144 is 30% of what number?

3. (a) What percent of 90 is 60?

(b) What percent of 1.44 is .36?

(c) What percent of 2500 is 2250?

(d) What percent of 80 is 120?

4. (a) The price of lumber has increased very quickly during the last year. In fact, the price is 60% higher than it was a year ago. Find last year's price for lumber now costing $1280.

(b) Enrollment in nursing classes at one school has increased 90% in the past few years. The enrollment now is 133 students. Find the previous enrollment.

Quick Check Answers
2. (a) 40 (b) 130 (c) 480
3. (a) 66 2/3% (b) 25%
(c) 90% (d) 150%
4. (a) $800 (b) 70

Let x be the number we need.

$$20\% \cdot x = 75$$
$$(.20) \cdot x = 75$$
$$x = \frac{75}{.20}$$
$$x = 375$$

Work Quick Check 2 at the side.

Example 5 What percent of 20 is 15?

If we use x to represent the unknown percent, then

what percent	of	20	is	15
↓	↓	↓	↓	↓
x	$\cdot$	20	=	15

equals $20x = 15.$

Multiply both sides of this equation by 1/20.

$$\frac{1}{20} \cdot 20x = \frac{1}{20} \cdot 15$$
$$x = \frac{15}{20}$$
$$x = .75$$
$$x = 75\%$$

Work Quick Check 3 at the side.

Example 6 Samantha Jones was given a 10% raise, bringing her salary to $770 per month. Find her previous salary.

Let x represent her previous salary. Her 10% raise is given by $10\% \cdot x$, or $.10x$. Finally,

previous salary	+	raise	=	new salary.
↓	↓	↓	↓	↓
x	+	$.10x$	=	770

Simplify:

$$x + .10x = 770$$
$$1.10x = 770$$

$x + .10x = (1 + .10)x = 1.10x.$

Multiply by 1/1.10.

$$\frac{1}{1.10} \cdot (1.10x) = \frac{1}{1.10} \cdot (770)$$
$$x = 700$$

Her previous salary was $700 per month.

Work Quick Check 4 at the side.

Example 7 The Thompson family has installed new insulation which has cut their monthly electric bill by 12% per month. The current bill is \$32 per month. Find the amount of the bill before the insulation was added.

Let x be the amount of the old bill. Then $12\% \cdot x$, or $.12x$, represents the amount of the savings. We have:

old bill	$-$	savings	$=$	new bill	
x	$-$	$.12x$	$=$	32	
		$.88x$	$=$	32	$x - .12x =$ $1x - .12x = .88x.$

Multiply both sides by $1/.88$:

$$\frac{1}{.88} \cdot (.88x) = \frac{1}{.88} \cdot 32$$

$$x = \frac{32}{.88}$$

$$x = 36.36. \quad \text{(Rounded off)}$$

The previous bill was \$36.36.

Work Quick Check 5 at the side.

5. (a) In a move to save money, the company that Sarah works for cut all salaries by 10%. Sarah's new salary is \$1080 per month. Find her original salary.

(b) An electric guitar is on sale at \$336, which is 16% off the original price. Find the original price.

Quick Check Answers

5. (a) \$1200 (b) \$400

WORK SPACE

name date hour

3.9 EXERCISES

Convert each percent to a decimal.

1. 27% = ______ 2. 39% = ______ 3. 96% = ______

4. 55% = ______ 5. 137% = ______ 6. 241% = ______

7. .2% = ______ 8. .7% = ______

Convert each decimal to a percent.

9. .7 = ______ 10. .3 = ______ 11. .42 = ______

12. .83 = ______ 13. 1.92 = ______ 14. 3.81 = ______

Solve. See Examples 1-4.

15. 625 is 125% of what number? ______

16. 300 is 75% of what number? ______

17. 8.3 is 10% of what number? ______

18. 72,000 is 180% of what number? ______

19. 32% of what number is 180? ______

20. 12% of what number is 375? ______

21. 75% of what number is 4800? ______

22. 3% of what number is 18? ______

23. 1318 is what percent of 2636? ______

24. 16 is what percent of 80? ______

25. 144 is what percent of 300? ______

26. 5 is what percent of 25? ______

Solve. See Examples 6 and 7.

27. At a certain amusement park, 10% of the customers ride the roller coaster. One hour, 796 people rode the roller coaster. How many customers were in the park that hour? ______

28. Mama Lulu's recipe for pizza sauce calls for 11% herbs and spices. If 97.9 gallons of herbs and spices are used, how much pizza sauce will be made? ______

29. Jeff McNett's overtime pay is $17.80, which is 12% of his total pay. Find his total pay. ______

30. At Goldie's Delicatessen, 16% of all customers order a dill pickle. In a recent week, 372 pickles were sold. Find the total number of customers. ______

31. Susan Wright estimates that the total sales at her Magic Mart food store next month will be $175,000, with total advertising expenses of $3500. What percent of total sales will be spent on advertising? ______

32. Sid's Pharmacy has a total monthly payroll of $6800, with $1496 of this amount going for fringe benefits. What percent of the total payroll goes for fringe benefits? ______

33. Mr. Block's income tax this year was $2013, an increase of 10% over last year's tax. Find the amount of last year's tax. ______

34. In the local election, 411,700 people voted. This was an increase of 15% over the number who voted last year. Find the number who voted last year. ______

35. Sara's Plant World collects 4% state sales tax on all sales. Total sales for one day, including tax, were $1648.40. Find the amount of the tax. ______

name date hour

CHAPTER 3 TEST

Simplify by combining terms.

1. $2x + 5 + 5x - 3x - 3 =$ ______

2. $k - 3k + 5k - 6k + 4k =$ ______

3. $9r + 3r - 4r - r - 8r =$ ______

4. $3z - 6z + 8 - 9 + 4z - 9 =$ ______

5. $4(2m + 1) - 3(m + 5) + 2m - 1 =$ ______

6. $3(2z - 8) - 2(2 - 3z) + 6 =$ ______

Solve each equation.

7. $x + 7 = 10$ ______

8. $2m - 5 = 3$ ______

9. $6v + 3 = 8v - 7$ ______

10. $3(a + 12) = 1 - 2(a - 5)$ ______

11. $6y - 3(y + 2) = 5$ ______

12. $4k - 6k + 8(k - 3) = -2(k + 12)$ ______

13. $4(p + 3) - 5 = 3(p + 4) - 10$ ______

14. $\frac{m}{5} = 2$ ______

15. $\frac{2}{3}z = 18$ ______

16. $4 - (3 - m) = 12 + 3m$ ______

17. $-(r + 4) = 2 + r$ ______

18. Solve the formula $I = prt$ for p. ______

19. Solve the formula $A = \frac{1}{2}(b + B)h$ for h. ______

Solve each inequality. Graph the solution.

20. $x + 4 \leqslant 8$

21. $5z \geqslant -10$

22. $-2m < -14$

23. $-3k < k - 8$

24. $5(k - 2) + 3 \leqslant 2(k - 3) + 2k$

25. $-4r + 2(r - 3) \geqslant 5r - (3 + 6r) + 1 - 8$

Write an equation for each problem, then solve.

26. A rectangle has a perimeter which is two inches less than three times the length. The perimeter is 190 inches. Find the dimensions of the rectangle. ______

27. Dick's lunch cost \$1.20. Sandwiches are 35¢ each and milk is 15¢ per glass. He bought one glass of milk. How many sandwiches did he buy? ______

28. Joe bicycled from here to there, a distance of 21 miles, in five hours. During the last two hours, he became tired and slowed down by two miles per hour. What was his speed for the first three hours? ______

29. Don is now twenty years older than Hank. In five years Don will be twice as old as Hank. What are their ages now? ______

30. 15 is what percent of 25? ______

31. What percent of 75 is 90? ______

4 Polynomials

4.1 EXPONENTS

OBJECTIVES

- Write and simplify numbers in exponential form
- Use the product rule for exponents
- Know the meaning of a 0 exponent
- Know and use the properties of exponents

Expressions like xx and xxx occur so frequently in algebra that it is convenient to write them in shortened form. For example, write

xx	as x^2,	read "x squared"
xxx	as x^3,	read "x cubed"
$xxxx$	as x^4,	read "x to the fourth power" or "x to the fourth"
$xxxxx$	as x^5,	read "x to the fifth"

and so on. In the symbol x^3, for example, x is called the **base** and 3 is called the **exponent** or **power**. The symbol x^3 itself is called an **exponential.**

Example 1 Write $3 \cdot 3 \cdot 3 \cdot 3 \cdot 3$ in exponential form. Evaluate the exponential.

Since 3 occurs as a factor five times, the base is 3 and the exponent is 5. The exponential is 3^5. The value is

$$3^5 = 3 \cdot 3 \cdot 3 \cdot 3 \cdot 3 = 243.$$

Work Quick Check 1 at the side.

Example 2 Evaluate each exponential. Name the base and the exponent.

(a) $5^4 = 5 \cdot 5 \cdot 5 \cdot 5 = 625$. The base is 5; the exponent is 4.

(b) $-(5^4) = -(5 \cdot 5 \cdot 5 \cdot 5) = -625$. The base is 5; the exponent is 4.

(c) $(-5)^4 = (-5)(-5)(-5)(-5) = 625$. The base is -5; the exponent is 4.

Work Quick Check 2 at the side.

By the definition of exponents,

$$\begin{aligned} 2^4 \cdot 2^3 &= (2 \cdot 2 \cdot 2 \cdot 2)(2 \cdot 2 \cdot 2) \\ &= 2 \cdot 2 \cdot 2 \cdot 2 \cdot 2 \cdot 2 \cdot 2 \\ &= 2^7. \end{aligned}$$

QUICK CHECKS

1. Write each in exponential form and evaluate.

(a) $2 \cdot 2 \cdot 2 \cdot 2$

(b) $7 \cdot 7 \cdot 7$

2. Evaluate each exponential. Name the base and the exponent.

(a) 6^3

(b) $(-2)^5$

(c) $-(3^2)$

Quick Check Answers

1. (a) $2^4 = 16$ (b) $7^3 = 343$
2. (a) 216; 6; 3 (b) -32; -2; 5 (c) -9; 3; 2

3. Find each product by the product rule, if possible.

(a) $8^2 \cdot 8^5$

(b) $3^5 \cdot 2^6$

(c) $(-7)^5 \cdot (-7)^3$

4. Multiply.

$5m^2 \cdot 2m^6$

Quick Check Answers
3. (a) 8^7 (b) cannot (c) $(-7)^8$
4. $10m^8$

In general, for positive integers m and n, we have the **product rule for exponentials.**

$$a^m \cdot a^n = a^{m+n}$$

The bases must be the same before the product rule for exponentials can be applied.

Example 3 Find each product by the product rule, if possible.

(a) $6^3 \cdot 6^5 = 6^{3+5} = 6^8$ by the product rule.

(b) $6^3 \cdot 4^5$ cannot be simplified by the product rule since the bases, 6 and 4, are different.

(c) $(-4)^5(-4)^3 = (-4)^{5+3} = (-4)^8$ by the product rule.

Work Quick Check 3 at the side.

Example 4 Multiply $2x^3$ by $3x^7$.

Since $2x^3$ means $2 \cdot x^3$ and $3x^7$ means $3 \cdot x^7$, we have

$$2x^3 \cdot 3x^7 = 2 \cdot 3 \cdot x^3 \cdot x^7 = 6x^{10}.$$

Work Quick Check 4 at the side.

The rule for division of exponentials is similar to the product rule. For example,

$$\frac{6^5}{6^2} = \frac{6 \cdot 6 \cdot 6 \cdot 6 \cdot 6}{6 \cdot 6} = 6 \cdot 6 \cdot 6 = 6^3.$$

Note that the difference of the exponents, $5 - 2$, gives the exponent of the answer, 3.

$$\frac{3^4}{3^6} = \frac{3 \cdot 3 \cdot 3 \cdot 3}{3 \cdot 3 \cdot 3 \cdot 3 \cdot 3 \cdot 3} = \frac{1}{3 \cdot 3} = \frac{1}{3^2}$$

In this example, the difference $6 - 4$ gives the new exponent, 2.

There is one other possibility. If the exponents in the numerator and denominator are equal, we have

$$\frac{6^5}{6^5} = \frac{6 \cdot 6 \cdot 6 \cdot 6 \cdot 6}{6 \cdot 6 \cdot 6 \cdot 6 \cdot 6} = 1.$$

However, if we subtract the exponents as we did above, we have

$$\frac{6^5}{6^5} = 6^{5-5} = 6^0.$$

This means that $6^0 = 1$. In general, for any number a, except 0, we define

$$a^0 = 1.$$

Example 5 Evaluate each exponential.

(a) $60^0 = 1$

(b) $(-60)^0 = 1$

(c) $-(60^0) = -(1) = -1$

Note the difference between Examples 5(b) and 5(c). In Example 5(b) the base is -60 and the exponent is 0. Any nonzero base raised to a zero exponent is 1. But in Example 5(c), the base is 60. Then $60^0 = 1$, so that $-60^0 = -1$.

Work Quick Check 5 at the side.

5. Evaluate.

(a) 28^0

(b) $(-16)^0$

(c) $-(7^0)$

Considering all three cases from above, the **quotient rule for exponents** can be expressed as follows. For positive integers m and n, and $a \neq 0$,

$$\frac{a^m}{a^n} = \begin{cases} a^{m-n} & \text{if } m \text{ is larger than } n \\ 1 & \text{if } m = n \\ \dfrac{1}{a^{n-m}} & \text{if } n \text{ is larger than } m. \end{cases}$$

Example 6 Find each quotient.

(a) $\dfrac{5^8}{5^3} = 5^{8-3} = 5^5$

(b) $\dfrac{4^2}{4^9} = \dfrac{1}{4^{9-2}} = \dfrac{1}{4^7}$

(c) $\dfrac{3^2x^5}{3^4x^3} = \dfrac{3^2}{3^4} \cdot \dfrac{x^5}{x^3}$

$= \dfrac{1}{3^2} \cdot \dfrac{x^2}{1}$

$= \dfrac{x^2}{3^2}$

$= \dfrac{x^2}{9}$

Work Quick Check 6 at the side.

6. Find each quotient.

(a) $\dfrac{12^5}{12^2}$

(b) $\dfrac{8^3}{8^7}$

(c) $\dfrac{4^2m^3}{4^5m}$

(Hint: $m = m^1$)

To simplify an expression such as $(8^3)^2$, we can use the definition of an exponential to write

$$\begin{aligned}(8^3)^2 &= (8^3)(8^3) \\ &= (8 \cdot 8 \cdot 8)(8 \cdot 8 \cdot 8) \\ &= 8^6.\end{aligned}$$

Looking just at the exponents, we see $3 \cdot 2 = 6$. In general, to evaluate $(a^m)^n$, where m and n are positive integers, we use the **power rule for exponentials.**

$$(a^m)^n = a^{mn}$$

Example 7 Use the power rule to evaluate the exponential $(2^5)^3$.

$$(2^5)^3 = 2^{5 \cdot 3} = 2^{15}$$

Quick Check Answers

5. (a) 1 (b) 1 (c) -1
6. (a) 12^3 (b) $1/8^4$ (c) $m^2/4^3$

7. Use the exponential rules to simplify.

(a) $(x^3)^4$

(b) $5(mn)^3$

(c) $(3a^2b^4)^5$

(d) $\left(\frac{5}{2}\right)^4$

We can use the properties studied in Chapter 1 to develop two more rules for exponentials. By definition,

$$(4 \cdot 8)^3 = (4 \cdot 8)(4 \cdot 8)(4 \cdot 8)$$

$$= 4 \cdot 4 \cdot 4 \cdot 8 \cdot 8 \cdot 8 \quad \text{(Commutative and associative properties)}$$

$$= 4^3 \cdot 8^3.$$

Based on this example, we have the following rule. For any positive integer m,

$$(ab)^m = a^m b^m.$$

Example 8 Simplify each exponential.

(a) $(3xy)^2 = 3^2x^2y^2 = 9x^2y^2$

(b) $9(pq)^2 = 9(p^2q^2) = 9p^2q^2$

(c) $(2m^2p^3)^4 = 2^4(m^2)^4(p^3)^4 = 2^4m^8p^{12} = 16m^8p^{12}$

Since a/b can be written as $a \cdot (1/b)$, the rule discussed above together with some of the properties of real numbers gives us the final rule for exponentials. For any positive integer m, and $b \neq 0$,

$$\left(\frac{a}{b}\right)^m = \frac{a^m}{b^m}$$

Example 9 $\left(\frac{2}{3}\right)^5 = \frac{2^5}{3^5} = \frac{32}{243}$

Work Quick Check 7 at the side.

Example 10 Use the exponential rules to simplify.

(a) $\frac{(4^2)^3}{4^5}$

Use the power rule and then the quotient rule.

$$\frac{(4^2)^3}{4^5} = \frac{4^6}{4^5} = 4^1 = 4$$

(b) $(2x)^3(2x)^2$

Use the product rule first.

$$(2x)^3(2x)^2 = (2x)^5 = 2^5x^5 = 32x^5$$

(c) $\left(\frac{2x^3}{5}\right)^4$

By the last two rules given above,

$$\left(\frac{2x^3}{5}\right)^4 = \frac{2^4x^{12}}{5^4} = \frac{16x^{12}}{625}.$$

Quick Check Answers

7. (a) x^{12} (b) $5m^3n^3$
(c) $3^5a^{10}b^{20}$ or $243a^{10}b^{20}$
(d) $5^4/2^4$ or 625/16

Summary of Rules for Exponentials

If m and n are any positive integers, then

$a^m \cdot a^n = a^{m+n}$ — Product rule

$$\frac{a^m}{a^n} = \begin{cases} a^{m-n} & \text{if } m \text{ is larger than } n \\ a^0 = 1 & \text{if } m = n \\ \dfrac{1}{a^{n-m}} & \text{if } n \text{ is larger than } m \end{cases} \quad (a \neq 0)$$ Quotient rule

$(a^m)^n = a^{mn}$ — Power rule

$(ab)^n = a^n b^n$

$\left(\frac{a}{b}\right)^m = \frac{a^m}{b^m} \quad (b \neq 0)$

WORK SPACE

name date hour

4.1 EXERCISES

Identify the base and exponent for each exponential. See Example 2.

1. 5^{12} ______
2. a^6 ______
3. $(3m)^4$ ______
4. -2^4 ______
5. -125^3 ______
6. $(-1)^8$ ______
7. $(-24)^2$ ______
8. $-(-3)^5$ ______
9. $3m^2$ ______
10. $5y^3$ ______

Write each expression using exponents. See Example 1.

11. $3 \cdot 3 \cdot 3 \cdot 3 \cdot 3$ ______
12. $4 \cdot 4 \cdot 4$ ______
13. $5 \cdot 5 \cdot 5 \cdot 5$ ______
14. $3 \cdot 3 \cdot 3 \cdot 3 \cdot 3 \cdot 3 \cdot 3 \cdot 3 \cdot 3$ ______
15. $(-2)(-2)(-2)(-2)(-2)$ ______
16. $(-1)(-1)(-1)(-1)$ ______
17. $\frac{1}{4 \cdot 4 \cdot 4 \cdot 4 \cdot 4}$ ______
18. $\frac{1}{(-2)(-2)(-2)}$ ______
19. $\frac{1}{3 \cdot 3 \cdot 3 \cdot 3}$ ______
20. $\frac{1}{2 \cdot 2 \cdot 2 \cdot 2 \cdot 2}$ ______

Evaluate each expression. For example, $5^2 + 5^3 = 25 + 125 = 150$.

21. $3^2 + 3^4$ ______
22. $2^8 - 2^6$ ______
23. $4^2 + 4^3$ ______
24. $3^3 + 3^4$ ______
25. $2^2 + 2^5$ ______
26. $4^2 + 4^1$ ______
27. $4^0 + 5^0$ ______
28. $3^0 + 8^0$ ______
29. $(-9)^0 + 9^0$ ______
30. $(-8)^0 + (-8)^0$ ______

Use the product rule to simplify each expression. Write each answer in exponential form. See Example 3.

31. $4^2 \cdot 4^3$ ________________

32. $3^5 \cdot 3^4$ ________________

33. $9^5 \cdot 9^3$ ________________

34. $8^6 \cdot 8^4$ ________________

35. $3^4 \cdot 3^7$ ________________

36. $2^5 \cdot 2^{15}$ ________________

37. $4^3 \cdot 4^5 \cdot 4$ ________________

38. $2^3 \cdot 2^4 \cdot 2^5$ ________________

39. $(-3)^3(-3)^2$ ________________

40. $(-4)^5(-4)^3$ ________________

41. $(-2)^3(-2)^6$ ________________

42. $(-3)^4(-3)^6$ ________________

Use the quotient rule to simplify each expression. Write each answer in exponential form. See Examples 5 and 6.

43. $\frac{4^3}{4^2}$ ________________

44. $\frac{11^5}{11^6}$ ________________

45. $\frac{4^2}{4^4}$ ________________

46. $\frac{14^{11}}{14^{15}}$ ________________

47. $\frac{8^9}{8^3}$ ________________

48. $\frac{5^{10}}{5^4}$ ________________

49. $\frac{6^3}{6^2}$ ________________

50. $\frac{7^{12}}{7^5}$ ________________

51. $\frac{(-14)^6}{(-14)^5}$ ________________

52. $\frac{(-3)^7}{(-3)^8}$ ________________

53. $\frac{-19^0}{(-18)^0}$ ________________

54. $\frac{14^0}{-16^0}$ ________________

Use the rules for exponentials to simplify each expression. Write each answer in exponential form. See Examples 4 and 7–10.

55. $x^4 \cdot x^5$ ________________

56. $m^2 \cdot m^7$ ________________

57. $r^3 \cdot r^8$ ________________

58. $p^4 \cdot p^{10}$ ________________

59. $\frac{(y^3)^3}{(y^2)^2}$ ________________

60. $\frac{(r^2)^4}{(r^3)^2}$ ________________

61. $\frac{a^{11}}{(a^2)^4}$ ________________

62. $\frac{s^{14}}{(s^5)^2}$ ________________

63. $\frac{(k^2)^9}{(k^6)^2}$ ________________

64. $\frac{(w^4)^2}{(w^7)^3}$ ________________

65. $(5m)^3$ ________________

66. $(2xy)^4$ ________________

name date hour

67. $(3mn)^4$ ________

68. $(-2ab)^5$ ________

69. $(-3x^5)^2$ ________

70. $(4m^3n^2)^4$ ________

71. $(5p^2q)^3$ ________

72. $(2^3a^4)^5$ ________

73. $\left(\frac{a}{5}\right)^3$ ________

74. $\left(\frac{9}{x}\right)^2$ ________

75. $\left(\frac{3mn}{2}\right)^5$ ________

76. $\left(\frac{2x^3}{3y^2}\right)^4$ ________

77. $\frac{x^7x^8(x^3)^2}{x^9x^7}$ ________

78. $\frac{(m^3)^2(m^2)^4m^8}{(m^9)^3}$ ________

79. $\frac{b^{11}(b^2)^4}{(b^3)^3(b^2)^6}$ ________

80. $\frac{(8m^2)^3(8m^4)^2}{(8m^3)^4}$ ________

WORK SPACE

OBJECTIVES

- Identify coefficients
- Combine like terms
- Know the various words about polynomials
- Find the value of a polynomial for a given value of the variable

Recall that in an expression such as

$$4x^3 + 6x^2 + 5x,$$

the quantities $4x^3$, $6x^2$, and $5x$ are called **terms.** In the term $4x^3$, the number 4 is called the **numerical coefficient,** or simply the **coefficient,** of x^3. In the same way, 6 is the coefficient of x^2 in the term $6x^2$, and 5 is the coefficient of x in the term $5x$.

Example 1 Name the coefficient in each term of these expressions.

(a) $4x^3$

The coefficient is 4.

(b) $x - 6x^4$

The coefficient of x is 1 because $x = 1 \cdot x$. The coefficient of x^4 is -6, since $x - 6x^4$ can be written as the sum $x + (-6x^4)$.

(c) $5 - v^3$

The coefficient of 5 is 5 since $5 = 5v^0$. If $5 - v^3$ is written as a sum, $5 + (-v^3)$, and if v^3 is written as $1 \cdot v^3$, we have

$$5 - v^3 = 5 + (-v^3) = 5 + (-1v^3).$$

Thus the coefficient of v^3 is -1.

Work Quick Check 1 at the side.

QUICK CHECKS

1. Name the coefficient in each term of these expressions.

(a) $3m^2$

(b) $2x^3 - x$

(c) $x + 8$

Like terms have exactly the same variable with the same exponent. Only the coefficients may be different. Examples include

$$19m^5 \quad \text{and} \quad 14m^5,$$

$$6y^9, \quad -37y^9, \quad \text{and} \quad y^9.$$

To add like terms, use the distributive property.

Example 2 Simplify each expression using the distributive property.

(a) $-4x^3 + 6x^3 = (-4 + 6)x^3 = 2x^3$

(b) $3x^4 + 5x^4 = (3 + 5)x^4 = 8x^4$

(c) $9x^6 - 14x^6 + x^6 = (9 - 14 + 1)x^6 = -4x^6$

(d) $12m^2 + 5m + 4m^2 = (12 + 4)m^2 + 5m = 16m^2 + 5m$

(e) $3x^2y + 4x^2y - x^2y = (3 + 4 - 1)x^2y = 6x^2y$

Example 2(d) shows that it is not possible to add $16m^2$ and $5m$. These two terms are unlike because the exponents on the variables are different. **Unlike terms** have different variables or different exponents on the same variables.

Quick Check Answers

1. (a) 3 (b) 2; -1 (c) 1; 8

2. Add any like terms in each expression.

(a) $5x^4 + 7x^4$

(b) $9pq + 3pq - 2pq$

(c) $r^2 + 3r + 5r^2$

(d) $8t + 6w$

Quick Check Answers
2. (a) $12x^4$ (b) $10pq$ (c) $6r^2 + 3r$ (d) cannot—unlike terms

Example 3 Simplify each expression, if possible.

(a) $4m^2 - 5m^3$ cannot be combined or simplified any further, because the exponents differ.

(b) $8x^4 + 9y^4$ cannot be combined or simplified, because the variables are not the same.

(c) $3z^2 + 3z + 3$ cannot be combined or simplified, because the exponents are different and the last term, 3, has no variable associated with it.

Work Quick Check 2 at the side.

One of the basic concepts in algebra is the polynomial. A **polynomial** is defined as any finite sum of terms which are the product of a number and a variable raised to a power, such as

$$4y^3 + 3x^2 - 2m.$$

[Recall that $4y^3 + 3x^2 - 2m = 4y^3 + 3x^2 + (-2)m$.] On the other hand,

$$2x^3 - x^2 + \frac{4}{x^4}$$

is not a polynomial because the last term is the quotient (not the product) of a number and a variable raised to a power.

In general, we shall be concerned only with polynomials containing a single variable, such as x. **A polynomial in x** is a polynomial whose terms contain only variables which are whole number powers of x (including the zero power of x). Thus

$$16x^8 - 7x^6 + 5x^5 + 5x^3 - 3x + 2$$

is a polynomial in x. (Note that $2 = 2x^0$.) This last polynomial is written in **descending powers** of the variable, since the exponents on x decrease from left to right.

Example 4 Examine these expressions.

$$x^2 + 3x^3 + 4x^4$$

$$2m + 5$$

$$p^3 - p^5$$

$$3k^2 - 2k + \frac{1}{k}$$

(a) Which are polynomials?

The first three are polynomials, since each one is the sum of terms which are the product of a number and a variable raised to a whole number power.

(b) Which are polynomials in x?

The first expression is a polynomial in x.

(c) Which are polynomials in descending powers?

$2m + 5$ is written in descending powers.

Work Quick Check 3 at the side.

The **degree** of a term with one variable is the exponent on the variable. Thus $3x^4$ has degree 4, $6x^{17}$ has degree 17, $5x$ has degree 1, and -7 has degree 0 (since -7 can be written as $-7x^0$). The **degree of a polynomial** in one variable is the highest exponent found in any nonzero term of the polynomial. Thus $3x^4 - 5x^2 + 6$ is of degree 4, while $5x$ is of degree 1, and $3(3x^0)$ is of degree 0.

Three types of polynomials are very common and are given special names. A polynomial with exactly three terms is called a **trinomial.** (Tri- means "three," as in triangle.) Examples are:

$$9m^3 - 4m^2 + 6,$$

$$19y^2 + 8y^9 + 5,$$

$$-3m^5 - 9m^2 + 2.$$

A polynomial with exactly two terms is called a **binomial.** (Bi- means "two," as in bicycle.) Examples are:

$$-9x^4 + 9x^3,$$

$$8m^2 + 6m,$$

$$3m^5 - 9m^2.$$

A polynomial with only one term is called a **monomial.** [Mon(o)- means "one," as in monaural.] Examples are:

$$9m, \quad -6y^5, \quad a^2, \quad 6.$$

Example 5 For each polynomial, first simplify if possible. Then give the degree and tell whether it is a monomial, a binomial, a trinomial, or none of these.

(a) $2x^3 + 5$

The polynomial cannot be simplified. The degree is 3. The polynomial is a binomial.

(b) $4x - 5x + 2x$

$$4x - 5x + 2x = x$$

The degree is 1. The polynomial is a monomial.

Work Quick Check 4 at the side.

A polynomial represents different numbers for different values of the variable as shown in the next examples.

Example 6 Find the value of $3x^4 - 5x^3 - 4x - 4$ when $x = 1$.

Substitute 1 for x.

3. From the given list, choose any expressions which are
(1) polynomials;
(2) polynomials in x;
(3) polynomials written in descending order.

(a) $3m^3 + 5m^2 - 2m + 1$

(b) $2p^4 + p^6$

(c) $\frac{1}{x} + 2x^2 + 3$

(d) $x - 3$

4. For each polynomial, first simplify if possible. Then give the degree and tell whether it is a monomial, binomial, trinomial, or none of these.

(a) $3x^2 + 2x - 4$

(b) $x^3 + 4x^3$

(c) $x^8 - x^7 + 2x^8$

Quick Check Answers

3. (a) (1); (3) (b) (1) only (c) none (d) (1); (2); (3)
4. (a) degree 2; trinomial (b) degree 3; monomial (simplify to $5x^3$) (c) degree 8; binomial (simplify to $3x^8 - x^7$)

5. (a) Find the value of $2x^3 + 3x - 4$ when $x = -2$.

(b) Find $P(-1)$ if $P(x) = 3x^5 + x^3$.

$$\begin{aligned} 3x^4 - 5x^3 - 4x - 4 &= 3(1)^4 - 5(1)^3 - 4(1) - 4 \\ &= 3(1) - 5(1) - 4 - 4 \\ &= 3 - 5 - 4 - 4 \\ &= -10 \end{aligned}$$

We sometimes use a capital letter to represent a polynomial. For example, if we let $P(x)$ represent the polynomial

$$3x^4 - 5x^3 - 4x - 4,$$

then $P(x) = 3x^4 - 5x^3 - 4x - 4$ where $P(x)$ is read "P of x." We sometimes express the fact that $P(x) = 92$ when $x = -2$ by writing $P(-2) = 92$. [Read $P(-2)$ as "P of -2."]

Example 7 If $P(x) = 9x^3 - 8x + 6$, find

(a) $P(-3)$

If we replace x by -3, we have

$$\begin{aligned} P(-3) &= 9(-3)^3 - 8(-3) + 6 \\ &= 9(-27) + 24 + 6 \\ &= -243 + 30 \\ P(-3) &= -213. \end{aligned}$$

(b) $P(1)$

$$P(1) = 9(1)^3 - 8(1) + 6 = 9 - 8 + 6 = 7$$

Work Quick Check 5 at the side.

Quick Check Answers

5. (a) −26 (b) −4

name date hour

4.2 EXERCISES

Find the value of each polynomial when $x = 2$ and when $x = -1$. See Example 6.

1. $2x^2 - 4x$ ________

2. $8x + 5x^2 + 2$ ________

3. $2x^5 - 4x^4 + 5x^3 - x^2$ ________

4. $9x + 1$ ________

5. $2x^2 + 5x + 1$ ________

6. $-3x^2 + 14x - 2$ ________

7. $-2x^2 + 3$ ________

8. $-3x^2 + 4x + 5$ ________

9. $-x^2 - x^3$ ________

10. $4x^2 - 3x + 2$ ________

Let $P(x) = x^3 - 3x^2 + 2x - 3$ and $Q(x) = x^4 - 1$. Find each value. See Example 7.

11. $P(-1)$ ________

12. $P(0)$ ________

13. $P(2)$ ________

14. $Q(2)$ ________

15. $P(-2)$ ________

16. $Q(1)$ ________

17. $Q(-2)$ ________

18. $P(-2) + Q(-2)$ ________

19. $P(-1) \cdot Q(-2)$ ________

20. $P(0) \cdot Q(0)$ ________

In each polynomial, combine terms whenever possible. See Examples 2 and 3.

21. $3m^5 + 5m^5$ ________

22. $-4y^3 + 3y^3$ ________

23. $2r^5 + (-3r^5)$ ________

24. $-19y^2 + 9y^2$ ________

25. $2m^5 - 5m^2$ ________

26. $-9y + 9y^2$ ________

27. $3x^5 + 2x^5 - 4x^5$ ________

28. $6x^3 + 8x^3 - 9x^3$ ________

29. $-4p^7 + 8p^7 - 5p^7$ ________

30. $-3a^8 + 4a^8 - 3a^8 + 2a^8$ ________

31. $4y^2 + 3y^2 - 2y^2 + y^2$ ________

32. $3r^5 - 8r^5 + r^5 - 2r^5$ ________

For each polynomial, first simplify, if possible, then give the degree of the polynomial and tell whether it is (a) a monomial, (b) a binomial, (c) a trinomial, (d) none of these. See Example 5.

33. $5x^4 - 8x$ ______

34. $4y - 8y$ ______

35. $23x^9 - \frac{1}{2}x^2 + x$ ______

36. $2m^7 - 3m^6 + 2m^5 + m$ ______

37. $x^8 + 3x^7 - 5x^4$ ______

38. $\frac{3}{5}x^5 + \frac{2}{5}x^5$ ______

39. $\frac{9}{11}x^2$ ______

40. -8 ______

For each statement, write always, sometimes, *or* never.

41. A binomial is a polynomial ______

42. A polynomial is a trinomial ______

43. A trinomial is a binomial ______

44. A monomial has no coefficient ______

45. A binomial is a trinomial ______

46. A polynomial of degree 4 has 4 terms ______

OBJECTIVES

- Add polynomials
- Subtract polynomials
- Work problems with both addition and subtraction

To add two polynomials, we add like terms as shown in Example 1.

Example 1 Add the polynomials $6x^3 - 4x^2 + 3$ and $-2x^3 + 7x^2 - 5$.

Write the sum

$$(6x^3 - 4x^2 + 3) + (-2x^3 + 7x^2 - 5).$$

Regroup to collect like terms and change all subtractions to additions of inverses.

$$[6x^3 + (-2)x^3] + [-4x^2 + 7x^2] + [3 + (-5)]$$

Now use the distributive property to combine each group of like terms. The result is the trinomial

$$4x^3 + 3x^2 - 2.$$

Work Quick Check 1 at the side.

The sum of two polynomials can also be found by placing one directly above the other with like terms lined up vertically in columns.

Example 2 Add $6x^3 - 4x^2 + 3$ and $-2x^3 + 7x^2 - 5$.

Write like terms in columns.

$$\begin{array}{r} 6x^3 - 4x^2 + 3 \\ \underline{-2x^3 + 7x^2 - 5} \end{array}$$

Now add, column by column.

$$\begin{array}{ccc} 6x^3 & -4x^2 & 3 \\ \underline{-2x^3} & \underline{7x^2} & \underline{-5} \\ 4x^3 & 3x^2 & -2 \end{array}$$

Add the three sums together. The result is the same answer we found in Example 1.

$$4x^3 + 3x^2 + (-2) = 4x^3 + 3x^2 - 2$$

Work Quick Check 2 at the side.

In Section 2.3, the difference $x - y$ was defined as $x + (-y)$. For example,

$$7 - 2 = 7 + (-2) = 5$$

and

$$-8 - (-2) = -8 + 2 = -6.$$

We use the same method to subtract polynomials.

QUICK CHECKS

1. Find each sum.

(a) $(2x^4 - 6x^2 + 7) + (-3x^4 + 5x^2 + 2)$

(b) $(3x^3 + 4x + 2) + (6x^3 - 5x - 7)$

2. Add each polynomial.

(a) $\begin{array}{r} 4x^3 - 3x^2 + 2x \\ \underline{6x^3 + 2x^2 - 3x} \end{array}$

(b) $\begin{array}{r} x^2 - 2x + 5 \\ \underline{4x^2 + 3x - 2} \end{array}$

Quick Check Answers

1. (a) $-x^4 - x^2 + 9$
(b) $9x^3 - x - 5$
2. (a) $10x^3 - x^2 - x$
(b) $5x^2 + x + 3$

3. Subtract and check your answers by addition.

(a) $(14y^3 - 6y^2 + 2y - 5) - (2y^3 - 7y^2 - 4y + 6)$

(b) $(7y^2 - 11y + 8) - (-3y^2 + 4y + 6)$

4. Use the method of subtracting by columns to solve each problem.

(a) $(14y^3 - 6y^2 + 2y) - (2y^3 - 7y^2 + 6)$

(b) $(6p^4 - 8p^3 + 2p - 1) - (-7p^4 + 6p^2 - 12)$

Quick Check Answers

3. (a) $12y^3 + y^2 + 6y - 11$
(b) $10y^2 - 15y + 2$

4. (a) $12y^3 + y^2 + 2y - 6$
(b) $13p^4 - 8p^3 - 6p^2 + 2p + 11$

Example 3 Subtract the polynomial $6x^3 - 4x^2 + 2$ from the polynomial $11x^3 + 2x^2 - 8$.

By the definition of subtraction,

$$(11x^3 + 2x^2 - 8) - (6x^3 - 4x^2 + 2) =$$
$$(11x^3 + 2x^2 - 8) + [-(6x^3 - 4x^2 + 2)].$$

To simplify $-(6x^3 - 4x^2 + 2)$ recall that

$$-(a + b) = -a + (-b).$$

Thus $-(6x^3 - 4x^2 + 2) = -6x^3 - (-4x^2) - 2$
$= -6x^3 + 4x^2 - 2.$

All the signs inside the parentheses, including the understood + on $6x^3$, have been changed. We now complete the subtraction.

$$\begin{aligned}
&(11x^3 + 2x^2 - 8) - (6x^3 - 4x^2 + 2) \\
&= (11x^3 + 2x^2 - 8) + [-(6x^3 - 4x^2 + 2)] \\
&= (11x^3 + 2x^2 - 8) + (-6x^3 + 4x^2 - 2) \\
&= [11x^3 + (-6x^3)] + (2x^2 + 4x^2) + [-8 + (-2)] \\
&= 5x^3 + 6x^2 - 10
\end{aligned}$$

To check a subtraction problem such as this, use the fact that if $a - b = c$, then $a = b + c$. For example, $6 - 2 = 4$. To check this, write $6 = 2 + 4$, which is correct. For the polynomials, to check the subtraction above, add $6x^3 - 4x^2 + 2$ and $5x^3 + 6x^2 - 10$. Since the sum is $11x^3 + 2x^2 - 8$, the subtraction was performed correctly.

Work Quick Check 3 at the side.

Subtraction can also be done in columns.

Example 4 Subtract $6x^3 - 4x^2 + 2$ from $11x^3 + 2x^2 - 8x$.

Step 1 Write the problem with like terms arranged in columns.

$$\begin{array}{rrrr}
11x^3 & + 2x^2 & - 8x & \\
6x^3 & - 4x^2 & & + 2 \\
\hline
\end{array}$$

Step 2 Take the inverse of each term in the second polynomial.

$$\begin{array}{rrrr}
11x^3 & + 2x^2 & - 8x & \\
-6x^3 & + 4x^2 & & - 2 \\
\hline
\end{array}$$

Step 3 Add column by column.

$$\begin{array}{rrrr}
11x^3 & + 2x^2 & - 8x & \\
-6x^3 & + 4x^2 & & - 2 \\
\hline
5x^3 & + 6x^2 & - 8x & - 2
\end{array}$$

Work Quick Check 4 at the side.

name date hour

4.3 EXERCISES

Add or subtract as indicated. See Examples 2 and 4.

1. Add:

$$\begin{array}{r} 3m^2 + 5m \\ 2m^2 - 2m \\ \hline \end{array}$$

2. Add:

$$\begin{array}{r} 4a^3 - 4a^2 \\ 6a^3 + 5a^2 \\ \hline \end{array}$$

3. Subtract:

$$\begin{array}{r} 12x^4 - x^2 \\ 8x^4 + 3x^2 \\ \hline \end{array}$$

4. Subtract:

$$\begin{array}{r} 2a + 5d \\ 3a - 6d \\ \hline \end{array}$$

5. Subtract:

$$\begin{array}{r} 2n^5 - 5n^3 + 6 \\ 3n^5 + 7n^3 + 8 \\ \hline \end{array}$$

6. Subtract:

$$\begin{array}{r} 3r^2 - 4r + 2 \\ 7r^2 + 2r - 3 \\ \hline \end{array}$$

7. Add:

$$\begin{array}{r} 9m^3 - 5m^2 + 4m - 8 \\ 3m^3 + 6m^2 + 8m - 6 \\ \hline \end{array}$$

8. Add:

$$\begin{array}{r} 12r^5 + 11r^4 - 7r^3 - 2r^2 - 5r - 3 \\ -8r^5 - 10r^4 + 3r^3 + 2r^2 - 5r + 7 \\ \hline \end{array}$$

9. Add:

$$\begin{array}{r} 12m^2 - 8m + 6 \\ 3m^2 + 5m - 2 \\ \hline \end{array}$$

10. Subtract:

$$\begin{array}{r} 5a^4 - 3a^3 + 2a^2 \\ a^3 - a^2 + a - 1 \\ \hline \end{array}$$

11. Add:

$$\begin{array}{r} 5b^2 + 6b + 2 \\ 3b^2 - 4b + 5 \\ \hline \end{array}$$

12. Add:

$$\begin{array}{r} 3w^2 - 5w + 2 \\ 4w^2 + 6w - 5 \\ 8w^2 + 7w - 2 \\ \hline \end{array}$$

Perform the indicated operations. See Examples 1 and 3.

13. $(2r^2 + 3r) - (3r^2 + 5r)$ ____________________

14. $(3r^2 + 5r - 6) + (2r - 5r^2)$ ____________________

15. $(8m^2 - 7m) - (3m^2 + 7m)$ ____________________

16. $(x^2 + x) - (3x^2 + 2x - 1)$ ____________________

17. $8 - (6s^2 - 5s + 7)$ ____________________

18. $2-[3-(4+s)]$ __________

19. $(8s-3s^2)+(-4s+5s^2)$ __________

20. $(3x^2+2x+5)+(8x^2-5x-4)$ __________

21. $(16x^3-x^2+3x)+(-12x^3+3x^2+2x)$ __________

22. $(-2b^6+3b^4-b^2)-(b^6+2b^4+2b^2)$ __________

23. $(7y^4+3y^2+2y)-(18y^4-5y^2-y)$ __________

24. $(3x^2+2x+5)+(-7x^2-8x+2)+(3x^2-4x+7)$ __________

25. $(9a^4-3a^2+2)+(4a^4-4a^2+2)+(-12a^4+6a^2-3)$ __________

26. $(4m^2-3m+2)+(5m^2+13m-4)-(16m^2+4m-3)$ __________

27. $[(8m^2+4m-7)-(2m^2-5m+2)]-(m^2+m+1)$ __________

28. $(9b^3-4b^2+3b+2)+(-2b^3-3b^2+b)-(8b^3+6b+4)$ __________

Write each statement as an equation or an inequality. Do not try to solve.

29. $4+x^2$ is larger than 8. __________

30. The difference between $5+2x$ and $6+3x$ is larger than $8x+x^2$. __________

31. The sum of $5+x^2$ and $3-2x$ is not equal to 5. __________

32. The sum of $3-2x+x^2$ and $8-9x+3x^2$ is negative. __________

OBJECTIVES

- Multiply a monomial and a polynomial
- Multiply two polynomials

The product of two monomials is found by using the rules for exponents and the commutative and associative properties. For example,

$$(6x^3)(4x^4) = 6 \cdot 4 \cdot x^3 \cdot x^4 = 24x^7.$$

Also,

$$(-8m^6)(-9n^6) = (-8)(-9)(m^6)(n^6) = 72m^6n^6.$$

To use this method to find the product of a monomial and a polynomial with more than one term, use the distributive property, as shown in Examples 1 and 2.

Example 1 Use the distributive property to multiply $4x^2(3x + 5)$.

$$\begin{aligned} 4x^2(3x + 5) &= (4x^2)(3x) + (4x^2)(5) \\ &= 12x^3 + 20x^2 \end{aligned}$$

Example 2 Use the distributive property to multiply $-8m^3(4m^3 + 3m^2 + 2m - 1)$.

$$\begin{aligned} &-8m^3(4m^3 + 3m^2 + 2m - 1) \\ &= (-8m^3)(4m^3) + (-8m^3)(3m^2) + (-8m^3)(2m) \\ &\quad + (-8m^3)(-1) \\ &= -32m^6 - 24m^5 - 16m^4 + 8m^3 \end{aligned}$$

Work Quick Check 1 at the side.

QUICK CHECKS

1. Find each product.

(a) $5m^3(2m + 7)$

(b) $-4y^2(3y^3 + 2y^2 - 4y + 8)$

The distributive property is also used to find the product of any two polynomials. Suppose we want to find the product of the polynomials $x + 1$ and $x - 4$. If we work with $x + 1$ as a single quantity, we can use the distributive property to write

$$(x + 1)(x - 4) = (x + 1)x + (x + 1)(-4).$$

Now use the distributive property to multiply $(x + 1)x$ and $(x + 1)(-4)$.

$$\begin{aligned} (x + 1)x + (x + 1)(-4) &= x(x) + 1(x) + x(-4) \\ &\quad + 1(-4) \\ &= x^2 + x + (-4x) + (-4) \\ &= x^2 - 3x - 4 \end{aligned}$$

Example 3 Multiply $(2x + 1)(3x + 5)$.

$$\begin{aligned} (2x + 1)(3x + 5) &= (2x + 1)(3x) + (2x + 1)(5) \\ &= (2x)(3x) + (1)(3x) + (2x)(5) \\ &\quad + (1)(5) \\ &= 6x^2 + 3x + 10x + 5 \\ &= 6x^2 + 13x + 5 \end{aligned}$$

Work Quick Check 2 at the side.

2. Multiply.

(a) $(4x + 3)(2x + 1)$

(b) $(3k - 2)(2k + 1)$

(c) $(m + 5)(3m - 4)$

Quick Check Answers

1. (a) $10m^4 + 35m^3$
(b) $-12y^5 - 8y^4 + 16y^3 - 32y^2$
2. (a) $8x^2 + 10x + 3$
(b) $6k^2 - k - 2$
(c) $3m^2 + 11m - 20$

3. Multiply.

(a) $\begin{array}{r} 2m + 3 \\ \underline{5m - 4} \end{array}$

(b) $\begin{array}{r} 4k - 6 \\ \underline{2k + 5} \end{array}$

4. Multiply.

(a) $\begin{array}{r} 3x^2 + 4x - 5 \\ \underline{x^2 + 4} \end{array}$

(b) $\begin{array}{r} a^3 + 3a - 4 \\ \underline{2a^2 + 6a + 5} \end{array}$

(c) $\begin{array}{r} k^3 - k^2 + k + 1 \\ \underline{k + 1} \end{array}$

Quick Check Answers
3. (a) $10m^2 + 7m - 12$ (b) $8k^2 + 8k - 30$
4. (a) $3x^4 + 4x^3 + 7x^2 + 16x - 20$ (b) $2a^5 + 6a^4 + 11a^3 + 10a^2 - 9a - 20$ (c) $k^4 + 2k + 1$

The work involved in multiplication can often be simplified by writing one polynomial above the other.

$$\begin{array}{r} 2x + 1 \\ \underline{3x + 5} \end{array}$$

We need not worry about lining up the like terms in columns, since we are multiplying, and any terms may be multiplied. To begin, multiply each of the terms in the top row by 5.

Step 1

$$\begin{array}{r} 2x + 1 \\ \underline{3x + 5} \\ 10x + 5 \end{array}$$

This is similar to ordinary multiplication. Then multiply $3x$ times each term in the top row. Be careful to place the like terms in columns, since the final step will involve addition (as in multiplying two whole numbers).

Step 2

$$\begin{array}{r} 2x + 1 \\ \underline{3x + 5} \\ 10x + 5 \\ \underline{6x^2 + 3x } \\ 6x^2 + 13x + 5 \end{array}$$

Thus $(2x + 1)(3x + 5) = 6x^2 + 13x + 5$.

Example 4 Multiply $(3p - 5)(2p + 6)$.

$$\begin{array}{r} 3p - 5 \\ \underline{2p + 6} \\ 18p - 30 \\ \underline{6p^2 - 10p } \\ 6p^2 + 8p - 30 \end{array}$$

Work Quick Check 3 at the side.

Example 5 Multiply $(4m^3 - 2m^2 + 4m)(m^2 + 5)$.

$$\begin{array}{r} 4m^3 - 2m^2 + 4m \\ \underline{m^2 + 5} \\ 20m^3 - 10m^2 + 20m \\ \underline{4m^5 - 2m^4 + 4m^3 } \\ 4m^5 - 2m^4 + 24m^3 - 10m^2 + 20m \end{array}$$

Work Quick Check 4 at the side.

Example 6 Find $(n + 2)^2$.

We know $(n + 2)^2 = (n + 2)(n + 2)$. We use the vertical method of multiplication.

$$\begin{array}{r} n + 2 \\ n + 2 \\ \hline 2n + 4 \\ n^2 + 2n \\ \hline n^2 + 4n + 4 \end{array}$$

Work Quick Check 5 at the side.

Example 7 Find $(x + 5)^3$.

Since $(x + 5)^3 = (x + 5)(x + 5)(x + 5)$, we begin by finding $(x + 5)(x + 5)$.

$$\begin{array}{r} x + 5 \\ x + 5 \\ \hline 5x + 25 \\ x^2 + 5x \\ \hline x^2 + 10x + 25 \end{array}$$

Now multiply this result times $x + 5$.

$$\begin{array}{r} x^2 + 10x + 25 \\ x + 5 \\ \hline 5x^2 + 50x + 125 \\ x^3 + 10x^2 + 25x \\ \hline x^3 + 15x^2 + 75x + 125 \end{array}$$

Work Quick Check 6 at the side.

5. Find each product.

(a) $(a + 4)^2$

(b) $(r - 5)^2$

(c) $(3k - 2)^2$

6. Find each product.

(a) $(m + 1)^3$

(b) $(r + 2)^4$

Quick Check Answers

5. (a) $a^2 + 8a + 16$ (b) $r^2 - 10r + 25$ (c) $9k^2 - 12k + 4$

6. (a) $m^3 + 3m^2 + 3m + 1$ (b) $r^4 + 8r^3 + 24r^2 + 32r + 16$

WORK SPACE

name date hour

4.4 EXERCISES

Find each product. See Examples 1 and 2.

1. $(-4x^5)(8x^2)$ ____________

2. $(-3x^7)(2x^5)$ ____________

3. $(5y^4)(3y^7)$ ____________

4. $(10p^2)(5p^3)$ ____________

5. $(15a^4)(2a^5)$ ____________

6. $(-3m^6)(-5m^4)$ ____________

7. $2m(3m + 2)$ ____________

8. $-5p(6 - 3p)$ ____________

9. $3p(-2p^3 + 4p^2)$ ____________

10. $4x(3 + 2x + 5x^3)$ ____________

11. $-8z(2z + 3z^2 + 3z^3)$ ____________

12. $7y(3 + 5y^2 - 2y^3)$ ____________

13. $2y(3 + 2y + 5y^4)$ ____________

14. $-2m^4(3m^2 + 5m + 6)$ ____________

Find each binomial product. See Examples 3 and 4.

15. $(m + 7)(m + 5)$ ____________

16. $(n - 1)(n + 4)$ ____________

17. $(x + 5)(x - 5)$ ____________

18. $(y + 8)(y - 8)$ ____________

19. $(t - 4)(t + 4)$ ____________

20. $(x - 4)(x + 2)$ ____________

21. $(6p + 5)(p - 1)$ ____________

22. $(2x + 3)(6x - 4)$ ____________

23. $(4m - 3)(4m + 3)$ ____________

24. $(3x - 2)(x + 5)$ ____________

25. $(b + 8)(6b - 2)$ ____________

26. $(5a + 1)(2a + 7)$ ____________

Find each product. See Example 5.

27. $(6x + 1)(2x^2 + 4x + 1)$ ____________

28. $(9y - 2)(8y^2 - 6y + 1)$ ____________

29. $(9a + 2)(9a^2 + a + 1)$ ____________________

30. $(2r - 1)(3r^2 + 4r - 4)$ ____________________

31. $(4m + 3)(5m^3 - 4m^2 + m - 5)$ ____________________

32. $(y + 4)(3y^4 - 2y^2 + 1)$ ____________________

33. $(2x - 1)(3x^5 - 2x^3 + x^2 - 2x + 3)$ ____________________

34. $(2a + 3)(a^4 - a^3 + a^2 - a + 1)$ ____________________

Find each product. See Examples 6 and 7.

35. $(x + 7)^2$ ____________________

36. $(m + 6)^2$ ____________________

37. $(a - 4)^2$ ____________________

38. $(b - 10)^2$ ____________________

39. $(m - 5)^3$ ____________________

40. $(p + 3)^3$ ____________________

41. $(k + 1)^4$ ____________________

42. $(r - 1)^4$ ____________________

4.5 PRODUCTS OF BINOMIALS

OBJECTIVES

- Multiply binomials by the shortcut method
- Square binomials
- Find the difference of two squares

The procedures described in Section 4.4 can be used to find the product of any two polynomials and are the only practical methods for multiplying polynomials with three or more terms. We can also use these methods to multiply two binomials. However, in practice, many of the polynomials that must be multiplied are binomials, so for binomials, we need a quick method that eliminates writing out all the steps. To find such a shortcut, let us look carefully at the process of multiplying two binomials. If we multiply $(x + 3)(x + 5)$ using the distributive property, we have

$$\begin{aligned}(x + 3)(x + 5) &= (x + 3)x + (x + 3)5 \\ &= (x)(x) + (3)(x) + (x)(5) + (3)(5) \\ &= x^2 + 3x + 5x + 15 \\ &= x^2 + 8x + 15.\end{aligned}$$

The first term in the second line, $(x)(x)$, is the product of the first term in each binomial.

$$(x + 3)(x + 5)$$

Multiply the first terms: $(x)(x)$.

The second term, $(3)(x)$, is the product of the last term of the first binomial and the first term of the second binomial. The product of the middle terms is sometimes called the **inner product.**

$$(x + 3)(x + 5)$$

Multiply the inner terms: $(3)(x)$.

The third term, $(x)(5)$, is the product of the first term of the first binomial and the last term of the second binomial. This is called the **outer product.**

$$(x + 3)(x + 5)$$

Multiply the outer terms: $(x)(5)$.

Finally, $(3)(5)$ is the product of the last term of each binomial.

$$(x + 3)(x + 5)$$

Multiply the last terms: $(3)(5)$.

In the third step of the multiplication, we add the inner product and the outer product. This step should be performed mentally, so that the three terms of the answer can be written down without any extra steps.

A summary of these steps is given below.

1. Multiply the two first terms of the binomials to get the first term of the answer.

QUICK CHECKS

1. Use the shortcut method to find each product.

(a) $(x + 4)(x - 3)$

(b) $(2x + 1)(5x - 4)$

2. Find the inner product and the outer product and add them to get the middle term of the answer.
3. Multiply the two last terms of the binomials to get the last term of the answer.

Example 1 Multiply $(x + 8)(x - 6)$ by the shortcut method.

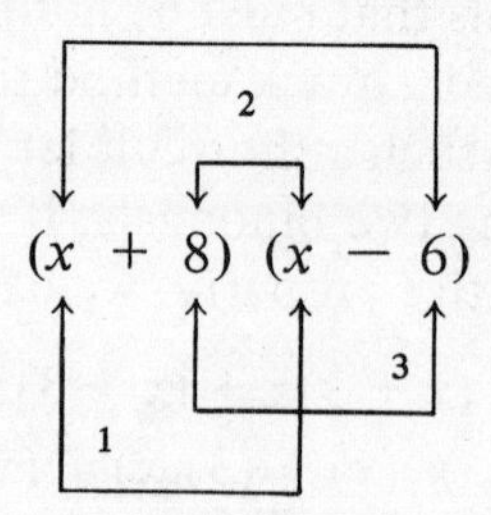

Step 1 Multiply $(x)(x)$ to get x^2.

Step 2 Multiply $(8)(x)$ and multiply $(-6)(x)$. Add to get $2x$.

Step 3 Multiply $(8)(-6)$ to get -48.

The result is $x^2 + 2x - 48$.

Example 2 Multiply $9x - 2$ and $3x + 1$ by the shortcut method.

Step 1 $(9x - 2)(3x + 1)$ to get $27x^2$

Step 2 $(9x - 2)(3x + 1)$ to get $9x - 6x = 3x$

Step 3 $(9x - 2)(3x + 1)$ to get -2

The result is $27x^2 + 3x - 2$.

Work Quick Check 1 at the side.

Special types of binomial products occur so often in practice that the form of the answers should be memorized. For example, we frequently need to find the square of a binomial.

Example 3 Find $(2m + 3)^2$.

Squaring $2m + 3$ by the shortcut method gives

$$(2m + 3)(2m + 3) = 4m^2 + 12m + 9.$$

In the result, note that we have the square of both the first and the last terms of the binomial:

$$(2m)^2 = 4m^2 \qquad \text{and} \qquad 3^2 = 9.$$

We also have twice the product of the two terms of the binomial, that is,

$$2(2m)(3) = 12m.$$

Quick Check Answers

1. (a) $x^2 + x - 12$
(b) $10x^2 - 3x - 4$

The general rule is that the square of a binomial is a trinomial composed of the square of the first term, plus twice the product of the two terms, plus the square of the second term of the binomial, or

$$(x + y)^2 = x^2 + 2xy + y^2.$$

Example 4 Use the formula to square each binomial.

(a) $(5z - 1)^2 = (5z)^2 + 2(5z)(-1) + (-1)^2$
$= 25z^2 - 10z + 1$

Recall that $(5z)^2 = 5^2 z^2 = 25z^2$.

(b) $(3b + 5r)^2 = (3b)^2 + 2(3b)(5r) + (5r)^2$
$= 9b^2 + 30br + 25r^2$

(c) $(2a - 9x)^2 = 4a^2 - 36ax + 81x^2$

Work Quick Check 2 at the side.

Binomial products of the form $(x + y)(x - y)$ also occur frequently. In these products, one binomial is the sum of two terms, while the other is the difference of the same two terms. For example, let us multiply $(a + 2)(a - 2)$ by the shortcut method.

$$(a + 2)(a - 2) = a^2 - 2a + 2a - 4$$
$$= a^2 - 4$$

In general,

$$(x + y)(x - y) = x^2 - y^2,$$

and the result, $x^2 - y^2$, is called the **difference of two squares.**

Example 5 Multiply $(5m + 3)(5m - 3)$.

Use the formula for the difference of two squares.

$$(5m + 3)(5m - 3) = (5m)^2 - 3^2$$
$$= 25m^2 - 9$$

Example 6 Multiply $(4x + y)(4x - y)$.

$$(4x + y)(4x - y) = (4x)^2 - y^2$$
$$= 16x^2 - y^2.$$

Work Quick Check 3 at the side.

2. Use the rule for $(x + y)^2$ to square each binomial.

(a) $(a + b)^2$

(b) $(2m - n)^2$

(c) $(4p + 3q)^2$

3. Use the formula for the difference of squares to multiply.

(a) $(6a + 3)(6a - 3)$

(b) $(10m + 7)(10m - 7)$

Quick Check Answers
2. (a) $a^2 + 2ab + b^2$ (b) $4m^2 - 4mn + n^2$ (c) $16p^2 + 24pq + 9q^2$
3. (a) $36a^2 - 9$ (b) $100m^2 - 49$

WORK SPACE

name date hour

4.5 EXERCISES

Find each product. See Examples 1 and 2.

1. $(r - 1)(r + 3)$ ____

2. $(x + 2)(x - 5)$ ____

3. $(x - 7)(x - 3)$ ____

4. $(r + 3)(r + 6)$ ____

5. $(2x - 1)(3x + 2)$ ____

6. $(4y - 5)(2y + 1)$ ____

7. $(6z + 5)(z - 3)$ ____

8. $(8a + 3)(6a + 1)$ ____

9. $(a + 4)(2a + 1)$ ____

10. $(3x - 1)(2x + 3)$ ____

11. $(2r - 1)(4r + 3)$ ____

12. $(5m + 2)(3m - 4)$ ____

13. $(2a + 4)(3a - 2)$ ____

14. $(11m - 10)(10m + 11)$ ____

15. $(4 + 5x)(5 - 4x)$ ____

16. $(8 + 3x)(2 - x)$ ____

17. $(-3 + 2r)(4 + r)$ ____

18. $(-5 + 6z)(2 - z)$ ____

19. $(-3 + a)(-5 - 2a)$ ____

20. $(-6 - 3y)(1 - 4y)$ ____

Find each square. See Examples 3 and 4.

21. $(m + 2)^2$ ____

22. $(x + 8)^2$ ____

23. $(5 + x)^2$ ____

24. $(2 - y)^2$ ____

25. $(x + 2y)^2$ ____

26. $(3m - n)^2$ ____

27. $(2z - 5x)^2$ ____

28. $(6a - b)^2$ ____

29. $(5p + 2q)^2$ ____

30. $(8a - 3b)^2$ ____

31. $(4a + 5b)^2$ ____

32. $(9y + z)^2$ ____

Find each product. See Examples 5 and 6.

33. $(m - n)(m + n)$ ______

34. $(p + q)(p - q)$ ______

35. $(r + z)(r - z)$ ______

36. $(a + b)(a - b)$ ______

37. $(6a - p)(6a + p)$ ______

38. $(5y + 3x)(5y - 3x)$ ______

39. $(2m - 5)(2m + 5)$ ______

40. $(3a - 5)(3a + 5)$ ______

41. $(7y + 10)(7y - 10)$ ______

42. $(6x + 3)(6x - 3)$ ______

Write each statement as an equation or an inequality using x to represent the unknown number. Do not try to solve.

43. The square of 3 more than a number is 5. ______

44. The square of the sum of a number and 6 is less than 3. ______

45. When 3 plus a number is multiplied by the number less 4, the result is greater than 7. ______

46. Twice a number plus 4, multiplied by 6 times the number, less 5, gives 8. ______

4.6 DIVIDING A POLYNOMIAL BY A MONOMIAL

OBJECTIVES

- Divide a polynomial by a monomial

To divide a monomial by another monomial, use the quotient rule for exponents. For example,

$$\frac{12x^2}{6x} = 2x, \quad \frac{25m^5}{5m^2} = 5m^3, \quad \text{and} \quad \frac{30a^2b^8}{15a^3b^3} = \frac{2b^5}{a}.$$

To divide a polynomial by a monomial, use the fact that

$$\frac{a}{b} = a \cdot \frac{1}{b}.$$

Example 1 Divide $5m^5 - 10m^3$ by $5m^2$.

We multiply by $1/5m^2$.

$$\frac{5m^5 - 10m^3}{5m^2} = (5m^5 - 10m^3)\frac{1}{5m^2}$$

Now use the distributive property and the quotient rule for exponents.

$$\begin{aligned}(5m^5 - 10m^3)\frac{1}{5m^2} &= (5m^5)\frac{1}{5m^2} - (10m^3)\frac{1}{5m^2} \\ &= \frac{5m^5}{5m^2} - \frac{10m^3}{5m^2} \\ &= m^3 - 2m\end{aligned}$$

Therefore,

$$\frac{5m^5 - 10m^3}{5m^2} = (5m^5 - 10m^3)\frac{1}{5m^2} = m^3 - 2m.$$

To check, multiply

$$5m^2(m^3 - 2m) = 5m^5 - 10m^3.$$

Since we cannot divide by 0, the quotient

$$(5m^5 - 10m^3)/5m^2$$

has no value if $m = 0$. In the rest of this chapter, we assume that no denominators are 0.

Work Quick Check 1 at the side.

QUICK CHECKS

1. Divide.

(a) $6p^4 + 18p^7 \div 3p^2$

(b) $12m^6 + 18m^5 + 30m^4 \div 6m^2$

Example 2 Divide $\dfrac{16a^5 - 12a^4 + 8a^2}{4a^3}$.

Quick Check Answers

1. (a) $2p^2 + 6p^5$
(b) $2m^4 + 3m^3 + 5m^2$

2. Divide.

(a) $\dfrac{20x^4 - 25x^3 + 5x}{5x^2}$

(b) $\dfrac{50m^4 - 30m^3 + 20m}{10m^3}$

3. Divide.

(a) $\dfrac{8y^7 - 9y^6 - 11y - 4}{y^2}$

(b) $\dfrac{12p^5 + 8p^4 + 6p^3 - 5p^2}{3p^3}$

Quick Check Answers

2. (a) $4x^2 - 5x + 1/x$

(b) $5m - 3 + \dfrac{2}{m^2}$

3. (a) $8y^5 - 9y^4 - \dfrac{11}{y} - \dfrac{4}{y^2}$

(b) $4p^2 + 8p/3 + 2 - 5/3p$

$$\begin{aligned}\frac{16a^5 - 12a^4 + 8a^2}{4a^3} &= (16a^5 - 12a^4 + 8a^2)\frac{1}{4a^3}\\ &= (16a^5)\frac{1}{4a^3} - (12a^4)\frac{1}{4a^3}\\ &\quad + (8a^2)\frac{1}{4a^3}\\ &= \frac{16a^5}{4a^3} - \frac{12a^4}{4a^3} + \frac{8a^2}{4a^3}\\ &= 4a^2 - 3a + \frac{2}{a}\end{aligned}$$

The result is not a polynomial because of the expression $2/a$. Note that although the sum and the product of two polynomials always result in a polynomial, the quotient of two polynomials may not.

Work Quick Check 2 at the side.

Some of the steps shown in Examples 1 and 2 may be dropped as shown in Example 3.

Example 3 Divide the polynomial

$$180y^{10} - 150y^8 + 120y^6 - 90y^4 + 100y$$

by the monomial $30y^2$.

Using the methods of this section, we have

$$\begin{aligned}&\frac{180y^{10} - 150y^8 + 120y^6 - 90y^4 + 100y}{30y^2}\\ &= \frac{180y^{10}}{30y^2} - \frac{150y^8}{30y^2} + \frac{120y^6}{30y^2} - \frac{90y^4}{30y^2} + \frac{100y}{30y^2}\\ &= 6y^8 - 5y^6 + 4y^4 - 3y^2 + \frac{10}{3y}.\end{aligned}$$

Work Quick Check 3 at the side.

name date hour

4.6 EXERCISES

Find each quotient.

1. $\dfrac{4x^2}{2x}$ ______

2. $\dfrac{8m^5}{2m}$ ______

3. $\dfrac{10a^3}{5a}$ ______

4. $\dfrac{36p^8}{4p^3}$ ______

5. $\dfrac{27k^4m^5}{3km^6}$ ______

6. $\dfrac{18x^5y^6}{3x^2y^2}$ ______

Divide each polynomial by 2m. See Examples 1–3.

7. $60m^4 - 20m^2$ ______

8. $16m^3 - 8m^2$ ______

9. $120m^6 - 60m^3 + 80m^2$ ______

10. $10m^5 - 16m^2 + 8m^3$ ______

11. $8m^3 - 4m^2 + 6m$ ______

12. $2m^5 - 4m^2 + 8m$ ______

13. $m^2 + m + 1$ ______

14. $2m^2 - 2m + 5$ ______

Divide each polynomial by 3x. See Examples 1–3.

15. $3x^4 + 9x^3 + 3x^2 + 6x$ ______

16. $15x^2 - 9x$ ______

17. $12x^4 - 3x^3 + 3x$ ______

18. $45x^3 + 15x^2 - 9x$ ______

19. $27x^3 - 9x^4 + 18x^5$ ______

20. $-12x^6 + 6x^5 + 3x^4 - 9x^3 + 3x^2$ ______

21. $x^3 + 6x^2 - x$ ______

22. $4x^4 - 3x^3 + 2x$ ______

Perform each division. See Examples 1–3.

23. $\dfrac{8k^4 - 12k^3 - 2k^2 + 7k - 3}{2k}$ ______

24. $\dfrac{27r^4 - 36r^3 - 6r^2 + 26r - 2}{3r}$ ______

25. $\dfrac{100p^5 - 50p^4 + 30p^3 - 30p}{10p^2}$ ______

26. $\dfrac{25 + 10p + p^2}{p}$ ______

27. $\dfrac{8x + 16x^2 + 10x^3}{4x^4}$ ______

28. $\dfrac{36m^5 - 24m^4 + 16m^3 - 8m^2}{4m^3}$ ______

29. $(16y^5 - 8y^2 + 12y) \div 4y^2$ ______

30. $(20a^4 - 15a^5 + 25a^3) \div 15a^4$ ______

31. $(120x^{11} - 60x^{10} + 140x^9 - 100x^8) \div 10x^{12}$ ______

Solve each problem.

32. What polynomial, when divided by $3x^2$, yields $4x^3 + 3x^2 - 4x + 2$ as a quotient? ______

33. What polynomial, when divided by $4m^3$, yields $-6m^2 + 4m$ as a quotient? ______

34. The quotient of a certain polynomial and $-7y^2$ is $9y^2 + 3y + 5 - 2/y$. Find the polynomial. ______

4.7 THE QUOTIENT OF TWO POLYNOMIALS

OBJECTIVES

- Divide a polynomial by a polynomial

To divide a polynomial by any other polynomial, we use a method of "long division." This is similar to long division of two whole numbers.

Step 1

Divide 27 into 6696.

$27\overline{)6696}$

Divide $2x + 3$ into $8x^3 - 4x^2 - 14x + 15$.

$2x + 3\overline{)8x^3 - 4x^2 - 14x + 15}$

Step 2

27 divides into 66 2 times; $2 \cdot 27 = 54$.

$$\begin{array}{r} 2 \\ 27\overline{)6696} \\ \underline{54} \end{array}$$

$2x$ divides into $8x^3$, $4x^2$ times; $4x^2(2x + 3) = 8x^3 + 12x^2$.

$$\begin{array}{rl} & 4x^2 \\ 2x + 3 & \overline{)8x^3 - 4x^2 - 14x + 15} \\ & \underline{8x^3 + 12x^2} \end{array}$$

Step 3

Subtract and bring down the next term.

$$\begin{array}{r} 2 \\ 27\overline{)6696} \\ \underline{54} \\ 129 \end{array}$$

Subtract and bring down the next term.

$$\begin{array}{rl} & 4x^2 \\ 2x + 3 & \overline{)8x^3 - 4x^2 - 14x + 15} \\ & \underline{8x^3 + 12x^2} \\ & \quad -16x^2 - 14x \end{array}$$

(To subtract two polynomials, change the sign of the second, and add.)

Step 4

27 divides into 129 4 times; $4 \cdot 27 = 108$.

$$\begin{array}{r} 24 \\ 27\overline{)6696} \\ \underline{54} \\ 129 \\ \underline{108} \end{array}$$

$2x$ divides into $-16x^2$, $-8x$ times; $-8x(2x + 3) = -16x^2 - 24x$.

$$\begin{array}{rl} & 4x^2 - 8x \\ 2x + 3 & \overline{)8x^3 - 4x^2 - 14x + 15} \\ & \underline{8x^3 + 12x^2} \\ & \quad -16x^2 - 14x \\ & \quad \underline{-16x^2 - 24x} \end{array}$$

Step 5

Subtract and bring down the next term.

$$\begin{array}{r} 24 \\ 27\overline{)6696} \\ \underline{54} \\ 129 \\ \underline{108} \\ 216 \end{array}$$

Subtract and bring down the next term.

$$\begin{array}{rl} & 4x^2 - 8x \\ 2x + 3 & \overline{)8x^3 - 4x^2 - 14x + 15} \\ & \underline{8x^3 + 12x^2} \\ & \quad -16x^2 - 14x \\ & \quad \underline{-16x^2 - 24x} \\ & \qquad\qquad 10x + 15 \end{array}$$

QUICK CHECKS

1. Divide.

$x^3 + x^2 + 4x - 6 \div x - 1$

Step 6

27 divides into 216 8 times; $8 \cdot 27 = 216$.

$$
\begin{array}{r}
248 \\
27\overline{)6696} \\
\underline{54} \\
129 \\
\underline{108} \\
216 \\
\underline{216}
\end{array}
$$

6696 divided by 27 is 248. There is no remainder.

$2x$ divides into $10x$ 5 times; $5(2x + 3) = 10x + 15$.

$$
\begin{array}{r}
4x^2 - 8x + 5 \\
2x + 3\overline{)8x^3 - 4x^2 - 14x + 15} \\
\underline{8x^3 + 12x^2} \qquad\qquad \\
-16x^2 - 14x \qquad \\
\underline{-16x^2 - 24x} \qquad \\
10x + 15 \\
\underline{10x + 15}
\end{array}
$$

$8x^3 - 4x^2 - 14x + 15$ divided by $2x + 3$ is $4x^2 - 8x + 5$. There is no remainder.

Step 7

Check by multiplication.

$27 \cdot 248 = 6696$

Check by multiplication.

$(2x + 3)(4x^2 - 8x + 5)$
$= 8x^3 - 4x^2 - 14x + 15$

Example 1 Divide $4x^3 - 4x^2 + 5x - 8$ by $2x - 1$.

$$
\begin{array}{r}
2x^2 - x + 2 \\
2x - 1\overline{)4x^3 - 4x^2 + 5x - 8} \\
\underline{4x^3 - 2x^2} \qquad\qquad \\
-2x^2 + 5x \qquad \\
\underline{-2x^2 + x} \qquad \\
4x - 8 \\
\underline{4x - 2} \\
-6
\end{array}
$$

1. $2x$ divides into $4x^3$, $2x^2$ times; $2x^2(2x - 1) = 4x^3 - 2x^2$.
2. Subtract; bring down the next term.
3. $2x$ divides into $-2x^2$, $-x$ times; $-x(2x - 1) = -2x^2 + x$.
4. Subtract; bring down the next term.
5. $2x$ divides into $4x$, 2 times; $2(2x - 1) = 4x - 2$.
6. Subtract. The remainder is -6.

Thus $2x - 1$ divides into $4x^3 - 4x^2 + 5x - 8$ with a quotient of $2x^2 - x + 2$ and a remainder of -6. The result is not a polynomial because of the remainder.

$$\frac{4x^3 - 4x^2 + 5x - 8}{2x - 1} = 2x^2 - x + 2 + \frac{-6}{2x - 1}$$

Check by multiplication.

$$(2x - 1)\left(2x^2 - x + 2 + \frac{-6}{2x - 1}\right)$$
$$= 4x^3 - 4x^2 + 5x - 8$$

Work Quick Check 1 at the side.

Quick Check Answers

1. $x^2 + 2x + 6$

Example 2 Divide $x^3 - 1$ by $x - 1$.

Here the polynomial $x^3 - 1$ is missing the x^2 term and the x term. When this is the case, the polynomial should be filled in with 0 as the coefficient for the missing terms.

$$x^3 - 1 = x^3 + 0x^2 + 0x - 1$$

Now we can rewrite the problem.

$$\begin{array}{r}
x^2 + x + 1 \\
x - 1\overline{)x^3 + 0x^2 + 0x - 1} \\
\underline{x^3 - x^2} \\
x^2 + 0x \\
\underline{x^2 - x} \\
x - 1 \\
\underline{x - 1}
\end{array}$$

There is no remainder. The quotient is $x^2 + x + 1$. Check by multiplication.

$$(x^2 + x + 1)(x - 1) = x^3 - 1$$

Work Quick Check 2 at the side.

2. Divide.

$x^3 - 8 \div x - 2$

Example 3 Divide $x^4 + 2x^3 + 2x^2 - x - 1$ by $x^2 + 1$.

The denominator of $x^2 + 1$ has a missing x term which we fill in with $0x$, since $x^2 + 1 = x^2 + 0x + 1$. We then proceed as usual through the division process.

$$\begin{array}{r}
x^2 + 2x + 1 \\
x^2 + 0x + 1\overline{)x^4 + 2x^3 + 2x^2 - x - 1} \\
\underline{x^4 + 0x^3 + x^2} \\
2x^3 + x^2 - x \\
\underline{2x^3 + 0x^2 + 2x} \\
x^2 - 3x - 1 \\
\underline{x^2 + 0x + 1} \\
-3x - 2
\end{array}$$

Since $-3x - 2$ is a polynomial of smaller degree than the divisor $(x^2 + 0x + 1)$, $-3x - 2$ is the remainder. We write the quotient as

$$x^2 + 2x + 1 + \frac{-3x - 2}{x^2 + 1}.$$

Work Quick Check 3 at the side.

3. Divide.

$2x^4 + 3x^3 - x^2 + 6x + 5$
$\div x^2 - 1$

Quick Check Answers

2. $x^2 + 2x + 4$

3. $2x^2 + 3x + 1 + \dfrac{9x + 6}{x^2 - 1}$

WORK SPACE

name date hour

4.7 EXERCISES

Perform each division. See Example 1.

1. $\dfrac{x^2 - x - 6}{x - 3}$ ______

2. $\dfrac{m^2 - 2m - 24}{m + 4}$ ______

3. $\dfrac{2y^2 + 9y - 35}{y + 7}$ ______

4. $\dfrac{y^2 + 2y + 1}{y + 1}$ ______

5. $\dfrac{p^2 + 2p - 24}{p + 6}$ ______

6. $\dfrac{x^2 + 11x + 24}{x + 8}$ ______

7. $\dfrac{r^2 - 8r + 15}{r - 3}$ ______

8. $\dfrac{t^2 - 3t - 10}{t - 5}$ ______

9. $\dfrac{12m^2 - 20m + 3}{2m - 3}$ ______

10. $\dfrac{2y^2 - 5y - 3}{2y + 1}$ ______

11. $\dfrac{2a^2 - 11a - 21}{2a + 3}$ ______

12. $\dfrac{9w^2 + 6w - 8}{3w - 2}$ ______

13. $\dfrac{2x^2 + 5x + 3}{2x + 1}$ ______

14. $\dfrac{4m^2 - 4m + 5}{2m - 1}$ ______

15. $\dfrac{2a^2 - 3a + 4}{2a + 1}$ ______

16. $\dfrac{4p^2 - 4p + 7}{2p - 1}$ ______

17. $\dfrac{2d^2 - 2d + 5}{2d + 4}$ ______

18. $\dfrac{4m^2 + 11m - 8}{m + 3}$ ______

19. $\dfrac{2x^3 - x^2 + 3x + 2}{2x + 1}$ ______

20. $\dfrac{12t^3 - 11t^2 + 9t + 18}{4t + 3}$ ______

21. $\dfrac{8k^4 - 12k^3 - 2k^2 + 7k - 6}{2k - 3}$ ______

22. $\dfrac{27r^4 - 36r^3 - 6r^2 + 26r - 24}{3r - 4}$ ______

Perform each division. See Examples 2 and 3.

23. $\dfrac{3y^3 + y^2 + 3y + 1}{y^2 + 1}$ ________

24. $\dfrac{2x^5 + 6x^4 - x^3 + 3x^2 - x}{2x^2 + 1}$ ________

25. $\dfrac{x^4 - x^2 - 6x}{x^2 - 2}$ ________

26. $\dfrac{x^4 - 2x^2 + 5}{x^2 - 1}$ ________

27. $\dfrac{x^3 + 1}{x + 1}$ ________

28. $\dfrac{x^4 - 1}{x^2 - 1}$ ________

29. $\dfrac{x^4 - 1}{x^2 + 1}$ ________

30. $\dfrac{x^5 - 1}{x^2 - 1}$ ________

OBJECTIVES

- Know the meaning of a negative exponent
- Use the properties of exponents with negative exponents

We want to define negative integer exponents so that all the rules for exponents are still valid. (See the list of rules in Section 4.1.) For example, by the quotient rule we subtract exponents.

$$\frac{3^8}{3^6} = 3^{8-6} = 3^2$$

However, consider the quotient $3^0/3^2$. Since $3^0 = 1$, we have

$$\frac{3^0}{3^2} = \frac{1}{3^2}.$$

Suppose we subtract exponents in this case.

$$\frac{3^0}{3^2} = 3^{0-2} = 3^{-2}$$

The exponential 3^{-2} is read "three to the negative two power." In order to have negative exponents satisfy the same properties as positive exponents, we must have

$$3^{-2} = \frac{1}{3^2}.$$

Thus, in general, for $a \neq 0$ and n = any integer, we define

$$a^{-n} = \frac{1}{a^n}.$$

Example 1 Simplify each term using the definition of negative integer exponents.

(a) $3^{-2} = \frac{1}{3^2} = \frac{1}{9}$

(b) $5^{-3} = \frac{1}{5^3} = \frac{1}{125}$

(c) $\left(\frac{1}{2}\right)^{-3} = \frac{1}{\left(\frac{1}{2}\right)^3} = \frac{1}{\frac{1}{8}} = 1 \cdot \frac{8}{1} = 8$

(d) $4^{-1} - 2^{-1} = \frac{1}{4} - \frac{1}{2} = \frac{1}{4} - \frac{2}{4} = -\frac{1}{4}$

Work Quick Check 1 at the side.

QUICK CHECKS

1. Write each of the following with a positive exponent.

(a) 4^{-3}

(b) 6^{-2}

(c) $\left(\frac{2}{3}\right)^{-2}$

(d) $2^{-1} + 5^{-1}$

By defining a^{-n} as we have done, all the rules for exponentials are still valid. For example, by the product rule, we have $a^m \cdot a^n = a^{m+n}$. Therefore,

$$5^3 \cdot 5^{-5} = 5^{3+(-5)} = 5^{-2} = \frac{1}{5^2} = \frac{1}{25}.$$

Quick Check Answers

1. (a) $1/4^3$ (b) $1/6^2$ (c) $(3/2)^2$ or $3^2/2^2$ (d) $1/2 + 1/5$ or $7/10$

2. Simplify.

(a) $(6^{-2})(6^5)$

(b) $(5^{-1})^3$

(c) $\left(\frac{5}{6}\right)^{-2}$

(d) $(2^{-5} \cdot 2^3)^2$

3. Simplify each of the following. Write all variables with positive exponents.

(a) $\frac{4^{-5}}{4^{-2}}$

(b) $\frac{6x^{-1}}{3x^2}$

(c) $\left(\frac{2x^{-2}}{3^2y^{-1}}\right)^{-3}$

Also,

$$(-8)^{-5} \cdot (-8)^7 = (-8)^{-5+7} = (-8)^2 = 64.$$

Example 2 Use the definition of negative exponents and the rules for exponents to simplify.

(a) $(3^{-4})^2 = 3^{(-4)2} = 3^{-8} = \frac{1}{3^8}$

(b) $\left(\frac{6}{5}\right)^{-2} = \frac{6^{-2}}{5^{-2}} = \frac{\frac{1}{6^2}}{\frac{1}{5^2}} = \frac{1}{6^2} \cdot \frac{5^2}{1} = \frac{5^2}{6^2} = \frac{25}{36}$

(c) $(2^3 \cdot 2^{-4})^2 = (2^{3+(-4)})^2 = (2^{-1})^2 = 2^{-2} = \frac{1}{4}$

Work Quick Check 2 at the side.

Using the definition of negative exponent presented above, the **quotient rule** can now be simplified.

For all integers m and n, and $a \neq 0$,

$$\frac{a^m}{a^n} = a^{m-n}.$$

Example 3 Use the quotient rule to simplify.

(a) $\frac{5^{-3}}{5^{-7}} = 5^{-3-(-7)} = 5^{-3+7} = 5^4$

(b) $\left(\frac{3x^{-2}}{4^{-1}y^3}\right)^{-3} = \frac{3^{-3}x^6}{4^3y^{-9}} = \frac{\frac{1}{3^3} \cdot x^6}{4^3 \cdot \frac{1}{y^9}} = \frac{\frac{x^6}{3^3}}{\frac{4^3}{y^9}} = \frac{x^6}{3^3} \cdot \frac{y^9}{4^3}$

$$= \frac{x^6y^9}{27 \cdot 64}$$

$$= \frac{x^6y^9}{1728}$$

Work Quick Check 3 at the side.

Quick Check Answers

2. (a) 6^3 (b) 5^{-3} or $1/5^3$ (c) $(6/5)^2$ or $6^2/5^2$ (d) 2^{-4} or $1/2^4$

3. (a) $1/4^3$ (b) $2/x^3$ (c) $3^6x^6/2^3y^3$

name date hour

4.8 EXERCISES

Evaluate each expression. See Example 1.

1. 3^{-3} ________

2. 4^{-2} ________

3. 5^{-2} ________

4. 2^{-5} ________

5. 9^{-1} ________

6. $(-12)^{-1}$ ________

7. $(-6)^{-2}$ ________

8. 8^{-3} ________

9. 7^{-1} ________

10. 12^{-2} ________

11. $\left(\frac{1}{2}\right)^{-5}$ ________

12. $\left(\frac{1}{5}\right)^{-2}$ ________

13. $\left(\frac{1}{2}\right)^{-1}$ ________

14. $\left(\frac{3}{4}\right)^{-1}$ ________

15. $\left(\frac{2}{3}\right)^{-3}$ ________

16. $\left(\frac{5}{4}\right)^{-2}$ ________

17. $2^{-1} + 3^{-1}$ ________

18. $3^{-1} - 4^{-1}$ ________

19. $5^{-1} + 4^{-1}$ ________

20. $3^{-1} + 6^{-1}$ ________

Perform the indicated operations. Write the answers without negative exponents and with each variable occurring only once. Any variable represents a positive real number. See Examples 2 and 3.

21. $3^4 \cdot 3^{-5}$ ________

22. $5^{-6} \cdot 5^7$ ________

23. $9^8 \cdot 9^{-7}$ ________

24. $4^3 \cdot 4^{-5}$ ________

25. $9^{-4} \cdot 9^{-2} \cdot 9^5$ ________

26. $6^{-2} \cdot 6^3 \cdot 6^{-4}$ ________

27. $\dfrac{8^9 \cdot 8^{-11}}{8^{-5}}$ ________

28. $\dfrac{7^4 \cdot 7^{-3}}{7^5}$ ________

29. $\dfrac{4^3 \cdot 4^{-5}}{4^7}$ ______________________

30. $\dfrac{2^5 \cdot 2^{-4}}{2^{-1}}$ ______________________

31. $\dfrac{5^{-3} \cdot 5^{-2}}{5^4}$ ______________________

32. $\dfrac{8^{-2} \cdot 8^5}{8^6}$ ______________________

33. $\dfrac{m^4 \cdot m^{-5}}{m^{-6}}$ ______________________

34. $\dfrac{p^3 \cdot p^{-5}}{p^5}$ ______________________

35. $\dfrac{m^{11} \cdot m^{-7}}{m^5}$ ______________________

36. $\dfrac{z^3 \cdot z^{-5}}{z^{10}}$ ______________________

37. $\dfrac{r^5 \cdot r^{-8}}{r^{-6} \cdot r^4}$ ______________________

38. $\dfrac{x^3 \cdot x^{-1}}{x^8 \cdot x^{-2}}$ ______________________

39. $\dfrac{a^6 \cdot a^{-3}}{a^{-5} \cdot a}$ ______________________

40. $\dfrac{b^{10} \cdot b^{-2}}{b^{-8} \cdot b^6}$ ______________________

41. $(3x^{-5})^2$ ______________________

42. $(-5p^{-4})^{-2}$ ______________________

43. $(9^{-1}y^5)^{-2}$ ______________________

44. $(4^{-2}m^{-3})^{-2}$ ______________________

45. $\left(\dfrac{a}{b}\right)^{-1}$ ______________________

46. $\left(\dfrac{2a}{3}\right)^{-2}$ ______________________

47. $\left(\dfrac{5m^{-2}}{m^{-1}}\right)^2$ ______________________

48. $\left(\dfrac{4x^3}{3^{-1}}\right)^{-1}$ ______________________

49. $\dfrac{(3x^2)^{-2}(5x^{-1})^3}{3x^{-5}}$ ______________________

50. $\dfrac{(2y^{-1}z^2)^2(3y^{-2}z^{-3})^3}{(y^3z^2)^{-1}}$ ______________________

51. $\dfrac{(4a^2b^3)^{-2}(2ab^{-1})^3}{(a^3b)^{-4}}$ ______________________

52. $\dfrac{(m^6n)^{-2}(m^2n^{-2})^3}{m^{-1}n^{-2}}$ ______________________

4.9 AN APPLICATION OF EXPONENTIALS: SCIENTIFIC NOTATION

OBJECTIVES

- Convert numbers to scientific notation
- Convert numbers out of scientific notation

One example of the use of exponentials comes from the field of science. The numbers occurring in science are often extremely large (such as the distance from the earth to the sun, which is 93,000,000 miles) or extremely small (the wavelength of yellow-green light is approximately .0000006 meters). Because of the difficulty of working with many zeros, scientists often express such numbers as exponentials. Each number is written as the product of a number a (where $1 \leqslant a < 10$) and some power of 10. This form is called **scientific notation.** There is always one nonzero integer before the decimal point. For example, 35 is written 3.5×10^1, or 3.5×10; while 56,200 is written 5.62×10^4, since

$$56{,}200 = 5.62 \times 10{,}000 = 5.62 \times 10^4.$$

Example 1 Express in scientific notation.

(a) $93{,}000{,}000 = 9.3 \times 10^7$

(b) $463{,}000{,}000{,}000{,}000 = 4.63 \times 10^{14}$

(c) $63{,}200{,}000{,}000 = 6.32 \times 10^{10}$

(d) $302{,}100 = 3.021 \times 10^5$

Work Quick Check 1 at the side.

Example 2 Express without exponents.

(a) 6.2×10^3

The number 6.2×10^3 is in scientific notation. We write

$$6.2 \times 10^3 = 6.2 \times 1000 = 6200.$$

(b) $4.283 \times 10^5 = 4.283 \times 100{,}000$
$= 428{,}300$

Work Quick Check 2 at the side.

To use scientific notation for very small numbers, note that $.1 = 1/10 = 10^{-1}$; $.01 = 1/100 = 10^{-2}$, and so on.

Example 3 Express in scientific notation.

(a) $.004 = 4 \times 10^{-3}$

(b) $.0000762 = 7.62 \times 10^{-5}$

(c) $.000000000834 = 8.34 \times 10^{-10}$

Work Quick Check 3 at the side.

QUICK CHECKS

1. Write each number in scientific notation.

(a) 63,000

(b) 5,870,000

2. Express without exponents.

(a) 4.2×10^3

(b) 8.7×10^5

3. Express in scientific notation.

(a) .0571

(b) .000062

Quick Check Answers
1. (a) 6.3×10^4 (b) 5.87×10^6
2. (a) 4200 (b) 870,000
3. (a) 5.71×10^{-2}
(b) 6.2×10^{-5}

4. Express without exponents.

(a) 6.42×10^{-3}

(b) $(2.6 \times 10^{4})(2 \times 10^{-6})$

(c) $\dfrac{4.8 \times 10^{2}}{2.4 \times 10^{-3}}$

Quick Check Answers

4. (a) .00642 (b) .052
(c) 200,000

Example 4 Express without exponents.

(a) $9.73 \times 10^{-2} = .0973$

(b) $(6 \times 10^{3})\ (5 \times 10^{-4})$

First find the product.

$$(6 \times 10^{3})\ (5 \times 10^{-4}) = (6 \times 5)\ (10^{3} \times 10^{-4})$$
$$= 30 \times 10^{-1}$$

Then express the result without exponents as 3.0.

(c) $$\frac{4 \times 10^{-5}}{2 \times 10^{3}} = \frac{4}{2} \times \frac{10^{-5}}{10^{3}} = 2 \times 10^{-8}$$
$$= .00000002$$

Work Quick Check 4 at the side.

name ____________________ date ____________________ hour ____________________

4.9 EXERCISES

Express each number in scientific notation. See Examples 1 and 3.

1. 6,835,000,000 ____________________
2. 321,000,000,000,000 ____________________
3. 8,360,000,000,000 ____________________
4. 6850 ____________________
5. 215 ____________________
6. 683 ____________________
7. 25,000 ____________________
8. 110,000,000 ____________________
9. .035 ____________________
10. .005 ____________________
11. .0101 ____________________
12. .0000006 ____________________
13. .000012 ____________________
14. .000000982 ____________________

Write each number without exponents. See Examples 2 and 4.

15. 8.1×10^9 ____________________
16. 3.5×10^2 ____________________
17. 9.132×10^6 ____________________
18. 2.14×10^0 ____________________
19. 3.24×10^8 ____________________
20. 4.35×10^4 ____________________
21. 3.2×10^{-4} ____________________
22. 5.76×10^{-5} ____________________
23. 4.1×10^{-2} ____________________
24. 1.79×10^{-3} ____________________
25. $(2 \times 10^8) \times (4 \times 10^{-3})$ ____________________
26. $(5 \times 10^4) \times (3 \times 10^{-2})$ ____________________
27. $(4 \times 10^{-1}) \times (1 \times 10^{-5})$ ____________________
28. $(6 \times 10^{-5}) \times (2 \times 10^4)$ ____________________

29. $\dfrac{9 \times 10^5}{3 \times 10^{-1}}$ ______________________

30. $\dfrac{12 \times 10^{-4}}{4 \times 10^4}$ ______________________

31. $\dfrac{4 \times 10^{-3}}{2 \times 10^{-2}}$ ______________________

32. $\dfrac{5 \times 10^{-1}}{1 \times 10^{-5}}$ ______________________

Write the numbers in each application in scientific notation. See Examples 1 and 3.

33. Light visible to the human eye has a wavelength between .0004 mm and .0008 mm. ______________________

34. In the ocean, the amount of oxygen per cubic mile of water is 4,037,000,000 tons, while the amount of radium is .0003 tons. ______________________

35. Each tide in the Bay of Fundy carries more than 3,680,000,000,000,000 cubic feet of water into the bay. ______________________

36. The mean (average) diameter of the sun is about 865,000 miles. ______________________

Write the numbers in each application without exponents. See Examples 2 and 4.

37. 1×10^3 cubic millimeters equals 6.102×10^{-2} cubic inch. ______________________

38. In the food chain which links the largest sea creature, the whale, to the smallest, the diatom, 4×10^{14} diatoms sustain a medium-sized whale for only a few hours. ______________________

name date hour

CHAPTER 4 TEST

If $P(x) = x^4 + 2x^2 - 7x + 2$, find each value.

1. $P(2)$ ________

2. $P(-1)$ ________

Write in exponential form; then evaluate.

3. $4 \cdot 4 \cdot 4 \cdot 4 \cdot 4$ ________

4. $\frac{6^4}{6^6}$ ________

5. $(-2)^2(-2)^3$ ________

6. $2^{-3} \cdot 2^7$ ________

7. $(2^3)^2$ ________

8. $(2/3)^2$ ________

9. 8^{-2} ________

10. $\frac{8^0 \cdot 8^{13}}{8^{12}}$ ________

11. Use the rules for exponentials to simplify $(2x^2y^3)^{-3}$, and write with only positive exponents. ________

For each polynomial, combine terms; then give the degree of the polynomial. Finally, select the most specific description from this list.

(a) trinomial (b) binomial (c) monomial (d) none of these

12. $3x^2 + 6x - 4x^2$ ________

13. $11m^3 - m^2 + m^4$ ________

14. $3x^3 - 4x^2 + 2x - 1$ ________

15. 7 ________

Perform the indicated operation.

16. $(3x^3 + 2x^2 - 5x + 3) + (7x^3 - 4x^2 - 3x - 3)$ ________

17. $(2x^5 + 3x^3 - 4x + 7) - (x^5 - 3x^3 + x^2 - 2x - 5)$ ________

18. $(y^2 - 5y - 3) + (3y^2 + 2y) - (y^2 - y - 1)$ ________________

19. $(10x^3)(-4x^2)$ ________________

20. $(r - 5)(r + 2)$ ________________

21. $(3t + 4)(2t - 3)$ ________________

22. $(2p + 5)^2$ ________________

23. $(x - 8)(x + 8)$ ________________

24. $(15r^4 - 10r^3 + 25r^2 - 15r) \div 5r$ ________________

25. $(3x^5 + 12x^4 - 9x^3 + 6x^2 - 60x - 120) \div (x - 2)$ ________________

Write in scientific notation.

26. 6,000,000 ________________

27. 245,000,000 ________________

Write without exponents.

28. 4.8×10^4 ________________

29. 2.91×10^8 ________________

30. 6.45×10^{-2} ________________

31. 1.03×10^{-5} ________________

32. $\dfrac{15 \times 10^5}{5 \times 10^{-2}}$ ________________

33. $\dfrac{4 \times 10^{-2}}{2 \times 10^3}$ ________________

5 Factoring

5.1 FACTORS

OBJECTIVES

- List the factors of a number
- Find the greatest common factor for a list of terms
- Factor out the greatest common factor

Since the product of 6 and 2 is 12, we call 6 and 2 **factors** of 12. Also, $6 \cdot 2$ is a **factored form** of 12. Other factored forms of 12 are $(-6)(-2)$, $3 \cdot 4$, $(-3)(-4)$, $12 \cdot 1$, and $(-12)(-1)$. Since $(1/2)(24) = 12$, and $(2/3)(18) = 12$, we might think of these as factored forms of 12. However, factors are usually limited to integers. Because of this, the factors of 12 are

$$1, -1, 2, -2, 3, -3, 4, -4, 6, -6, 12, -12.$$

In general, an integer a is a **factor** of an integer b if b can be divided by a with no remainder.

Example 1 The factors of 36 are 1, −1, 2, −2, 3, −3, 4, −4, 6, −6, 9, −9, 12, −12, 18, −18, 36, −36.

Example 2 The factors of 11 are 11, −11, 1, and −1.

Work Quick Check 1 at the side.

QUICK CHECKS

1. List all factors of each integer.

(a) 24

(b) 40

(c) 7

(d) 19

An integer which is a factor of two or more other integers is called a **common factor** of those integers. For example, 6 is a common factor of 18 and 24, since 6 is a factor of both 18 and 24. Other common factors of 18 and 24 are −6, −3, −2, −1, 1, 2, and 3. The **greatest common factor** of a set of integers is the largest common factor of the set. Thus, 6 is the greatest common factor of 18 and 24 because 6 is the largest common factor of these numbers.

Example 3

(a) The greatest common factor of 7, 11, and 14 is 1, since 1 is the largest number that divides into all three numbers.

(b) The greatest common factor of 2, 4, 6, and 8 is 2.

(c) The number 3 is the greatest common factor of 9, 27, and 6.

(d) The number 4 is the greatest common factor of the numbers 8, 12, 16, and 32.

Quick Check Answers

1. (a) 1, −1, 2, −2, 3, −3, 4, −4, 6, −6, 8, −8, 12, −12, 24, −24 (b) 1, −1, 2, −2, 4, −4, 5, −5, 8, −8, 10, −10, 20, −20, 40, −40 (c) 1, −1, 7, −7 (d) 1, −1, 19, −19

2. Find the greatest common factor for each list of numbers.

(a) 30, 20, 15

(b) 42, 28, 35

(c) 12, 18, 26, 32

(d) 10, 15, 21

3. Find the greatest common factor of each set of terms.

(a) $6m^4, 9m^2, 12m^5$

(b) $25y^{11}, 30y^7$

(c) $12p^5, 18q^4$

(d) $11r^9, 10r^{15}, 8r^{12}$

(e) y^4z^2, y^6z^8, z^9

(f) $12p^{11}, 17q^5$

Quick Check Answers

2. (a) 5 (b) 7 (c) 2 (d) 1
3. (a) $3m^2$ (b) $5y^7$ (c) 6 (d) r^9 (e) z^2 (f) 1

Work Quick Check 2 at the side.

You can also find the greatest common factor of a collection of terms. For example, the terms x^4, x^5, x^6, and x^7 have x^4 as the greatest common factor, because x^4 is the highest power of x that is a factor of each of the terms x^4, x^5, x^6, and x^7. To see this, write the terms in factored form.

$$x^4 = x^4 \cdot 1$$
$$x^5 = x^4 \cdot x$$
$$x^6 = x^4 \cdot x^2$$
$$x^7 = x^4 \cdot x^3$$

Generally, the exponent on the greatest common factor is the *smallest* exponent that appears on the terms. In the example above, 4 is the smallest exponent on x^4, x^5, x^6, x^7, so x^4 is the greatest common factor for these terms.

Example 4 Find the greatest common factor of the terms

$$y^2, y^5, y^7, \text{ and } y^{15}.$$

Here 2 is the smallest exponent on y, so y^2 is the greatest common factor.

Example 5 Find the greatest common factor of the terms

$$21m^7, -18m^6, 45m^8, \text{ and } -24m^5.$$

First, 3 is the greatest common factor of the coefficients 21, −18, 45, and −24. The smallest exponent on m is 5, so the greatest common factor of the terms is $3m^5$.

Example 6 Find the greatest common factor of the terms

$$x^4y^2, x^7y^5, x^3y^7, \text{ and } y^{15}.$$

There is no x in the last term, y^{15}, so that x will not appear in the greatest common factor. There is a y in each term, however, with 2 the smallest exponent on y. The greatest common factor is y^2.

Work Quick Check 3 at the side.

We can use the idea of a greatest common factor to write a polynomial in factored form. For example, the polynomial

$$3m + 12$$

is made up of the two terms $3m$ and 12. The greatest common factor for these two terms is 3. Write $3m + 12$ so that each term is a product with 3 as one factor.

$$3m + 12 = 3 \cdot m + 3 \cdot 4$$

Now use the distributive property.

$$3m + 12 = 3 \cdot m + 3 \cdot 4 = 3(m + 4)$$

The factored form of $3m + 12$ is $3(m + 4)$. This process is called **factoring out the greatest common factor.**

Example 7 Factor out the greatest common factor.

(a) $20m^5 + 10m^4 + 15m^3$

The greatest common factor for the terms of this polynomial is $5m^3$.

$$\begin{aligned} 20m^5 + 10m^4 + 15m^3 &= (5m^3)(4m^2) \\ &\quad + (5m^3)(2m) + (5m^3)3 \\ &= 5m^3(4m^2 + 2m + 3) \end{aligned}$$

To check your work, multiply $5m^3$ and $(4m^2 + 2m + 3)$. You should get the original polynomial as your answer.

(b) $48y^{12} - 36y^{10} + 12y^7$
$= (12y^7)(4y^5) - (12y^7)(3y^3) + (12y^7)1$
$= 12y^7(4y^5 - 3y^3 + 1)$

(c) $x^5 - x^3 = (x^3)x^2 - (x^3)1$
$= x^3(x^2 - 1)$

(d) $20m^7p^2 - 36m^3p^4 = 4m^3p^2(5m^4 - 9p^2)$

Work Quick Check 4 at the side.

4. Factor out the greatest common factor.

(a) $32p^2 + 16p + 48$

(b) $10y^5 - 8y^4 + 6y^2$

(c) $27a^5 + 9a^4$

(d) $m^7 + m^9$

(e) $8p^5q^2 + 16p^6q^3 - 12p^4q^7$

Quick Check Answers

4. (a) $16(2p^2 + p + 3)$
(b) $2y^2(5y^3 - 4y^2 + 3)$
(c) $9a^4(3a + 1)$ (d) $m^7(1 + m^2)$
(e) $4p^4q^2(2p + 4p^2q - 3q^5)$

WORK SPACE

name date hour

5.1 EXERCISES

Find all the factors of each number. See Examples 1 and 2.

1. 14 ______ 2. 18 ______ 3. 27 ______

4. 35 ______ 5. 45 ______ 6. 50 ______

7. 60 ______ 8. 72 ______ 9. 100 ______

10. 130 ______ 11. 29 ______ 12. 37 ______

Find the greatest common factor for each set of terms. See Examples 3–6.

13. $12y$, 24 ______ 14. $72m$, 12 ______

15. $30p^2, 20p^3, 40p^5$ ______ 16. $14r^5, 28r^2, 56r^8$ ______

17. $18r, 32y, 11z$ ______ 18. $45m^2, 12n, 7p^2$ ______

19. $18m^2n^2, 36m^4n^5, 12m^3n$ ______ 20. $50p^5r^2, 25p^4r^7, 30p^7r^8$ ______

Complete the factoring.

21. $12 = 6(\quad)$ 22. $18 = 9(\quad)$ 23. $3x^2 = 3x(\quad)$

24. $8x^3 = 8x(\quad)$ 25. $9m^4 = 3m^2(\quad)$ 26. $12p^5 = 6p^3(\quad)$

27. $-8z^9 = -4z^5(\quad)$ 28. $-15k^{11} = -5k^8(\quad)$ 29. $x^2y^3 = xy(\quad)$

30. $a^3b^2 = a^2b(\quad)$ 31. $6x^2y^3 = 6xy(\quad)$ 32. $27a^3b^2 = 9a^2b(\quad)$

33. $14x^4y^3 = 2xy(\quad)$ 34. $-16m^3n^3 = 4mn^2(\quad)$

Factor out the greatest common factor. See Example 7.

35. $12x + 24$ ______ 36. $18m - 9$ ______

37. $3 + 36d$ ______

38. $15 + 25r$ ______

39. $9a^2 - 18a$ ______

40. $21m^5 - 14m^4$ ______

41. $65y^9 - 35y^5$ ______

42. $100a^4 - 16a^2$ ______

43. $121p^5 - 33p^4$ ______

44. $8p^2 - 4p^4$ ______

45. $11z^2 - 100$ ______

46. $12z^2 - 11y^4$ ______

47. $9m^2 + 90m^3$ ______

48. $16r^2s + 64rs^2$ ______

49. $19y^3p^2 + 38y^2p^3$ ______

50. $4mn^2 - 12m^2n$ ______

51. $18x^2y^3 - 24x^4y$ ______

52. $100m^5 - 50m^3 + 100m^2$ ______

53. $13y^6 + 26y^5 - 39y^3$ ______

54. $5x^4 + 25x^3 - 20x^2$ ______

55. $16a^3 + 8a^2 + 24a$ ______

56. $6a^2 + 8c^2 - 4b^2$ ______

57. $45q^4p^5 - 36qp^6 + 81q^2p^3$ ______

58. $a^5 + 2a^5b + 3a^5b^2 - 4a^5b^3$ ______

59. $a^3b^5 - a^2b^7 + ab^3$ ______

60. $m^6n^5 - 2m^5 + 5m^3n^5$ ______

OBJECTIVES

- Factor trinomials with a coefficient of 1 for the squared term
- Factor such polynomials after factoring out the greatest common factor

The product of two binomials is usually a trinomial. For example,

$$(x + 4)(x - 6) = x^2 - 2x - 24;$$

$$(y + 5)(y - 3) = y^2 + 2y - 15.$$

Exceptions to this rule include $(x + 2)(x - 2) = x^2 - 4$, which is not a trinomial.

In this section our goal is to factor a trinomial as the product of two binomial factors. We limit ourselves to trinomials like $x^2 - 2x - 24$ or $y^2 + 2y - 15$, where the coefficient of the squared term is 1.

Let's try to factor $x^2 + 5x + 6$. We want to find integers a and b such that

$$x^2 + 5x + 6 = (x + a)(x + b).$$

To find these integers a and b, first multiply the right-hand side of the equation.

$$(x + a)(x + b) = x^2 + ax + bx + ab$$

By the distributive property,

$$x^2 + ax + bx + ab = x^2 + (a + b)x + ab.$$

Thus, we want integers a and b such that

$$x^2 + 5x + 6 = x^2 + (a + b)x + ab.$$

The integers a and b must satisfy the conditions

$$a + b = 5$$
$$ab = 6.$$

Can we find two integers whose sum is 5 ($a + b = 5$) and whose product is 6 ($ab = 6$)? Since many pairs of integers can be found which have a sum of 5, it is best to list first those pairs of integers whose product is 6.

$$1 \cdot 6 = 6 \qquad 1 + 6 = 7$$
$$2 \cdot 3 = 6 \qquad 2 + 3 = 5$$
$$(-1)(-6) = 6 \qquad -1 + (-6) = -7$$
$$(-2)(-3) = 6 \qquad -2 + (-3) = -5$$

All four pairs have a product of 6, but only the pair with 2 and 3 has a sum of 5. Then 2 and 3 are the integers we are looking for.

$$x^2 + 5x + 6 = (x + 2)(x + 3)$$

The trinomial $x^2 + 5x + 6$ has been factored into the product of the binomials $x + 2$ and $x + 3$.

To check, multiply the binomials.

$$(x + 2)(x + 3) = x^2 + 5x + 6$$

Note that the method we used can only be used on trinomials where the coefficient of the squared term is 1. Methods for other trinomials will be presented later.

QUICK CHECKS

1. Complete the given lists of numbers, then factor the given trinomial.

(a) $m^2 + 11m + 30$

$ab = 30, a + b = 11$

$30, 1 \quad 30 + 1 = 31$

$15, 2 \quad 15 + 2 = \_\_$

$10, 3 \quad 10 + 3 = \_\_$

$6, 5 \quad 6 + 5 = \_\_$

(b) $y^2 + 12y + 20$

$ab = 20, a + b = 12$

$20, 1 \quad 20 + 1 = 21$

$10, \_ \quad 10 + \_ = \_\_$

$5, \_ \quad 5 + \_ = \_\_$

2. Factor each trinomial.

(a) $p^2 + 7p + 6$

(b) $y^2 + 4y + 3$

(c) $a^2 - 9a - 22$

(d) $r^2 - 6r - 16$

(e) $m^2 - m - 12$

(f) $r^2 - 6r + 5$

3. Factor each trinomial, where possible.

(a) $x^2 + x + 1$

(b) $r^2 - 3r - 4$

(c) $m^2 - 2m + 5$

(d) $y^2 - 11y + 30$

Quick Check Answers

1. (a) 17; 13; 11; $(m + 6)(m + 5)$ (b) 2; 2; 12; 4; 4; 9; $(y + 10)(y + 2)$
2. (a) $(p + 6)(p + 1)$ (b) $(y + 3)(y + 1)$ (c) $(a - 11)(a + 2)$ (d) $(r - 8)(r + 2)$ (e) $(m - 4)(m + 3)$ (f) $(r - 5)(r - 1)$
3. (a) cannot be factored (b) $(r - 4)(r + 1)$ (c) cannot be factored (d) $(y - 5)(y - 6)$

Example 1 To factor $m^2 + 9m + 14$, look for two integers whose product is 14, and whose sum is 9. List the pairs of integers whose products are 14. Then examine the sums.

$$\begin{array}{rr} 14, 1 & 14 + 1 = 15 \\ 7, 2 & 7 + 2 = 9 \\ -14, -1 & -14 + (-1) = -15 \\ -7, -2 & -7 + (-2) = -9 \end{array}$$

From the list, 7 and 2 are the integers we need, since $7 \cdot 2 = 14$ and $7 + 2 = 9$. Thus the binomial factors of $m^2 + 9m + 14$ are $(m + 2)(m + 7)$.

Work Quick Check 1 at the side.

Example 2 To factor $p^2 - 2p - 15$, find two integers whose product is -15 and whose sum is -2. If these numbers do not come to mind right away, we can always find them (if they exist) by listing all the pairs of integers whose product is -15.

$$\begin{array}{rr} 15, -1 & 15 + (-1) = 14 \\ 5, -3 & 5 + (-3) = 2 \\ -15, 1 & -15 + 1 = -14 \\ -5, 3 & -5 + 3 = -2 \end{array}$$

The integers we need are -5 and 3. The factored trinomial is

$$p^2 - 2p - 15 = (p - 5)(p + 3).$$

Work Quick Check 2 at the side.

Example 3 To try to factor $x^2 + 5x + 12$, we first list all pairs of integers whose product is 12.

$$\begin{array}{rr} 12, 1 & 12 + 1 = 13 \\ 6, 2 & 6 + 2 = 8 \\ 3, 4 & 3 + 4 = 7 \\ -12, -1 & -12 + (-1) = -13 \\ -6, -2 & -6 + (-2) = -8 \\ -3, -4 & -3 + (-4) = -7 \end{array}$$

None of the pairs of integers has a sum of 5. Because of this, the trinomial $x^2 + 5x + 12$ *cannot be factored.*

Example 4 There are no integers whose product is 11 and whose sum is -8, so that $k^2 - 8k + 11$ cannot be factored.

Work Quick Check 3 at the side.

Example 5 Factor $4x^5 - 28x^4 + 40x^3$.

First, factor out the greatest common factor, $4x^3$.

$$4x^5 - 28x^4 + 40x^3 = 4x^3(x^2 - 7x + 10)$$

Now factor $x^2 - 7x + 10$. The integers -5 and -2 have a product of 10 and a sum of -7. The complete factored form is

$$4x^5 - 28x^4 + 40x^3 = 4x^3(x - 5)(x - 2).$$

Work Quick Check 4 at the side.

Example 6 To factor $z^2 - 2bz - 3b^2$, look for two numbers whose product is $-3b^2$ and whose sum is $-2b$. The numbers we need are $-3b$ and b. Thus

$$z^2 - 2bz - 3b^2 = (z - 3b)(z + b).$$

Work Quick Check 5 at the side.

4. Factor each trinomial as completely as possible.

(a) $2p^3 + 6p^2 - 8p$

(b) $3x^4 - 15x^3 + 18x^2$

(c) $4m^7 + 28m^6 + 48m^5$

5. Factor each trinomial.

(a) $b^2 - 3ab - 4a^2$

(b) $p^2 + 6pq + 5q^2$

(c) $r^2 - 6rs + 8s^2$

Quick Check Answers

4. (a) $2p(p + 4)(p - 1)$
(b) $3x^2(x - 3)(x - 2)$
(c) $4m^5(m + 4)(m + 3)$

5. (a) $(b - 4a)(b + a)$
(b) $(p + 5q)(p + q)$
(c) $(r - 4s)(r - 2s)$

WORK SPACE

name date hour

5.2 EXERCISES

Complete the factoring.

1. $x^2 + 10x + 21 = (x + 7)(\qquad)$
2. $p^2 + 11p + 30 = (p + 5)(\qquad)$
3. $r^2 + 15r + 56 = (r + 7)(\qquad)$
4. $x^2 + 15x + 44 = (x + 4)(\qquad)$
5. $t^2 - 14t + 24 = (t - 2)(\qquad)$
6. $x^2 - 9x + 8 = (x - 1)(\qquad)$
7. $x^2 - 12x + 32 = (x - 4)(\qquad)$
8. $y^2 - 2y - 15 = (y + 3)(\qquad)$
9. $m^2 + 2m - 24 = (m - 4)(\qquad)$
10. $x^2 + 9x - 22 = (x - 2)(\qquad)$
11. $p^2 + 7p - 8 = (p + 8)(\qquad)$
12. $y^2 - 7y - 18 = (y + 2)(\qquad)$
13. $x^2 - 7x - 30 = (x - 10)(\qquad)$
14. $k^2 - 3k - 28 = (k - 7)(\qquad)$

Factor as completely as possible. If a polynomial cannot be factored, write "cannot be factored." See Examples 1-6.

15. $x^2 + 6x + 5$ ____________
16. $y^2 + y - 72$ ____________
17. $a^2 + 9a + 20$ ____________
18. $b^2 + 8b + 15$ ____________
19. $x^2 - 8x + 7$ ____________
20. $m^2 + m - 20$ ____________
21. $p^2 + 4p + 5$ ____________
22. $n^2 - 4n - 12$ ____________
23. $y^2 - 6y + 8$ ____________
24. $r^2 - 11r + 30$ ____________
25. $s^2 + 2s - 35$ ____________
26. $h^2 + 11h + 12$ ____________
27. $n^2 - 12n - 35$ ____________
28. $a^2 - 2a - 99$ ____________
29. $b^2 - 11b + 24$ ____________
30. $x^2 - 9x + 20$ ____________

31. $y^2 - 4y - 21$ ______________________

32. $z^2 - 14z + 49$ ______________________

33. $y^2 - 12y + 8$ ______________________

34. $r^2 + r - 42$ ______________________

35. $z^2 - 3z - 40$ ______________________

36. $p^2 + 5p - 66$ ______________________

37. $3m^3 + 12m^2 + 9m$ ______________________

38. $3y^5 - 18y^4 + 15y^3$ ______________________

39. $6a^2 - 48a - 120$ ______________________

40. $h^7 - 5h^6 - 14h^5$ ______________________

41. $3j^3 - 30j^2 + 72j$ ______________________

42. $2x^6 - 8x^5 - 42x^4$ ______________________

43. $3x^4 - 3x^3 - 90x^2$ ______________________

44. $2y^3 - 8y^2 - 10y$ ______________________

45. $x^2 + 4ax + 3a^2$ ______________________

46. $x^2 - mx - 6m^2$ ______________________

47. $y^2 - by - 30b^2$ ______________________

48. $z^2 + 2zx - 15x^2$ ______________________

49. $x^2 + xy - 30y^2$ ______________________

50. $a^2 - ay - 56y^2$ ______________________

OBJECTIVES

- Factor trinomials not having 1 as the coefficient of the squared term

In this section we factor trinomials where the coefficient of the squared term is *not* 1, such as

$$2x^2 + 7x + 6.$$

There is no shortcut way to factor this trinomial—the only method that we can use is the method of trial and error.

The possible factors of $2x^2$ are $2x$ and x, or $-2x$ and $-x$. We normally use only positive coefficients for the first factors. If we use $2x$ and x, we can start the factoring of $2x^2 + 7x + 6$ by writing

$$2x^2 + 7x + 6 = (2x + \quad)(x + \quad).$$

The last term of $2x^2 + 7x + 6$ is 6. There are several possible factors of 6, including 6 and 1. We don't know if these are the correct factors or not, so we try them.

$$(2x + 1)(x + 6) = 2x^2 + 13x + 6$$
$$\neq 2x^2 + 7x + 6 \quad \text{Wrong}$$

We can exchange the 6 and 1.

$$(2x + 6)(x + 1) = 2x^2 + 8x + 6$$
$$\neq 2x^2 + 7x + 6 \quad \text{Wrong}$$

No matter how we use the 6 and 1, we can't get the correct answer. Let us try 2 and 3, which are also factors of 6.

$$(2x + 2)(x + 3) = 2x^2 + 8x + 6$$
$$\neq 2x^2 + 7x + 6 \quad \text{Wrong}$$

We could have noticed earlier that $(2x + 2)(x + 3)$ couldn't be correct. This is because $2x + 2$ has a common factor, and the given trinomial $2x^2 + 7x + 6$ does not. *If the original polynomial has no common factor, none of its factors can have one either.*

Let us exchange the 2 and 3.

$$(2x + 3)(x + 2) = 2x^2 + 7x + 6 \quad \text{Correct}$$

We finally found the answer—the trinomial $2x^2 + 7x + 6$ can be factored as $(2x + 3)(x + 2)$.

Example 1 Factor $3y^2 + 14y + 8$.

The only positive factors of $3y^2$ are $3y$ and y, so we start with

$$(3y + \quad)(y + \quad).$$

Now use trial and error. The factors of 8 are 8 and 1, or 4 and 2. Try 4 and 2.

$$(3y + 4)(y + 2) = 3y^2 + 10y + 8 \quad \text{Wrong}$$

Try 2 and 4.

$$(3y + 2)(y + 4) = 3y^2 + 14y + 8 \quad \text{Correct}$$

Work Quick Check 1 at the side.

QUICK CHECKS

1. Use trial and error to factor each trinomial.

(a) $2m^2 + 7m + 3$

(b) $5p^2 + 16p + 3$

(c) $3p^2 + 4p + 1$

(d) $3r^2 + 13r + 4$

Quick Check Answers

1. (a) $(2m + 1)(m + 3)$
(b) $(5p + 1)(p + 3)$
(c) $(3p + 1)(p + 1)$
(d) $(3r + 1)(r + 4)$

2. Factor each trinomial.

(a) $6x^2 + 5x - 4$

(b) $3x^2 - 7x - 6$

(c) $5p^2 + 13p - 6$

(d) $6m^2 - 11m - 10$

Example 2 Factor $8p^2 + 14p + 5$.

The number 8 has several possible pairs of factors, while 5 has only 1 and 5 or −1 and −5. For this reason, it is easiest to begin by considering the factor 5. We can ignore the negative factors since all terms of the trinomial have positive coefficients. Thus, if $8p^2 + 14p + 5$ can be factored, it will have to be factored as

$$(\quad + 5)(\quad + 1).$$

Now we must use trial and error to decide on the factors of $8p^2$. The possibilities are $8p$ and p, or $4p$ and $2p$.

$$(8p + 5)(p + 1) = 8p^2 + 13p + 5 \quad \text{Wrong}$$
$$(p + 5)(8p + 1) = 8p^2 + 41p + 5 \quad \text{Wrong}$$
$$(4p + 5)(2p + 1) = 8p^2 + 14p + 5 \quad \text{Correct}$$

By trial and error, we have found that $8p^2 + 14p + 5$ factors as

$$(4p + 5)(2p + 1).$$

Example 3 Factor $6x^2 - 11x + 3$.

There are several possible factors for 6, while 3 has only 1 and 3 or −1 and −3. Thus, we begin by factoring 3. Since the middle term of $6x^2 - 11x + 3$ has a negative coefficient, we have to consider negative factors. Let's try −3 and −1 as factors of 3, so that our first step is

$$(\quad - 3)(\quad - 1).$$

The factors of $6x^2$ are $6x$ and x, or $3x$ and $2x$. Begin the process of trial and error.

$$(2x - 3)(3x - 1) = 6x^2 - 11x + 3 \quad \text{Correct}$$

We got our answer on the first attempt.

Example 4 Factor $8x^2 + 6x - 9$.

The integer 8 has several possible factors, as does −9. Since the coefficient of the middle term is small, it probably would be wise to avoid large factors such as 8 or 9. Let us begin by trying 4 and 2 as factors of 8, and 3 and −3 as factors of −9.

$$(4x + 3)(2x - 3) = 8x^2 - 6x - 9 \quad \text{Wrong}$$
$$(4x - 3)(2x + 3) = 8x^2 + 6x - 9 \quad \text{Correct}$$

Work Quick Check 2 at the side.

Example 5 Factor $12a^2 - ab - 20b^2$.

There are several possible factors of $12a^2$, including $12a$ and a, $6a$ and $2a$, and $3a$ and $4a$, just as there are many possible factors of $-20b^2$, including $-20b$ and b, $10b$ and $-2b$, $-10b$ and $2b$, $4b$ and $-5b$, and $-4b$ and $5b$. Once again, since our desired middle term is small we avoid the larger factors. Let us try as factors $6a$ and $2a$ and $4b$ and $-5b$.

Quick Check Answers

2. (a) $(3x + 4)(2x - 1)$
(b) $(3x + 2)(x - 3)$
(c) $(5p - 2)(p + 3)$
(d) $(2m - 5)(3m + 2)$

$$(6a + 4b)(2a - 5b)$$

This cannot be correct, as we mentioned before, since $6a + 4b$ has a common factor, while the trinomial itself has none. For the same reason, $(6a - 5b)(2a + 4b)$ cannot be correct.

Let's try $3a$ and $4a$ with $4b$ and $-5b$.

$$(3a + 4b)(4a - 5b) = 12a^2 + ab - 20b^2 \quad \text{Wrong}$$

Here the middle term has the wrong sign. We try $3a$ and $4a$ with $-4b$ and $5b$.

$$(3a - 4b)(4a + 5b) = 12a^2 - ab - 20b^2 \quad \text{Correct}$$

Work Quick Check 3 at the side.

Example 6 Factor $28x^5 - 58x^4 - 30x^3$.

First factor out $2x^3$, the greatest common factor.

$$28x^5 - 58x^4 - 30x^3 = 2x^3(14x^2 - 29x - 15)$$

We must use trial and error to factor $14x^2 - 29x - 15$. Let's try $7x$ and $2x$ as factors of $14x^2$, and -3 and 5 as factors of -15.

$$(7x - 3)(2x + 5) = 14x^2 + 29x - 15 \quad \text{Wrong}$$

The middle term differs only in sign, and so we try

$$(7x + 3)(2x - 5) = 14x^2 - 29x - 15. \quad \text{Correct}$$

The factored form of $28x^5 - 58x^4 - 30x^3$ is

$$2x^3(7x + 3)(2x - 5).$$

Work Quick Check 4 at the side.

3. Factor each trinomial.

(a) $2x^2 - 5xy - 3y^2$

(b) $8a^2 + 2ab - 3b^2$

(c) $3r^2 + 8rs + 5s^2$

(d) $6m^2 + 11mn - 10n^2$

4. Factor each polynomial as completely as possible.

(a) $4x^2 - 2x - 30$

(b) $15y^3 + 55y^2 + 30y$

(c) $18p^4 + 63p^3 + 27p^2$

(d) $16m^5 - 8m^4 - 168m^3$

Quick Check Answers

3. (a) $(2x + y)(x - 3y)$
(b) $(4a + 3b)(2a - b)$
(c) $(3r + 5s)(r + s)$
(d) $(3m - 2n)(2m + 5n)$

4. (a) $2(2x + 5)(x - 3)$
(b) $5y(3y + 2)(y + 3)$
(c) $9p^2(2p + 1)(p + 3)$
(d) $8m^3(2m - 7)(m + 3)$

WORK SPACE

name date hour

5.3 EXERCISES

Complete the factoring.

1. $2x^2 - x - 1 = (2x + 1)(\qquad)$
2. $3a^2 + 5a + 2 = (3a + 2)(\qquad)$
3. $5b^2 - 16b + 3 = (5b - 1)(\qquad)$
4. $2x^2 + 11x + 12 = (2x + 3)(\qquad)$
5. $4y^2 + 17y - 15 = (y + 5)(\qquad)$
6. $7z^2 + 10z - 8 = (z + 2)(\qquad)$
7. $15x^2 + 7x - 4 = (3x - 1)(\qquad)$
8. $12c^2 - 7c - 12 = (4c + 3)(\qquad)$
9. $2m^2 + 19m - 10 = (2m - 1)(\qquad)$
10. $6x^2 + x - 12 = (2x + 3)(\qquad)$
11. $6a^2 + 7ab - 20b^2 = (2a + 5b)(\qquad)$
12. $9m^2 - 3mn - 2n^2 = (3m - 2n)(\qquad)$
13. $4k^2 + 13km + 3m^2 = (4k + m)(\qquad)$
14. $6x^2 - 13xy - 5y^2 = (3x + y)(\qquad)$
15. $4x^3 - 10x^2 - 6x = 2x(\qquad) = 2x(2x + 1)(\qquad)$
16. $15r^3 - 39r^2 - 18r = 3r(\qquad) = 3r(5r + 2)(\qquad)$
17. $6m^6 + 7m^5 - 20m^4 = m^4(\qquad) = m^4(3m - 4)(\qquad)$
18. $16y^5 - 4y^4 - 6y^3 = 2y^3(\qquad) = 2y^3(4y - 3)(\qquad)$

Factor as completely as possible. See Examples 1-6.

19. $2x^2 + 7x + 3$ __________
20. $3y^2 + 13y + 4$ __________
21. $3a^2 + 10a + 7$ __________
22. $7r^2 + 8r + 1$ __________
23. $4r^2 + r - 3$ __________
24. $3p^2 + 2p - 8$ __________
25. $15m^2 + m - 2$ __________
26. $6x^2 + x - 1$ __________

27. $8m^2 - 10m - 3$ ______

28. $2a^2 - 17a + 30$ ______

29. $5a^2 - 7a - 6$ ______

30. $12s^2 + 11s - 5$ ______

31. $3r^2 + r - 10$ ______

32. $20x^2 - 28x - 3$ ______

33. $4y^2 + 69y + 17$ ______

34. $21m^2 + 13m + 2$ ______

35. $38x^2 + 23x + 2$ ______

36. $20y^2 + 39y - 11$ ______

37. $10x^2 + 11x - 6$ ______

38. $6b^2 + 7b + 2$ ______

39. $6w^2 + 19w + 10$ ______

40. $20q^2 - 41q + 20$ ______

41. $6q^2 + 23q + 21$ ______

42. $8x^2 + 47x - 6$ ______

43. $10m^2 - 23m + 12$ ______

44. $4t^2 - 5t - 6$ ______

45. $8k^2 + 2k - 15$ ______

46. $15p^2 - p - 6$ ______

47. $10m^2 - m - 24$ ______

48. $16a^2 + 30a + 9$ ______

49. $8x^2 - 14x + 3$ ______

50. $24b^2 - 37b - 5$ ______

51. $40m^2 + m - 6$ ______

52. $15a^2 + 22a + 8$ ______

53. $2m^3 + 2m^2 - 40m$ ______

54. $15n^4 - 39n^3 + 18n^2$ ______

55. $24a^4 + 10a^3 - 4a^2$ ______

56. $18x^5 + 15x^4 - 75x^3$ ______

57. $32z^5 - 20z^4 - 12z^3$ ______

58. $15x^2y^2 - 7xy^2 - 4y^2$ ______

59. $12p^2 + 7pq - 12q^2$ ______

60. $6m^2 - 5mn - 6n^2$ ______

61. $25a^2 + 25ab + 6b^2$ ______

62. $6x^2 - 5xy - y^2$ ______

63. $6a^2 - 7ab - 5b^2$ ______

64. $25g^2 - 5gh - 2h^2$ ______

OBJECTIVES

- Factor the difference of two squares
- Factor a perfect square trinomial

In this section we look at methods for factoring two special types of polynomials. First, we factor a binomial which is the **difference of two squares**, such as

$$x^2 - 25 = x^2 - 5^2, \quad \text{or} \quad 9y^2 - 64 = (3y)^2 - 8^2.$$

To factor $x^2 - 25$, rewrite it as $x^2 + 0x - 25$. We can factor this trinomial by finding two integers whose product is -25 and whose sum is 0. The integers we need are 5 and -5, so that $x^2 - 25$ factors as

$$x^2 - 25 = (x + 5)(x - 5).$$

A binomial which is the difference of two squares always factors this way, as shown in Examples 1–3.

Example 1

(a) $x^2 - 49 = (x + 7)(x - 7)$

(b) $z^2 - 4 = (z + 2)(z - 2)$

(c) $y^2 - m^2 = (y + m)(y - m)$

(d) $p^2 + 16$ cannot be factored

Example 2 Factor $25m^2 - 16$.

This is the difference of two squares, since

$$25m^2 - 16 = (5m)^2 - 4^2.$$

Factor this as

$$(5m + 4)(5m - 4).$$

Example 3

(a) $49z^2 - 64 = (7z)^2 - 8^2 = (7z + 8)(7z - 8)$

(b) $9a^2 - 4b^2 = (3a)^2 - (2b)^2 = (3a + 2b)(3a - 2b)$

(c) $81y^2 - 36 = 9(9y^2 - 4) = 9(3y + 2)(3y - 2)$

(d) $m^4 - 16 = (m^2)^2 - 4^2$
$= (m^2 + 4)(m^2 - 4)$
$= (m^2 + 4)(m + 2)(m - 2)$

Work Quick Check 1 at the side.

QUICK CHECKS

1. Factor.

(a) $p^2 - 100$

(b) $9m^2 - 49$

(c) $64a^2 - 25$

(d) $50r^2 - 32$

(e) $27y^2 - 75$

(f) $k^4 - 49$

(g) $9r^4 - 100$

A quantity is a perfect square if it can be factored as the square of another quantity. Thus, 144, $4x^2$, and $81m^6$ are all perfect squares, since

$$144 = 12^2, \quad 4x^2 = (2x)^2, \quad 81m^6 = (9m^3)^2.$$

A **perfect square trinomial** is a trinomial that is the square of a binomial. For example,

$$x^2 + 8x + 16 = (x + 4)^2.$$

Quick Check Answers

1. (a) $(p + 10)(p - 10)$
(b) $(3m + 7)(3m - 7)$
(c) $(8a + 5)(8a - 5)$
(d) $2(5r + 4)(5r - 4)$
(e) $3(3y + 5)(3y - 5)$
(f) $(k^2 + 7)(k^2 - 7)$
(g) $(3r^2 + 10)(3r^2 - 10)$

2. Factor each trinomial which is a perfect square.

(a) $y^2 + 6y + 9$

(b) $p^2 - 8p + 16$

(c) $r^2 + 12r + 36$

(d) $4m^2 + 20m + 25$

(e) $16a^2 + 56a + 49$

(f) $121p^2 + 110p + 100$

For a trinomial to be a perfect square, two of its terms must be perfect squares. Thus $16x^2 + 4x + 8$ is not a perfect square trinomial since only the term $16x^2$ is a perfect square.

On the other hand, just because two of the terms are perfect squares, we cannot be sure that the trinomial is a perfect square trinomial. For example, $x^2 + 6x + 36$ has two perfect square terms, but it is not a perfect square trinomial. (Try to find a binomial that can be squared to give $x^2 + 6x + 36$.)

In general, the square of a binomial is of the form

$$(a + b)^2 = a^2 + 2ab + b^2.$$

The middle term of a perfect square trinomial is always twice the product of the two terms in the squared binomial. We can use this fact to check any attempt to factor a trinomial that appears to be a perfect square.

Factor $x^2 + 10x + 25$. The term x^2 is a perfect square, and so is 25. We can try to factor the trinomial as

$$x^2 + 10x + 25 = (x + 5)^2.$$

To check, take twice the product of the two terms in the squared binomial.

$$2 \cdot x \cdot 5 = 10x$$

Since $10x$ is the middle term of the trinomial, the trinomial is a perfect square, and can be factored as $(x + 5)^2$.

Example 4 Factor each perfect square trinomial.

(a) $x^2 - 22x + 121$

The first and last terms are perfect squares $(121 = 11^2)$. We need to check and see if the middle term of $x^2 - 22x + 121$ is twice the product of the two terms of the binomial.

$$2 \cdot x \cdot 11 = 22x$$

first term of binomial ↑ (under x); second term of binomial ↑ (under 11)

Since this is true, $x^2 - 22x + 121$ is a perfect square trinomial, and

$$x^2 - 22x + 121 = (x - 11)^2.$$

The middle sign in the binomial, a minus sign in this case, is always the same as the middle sign in the trinomial.

(b) $9m^2 - 24m + 16 = (3m - 4)^2$

(c) $16r^2 + 8r + 1 = (4r + 1)^2$

(d) $25y^2 + 20y + 16$

The first and last terms are perfect squares.

$$25y^2 = (5y)^2 \quad \text{and} \quad 16 = 4^2$$

Twice the product of the first and last terms is

$$2 \cdot 5y \cdot 4 = 40y,$$

which is not the middle term of $25y^2 + 20y + 16$. This polynomial is not a perfect square.

Work Quick Check 2 at the side.

Quick Check Answers

2. (a) $(y + 3)^2$ (b) $(p - 4)^2$ (c) $(r + 6)^2$ (d) $(2m + 5)^2$ (e) $(4a + 7)^2$ (f) not a perfect square trinomial

name date hour

5.4 EXERCISES

Factor each binomial completely. The table of squares and square roots in the back of the book may be helpful. See Examples 1-3.

1. $x^2 - 16$ __________

2. $m^2 - 25$ __________

3. $p^2 - 4$ __________

4. $r^2 - 9$ __________

5. $m^2 - n^2$ __________

6. $p^2 - q^2$ __________

7. $a^2 - b^2$ __________

8. $r^2 - t^2$ __________

9. $9m^2 - 1$ __________

10. $16y^2 - 9$ __________

11. $25m^2 - 16$ __________

12. $144y^2 - 25$ __________

13. $36t^2 - 16$ __________

14. $9 - 36a^2$ __________

15. $25a^2 - 16r^2$ __________

16. $100k^2 - 49m^2$ __________

17. $x^2 + 16$ __________

18. $m^2 + 100$ __________

19. $p^4 - 36$ __________

20. $r^4 - 9$ __________

21. $a^4 - 1$ __________

22. $x^4 - 16$ __________

23. $m^4 - 81$ __________

24. $p^4 - 256$ __________

Factor any expressions that are perfect square trinomials. See Example 4.

25. $a^2 + 4a + 4$ __________

26. $p^2 + 2p + 1$ __________

27. $x^2 - 10x + 25$ __________

28. $y^2 - 8y + 16$ __________

29. $a^2 + 14a + 49$ __________

30. $m^2 - 20m + 100$ __________

31. $k^2 + 22k + 121$ ____________________

32. $r^2 + 24r + 144$ ____________________

33. $y^2 - 10y + 100$ ____________________

34. $z^2 + 7z + 49$ ____________________

35. $9y^2 + 14y + 25$ ____________________

36. $16m^2 + 42m + 49$ ____________________

37. $16a^2 - 40ab + 25b^2$ ____________________

38. $36y^2 - 60yp + 25p^2$ ____________________

39. $100m^2 + 100m + 25$ ____________________

40. $100a^2 - 140ab + 49b^2$ ____________________

41. $49x^2 + 28xy + 4y^2$ ____________________

42. $64y^2 - 48ya + 9a^2$ ____________________

43. $4c^2 + 12cd + 9d^2$ ____________________

44. $16t^2 - 40tr + 25r^2$ ____________________

45. $25h^2 - 20hy + 4y^2$ ____________________

46. $9x^2 + 24xy + 16y^2$ ____________________

OBJECTIVES

- Solve quadratic equations by factoring
- Solve other equations by factoring

In this section we study **quadratic equations,** which are equations that contain a squared term and no terms of a higher degree. Examples of quadratic equations include

$$x^2 + 5x + 6 = 0, \qquad 2a^2 - 5a = 3, \qquad \text{and } y^2 = 4.$$

Some quadratic equations can be solved by factoring. A more general method for those equations that cannot be solved by factoring is given in Chapter 10.

To solve a quadratic equation by factoring, we use the **zero-factor** property:

> If a and b represent real numbers, and if $ab = 0$, then $a = 0$ or $b = 0$.

In other words, if the product of two numbers is zero, then at least one of the numbers must be zero.

QUICK CHECKS

1. Solve each equation.

(a) $(x - 5)(x + 2) = 0$

Example 1 Solve the equation $(x + 3)(2x - 1) = 0$.

Here we are told that the product $(x + 3)(2x - 1)$ is equal to zero. By the zero-factor property, the only way that the product of these two factors can be zero is if at least one of the factors is zero. Therefore, either $x + 3$ or $2x - 1$ is 0. Solve each of these two equations.

$$x + 3 = 0 \qquad\qquad 2x - 1 = 0$$
$$x = -3 \qquad\qquad 2x = 1$$
$$x = \frac{1}{2}$$

The given equation $(x + 3)(2x - 1) = 0$ has two solutions, $x = -3$ and $x = 1/2$. To check these answers, substitute -3 for x in the original equation, and then substitute $1/2$ for x.

(b) $(3x - 2)(x + 6) = 0$

(c) $(5x + 7)(2x + 3) = 0$

Work Quick Check 1 at the side.

In Example 1, the equation that we were to solve was presented in factored form. If the equation is not already factored, we must factor it ourselves before attempting to solve it.

(d) $(2x + 9)(3x - 5) = 0$

Example 2 Solve the equation $x^2 - 5x = -6$.

First, rewrite the equation so that it is equal to 0. In this case, add 6 to both sides. This gives

$$x^2 - 5x + 6 = 0.$$

Now factor $x^2 - 5x + 6$. We need two numbers whose product is 6 and whose sum is -5. These two numbers are -2 and -3. This gives

$$x^2 - 5x + 6 = (x - 2)(x - 3).$$

Quick Check Answers
1. (a) 5, −2 (b) 2/3, −6 (c) −7/5, −3/2 (d) −9/2, 5/3

2. Solve each equation.

(a) $m^2 - 3m - 10 = 0$

(b) $a^2 + 6a + 8 = 0$

(c) $y^2 - y = 6$

(d) $r^2 = 8 - 2r$

3. Solve each equation.

(a) $2a^2 - a - 3 = 0$

(b) $3r^2 + 10r - 8 = 0$

(c) $3x^2 = 11x + 4$

(d) $12m^2 - 17m = 5$

Quick Check Answers

2. (a) 5, −2 (b) −4, −2 (c) 3, −2 (d) −4, 2
3. (a) 3/2, −1 (b) 2/3, −4 (c) −1/3, 4 (d) −1/4, 5/3

The original equation is now

$$(x - 2)(x - 3) = 0.$$

Proceed as in Example 1. Make each factor equal to 0.

$$x - 2 = 0 \quad \text{or} \quad x - 3 = 0$$

To solve the equation on the left, add 2 to both sides. On the right, add 3 to both sides. Doing this, we get the solutions

$$x = 2 \quad \text{or} \quad x = 3.$$

Check by substituting 2 and 3 for x in the original equation.

Example 3 Solve the equation $y^2 = y + 20$.

We need 0 alone on one side of the equals sign. We can get 0 alone if we add $-y$ and -20 to both sides of the equals sign.

$$y^2 - y - 20 = 0$$

Factor $y^2 - y - 20$.

$$(y - 5)(y + 4) = 0$$

This product of two factors can equal 0 only if at least one of the factors is 0. This gives two equations,

$$y - 5 = 0 \quad \text{or} \quad y + 4 = 0.$$

Solve each of these two equations to get the final solutions.

$$y = 5 \quad \text{or} \quad y = -4$$

Work Quick Check 2 at the side.

Example 4 Solve the equation $2p^2 - 13p + 20 = 0$.

Factor $2p^2 - 13p + 20$ by the method of trial and error.

$$(2p - 5)(p - 4) = 0$$

Make each of these two factors equal to 0.

$$2p - 5 = 0 \quad \text{or} \quad p - 4 = 0$$

To solve the equation on the left, first add 5 to both sides of the equation. Then multiply both sides by 1/2. To solve the equation on the right, add 4 to both sides.

$$2p = 5 \qquad p = 4$$

$$p = \frac{5}{2}$$

The solutions of $2p^2 - 13p + 20 = 0$ are 5/2 and 4.

Work Quick Check 3 at the side.

We can also use the zero-factor property to solve equations which have more than two factors, as shown in Example 5.

Example 5 Solve the equation $6z^3 - 6z = 0$.

First, factor out the greatest common factor in $6z^3 - 6z$.

$$6z^3 - 6z = 6z(z^2 - 1)$$

Now factor $z^2 - 1$ as $(z + 1)(z - 1)$. Doing this, we get the final result,

$$6z^3 - 6z = 6z(z + 1)(z - 1).$$

The original equation can now be written as

$$6z(z + 1)(z - 1) = 0.$$

This product can equal 0 only if at least one of the factors is 0. This means that we can now write three equations, one for each factor.

$$6z = 0 \quad \text{or} \quad z + 1 = 0 \quad \text{or} \quad z - 1 = 0$$

After solving all three of these equations, we get the final solutions.

$$z = 0 \quad \text{or} \quad z = -1 \quad \text{or} \quad z = 1$$

Example 6 Solve the equation $(2x - 1)(x^2 - 9x + 20) = 0$.

Factor $x^2 - 9x + 20$ as $(x - 5)(x - 4)$. Then we can rewrite the original equation as

$$(2x - 1)(x - 5)(x - 4) = 0.$$

Make each of these three factors equal to 0.

$$2x - 1 = 0 \quad \text{or} \quad x - 5 = 0 \quad \text{or} \quad x - 4 = 0$$

After solving all three of these equations, we end up with

$$x = 1/2 \quad \text{or} \quad x = 5 \quad \text{or} \quad x = 4$$

as the solutions of the original equation.

Work Quick Check 4 at the side.

4. Solve each equation.

(a) $2a(a - 1)(a + 3) = 0$

(b) $r^3 - 16r = 0$

(c) $(m + 3)(m^2 - 11m + 10) = 0$

(d) $(2x + 1)(2x^2 + 7x - 15) = 0$

Quick Check Answers

4. (a) 0, 1, −3 (b) 0, 4, −4
(c) −3, 1, 10 (d) −1/2, 3/2, −5

WORK SPACE

name date hour

5.5 EXERCISES

Solve each equation. See Example 1.

1. $(x-2)(x+4)=0$ ________

2. $(y-3)(y+5)=0$ ________

3. $(3x+5)(2x-1)=0$ ________

4. $(2a+3)(a-2)=0$ ________

5. $(5p+1)(2p-1)=0$ ________

6. $(3k-8)(k+7)=0$ ________

7. $(2m+9)(3m-1)=0$ ________

8. $(9a-2)(3a+1)=0$ ________

9. $(x-1)(3x+5)=0$ ________

10. $(k-3)(k+5)=0$ ________

11. $(3r-7)(2r+8)=0$ ________

12. $(5a+2)(3a-1)=0$ ________

Solve each equation. See Examples 2 and 3.

13. $x^2+5x+6=0$ ________

14. $y^2-3y+2=0$ ________

15. $r^2-5r-6=0$ ________

16. $y^2-y-12=0$ ________

17. $m^2+3m-28=0$ ________

18. $p^2-p-6=0$ ________

19. $a^2=24-5a$ ________

20. $r^2=2r+15$ ________

21. $x^2=3+2x$ ________

22. $m^2=3m+4$ ________

23. $z^2=-2-3z$ ________

24. $p^2=2p+3$ ________

25. $m^2+8m+16=0$ ________

26. $b^2-6b+9=0$ ________

Solve each equation. See Example 4.

27. $3a^2+5a-2=0$ ________

28. $6r^2-r-2=0$ ________

29. $2k^2-k-10=0$ ________

30. $6x^2-7x-5=0$ ________

31. $6p^2 = 4 - 5p$ __________

32. $6x^2 - 5x = 4$ __________

33. $6a^2 = 5 - 13a$ __________

34. $9s^2 + 12s = -4$ __________

35. $3a^2 + 7a = 20$ __________

36. $6z^2 + 11z + 3 = 0$ __________

37. $15r^2 = r + 2$ __________

38. $3m^2 = 5m + 28$ __________

39. $2b^2 + 3b - 9 = 0$ __________

40. $5b^2 = 8b + 4$ __________

41. $16r^2 - 25 = 0$ __________

42. $4k^2 - 9 = 0$ __________

43. $9m^2 - 36 = 0$ __________

44. $16x^2 - 64 = 0$ __________

Solve each equation. See Examples 5 and 6.

45. $(2r - 5)(3r^2 - 16r + 5) = 0$ __________

46. $(3m - 4)(6m^2 + m - 2) = 0$ __________

47. $x^3 - 25x = 0$ __________

48. $m^3 - 4m = 0$ __________

49. $9y^3 - 49y = 0$ __________

50. $16r^3 - 9r = 0$ __________

5.6 APPLICATIONS OF QUADRATIC EQUATIONS

OBJECTIVES

- Convert word problems to quadratic equations and solve them

In this section we look at problems whose solutions involve quadratic equations.

Example 1 The width of a rectangle is 4 centimeters less than the length. The area is 96 square centimeters. Find the length and width.

We can use x to represent the length of the rectangle. Then, according to the statement of the problem, we can write the width as $x - 4$ (the width is less than the length.) (See Figure 1.) The area of a rectangle is given by the formula

$$\text{area} = lw = \text{length} \times \text{width}.$$

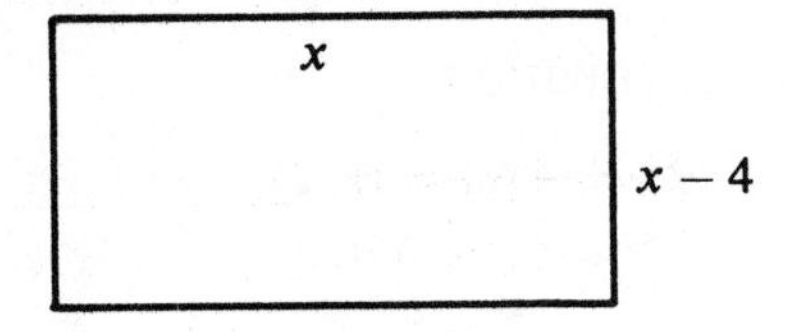

Figure 1

In our problem, the area is 96, the length is x, and the width is $x - 4$. We substitute these values into the formula.

$$96 = x(x - 4)$$
$$96 = x^2 - 4x$$

This is the quadratic equation

$$x^2 - 4x - 96 = 0.$$

Factor.

$$(x - 12)(x + 8) = 0$$

The solutions of the equation are $x = 12$ or $x = -8$.

However, the equation is only an algebraic representation of the area of a rectangle in the physical world. We must always be careful to check solutions against known physical facts. Since a rectangle cannot have a negative length, we discard the solution -8. Then 12 centimeters is the length of the rectangle, and $12 - 4 = 8$ centimeters is the width.

Work Quick Check 1 at the side.

QUICK CHECKS

1. The length of a rectangle is 2 meters more than the width. The area is 48 square meters. Find the length and width of the rectangle.

Example 2 The length of a rectangle is 4 more than the width. The area of the rectangle is numerically 1 more than the perimeter. (See Figure 2.) Find the length and width of the rectangle.

$x + 4$

x

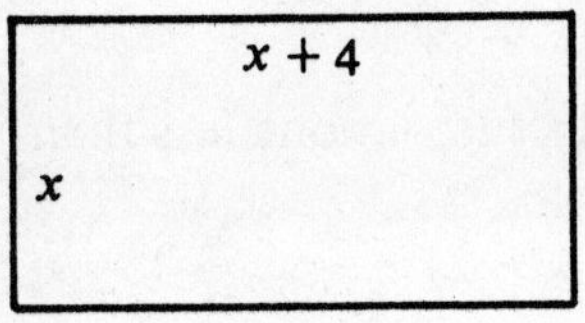

Figure 2

$$\text{area} = x(x + 4)$$

$$\text{perimeter} = 2x + 2(x + 4)$$

Quick Check Answers

1. 8, 6

2. The length of a rectangle is 5 more than the width. The area is numerically 32 more than the perimeter. Find the length and width of the rectangle.

Let x be the width of the rectangle. Then the length is $x + 4$. The area is the product of the length and width or

$$\text{area} = x(x + 4).$$

The perimeter of a rectangle is given by the formula

$$P = 2l + 2w.$$

In this problem, $l = x + 4$ and $w = x$, so that the perimeter is

$$P = 2(x + 4) + 2x. \qquad \text{Let } l = x + 4 \text{ and } w = x$$

According to the problem, the area is numerically 1 more than the perimeter, or

area	is	1	more than	perimeter.
↓	↓	↓	↓	↓
$x(x + 4)$	=	1	+	$2(x + 4) + 2x$

Simplify and solve this equation.

$$x^2 + 4x = 1 + 2x + 8 + 2x$$
$$x^2 + 4x = 9 + 4x$$
$$x^2 = 9 \qquad \text{Add } -4x \text{ to both sides}$$
$$x^2 - 9 = 0 \qquad \text{Add } -9 \text{ to both sides}$$
$$(x + 3)(x - 3) = 0 \qquad \text{Factor}$$

$$x + 3 = 0 \quad \text{or} \quad x - 3 = 0$$
$$x = -3 \quad \text{or} \quad x = 3$$

A rectangle cannot have a negative width, so we ignore $x = -3$. The answer is $x = 3$, so that

$$\text{width} = x = 3; \;\; \text{length} = x + 4 = 3 + 4 = 7.$$

The rectangle is 3 by 7.

Work Quick Check 2 at the side.

Quadratic equations often come up when we work problems about **consecutive integers.** These integers are in a row, such as 8 and 9, or 15 and 16. Consecutive **odd** integers are odd integers in a row, such as 5 and 7, or 19 and 21.

Example 3 The product of two consecutive odd integers is one less than 5 times their sum. Find the integers.

Let's use s to represent the smaller of the two integers. Since the problem mentions consecutive *odd* integers, we use $s + 2$ for the larger of the two integers.

According to the problem, the product is 1 less than 5 times the sum.

product	is	5 times sum	one less than
↓	↓	↓	↓
$s(s + 2)$	=	$5(s + s + 2)$	-1

Quick Check Answers
2. 11, 6

Simplify this equation and solve it.

$$s^2 + 2s = 5s + 5s + 10 - 1$$
$$s^2 + 2s = 10s + 9$$
$$s^2 - 8s - 9 = 0$$
$$(s - 9)(s + 1) = 0$$
$$s - 9 = 0 \quad \text{or} \quad s + 1 = 0$$
$$s = 9 \quad \text{or} \quad s = -1.$$

We need to find two consecutive odd integers.

If $s = 9$ is the first, then $s + 2 = 11$ is the second.

If $s = -1$ is the first, then $s + 2 = 1$ is the second.

There are two pairs of integers satisfying our problem: 9 and 11 or -1 and 1.

Work Quick Check 3 at the side.

3. (a) The product of two consecutive even integers is 4 more than two times their sum. Find the integers.

(b) Find three consecutive integers such that the product of the first two is 2 more than 6 times the third.

Quick Check Answers

3. (a) 4, 6 or $-2, 0$ (b) 7, 8, 9 or $-2, -1, 0$

WORK SPACE

name date hour

5.6 EXERCISES

Solve each problem.

1. The length of a rectangle is 5 centimeters more than the width. The area is 66 square centimeters. Find the length and width of the rectangle. ____________

2. The length of a rectangle is 1 foot more than the width. The area is 56 square feet. Find the length and width of the rectangle. ____________

3. The width of a rectangle is 3 meters less than its length. The area of the rectangle is 70 square meters. Find the dimensions of the rectangle. ____________

4. The width of a rectangle is 7 meters less than its length. The area is 8 square meters. Find the dimensions of the rectangle. ____________

5. The length of a rectangle is twice its width. If the width were increased by 2 inches, while the length remained the same, the resulting rectangle would have an area of 48 square inches. Find the dimensions of the original rectangle. ____________

6. The length of a rectangle is 3 times its width. If the length were decreased by 1, while the width stayed the same, the area of the new figure would be 44 square centimeters. Find the length and width of the original rectangle. ____________

7. The length of a rectangle is 3 more than the width. The area is numerically 4 less than the perimeter. Find the dimensions of the rectangle. ____________

8. The width of a rectangle is 5 less than the length. The area is numerically 10 more than the perimeter. Find the dimensions of the rectangle. ____________

Problems 9 and 10 require the formula for the area of a triangle.

Area $= \frac{1}{2}bh$

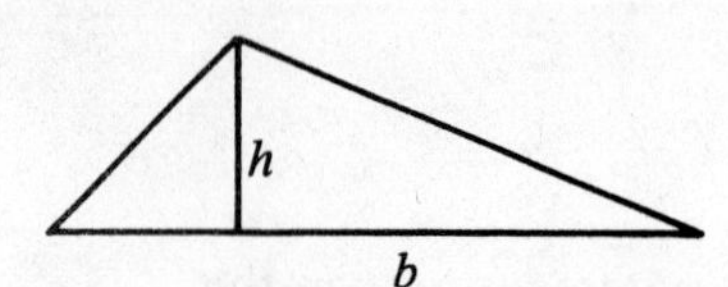

9. The area of a triangle is 25 square centimeters. The base is twice the height. Find the length of the base and the height of the triangle.

10. The height of a triangle is 3 inches more than the base. The area of the triangle is 27 square inches. Find the length of the base and the height of the triangle.

Problems 11 and 12 require the formula for the volume of a pyramid.

$V = \frac{1}{3}Bh$, where B is the area of the base

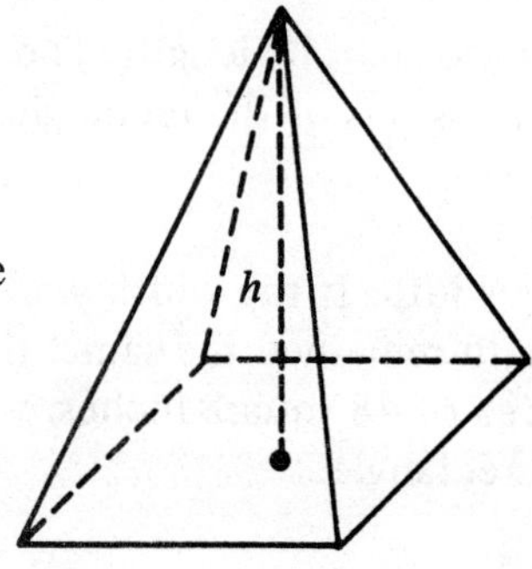

11. The volume of a pyramid is 32 cubic meters. Suppose the numerical value of the height is 10 meters less than the numerical value of the area of the base. Find the height and the area of the base.

12. Suppose a pyramid has a rectangular base whose width is three centimeters less than the length. If the height is 8 and the volume is 144, find the dimensions of the base.

13. One square has sides one foot less than the length of the sides of a second square. If the difference of the areas of the two squares is 37 square feet, find the lengths of the sides of the two squares.

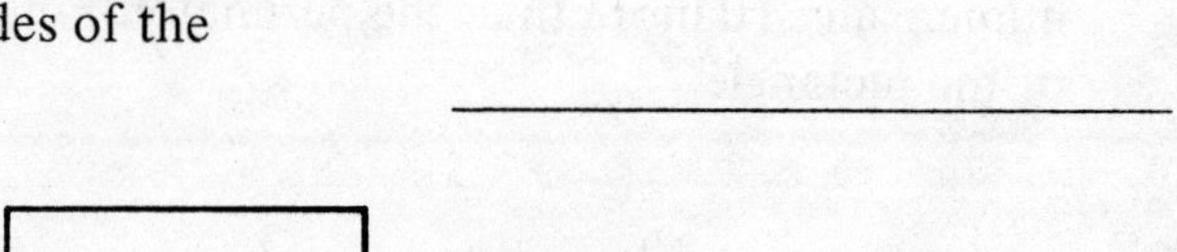

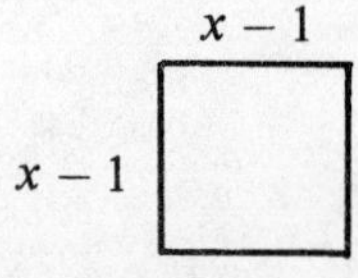

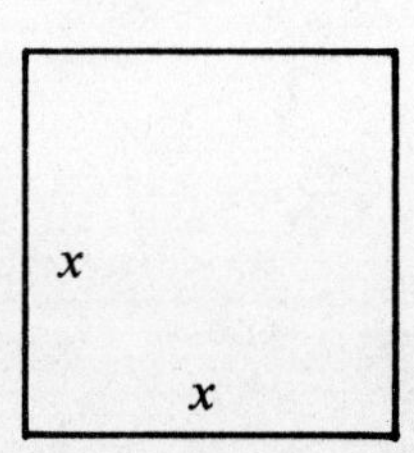

name | date | hour

14. John wishes to build a box to hold his boats. The box is to be 4 feet high, and the width of the box is to be one foot less than the length. The volume of the box will be 120 cubic feet. Find the dimensions of the box. (Hint: The formula for the volume of a box is given in the back of the book.)

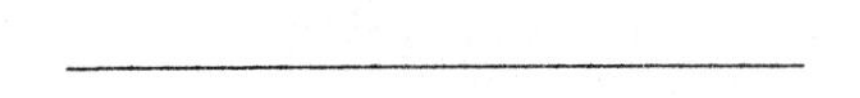

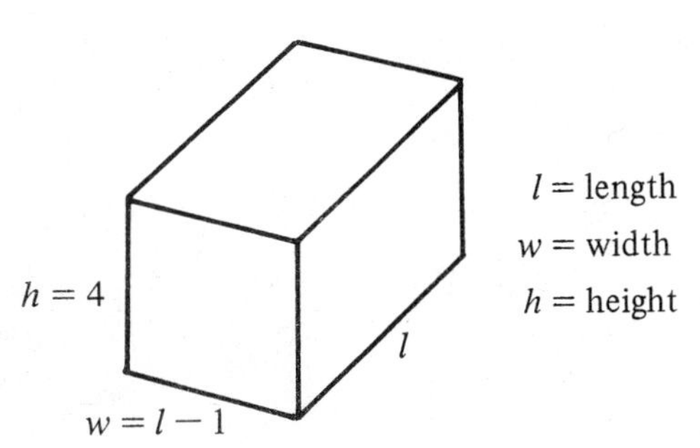

15. The product of two consecutive integers is two more than twice their sum. Find the integers.

16. The product of two consecutive even integers is 60 more than twice the larger. Find the integers.

17. Find three consecutive even integers such that four times the sum of all three equals the product of the smaller two.

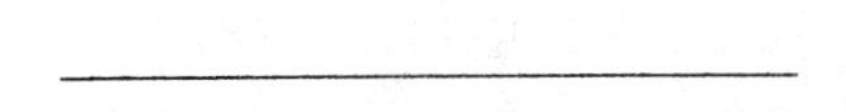

18. One number is four more than another. The square of the smaller increased by three times the larger is 66. Find the numbers.

19. If the square of the sum of two consecutive integers is diminished by three times their product, the result is 31. Find the integers.

20. If the square of the larger of two numbers is diminished by six times the smaller, the result is five times the larger. The larger is twice the smaller. Find the numbers.

OBJECTIVES

- Solve quadratic inequalities and graph the answers

A **quadratic inequality** is an inequality that involves a second degree polynomial. Examples of quadratic inequalities include

$$2x^2 + 3x - 5 < 0, \quad x^2 \leqslant 4, \quad \text{and } x^2 + 5x + 6 > 0.$$

Examples 1 and 2 show how to solve these inequalities.

Example 1 Solve $x^2 - x - 6 > 0$.

Step 1 Solve the corresponding equation, $x^2 - x - 6 = 0$. Factor $x^2 - x - 6$ as $(x - 3)(x + 2)$, so that $x^2 - x - 6 = 0$ becomes

$$(x - 3)(x + 2) = 0.$$

$$x - 3 = 0 \quad \text{or} \quad x + 2 = 0$$

$$x = 3 \quad \text{or} \quad x = -2$$

The numbers 3 and −2 are the only values of x that make $x^2 - x - 6$ equal 0. Any other values of x will make $x^2 - x - 6$ either more than 0 or less than 0.

Step 2 Draw a number line, and mark 3 and −2 on it. (Be sure to place the numbers on the number line in numerical order. See Figure 3.)

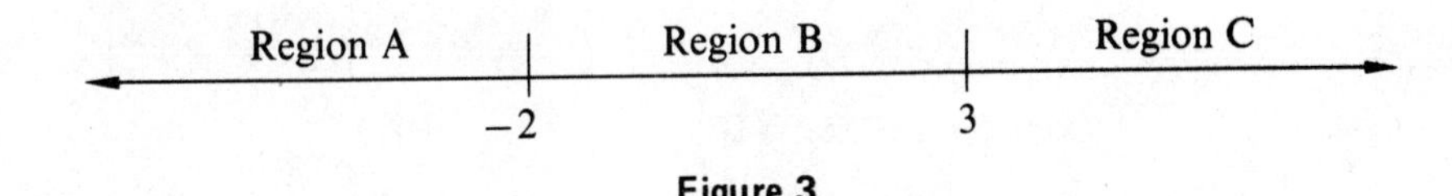

Figure 3

The number line is divided into 3 regions, A, B, and C. If one number in region A satisfies the original inequality, then all numbers in region A will satisfy it; the same is true for regions B and C.

Step 3 Select any number in region A and see if it satisfies the original inequality. Let us choose $x = -4$.

$x^2 - x - 6 > 0$	Original inequality
$(-4)^2 - (-4) - 6 > 0$	Let $x = -4$
$16 + 4 - 6 > 0$	
$14 > 0$	True

Since one number in region A makes the inequality true, all numbers in region A make the inequality true.

Step 4 Try a number from region B, such as $x = 0$.

$x^2 - x - 6 > 0$	Original inequality
$0^2 - 0 - 6 > 0$	Let $x = 0$
$-6 > 0$	False

No number in region B is part of the solution.

QUICK CHECKS

1. Solve each quadratic inequality. Graph each solution.

(a) $x^2 - 3x - 4 > 0$

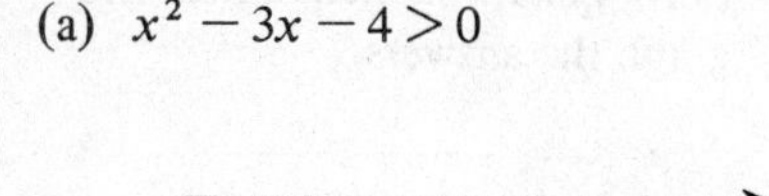

(b) $2k^2 - k - 6 < 0$

(c) $6x^2 + 13x - 5 > 0$

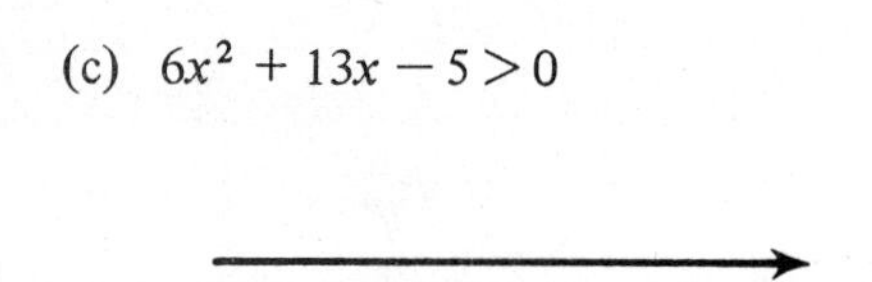

Step 5 For region C, try $x = 5$.

$$x^2 - x - 6 > 0 \qquad \text{Original inequality}$$
$$5^2 - 5 - 6 > 0 \qquad \text{Let } x = 5$$
$$25 - 5 - 6 > 0$$
$$14 > 0 \qquad \text{True}$$

All points in region C belong to the solution.

In summary, the points of regions A and C make the inequality true. The solution can be written

$$x < -2 \quad \text{or} \quad x > 3.$$

The solution can be graphed as in Figure 4.

−2 3

Figure 4

Work Quick Check 1 at the side.

Example 2 Solve $x^2 - 3x - 10 \leq 0$.

Step 1 Solve the equation $x^2 - 3x - 10 = 0$.

$$x^2 - 3x - 10 = 0$$
$$(x - 5)(x + 2) = 0$$
$$x - 5 = 0 \quad \text{or} \quad x + 2 = 0$$
$$x = 5 \quad \text{or} \quad x = -2$$

Step 2 Draw a number line, and label −2 and 5 on it. (See Figure 5.) This time, the numbers −2 and 5 are part of the solution, since we want all values of x which make $x^2 - 3x - 10 < 0$ or $x^2 - 3x - 10 = 0$ true.

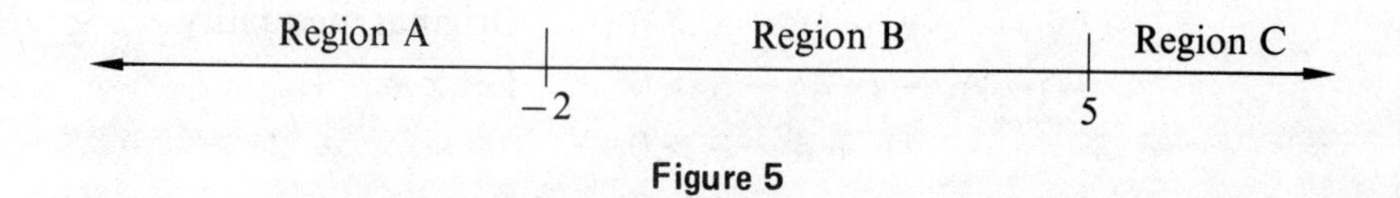

Figure 5

Step 3 Choose a value from region A, such as $x = -6$.

$$x^2 - 3x - 10 \leq 0 \qquad \text{Original inequality}$$
$$(-6)^2 - 3(-6) - 10 \leq 0 \qquad \text{Let } x = -6$$
$$36 + 18 - 10 \leq 0$$
$$44 \leq 0 \qquad \text{False}$$

No point of region A belongs to the solution.

Quick Check Answers

1.(a) −1 4

(b) −3/2 2

(c) −5/2 1/3

Step 4 What about region B? Let us try $x = 0$.

$$x^2 - 3x - 10 \leq 0 \quad \text{Original inequality}$$
$$0^2 - 3(0) - 10 \leq 0 \quad \text{Let } x = 0$$
$$-10 \leq 0 \quad \text{True}$$

The points of region B belong to the solution.

Step 5 Try $x = 7$ to check region C.

$$x^2 - 3x - 10 \leq 0 \quad \text{Original inequality}$$
$$7^2 - 3(7) - 10 \leq 0 \quad \text{Let } x = 7$$
$$49 - 21 - 10 \leq 0$$
$$18 \leq 0 \quad \text{False}$$

The points of region C do not belong to the solution. Only the points of region B, together with the end points -2 and 5, satisfy the original inequality. The solution is written

$$-2 \leq x \leq 5,$$

and is graphed as shown in Figure 6.

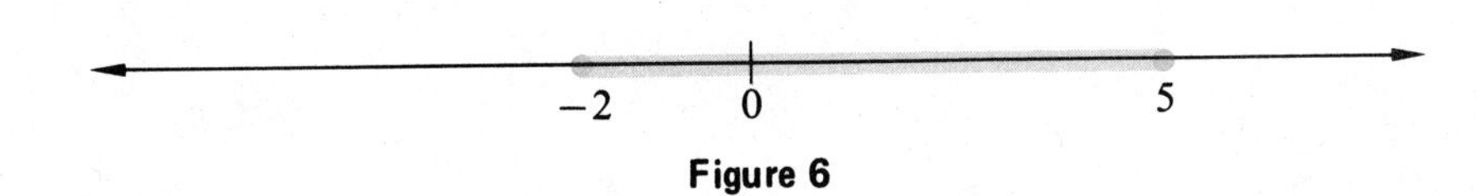

Figure 6

Work Quick Check 2 at the side.

2. Solve and graph each solution.

(a) $x^2 + 2x - 15 \geq 0$

(b) $2p^2 + p - 3 \leq 0$

(c) $4m^2 - 17m - 15 \leq 0$

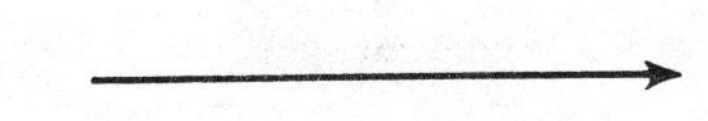

Quick Check Answers

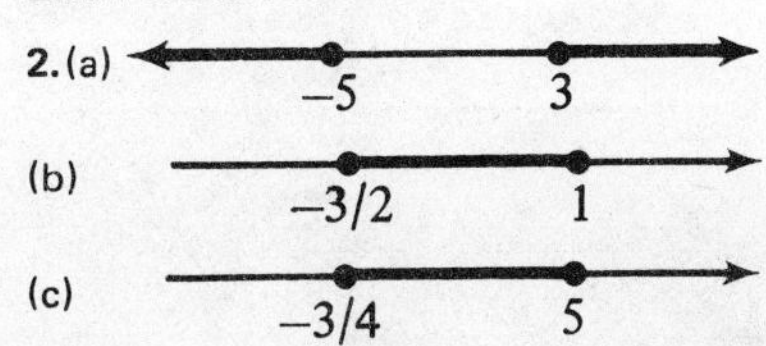

WORK SPACE

name date hour

5.7 EXERCISES

Solve each inequality and graph each solution. See Examples 1 and 2.

1. $(m + 2)(m - 5) < 0$

2. $(k - 1)(k + 3) > 0$

3. $(t + 6)(t + 5) \geqslant 0$

4. $(g - 2)(g - 4) \leqslant 0$

5. $(a + 3)(a - 3) < 0$

6. $(b - 2)(b + 2) > 0$

7. $(a + 6)(a - 7) \geqslant 0$

8. $(z - 5)(z - 4) \leqslant 0$

9. $m^2 + 5m + 6 > 0$

10. $y^2 - 3y + 2 < 0$

11. $z^2 - 4z - 5 \leqslant 0$

12. $3p^2 - 5p - 2 \leqslant 0$

13. $5m^2 + 3m - 2 < 0$

14. $2k^2 + 7k - 4 > 0$

15. $6r^2 - 5r - 4 < 0$

16. $6r^2 + 7r - 3 > 0$

17. $a^2 - 16 < 0$

18. $9m^2 - 36 > 0$

19. $r^2 - 100 \geqslant 0$

20. $q^2 - 7q + 6 < 0$

21. $2k^2 - 7k - 15 \leqslant 0$

22. $6m^2 + m - 1 > 0$

23. $30r^2 + 3r - 6 \leqslant 0$

24. $12p^2 + 11p + 2 < 0$

25. $r^2 > 16$

26. $m^2 \geqslant 25$

name date hour

CHAPTER 5 TEST

Factor as completely as possible.

1. $16m^2 - 24m$ ____

2. $6xy + 12y^2$ ____

3. $28pq + 14p + 56p^2$ ____

4. $3m^2n + 9mn + 6mn^2$ ____

5. $12p + 11r$ ____

6. $x^2 + 11x + 30$ ____

7. $p^2 + 6p - 7$ ____

8. $2y^2 - 7y - 15$ ____

9. $4m^2 + 4m - 3$ ____

10. $3x^2 + 13x - 10$ ____

11. $10z^2 + 7z + 1$ ____

12. $10a^2 - 23a - 5$ ____

13. $12r^2 + 19r + 5$ ____

14. $m^2 + 11m + 14$ ____

15. $a^2 + 3ab - 10b^2$ ____

16. $6r^2 - rs - 2s^2$ ____

17. $x^2 - 25$ ____

18. $25m^2 - 49$ ____

19. $4p^2 + 12p + 9$ ____

20. $25z^2 - 10z + 1$ ____

21. $4p^3 + 16p^2 + 16p$ ____

22. $10m^4 + 55m^3 + 25m^2$ ____

Solve each equation.

23. $y^2 + 3y + 2 = 0$ ____

24. $3x^2 + 5x = 2$ ____

25. $2p^2 + 3p = 20$ ____

26. $x^2 - 4 = 0$ ____

27. $z^3 = 16z$ ____________________

Write an equation for each problem and solve it.

28. The length of a certain rectangle is one inch less than twice the width. The area is 15 square inches. Find the dimensions. ____________________

29. The length of a rectangle is 7 more than the width. The area is numerically 4 more than the perimeter. Find the length and width of the rectangle. ____________________

30. One number is nine larger than another. Their product is eleven more than five times their sum. Find the numbers. ____________________

Graph the solution of each quadratic inequality.

31. $m^2 + 2m - 24 > 0$

32. $2p^2 + 5p - 3 \leqslant 0$

33. $6r^2 + r - 2 \geqslant 0$

6 Rational Expressions

6.1 THE FUNDAMENTAL PROPERTY OF RATIONAL EXPRESSIONS

OBJECTIVES

- Find the domain of a rational expression
- Find the numerical value of a rational expression
- Reduce a rational expression to lowest terms

The quotient of two integers is called a fraction or a rational number. Similarly, the quotient of two polynomials is called a **rational expression.**

> If P and Q are polynomials with $Q \neq 0$,
> then P/Q is a rational expression.

Examples of rational expressions include

$$\frac{-6x}{x^3+8}, \quad \frac{9x}{y+3}, \quad \frac{2m^3}{8}.$$

However, $8x/0$ is not a rational expression since division by zero is not possible. For that reason, be careful when substituting a number for the variable in the denominator of a rational expression. For example, in the rational expression

$$\frac{8x^2}{x-3}$$

x can take on any numerical value except 3. When $x = 3$, the denominator of the rational expression becomes zero, making the expression meaningless. The set of all permissible numerical values for the variable in a rational expression is called the **domain** of the rational expression. The domain of

$$\frac{8x^2}{x-3}$$

is all real numbers except 3.

In this book, unless stated otherwise, the domain of a rational expression will be the set of all real numbers except those which make the denominator zero.

Example 1 Find the domain of the rational expressions.

(a) $\dfrac{p+5}{p+2}$

QUICK CHECKS

1. Find the domain of each rational expression.

(a) $\frac{x+2}{x-5}$

(b) $\frac{3r}{r^2+6r+8}$

2. Find the value of each rational expression when $x = 3$.

(a) $\frac{x}{2x+1}$

(b) $\frac{2x+6}{x-3}$

Quick Check Answers

1. (a) $x \neq 5$ (b) $r \neq -4$, $r \neq -2$
2. (a) $\frac{3}{7}$ (b) not a real number

Since $p = -2$ will make the denominator equal to 0, the domain is $p \neq -2$.

(b) $\frac{9m^2}{m^2 - 5m + 6}$

To find the numbers which will make the denominator 0, we must solve

$$m^2 - 5m + 6 = 0.$$

We can factor the polynomial and set each factor equal to 0.

$$(m-2)(m-3) = 0$$

$$m - 2 = 0 \quad \text{or} \quad m - 3 = 0$$

$$m = 2 \quad \text{or} \quad m = 3$$

The domain is $m \neq 2$ and $m \neq 3$.

Work Quick Check 1 at the side.

Example 2 Find the numerical value of $\frac{3x+6}{2x-4}$ for each value of x.

(a) $x = 1$

We find the value of the rational expression by substitution.

$$\frac{3x+6}{2x-4} = \frac{3(1)+6}{2(1)-4} = \frac{9}{-2} = -\frac{9}{2}$$

(b) $x = 2$

If we substitute 2 for x, the denominator is zero. Thus there is no value for the rational expression when $x = 2$. The domain is $x \neq 2$.

Work Quick Check 2 at the side.

A rational expression represents a number for each value of the variable in its domain. Thus, the properties of rational numbers also apply to rational expressions. In order to reduce a rational expression to lowest terms, we use the **fundamental property of rational expressions**:

> If P/Q is a rational expression, and if K represents any rational expression, $K \neq 0$, then
>
> $$\frac{PK}{QK} = \frac{P}{Q}.$$

Example 3 Reduce $\frac{14k^2}{2k^3}$ to lowest terms.

First factor the rational expression. Then use the fundamental property.

$$\frac{14k^2}{2k^3} = \frac{2 \cdot 7 \cdot k \cdot k}{2 \cdot k \cdot k \cdot k} = \frac{7(2kk)}{k(2kk)} = \frac{7}{k}$$

Work Quick Check 3 at the side.

Example 4 Reduce $\frac{3x-12}{5x-20}$ to lowest terms.

Begin by factoring both numerator and denominator. Then use the fundamental property.

$$\frac{3x-12}{5x-20} = \frac{3(x-4)}{5(x-4)} = \frac{3}{5}$$

Work Quick Check 4 at the side.

Example 5 Reduce $\frac{m^2+2m-8}{2m^2-m-6}$ to lowest terms.

Always begin by factoring both numerator and denominator, if possible. Then use the fundamental property.

$$\frac{m^2+2m-8}{2m^2-m-6} = \frac{(m+4)(m-2)}{(2m+3)(m-2)} = \frac{m+4}{2m+3}$$

Work Quick Check 5 at the side.

Example 6 Reduce $\frac{x-y}{y-x}$ to lowest terms.

At first glance, there does not seem to be any way in which we can factor $x-y$ and $y-x$ to get a common factor. However, note that

$$y-x = -1(-y+x) = -1(x-y).$$

With these factors, use the fundamental property to simplify the rational expression.

$$\frac{x-y}{y-x} = \frac{1(x-y)}{-1(x-y)} = \frac{1}{-1} = -1$$

Work Quick Check 6 at the side.

3. Use the fundamental property to reduce the rational expressions.

(a) $\frac{5x^4}{15x^2}$

(b) $\frac{6p^3}{2p^2}$

4. Reduce to lowest terms.

(a) $\frac{8p+8q}{5p+5q}$

(b) $\frac{4y+2}{6y+3}$

5. Reduce to lowest terms.

(a) $\frac{a^2-b^2}{a^2+2ab+b^2}$

(b) $\frac{x^2+4x+4}{4x+8}$

6. Reduce to lowest terms.

(a) $\frac{m-n}{n-m}$

(b) $\frac{x-3}{3-x}$

Quick Check Answers

3. (a) $\frac{x^2}{3}$ (b) $3p$
4. (a) $\frac{8}{5}$ (b) $\frac{2}{3}$
5. (a) $\frac{a-b}{a+b}$ (b) $\frac{x+2}{4}$
6. (a) -1 (b) -1

WORK SPACE

name date hour

6.1 EXERCISES

Find the domain of each rational expression. See Example 1.

1. $\frac{3}{4x}$ ______

2. $\frac{5}{2x}$ ______

3. $\frac{8}{x-4}$ ______

4. $\frac{6}{x+3}$ ______

5. $\frac{x^2}{x+5}$ ______

6. $\frac{3x^2}{2x-1}$ ______

Find the numerical value of each rational expression when x = −3. See Example 2.

7. $\frac{x^3}{2x^2}$ ______

8. $\frac{-5x+1}{2x}$ ______

9. $\frac{4x^2-2x}{3x}$ ______

10. $\frac{2x+5}{3+x}$ ______

11. $\frac{(-8x)^2}{3x+9}$ ______

12. $\frac{x^2-1}{x}$ ______

Reduce to lowest terms. If possible, factor the numerator and denominator first. See Examples 3-5.

13. $\frac{2y}{3y}$ ______

14. $\frac{5m}{10}$ ______

15. $\frac{12k^2}{6k}$ ______

16. $\frac{9m^3}{3m}$ ______

17. $\frac{-8y^6}{6y^3}$ ______

18. $\frac{16x^4}{-8x^2}$ ______

19. $\frac{12m^2p}{9mp^2}$ ______

20. $\frac{6a^2b^3}{24a^3b^2}$ ______

21. $\frac{6(y+2)}{8(y+2)}$ ______

22. $\frac{9(m+2)}{5(m+2)}$ ______

23. $\frac{(x+1)(x-1)}{(x+1)^2}$ ______

24. $\frac{3(t+5)}{(t+5)(t-1)}$ ______

25. $\dfrac{m^2 - n^2}{m + n}$ ______

26. $\dfrac{a^2 - b^2}{a - b}$ ______

27. $\dfrac{5m^2 - 5m}{10m - 10}$ ______

28. $\dfrac{3y^2 - 3y}{2(y - 1)}$ ______

29. $\dfrac{16r^2 - 4s^2}{4r - 2s}$ ______

30. $\dfrac{11s^2 - 22s^3}{6 - 12s}$ ______

31. $\dfrac{m^2 - 4m + 4}{m^2 + m - 6}$ ______

32. $\dfrac{a^2 - a - 6}{a^2 + a - 12}$ ______

33. $\dfrac{x^2 + 3x - 4}{x^2 - 1}$ ______

34. $\dfrac{8m^2 + 6m - 9}{16m^2 - 9}$ ______

Reduce to lowest terms. See Example 6.

35. $\dfrac{-a + b}{b - a}$ ______

36. $\dfrac{b - a}{a - b}$ ______

37. $\dfrac{x^2 - 1}{1 - x}$ ______

38. $\dfrac{p^2 - q^2}{q - p}$ ______

39. $\dfrac{m^2 - 4m}{4m - m^2}$ ______

40. $\dfrac{s^2 - r^2}{r^2 - s^2}$ ______

6.2 MULTIPLICATION AND DIVISION OF RATIONAL EXPRESSIONS

OBJECTIVES

- Multiply rational expressions
- Divide rational expressions

To multiply two fractions, we multiply the numerators and multiply the denominators. The same rule applies to multiplication of rational expressions.

> The product of the rational expressions P/Q and R/S is
>
> $$\frac{P}{Q} \cdot \frac{R}{S} = \frac{PR}{QS}.$$

Example 1 Find the product of $6/x$ and $x^2/12$.

Using the definition of multiplication, we find the product of the numerators and the product of the denominators.

$$\frac{6}{x} \cdot \frac{x^2}{12} = \frac{6 \cdot x^2}{x \cdot 12}$$

$$= \frac{x}{2}$$

In the last step, use the fundamental property to reduce to lowest terms.

Work Quick Check 1 at the side.

Example 2 Find the product $\dfrac{x+y}{2x} \cdot \dfrac{x^2}{(x+y)^2}$.

Using the definition of multiplication, we have

$$\frac{x+y}{2x} \cdot \frac{x^2}{(x+y)^2} = \frac{(x+y)x^2}{2x(x+y)^2}$$

$$= \frac{(x+y)x^2}{2x(x+y)(x+y)}$$

$$= \frac{x}{2(x+y)}.$$

In the last two steps, factor and use the fundamental property of rational expressions to reduce the answer to lowest terms.

Work Quick Check 2 at the side.

Example 3 Find the product.

$$\frac{x^2+3x}{x^2-3x-4} \cdot \frac{x^2-5x+4}{x^2+2x-3}$$

Use the definition of multiplication. However, before multiplying, factor the numerators and denominators wherever possible. Then use the fundamental theorem to reduce to lowest terms.

QUICK CHECKS

1. Find the products.

(a) $\dfrac{3m^2}{2} \cdot \dfrac{10}{m}$

(b) $\dfrac{8p^2q}{3} \cdot \dfrac{9}{pq^2}$

2. Find the products.

(a) $\dfrac{a+b}{5} \cdot \dfrac{30}{2(a+b)}$

(b) $\dfrac{3(p-q)}{p} \cdot \dfrac{q}{2(p-q)}$

Quick Check Answers

1. (a) $15m$ (b) $\dfrac{24p}{q}$

2. (a) 3 (b) $\dfrac{3q}{2p}$

$$\frac{x^2+3x}{x^2-3x-4} \cdot \frac{x^2-5x+4}{x^2+2x-3}$$

$$= \frac{x(x+3)}{(x-4)(x+1)} \cdot \frac{(x-4)(x-1)}{(x+3)(x-1)}$$

$$= \frac{x(x+3)(x-4)(x-1)}{(x-4)(x+1)(x+3)(x-1)}$$

$$= \frac{x}{x+1}$$

Notice in the second step above that we do not multiply the factors together, since we want to look for common factors to reduce the product to lowest terms.

Work Quick Check 3 at the side.

To *divide* the fraction a/b by the nonzero fraction c/d, we multiply a/b by the reciprocal of c/d, which is d/c. Division of rational expressions is defined in the same way.

> If P/Q and R/S are any two rational expressions, with $R/S \neq 0$, then their quotient is
>
> $$\frac{P}{Q} \div \frac{R}{S} = \frac{P}{Q} \cdot \frac{S}{R} = \frac{PS}{QR}.$$

Example 4 Find the quotient.

$$\frac{(3m)^2}{(2n)^3} \div \frac{6m^3}{16n^2}$$

$$= \frac{9m^2}{8n^3} \div \frac{6m^3}{16n^2}$$

$$= \frac{9m^2}{8n^3} \cdot \frac{16n^2}{6m^3}$$

$$= \frac{9 \cdot 16m^2n^2}{8 \cdot 6n^3m^3}$$

$$= \frac{3}{mn}$$

Work Quick Check 4 at the side.

Example 5 Find the quotient.

$$\frac{x^2-4}{(x+3)(x-2)} \div \frac{(x+2)(x+3)}{2x}$$

First, use the definition of division.

$$\frac{x^2-4}{(x+3)(x-2)} \div \frac{(x+2)(x+3)}{2x}$$

$$= \frac{x^2-4}{(x+3)(x-2)} \cdot \frac{2x}{(x+2)(x+3)}$$

3. Find the products.

(a) $\frac{x^2+7x+10}{3x+6} \cdot \frac{6x-6}{x^2+2x-15}$

(b) $\frac{m^2+4m-5}{m+5} \cdot \frac{m^2+8m+15}{m-1}$

4. Find the quotients.

(a) $\frac{5a^2b}{2} \div \frac{10ab^2}{8}$

(b) $\frac{(3t)^2}{w} \div \frac{3t^2}{5w^4}$

Quick Check Answers

3. (a) $\frac{2(x-1)}{x-3}$ (b) $(m+5)(m+3)$

4. (a) $\frac{2a}{b}$ (b) $15w^3$

Next, be sure all numerators and all denominators are factored. Here, we factor $x^2 - 4$. Recall that $x^2 - 4 = (x + 2)(x - 2)$.

$$= \frac{(x + 2)(x - 2)}{(x + 3)(x - 2)} \cdot \frac{2x}{(x + 2)(x + 3)}$$

Now multiply the numerators and the denominators and simplify.

$$= \frac{(x + 2)(x - 2)(2x)}{(x + 3)(x - 2)(x + 2)(x + 3)}$$

$$= \frac{2x}{(x + 3)^2}$$

Work Quick Check 5 at the side.

Example 6 Find the quotient.

$$\frac{m^2 - 4}{m^2 - 1} \div \frac{2m^2 + 4m}{m - 1}$$

Use the definition of division.

$$\frac{m^2 - 4}{m^2 - 1} \div \frac{2m^2 + 4m}{m - 1} = \frac{m^2 - 4}{m^2 - 1} \cdot \frac{m - 1}{2m^2 + 4m}$$

Before multiplying, factor all numerators and denominators. Then multiply and simplify.

$$= \frac{(m + 2)(m - 2)}{(m + 1)(m - 1)} \cdot \frac{m - 1}{2m(m + 2)}$$

$$= \frac{(m + 2)(m - 2)(m - 1)}{(m + 1)(m - 1)(2m)(m + 2)}$$

$$= \frac{m - 2}{2m(m + 1)}$$

Work Quick Check 6 at the side.

5. Find the quotients.

(a) $\frac{y^2 + 4y + 3}{y + 3} \div \frac{y^2 - 4y - 5}{y - 3}$

(b) $\frac{4x(x + 3)}{2x + 1} \div \frac{x^2(x + 3)}{4x^2 - 1}$

6. Find the quotients.

(a) $\frac{x^2 - y^2}{x^2 - 1} \div \frac{x^2 + 2xy + y^2}{x^2 + x}$

(b) $\frac{ab - a^2}{a^2 - 1} \div \frac{b - a}{a^2 + 2a + 1}$

Quick Check Answers

5. (a) $\frac{y - 3}{y - 5}$ (b) $\frac{4(2x - 1)}{x}$

6. (a) $\frac{x(x - y)}{(x - 1)(x + y)}$

(b) $\frac{a(a + 1)}{a - 1}$

WORK SPACE

name date hour

6.2 EXERCISES

Find each product or quotient. Write each answer in lowest terms. See Examples 1 and 4.

1. $\dfrac{9m^2}{16} \cdot \dfrac{4}{3m}$ ______

2. $\dfrac{21z^4}{8y} \cdot \dfrac{4y^3}{7z^5}$ ______

3. $\dfrac{4p^2}{8p} \cdot \dfrac{3p^3}{16p^4}$ ______

4. $\dfrac{6x^3}{9x} \cdot \dfrac{12x}{x^2}$ ______

5. $\dfrac{8a^4}{12a^3} \cdot \dfrac{9a^5}{3a^2}$ ______

6. $\dfrac{14p^5}{2p^2} \cdot \dfrac{8p^6}{28p^9}$ ______

7. $\dfrac{3r^2}{9r^3} \div \dfrac{8r^4}{6r^5}$ ______

8. $\dfrac{15m^{10}}{9m^5} \div \dfrac{6m^6}{10m^4}$ ______

9. $\dfrac{3m^2}{(4m)^3} \div \dfrac{9m^3}{32m^4}$ ______

10. $\dfrac{5x^3}{(4x)^2} \div \dfrac{15x^2}{8x^4}$ ______

Find each product or quotient. Write each answer in lowest terms. See Examples 2, 3, 5, and 6.

11. $\dfrac{a+b}{2} \cdot \dfrac{12}{(a+b)^2}$ ______

12. $\dfrac{3(x-1)}{y} \cdot \dfrac{2y}{5(x-1)}$ ______

13. $\dfrac{a-3}{16} \div \dfrac{a-3}{32}$ ______

14. $\dfrac{9}{8-2y} \div \dfrac{3}{4-y}$ ______

15. $\dfrac{2k+8}{6} \div \dfrac{3k+12}{2}$ ______

16. $\dfrac{5m+25}{10} \cdot \dfrac{12}{6m+30}$ ______

17. $\dfrac{9y-18}{6y+12} \cdot \dfrac{3y+6}{15y-30}$ ______

18. $\dfrac{12p+24}{36p-36} \div \dfrac{6p+12}{8p-8}$ ______

19. $\dfrac{3r+12}{8} \cdot \dfrac{16r}{9r+36}$ ______

20. $\dfrac{2r+2p}{8z} \div \dfrac{r^2+rp}{72}$ ______

21. $\dfrac{y^2-16}{y+3} \div \dfrac{y-4}{y^2-9}$ ______

22. $\dfrac{9(y-4)^2}{8(z+3)^2} \cdot \dfrac{16(z+3)}{3(y-4)}$ ______

23. $\dfrac{6(m+2)}{3(m-1)^2} \div \dfrac{(m+2)^2}{9(m-1)}$ ______

24. $\dfrac{4y+12}{2y-10} \div \dfrac{y^2-9}{y^2-y-20}$ ______

25. $\dfrac{2-y}{8} \cdot \dfrac{7}{y-2}$ ______

(Hint: Recall Example 6, Section 6.1)

26. $\dfrac{9-2z}{3} \cdot \dfrac{9}{2z-9}$ ______

27. $\dfrac{8-r}{8+r} \div \dfrac{r-8}{r+8}$ ______

28. $\dfrac{6r-18}{3r^2+2r-8} \cdot \dfrac{12r-16}{4r-12}$ ______

29. $\dfrac{k^2-k-6}{k^2+k-12} \div \dfrac{k^2+2k-3}{k^2+3k-4}$ ______

30. $\dfrac{m^2+3m+2}{m^2+5m+4} \cdot \dfrac{m^2+10m+24}{m^2+5m+6}$ ______

31. $\dfrac{n^2-n-6}{n^2-2n-8} \cdot \dfrac{n^2+7n+12}{n^2-9}$ ______

32. $\dfrac{6n^2-5n-6}{6n^2+5n-6} \cdot \dfrac{12n^2-17n+6}{12n^2-n-6}$ ______

33. $\dfrac{16-r^2}{r^2+2r-8} \div \dfrac{r^2-2r-8}{4-r^2}$ ______

34. $\dfrac{y^2+y-2}{y^2+3y-4} \div \dfrac{y+2}{y+3}$ ______

35. $\dfrac{2m^2-5m-12}{m^2-10m+24} \div \dfrac{4m^2-9}{m^2-9m+18}$ ______

36. $\dfrac{9z^2+27zm}{9m^2+27zm} \div \dfrac{8zm+24m^2}{16zm+48z^2}$ ______

37. $\dfrac{21p^2-20pq-q^2}{p^2+pq-2q^2} \div \dfrac{21p^2+22pq+q^2}{p^2+pq-2q^2}$ ______

38. $\dfrac{2m^2+7m+3}{m^2-9} \cdot \dfrac{m^2-3m}{2m^2+11m+5}$ ______

39. $\dfrac{(x+1)^3(x+4)}{x^2+5x+4} \div \dfrac{x^2+2x+1}{x^2+3x+2}$ ______

40. $\dfrac{m^2-m-6}{3m^2+10m+8} \cdot \dfrac{6m^2+17m+12}{4m^2-9}$ ______

41. $\left(\dfrac{x^2+10x+25}{x^2+10x} \cdot \dfrac{10x}{x^2+15x+50}\right) \div \dfrac{x+5}{x+10}$ ______

42. $\left(\dfrac{m^2-12m+32}{8m} \cdot \dfrac{m^2-8m}{m^2-8m+16}\right) \div \dfrac{m-8}{m-4}$ ______

6.3 ADDITION AND SUBTRACTION OF RATIONAL EXPRESSIONS

OBJECTIVES

- Add rational expressions having the same denominator
- Add rational expressions having different denominators
- Subtract rational expressions

To find the sum of two rational expressions, we use a procedure similar to adding two fractions.

> If P and Q are rational expressions, then
> $$\frac{P}{Q}+\frac{R}{Q}=\frac{P+R}{Q}.$$

Example 1 Add $\dfrac{3x}{x+1}+\dfrac{2x}{x+1}$.

Since the denominators are the same, the sum is found by adding the two numerators and keeping the same (common) denominator.

$$\frac{3x}{x+1}+\frac{2x}{x+1}=\frac{3x+2x}{x+1}$$
$$=\frac{5x}{x+1}$$

Work Quick Check 1 at the side.

QUICK CHECKS

1. Find each sum.

(a) $\dfrac{3}{y+4}+\dfrac{2}{y+4}$

(b) $\dfrac{x}{x+y}+\dfrac{1}{x+y}$

If the denominators of the two rational expressions to be added are different, the following steps are required. These are the same steps used to add fractions with different denominators.

> *Step 1* Find the lowest common denominator (LCD).
>
> *Step 2* Rewrite each rational expression as a fraction with the LCD as denominator.
>
> *Step 3* Add the numerators to get the numerator of the sum. The LCD is the denominator of the sum.

Example 2 Add $\dfrac{2}{3y}+\dfrac{1}{4y}$.

The lowest common denominator is the smallest number that both denominators can be divided into evenly. In this example, we use $12y$ as the LCD, since it is the smallest number that both $3y$ and $4y$ will divide into evenly. In the first term, $\dfrac{2}{3y}$, multiply both the numerator and denominator by 4 to get the LCD of $12y$ and still leave the value of the fraction unchanged. Multiply both the numerator and denominator in the second term, $\dfrac{1}{4y}$, by 3 to get $12y$ as the LCD. Then we have

$$\frac{2}{3y}+\frac{1}{4y}=\frac{2(4)}{3y(4)}+\frac{1(3)}{4y(3)}=\frac{8}{12y}+\frac{3}{12y}=\frac{11}{12y}.$$

Work Quick Check 2 at the side.

2. Find each sum.

(a) $\dfrac{6}{5x}+\dfrac{9}{2x}$

(b) $\dfrac{m}{3n}+\dfrac{2}{7n}$

Quick Check Answers

1. (a) $\dfrac{5}{y+4}$ (b) $\dfrac{x+1}{x+y}$
2. (a) $\dfrac{57}{10x}$ (b) $\dfrac{7m+6}{21n}$

3. Find the sums.

(a) $\dfrac{2p}{3p+3} + \dfrac{5p}{2p+2}$

(b) $\dfrac{4}{y^2-1} + \dfrac{6}{y+1}$

(c) $\dfrac{-9}{p+1} + \dfrac{p}{p^2-1}$

4. Find the sum of

$$\frac{4m}{m^2+3m+2} + \frac{2m-1}{m^2+6m+5}.$$

Quick Check Answers

3. (a) $\dfrac{19p}{6(p+1)}$

(b) $\dfrac{6y-2}{(y+1)(y-1)}$

(c) $\dfrac{-8p+9}{(p+1)(p-1)}$

4. $\dfrac{6m^2+23m-2}{(m+2)(m+1)(m+5)}$

Example 3 Add $\dfrac{x}{x^2-1} + \dfrac{x}{x+1}$.

To find the LCD, factor both denominators.

$$x^2-1 = (x+1)(x-1); \; x+1 \text{ cannot be factored.}$$

The expression is now written as

$$\frac{x}{(x+1)(x-1)} + \frac{x}{(x+1)}.$$

Since the term $(x+1)$ is already contained in the denominators of both fractions, we need to multiply only the second fraction by $(x-1)$ to get the LCD.

$$\frac{x}{(x+1)(x-1)} + \frac{x(x-1)}{(x+1)(x-1)}$$

With both denominators now being the same, we add the numerators.

$$\frac{x+x(x-1)}{(x+1)(x-1)} = \frac{x+x^2-x}{(x+1)(x-1)} = \frac{x^2}{(x+1)(x-1)}$$

Work Quick Check 3 at the side.

Example 4 Add $\dfrac{2x}{x^2+5x+6} + \dfrac{x+1}{x^2+2x-3}$.

To begin, we factor the denominators completely.

$$\frac{2x}{(x+2)(x+3)} + \frac{x+1}{(x+3)(x-1)}$$

The LCD is $(x+2)(x+3)(x-1)$. By the fundamental property,

$$\frac{2x}{(x+2)(x+3)} + \frac{x+1}{(x+3)(x-1)}$$
$$= \frac{2x(x-1)}{(x+2)(x+3)(x-1)} + \frac{(x+1)(x+2)}{(x+3)(x-1)(x+2)}.$$

Since the two rational expressions above have the same denominator, we add their numerators, just as with fractions.

$$= \frac{2x(x-1)+(x+1)(x+2)}{(x+2)(x+3)(x-1)}$$
$$= \frac{2x^2-2x+x^2+3x+2}{(x+2)(x+3)(x-1)}$$
$$= \frac{3x^2+x+2}{(x+2)(x+3)(x-1)}$$

In a problem of this type, it is often convenient to leave the denominator in factored form.

Work Quick Check 4 at the side.

To subtract rational expressions, we use the following rule.

> If P/Q and R/S are rational expressions, then
>
> $$\frac{P}{Q}-\frac{R}{S}=\frac{P}{Q}+\left(\frac{-R}{S}\right).$$

5. Find each difference. Reduce the answers to lowest terms.

(a) $\frac{5}{2y}-\frac{9}{2y}$

(b) $\frac{3}{m^2}-\frac{-2}{m^2}$

Example 5 Find $\frac{12}{x^2}-\frac{-8}{x^2}$.

$$\frac{12}{x^2}-\frac{-8}{x^2}=\frac{12}{x^2}+\frac{-(-8)}{x^2}=\frac{12}{x^2}+\frac{8}{x^2}=\frac{20}{x^2}$$

Work Quick Check 5 at the side.

Example 6 Find $\frac{6x}{(x-1)^2}-\frac{2}{x^2-1}$.

$$\frac{6x}{(x-1)^2}-\frac{2}{x^2-1}$$

$$=\frac{6x}{(x-1)(x-1)}+\frac{-2}{(x-1)(x+1)}$$

We change subtraction to addition and change the numerator of the second fraction from 2 to -2. We also factor the two denominators, so we can identify a common denominator as $(x-1)(x-1)(x+1)$. We use the factor $x-1$ twice, since it appears twice in the first denominator.

$$=\frac{6x(x+1)}{(x-1)(x-1)(x+1)}+\frac{-2(x-1)}{(x-1)(x-1)(x+1)}$$

$$=\frac{6x(x+1)+[-2(x-1)]}{(x-1)(x-1)(x+1)}$$

$$=\frac{6x^2+6x-2x+2}{(x-1)(x-1)(x+1)}$$

$$=\frac{6x^2+4x+2}{(x-1)(x-1)(x+1)}$$

6. Find $\frac{4y}{y^2-1}-\frac{5}{(y+1)^2}$.

Work Quick Check 6 at the side.

Quick Check Answers

5. (a) $\frac{-2}{y}$ (b) $\frac{5}{m^2}$

6. $\frac{4y^2-y+5}{(y+1)^2(y-1)}$

WORK SPACE

name date hour

6.3 EXERCISES

Find sums or differences. When possible, reduce the answers to lowest terms. See Examples 1 and 5.

1. $\frac{2}{p}+\frac{5}{p}$ __________

2. $\frac{3}{r}+\frac{6}{r}$ __________

3. $\frac{9}{k}-\frac{12}{k}$ __________

4. $\frac{15}{z}-\frac{25}{z}$ __________

5. $\frac{y}{y+1}+\frac{1}{y+1}$ __________

6. $\frac{3m}{m-4}+\frac{-12}{m-4}$ __________

7. $\frac{m^2}{m-n}-\frac{n^2}{m-n}$ __________

8. $\frac{a+b}{2}-\frac{a-b}{2}$ __________

9. $\frac{m^2}{m+6}+\frac{6m}{m+6}$ __________

10. $\frac{y^2}{y-1}+\frac{-y}{y-1}$ __________

11. $\frac{2}{2r+2}+\frac{2r}{2r+2}$ __________

12. $\frac{a^2}{a+b}+\frac{ab}{a+b}$ __________

Find sums or differences. When possible, reduce the answers to lowest terms. See Examples 2, 3, 4, and 6.

13. $\frac{3}{m}+\frac{1}{2}$ __________

14. $\frac{6}{p}-\frac{2}{3}$ __________

15. $\frac{9}{m}+\frac{3}{2}$ __________

16. $\frac{9}{10}+\frac{r}{2}$ __________

17. $\frac{3}{5}-\frac{1}{y}$ __________

18. $\frac{9y}{7}-\frac{3y}{8}$ __________

19. $\frac{5m}{6}-\left(\frac{2m}{3}-\frac{m}{6}\right)$ __________

20. $\left(\frac{3}{x}+\frac{4}{2x}\right)-\frac{5}{4x}$ __________

21. $\frac{4+2k}{5}+\frac{2+k}{10}$ __________

22. $\frac{5-4r}{8}-\frac{2-3r}{6}$ __________

23. $\frac{6}{y^2}-\frac{2}{y}$ __________

24. $\frac{3}{p}+\frac{5}{p^2}$ __________

25. $\frac{9}{2p}+\frac{4}{p^2}$ __________

26. $\frac{15}{4k^2}-\frac{3}{k}$ __________

27. $\frac{3m+n}{3}+\frac{m+2n}{6}$ ____

28. $\frac{5r+s}{3}-\frac{2r-s}{9}$ ____

29. $\frac{-1}{x^2}+\frac{-3}{xy}$ ____

30. $\frac{9}{p^2}+\frac{p}{x}$ ____

31. $\frac{m+2}{m}+\frac{m}{m+2}$ ____

32. $\frac{2x-5}{x-2}+\frac{x}{2x-4}$ ____

33. $\frac{8}{x-2}-\frac{4}{x+2}$ ____

34. $\frac{6}{m-n}-\frac{2}{m+n}$ ____

35. $\frac{2x}{x+y}-\frac{3x}{2x+2y}$ ____

36. $\frac{1}{a+b}-\frac{a}{a^2-b^2}$ ____

37. $\frac{1}{m^2-1}-\frac{1}{m^2+3m+2}$ ____

38. $\frac{2}{4y^2-16}+\frac{3}{4+2y}$ ____

39. $\frac{1}{m^2-9}+\frac{1}{m+3}$ ____

40. $\frac{2}{y^2-4}-\frac{3}{2y+4}$ ____

41. $\frac{4}{2-m}+\frac{7}{m-2}$ ____

42. $\frac{9}{8-y}+\frac{6}{y-8}$ ____

43. $\frac{-1}{-3+y}-\frac{2}{y-3}$ ____

44. $\frac{-8}{11+p}-\frac{6}{p+11}$ ____

45. $\frac{5m}{m+2n}-\frac{3m}{-m-2n}$ ____

46. $\frac{6k}{2k+3m}-\frac{4k}{-2k-3m}$ ____

47. $\frac{x+3y}{x^2+2xy+y^2}+\frac{x-y}{x^2+4xy+3y^2}$ ____

48. $\frac{m+1}{m^2-1}+\frac{m-1}{m^2+2m+1}$ ____

OBJECTIVES

- Solve equations involving rational expressions

To solve equations with fractions, we first simplify the equation by using the multiplication property of equality. The goal is to replace the equation having fractions with another equation which does not have fractions. To do this, we choose as a multiplier the LCD of all denominators in the fractions of the equation.

Example 1 Solve $\frac{x}{3} + \frac{x}{4} = 10 + x$.

The LCD of the two fractions is 12. Therefore, we begin by multiplying both sides of the equation by 12.

$$12\left(\frac{x}{3} + \frac{x}{4}\right) = 12(10 + x)$$

$$12\left(\frac{x}{3}\right) + 12\left(\frac{x}{4}\right) = 12(10) + 12x$$

$$\frac{12x}{3} + \frac{12x}{4} = 120 + 12x$$

$$4x + 3x = 120 + 12x$$

$$7x = 120 + 12x$$

$$-5x = 120$$

$$x = -24$$

Check the solution $x = -24$ by substituting -24 for x in each side of the equation separately.

$$\frac{x}{3} + \frac{x}{4} = \frac{-24}{3} + \frac{-24}{4} = -8 + (-6) = -14$$

and

$$10 + x = 10 + (-24) = -14$$

Therefore, -24 is the solution.

Work Quick Check 1 at the side.

1. Solve each equation and check your answers.

(a) $\frac{x}{5} + 3 = \frac{3}{5}$

(b) $\frac{x}{2} - \frac{x}{3} = \frac{5}{6}$

(c) $\frac{k}{7} - \frac{k}{2} = -5$

When solving equations which have a variable in the denominator, we must remember that the number 0 cannot be used as a denominator. Therefore, the solution cannot be a number which will make the denominator equal to zero.

Example 2 Solve $\frac{x}{x-2} = \frac{2}{x-2} + 2$.

The common denominator is $x - 2$. Multiply both sides of the equation by $x - 2$.

$$(x-2)\left(\frac{x}{x-2}\right) = (x-2)\left(\frac{2}{x-2}\right) + (x-2)(2)$$

$$x = 2 + 2x - 4$$

$$x = -2 + 2x$$

$$0 = -2 + x$$

$$2 = x$$

Quick Check Answers

1. (a) -12 (b) 5 (c) 14

QUICK CHECKS

2. Solve $1 - \frac{2}{x+1} = \frac{2x}{x+1}$ and check your answer.

3. Solve each equation and check your answers.

(a) $\frac{2p}{p^2 - 1} = \frac{2}{p+1} - \frac{1}{p-1}$

(b) $\frac{8r}{4r^2 - 1} = \frac{3}{2r+1} + \frac{3}{2r-1}$

Quick Check Answers

2. When the equation is solved, -1 is found. However, $x = -1$ leads to a 0 denominator in the original equation, so that there is no solution.

3. (a) -3 (b) 0

The solution is 2. However, we cannot have a solution of 2 in this equation because this x-value makes both denominators zero, so the equation is meaningless. Therefore, this equation has no solution.

Work Quick Check 2 at the side.

Example 3 Solve $\frac{2m}{m^2 - 4} + \frac{1}{m-2} = \frac{2}{m+2}$.

Since $m^2 - 4 = (m+2)(m-2)$, use $(m+2)(m-2)$ as the common denominator.

$$(m+2)(m-2)\left(\frac{2m}{m^2-4} + \frac{1}{m-2}\right) = (m+2)(m-2)\frac{2}{m+2}$$

$$(m+2)(m-2)\frac{2m}{m^2-4} + (m+2)(m-2)\frac{1}{m-2} = (m+2)(m-2)\frac{2}{m+2}$$

$$2m + m + 2 = 2(m-2)$$

$$3m + 2 = 2m - 4$$

$$m = -6$$

Check that -6 is indeed a solution by replacing m with -6 in the original equation.

Work Quick Check 3 at the side.

Example 4 Solve $\frac{2}{x^2 - x} = \frac{1}{x^2 - 1}$.

To solve the equation, we begin by finding a common denominator. We can factor $x^2 - x$ as $x(x-1)$, while $x^2 - 1$ can be factored as $(x+1)(x-1)$. The least common denominator of the two rational expressions is $x(x+1)(x-1)$. We multiply both sides of the equation by $x(x+1)(x-1)$.

$$x(x+1)(x-1)\frac{2}{x(x-1)} = x(x+1)(x-1)\frac{1}{(x+1)(x-1)}$$

$$2(x+1) = x$$

$$2x + 2 = x$$

$$x + 2 = 0$$

$$x = -2.$$

To be sure that $x = -2$ is a solution, substitute -2 for x in the original equation.

$$\frac{2}{x^2 - x} = \frac{2}{(-2)^2 - (-2)} = \frac{2}{4 + 2} = \frac{2}{6} = \frac{1}{3}$$

$$\frac{1}{x^2 - 1} = \frac{1}{(-2)^2 - 1} = \frac{1}{4 - 1} = \frac{1}{3}$$

Since -2 satisfies the equation, the solution is -2.

Work Quick Check 4 at the side.

Example 5 Solve $\frac{1}{x-1} + \frac{1}{2} = \frac{2}{x^2 - 1}$.

The least common denominator is $2(x + 1)(x - 1)$. We multiply both sides of the equation by this common denominator.

$$2(x + 1)(x - 1)\left(\frac{1}{x - 1} + \frac{1}{2}\right)$$
$$= 2(x + 1)(x - 1)\frac{2}{(x + 1)(x - 1)}$$

$$2(x + 1)(x - 1)\frac{1}{x - 1} + 2(x + 1)(x - 1)\frac{1}{2}$$
$$= 2(x + 1)(x - 1)\frac{2}{(x + 1)(x - 1)}$$

$$2(x + 1) + (x + 1)(x - 1) = 4$$
$$2x + 2 + x^2 - 1 = 4$$
$$x^2 + 2x + 1 = 4$$
$$x^2 + 2x - 3 = 0$$

Factoring, we have

$$(x + 3)(x - 1) = 0.$$

Therefore, it seems $x = -3$ or $x = 1$. But 1 makes two of the denominators from the original equation equal to zero and thus is not a solution. However, -3 is a solution, as you can show by substituting -3 for x in the original equation.

Work Quick Check 5 at the side.

4. Solve each equation and check your answers.

(a) $\frac{4}{3m + 3} = \frac{m + 1}{m^2 + m}$

(b) $\frac{2}{p^2 - 2p} = \frac{3}{p^2 - p}$

5. Solve each equation and check your answers.

(a) $\frac{1}{x - 2} + \frac{1}{5} = \frac{2}{5(x^2 - 4)}$

(b) $\frac{2}{k - 1} - 1 = \frac{3}{k^2 - 1}$

Quick Check Answers

4. (a) 3 (b) 4
5. (a) $-4, -1$ (b) 0, 2

name date hour

6.4 EXERCISES

Solve each equation and check your answers. See Examples 1 and 2.

1. $\frac{1}{4}=\frac{x}{2}$ ________
2. $\frac{2}{m}=\frac{5}{12}$ ________
3. $\frac{9}{k}=\frac{3}{4}$ ________
4. $\frac{p}{15}=\frac{4}{15}$ ________
5. $\frac{3}{4}-m=2m$ ________
6. $3r-\frac{1}{2}=\frac{11}{2}$ ________
7. $\frac{6}{x}-\frac{4}{x}=5$ ________
8. $\frac{3}{x}+\frac{2}{x}=5$ ________
9. $\frac{x}{2}-\frac{x}{4}=6$ ________
10. $\frac{4}{y}+\frac{2}{3}=1$ ________
11. $\frac{9}{m}=5-\frac{1}{m}$ ________
12. $\frac{3x}{5}+2=\frac{1}{4}$ ________
13. $\frac{2t}{7}-5=t$ ________
14. $\frac{1}{2}+\frac{2}{m}=1$ ________
15. $\frac{x+1}{2}=\frac{x+2}{3}$ ________
16. $\frac{t-4}{3}=t+2$ ________
17. $\frac{3m}{2}+m=5$ ________
18. $\frac{9}{x+1}=3$ ________
19. $\frac{9}{x-2}=3$ ________
20. $\frac{2y-1}{y}+2=\frac{1}{2}$ ________
21. $\frac{2k+3}{k}=\frac{3}{2}$ ________
22. $\frac{a}{2}-\frac{17+a}{5}=2a$ ________
23. $\frac{5-y}{y}+\frac{3}{4}=\frac{7}{y}$ ________
24. $\frac{x}{x-4}=\frac{2}{x-4}+5$ ________
25. $\frac{a-4}{4}=\frac{a+8}{16}$ ________
26. $\frac{m-2}{5}=\frac{m+8}{10}$ ________

27. $\dfrac{2p+8}{9} = \dfrac{10p+4}{27}$ ____________

28. $\dfrac{5r-3}{7} = \dfrac{15r-2}{28}$ ____________

Solve each equation and check your answers. See Examples 3-5.

29. $\dfrac{8x-1}{6x+8} = \dfrac{3}{4}$ ____________

30. $\dfrac{6m+9}{5m+10} = \dfrac{3}{5}$ ____________

31. $\dfrac{2}{y} = \dfrac{y}{5y-12}$ ____________

32. $\dfrac{8x+3}{x} = 3x$ ____________

33. $\dfrac{m}{2m+2} = \dfrac{-2m}{4m+4} + \dfrac{2m-3}{m+1}$ ____________

34. $\dfrac{5p+1}{3p+3} = \dfrac{5p-5}{5p+5} + \dfrac{3p-1}{p+1}$ ____________

35. $\dfrac{1}{x^2+5x+6} + \dfrac{1}{x^2-2x-8} = \dfrac{-1}{12(x+2)}$ ____________

36. $\dfrac{x+4}{x^2-3x+2} - \dfrac{5}{x^2-4x+3} = \dfrac{x-4}{x^2-5x+6}$ ____________

OBJECTIVES

- Solve problems about numbers involving rational expressions
- Solve distance problems involving rational expressions
- Solve work problems involving rational expressions

We are now ready to discuss some applications which involve rational expressions.

Example 1 If the same number is added to both the numerator and denominator of the fraction 3/4, the result is 5/6. Find the number.

If x represents the number that is added to the numerator and denominator, we can write

$$\frac{3+x}{4+x}$$

to represent the result of adding the same number to both the numerator and denominator. Since this result is $\frac{5}{6}$,

$$\frac{3+x}{4+x} = \frac{5}{6}.$$

If we multiply both sides of the equation by the common denominator $6(4 + x)$, we have

$$6(4+x)\frac{3+x}{4+x} = 6(4+x)\frac{5}{6}$$
$$6(3+x) = 5(4+x)$$
$$18 + 6x = 20 + 5x$$
$$x = 2.$$

Work Quick Check 1 at the side.

QUICK CHECKS

1. A certain number is added to the numerator and subtracted from the denominator of 5/8. The new number equals the reciprocal of 5/8. Find the number.

Example 2 The Big Muddy River has a current of 3 miles per hour. A motorboat takes as long to go 12 miles downstream as to go 8 miles upstream. What is the speed of the boat in still water?

This problem requires the distance formula, $d = rt$ (distance = rate · time). For our problem, let's use x to represent the speed of the boat in still water. Since the current pushes the boat when the boat is going downstream, the speed of the boat downstream will be the sum of the speed of the boat and the speed of the current, or $x + 3$ miles per hour. Similarly, the boat's speed going upstream is given by $x - 3$ miles per hour.

We can summarize the information in the problem in a chart.

	d	r	t
downstream	12	$x+3$	
upstream	8	$x-3$	

Quick Check Answers
1. 3

To fill in the last column, representing time, we solve the formula $d = rt$ for t.

$$d = rt$$
$$\frac{d}{r} = t$$

Then the time upstream is

$$\frac{d}{r} = \frac{8}{x - 3},$$

while the time downstream is

$$\frac{d}{r} = \frac{12}{x + 3}.$$

Now we can complete the chart.

	d	r	t
downstream	12	$x + 3$	$\frac{12}{x + 3}$
upstream	8	$x - 3$	$\frac{8}{x - 3}$

The problem states that the time upstream equals the time downstream. Thus the two times from the chart are equal.

$$\frac{12}{x + 3} = \frac{8}{x - 3}$$

To solve this equation, multiply both sides by $(x + 3)(x - 3)$.

$$(x + 3)(x - 3)\frac{12}{x + 3} = (x + 3)(x - 3)\frac{8}{x - 3}$$
$$12(x - 3) = 8(x + 3)$$
$$12x - 36 = 8x + 24$$
$$4x = 60$$
$$x = 15$$

The speed of the boat in still water is 15 miles per hour.

To check, note that the speed of the boat downstream is $15 + 3 = 18$ miles per hour, and traveling 12 miles takes

$$t = 12/18,$$
$$t = 2/3 \text{ hour}.$$

On the other hand, the speed of the boat upstream is $15 - 3 = 12$ miles per hour, and traveling 8 miles takes

$$t = 8/12,$$
$$t = 2/3 \text{ hour}.$$

The time upstream equals the time downstream, as required.

Work Quick Check 2 at the side.

Example 3 Working alone, John can cut his lawn in 8 hours. If John's pet sheep is released to eat the grass, the lawn can be cut in 14 hours. If both John and the sheep work on the lawn, how long will it take to cut it?

Let x be the number of hours that it takes John and the sheep to cut the lawn, working together. Certainly x will be less than 8, since John alone can cut the lawn in 8 hours. In one hour, John can do $1/8$ of the lawn, and in one hour the sheep can do $1/14$ of the lawn. Since it takes them x hours to cut the lawn when working together, in one hour together they can do $1/x$ of the lawn. The amount of the lawn cut by John in one hour plus the amount cut by the sheep in one hour must equal the amount they can do together in one hour. In symbols,

$$\frac{1}{8} + \frac{1}{14} = \frac{1}{x}.$$

The quantity $56x$ is the least common denominator for 8, 14, and x, so we multiply both sides of the equation by $56x$.

$$\begin{aligned} 56x\left(\frac{1}{8} + \frac{1}{14}\right) &= 56x \cdot \frac{1}{x} \\ 56x \cdot \frac{1}{8} + 56x \cdot \frac{1}{14} &= 56x \cdot \frac{1}{x} \\ 7x + 4x &= 56 \\ 11x &= 56 \\ x &= \frac{56}{11}. \end{aligned}$$

Working together, John and his sheep can cut the lawn in $56/11$ hours, or $5\ 1/11$ hours, about 5 hours and 5 minutes.

Work Quick Check 3 at the side.

2. An airplane, maintaining a constant airspeed, takes as long to go 450 miles with the wind as it does to go 375 miles against the wind. If the wind is blowing at 15 mph, what is the speed of the plane?

3. Jerry and Louise operate a small roofing company. Louise can roof an average house alone in 9 hours. Jerry can roof a house alone in 8 hours. How long will it take them to do the job if they work together?

Quick Check Answers

2. 165 mph

3. $\frac{72}{17}$ hours

WORK SPACE

name date hour

6.5 EXERCISES

Solve each problem. See Example 1.

1. One half of a number is three more than one-sixth of the same number. What is the number? ____________

2. The numerator of the fraction 4/7 is increased by an amount so that the value of the resulting fraction is 27/21. By what amount was the numerator increased? ____________

3. In a certain fraction, the denominator is 5 larger than the numerator. If 3 is added to both the numerator and the denominator, the result is 3/4. Find the original fraction. ____________

4. The denominator of a certain fraction is three times the numerator. If 1 is added to the numerator and subtracted from the denominator, the result equals 1/2. Find the original fraction. ____________

5. One number is three more than another. If the smaller is added to two thirds the larger, the result is four fifths the sum of the original numbers. Find the numbers. ____________

6. The sum of a number and its reciprocal is 5/2. Find the number. ____________

7. If twice the reciprocal of a number is subtracted from the number, the result is −7/3. Find the number. ____________

8. The sum of the reciprocals of two consecutive integers is 5/6. Find the integers. ____________

9. If three times a number is added to twice its reciprocal, the answer is 5. Find the number. ____________

10. If twice a number is subtracted from 3 times its reciprocal, the result is 1. Find the number. ____________

11. A man and his son worked four days at a job. The son's daily wage was 2/5 that of the father. If together they earned \$336, what were their daily wages? ____________

12. The profits from a student show are to be given to two scholarships in the ratio 2/3 (2 to 3). If the fund receiving the larger amount was given \$390, how much was given to the other fund? ____________

Solve each problem. See Example 2.

13. Sam can row four miles per hour in still water. It takes as long rowing 8 miles upstream as 24 miles downstream. How fast is the current? ____________

14. Mary flew from Philadelphia to Des Moines at 180 miles per hour, and from Des Moines to Philadelphia at 150 miles per hour. The trip at the slower speed took one hour longer than the trip at the higher speed. Find the distance between the two cities. (Assume there was no wind in either direction.) ____________

15. On a business trip, Arlene traveled to her destination at an average speed of 60 m.p.h. Coming home, her average speed was 50 m.p.h. and the trip took 1/2 hour longer. How far did she travel each way? ____________

16. Rae flew her airplane 500 miles against the wind in the same time it took her to fly it 600 miles with the wind. If the speed of the wind was 10 m.p.h., what was the average speed of her plane? ____________

17. The distance from Seattle, Washington to Victoria, British Columbia is about 148 miles by ferry. It takes about 2 hours less to travel by ferry from Victoria to Vancouver, British Columbia, a distance of about 74 miles. What is the average speed of the ferry? ____________

18. Ron flew his plane 800 miles in 1.5 hours more than it took him to fly 500 miles. What was his average speed? ____________

Solve each problem. See Example 3.

19. Paul can tune his Toyota engine in 2 hours. His friend Marco can do the job in 3 hours. How long would it take them if they worked together? ____________

name date hour

20. George can paint a room, working alone, in 8 hours. Jenny can paint the same room, working alone, in 6 hours. How long will it take them if they work together? ____________

21. Machine A can do a certain job in 7 hours, while machine B takes 12 hours. How long will it take the two machines working together? ____________

22. One pipe can fill a swimming pool in 6 hours, while another pipe can do it in 9 hours. How long will it take the two pipes working together to fill the pool 3/4 full? ____________

23. Dennis can do a job in 4 days. When Dennis and Sue work together, the job takes 2 1/3 days. How long would the job take Sue if she worked alone? ____________

24. An inlet pipe can fill a swimming pool in nine hours, while an outlet pipe can empty the pool in 12 hours. Through an error, both pipes are left open. How long will it take to fill the pool? ____________

25. A cold water faucet can fill a sink in 12 minutes, and a hot water faucet in 15. The drain can empty the sink in 25 minutes. If both faucets are on and the drain is open, how long will it take to fill the sink? ____________

26. Refer to Exercise 24. Assume the error was discovered after both pipes had been running for 3 hours, and the outlet pipe was then closed. How much more time would then be required to fill the pool? (Hint: How much of the job had been done when the error was discovered?) ____________

WORK SPACE

OBJECTIVES

- Simplify complex fractions

A rational expression containing fractions in the numerator, denominator, or both, is called a **complex fraction**. Examples of complex fractions include

$$\frac{3+\frac{4}{x}}{5}, \quad \frac{\frac{3x^2-5x}{6x^2}}{2x-\frac{1}{x}}, \quad \frac{3+x}{5-\frac{2}{x}}.$$

Complex fractions are simplified by rewriting the numerator and denominator as single fractions and then performing the indicated division.

Example 1 Simplify the complex fraction

$$\frac{6+\frac{3}{x}}{\frac{2x+1}{8}}.$$

First write the numerator as a rational expression by adding 6 and $3/x$.

$$6+\frac{3}{x}=\frac{6}{1}+\frac{3}{x}$$
$$=\frac{6x}{x}+\frac{3}{x}$$
$$=\frac{6x+3}{x}$$

The complex fraction can now be written as

$$\frac{\frac{6x+3}{x}}{\frac{2x+1}{8}}.$$

Now use the rule for division and the fundamental property.

$$\frac{6x+3}{x}\div\frac{2x+1}{8}=\frac{6x+3}{x}\cdot\frac{8}{2x+1}$$
$$=\frac{3(2x+1)}{x}\cdot\frac{8}{2x+1}$$
$$=\frac{24}{x}$$

Work Quick Check 1 at the side.

QUICK CHECKS

1. Simplify the complex fractions.

(a) $\frac{m+\frac{1}{2}}{\frac{6m+3}{4m}}$

(b) $\frac{6+\frac{1}{x}}{5-\frac{2}{x}}$

Quick Check Answers

1. (a) $\frac{2m}{3}$ (b) $\frac{6x+1}{5x-2}$

2. Simplify the complex fractions.

(a) $\dfrac{\dfrac{rs^2}{t}}{\dfrac{r^2s}{t^2}}$

(b) $\dfrac{\dfrac{m^2n^3}{p}}{\dfrac{m^4n}{p^2}}$

3. Simplify.

$\dfrac{\dfrac{m+n}{m}}{\dfrac{m}{n}-\dfrac{n}{m}}$

Quick Check Answers

2. (a) $\dfrac{st}{r}$ (b) $\dfrac{n^2p}{m^2}$

3. $\dfrac{n}{m-n}$

Example 2 Simplify the complex fraction

$$\frac{\dfrac{xp}{q^3}}{\dfrac{p^2}{qx^2}}.$$

Here, the numerator and denominator are already single fractions, so we use the division rule and then the fundamental property.

$$\frac{xp}{q^3} \div \frac{p^2}{qx^2} = \frac{xp}{q^3} \cdot \frac{qx^2}{p^2}$$
$$= \frac{x^3}{q^2p}$$

Work Quick Check 2 at the side.

Example 3 Simplify the complex fraction

$$\frac{\dfrac{x}{x+y}}{\dfrac{1}{x}+\dfrac{1}{y}}.$$

First simplify the denominator by adding $1/x$ and $1/y$.

$$\frac{1}{x}+\frac{1}{y} = \frac{y}{xy}+\frac{x}{xy}$$
$$= \frac{x+y}{xy}$$

Now use the division rule and the fundamental property.

$$\frac{x}{x+y} \div \frac{x+y}{xy} = \frac{x}{x+y} \cdot \frac{xy}{x+y}$$
$$= \frac{x^2y}{(x+y)^2}$$

Work Quick Check 3 at the side.

name date hour

6.6 EXERCISES

Simplify each complex fraction. See Examples 1–3.

1. $\dfrac{\dfrac{p}{q^2}}{\dfrac{p^2}{q}}$ ______

2. $\dfrac{\dfrac{ab}{x}}{\dfrac{a^2}{2x}}$ ______

3. $\dfrac{\dfrac{x}{y}}{\dfrac{x^2}{y}}$ ______

4. $\dfrac{\dfrac{pq}{r}}{\dfrac{p^2q}{r^2}}$ ______

5. $\dfrac{\dfrac{x+1}{y}}{\dfrac{y+1}{x}}$ ______

6. $\dfrac{\dfrac{m+n}{m}}{\dfrac{m-n}{n}}$ ______

7. $\dfrac{\dfrac{a}{b+1}}{\dfrac{a^2}{b}}$ ______

8. $\dfrac{\dfrac{1}{x}}{\dfrac{1+x}{1-x}}$ ______

9. $\dfrac{\dfrac{2x+3}{y}}{\dfrac{x-1}{2y}}$ ______

10. $\dfrac{\dfrac{k}{k+1}}{\dfrac{5}{2(k+1)}}$ ______

11. $\dfrac{\dfrac{3}{y}+1}{\dfrac{3+y}{2}}$ ______

12. $\dfrac{y+\dfrac{2}{y}}{\dfrac{y^2+2}{3}}$ ______

13. $\dfrac{\dfrac{1}{x}+\dfrac{1}{y}}{\dfrac{1}{x+y}}$ ______

14. $\dfrac{m+\dfrac{1}{m}}{\dfrac{3}{m}-m}$ ______

15. $\dfrac{x + \dfrac{1}{y}}{\dfrac{1}{x} + y}$ ____________________

16. $\dfrac{y - \dfrac{1}{y}}{y + \dfrac{1}{y}}$ ____________________

17. $\dfrac{\dfrac{p+q}{p}}{\dfrac{1}{p} + \dfrac{1}{q}}$ ____________________

18. $\dfrac{r + \dfrac{1}{r}}{\dfrac{1}{r} - r}$ ____________________

19. $\dfrac{\dfrac{1}{m+n}}{\dfrac{4}{m^2 - n^2}}$ ____________________

20. $\dfrac{\dfrac{a}{a+1}}{\dfrac{2}{a^2 - 1}}$ ____________________

21. $\dfrac{\dfrac{1}{m+1} - 1}{\dfrac{1}{m-1} + 1}$ ____________________

22. $\dfrac{\dfrac{2}{x-1} + 2}{\dfrac{2}{x+1} - 2}$ ____________________

23. $\dfrac{\dfrac{y+1}{y-1}}{\dfrac{1}{y+1}}$ ____________________

24. $\dfrac{\dfrac{a-b}{a+b}}{\dfrac{a}{a-b}}$ ____________________

name ________ date ________ hour ________

CHAPTER 6 TEST

Reduce to lowest terms.

1. $\frac{8m^2n^2}{4m^3n^5}$ ________

2. $\frac{5s^3 - 5s}{2s + 2}$ ________

Perform the indicated operations. Write all answers in lowest terms.

3. $\frac{x^6y}{x^3} \cdot \frac{y^2}{x^2y^3}$ ________

4. $\frac{m^3}{n} \div \frac{m^2}{n^3}$ ________

5. $\frac{5}{x} - \frac{6}{x}$ ________

6. $\frac{1}{a+1} + \frac{5}{6a+6}$ ________

7. $1 - \frac{2}{t}$ ________

8. $\frac{-3}{2k} + \frac{1}{k+2}$ ________

9. $\frac{6m^2 - m - 2}{8m^2 + 10m + 3} \cdot \frac{4m^2 + 7m + 3}{3m^2 + 5m + 2}$ ________

10. $\frac{5a^2 + 7a - 6}{2a^2 + 3a - 2} \div \frac{5a^2 + 17a - 12}{2a^2 + 5a - 3}$ ________

11. $\frac{\frac{3}{x}}{\frac{1}{1+x}}$ ________

Solve each equation.

12. $\frac{1}{8} = \frac{x}{12}$ ________

13. $\frac{3}{t-1} + \frac{1}{t+1} = \frac{6}{5}$ ________

For each problem, write an equation and solve it.

14. If four times a number is added to the reciprocal of twice the number, the result is 3. Find the number. ________

15. If the numerator of $3/x$ is decreased by x, the result is $2/3$. Find x. ________________

16. Harry can paint his house, working alone, in five hours. His wife Gertie can do it in four hours. How long will it take them working together? ________________

7 Graphing Linear Equations

7.1 LINEAR EQUATIONS IN TWO VARIABLES

OBJECTIVES

- Find the value of one variable in an equation, given the value of the other
- Complete ordered pairs for a given equation
- Decide whether a given ordered pair is a solution of a given equation

All the equations we have studied so far have contained only one variable, such as

$$3x + 5 = 12 \quad \text{or} \quad 2x^2 + x + 5 = 0.$$

In this chapter, we begin a study of equations in *two* variables, such as

$$y = 4x + 5 \quad \text{or} \quad 2x + 3y = 6.$$

Example 1 Suppose $y = 4x + 5$. Find y for the given values of x.

(a) $x = 3$

Substitute 3 for x.

$$y = 4x + 5$$
$$y = 4(3) + 5 \qquad \text{Let } x = 3$$
$$y = 12 + 5$$
$$y = 17$$

If $x = 3$, then $y = 17$.

(b) $x = -5$

$$y = 4(-5) + 5 = -20 + 5 = -15$$

If $x = -5$, then $y = -15$.

In part (a) of Example 1, when $x = 3$, then $y = 17$. This statement is usually abbreviated

$$(3, 17).$$

This abbreviation gives the x-value, 3, and the y-value, 17, as a pair of numbers, written inside parentheses. The x-value is always given first. A pair of numbers written in this order is called an **ordered pair.**

Example 2 Complete the ordered pairs for the equation $y = 4x + 5$.

QUICK CHECKS

1. Complete the ordered pairs for the equation $y = 2x - 9$.

(a) (5,)

(b) (2,)

(c) (, 7)

(d) (, −13)

Quick Check Answers

1. (a) (5, 1) (b) (2, −5) (c) (8, 7) (d) (−2, −13)

(a) (7,)

In this ordered pair, $x = 7$. To find the corresponding value of y, replace x with 7 in the equation $y = 4x + 5$.

$$y = 4(7) + 5 = 28 + 5 = 33$$

This gives the ordered pair (7, 33).

(b) (−9,)

To find the value of y, replace x with −9 in the equation.

$$y = 4(-9) + 5 = -36 + 5 = -31$$

This gives the ordered pair (−9, −31).

(c) (, 13)

In this ordered pair, $y = 13$. To find the value of x, replace y with 13 in the equation, and then solve for x.

$$\begin{aligned} y &= 4x + 5 & & \\ 13 &= 4x + 5 & & \text{Let } y = 13 \\ 8 &= 4x & & \text{Add } -5 \text{ to both sides} \\ 2 &= x & & \text{Multiply both sides by } \frac{1}{2} \end{aligned}$$

This gives the ordered pair (2, 13).

Work Quick Check 1 at the side.

Example 3 Complete the ordered pairs for the equation $5x - y = 24$.

Equation	Ordered pairs
$5x - y = 24$	(5,) (−3,) (0,)

To find the y-value of the ordered pair (5,), replace x with 5 in the equation and solve the resulting equation for y.

$$\begin{aligned} 5x - y &= 24 & & \\ 5(5) - y &= 24 & & \text{Let } x = 5 \\ 25 - y &= 24 & & \\ -y &= -1 & & \text{Add } -25 \text{ to both sides} \\ y &= 1 & & \end{aligned}$$

This gives the ordered pair (5, 1).

We complete the ordered pair (−3,) by letting $x = -3$ in the equation. Likewise, we complete (0,) by letting $x = 0$.

If $x = -3$,	If $x = 0$,
then $5x - y = 24$	then $5x - y = 24$
becomes $5(-3) - y = 24$	becomes $5(0) - y = 24$
$-15 - y = 24$	$0 - y = 24$
$-y = 39$	$-y = 24$
$y = -39$	$y = -24$

The completed ordered pairs are as follows.

Equation	Ordered pairs
$5x - y = 24$	$(5, 1)$ $(-3, -39)$ $(0, -24)$

2. Complete the ordered pairs for the equation $2x - 3y = 12$.

(a) $(0, \ )$

(b) $(\ , 0)$

(c) $(3, \ )$

(d) $(-6, \ )$

(e) $(\ , -3)$

(f) $(4, \ )$

Example 4 Complete the ordered pairs for the equation $x - 2y = 8$.

Equation	Ordered pairs
$x - 2y = 8$	$(2, \ )$ $(\ , 0)$
	$(10, \ )$ $(\ , -2)$

Complete the two ordered pairs on the left by letting $x = 2$ and $x = 10$, respectively.

If	$x = 2,$	If	$x = 10,$
then	$x - 2y = 8$	then	$x - 2y = 8$
becomes	$2 - 2y = 8$	becomes	$10 - 2y = 8$
	$-2y = 6$		$-2y = -2$
	$y = -3$		$y = 1$

Now complete the two ordered pairs on the right by letting $y = 0$ and $y = -2$, respectively.

If	$y = 0$	If	$y = -2,$
then	$x - 2y = 8$	then	$x - 2y = 8$
becomes	$x - 2(0) = 8$	becomes	$x - 2(-2) = 8$
	$x - 0 = 8$		$x + 4 = 8$
	$x = 8$		$x = 4$

The completed ordered pairs are as follows.

Equation	Ordered pairs
$x - 2y = 8$	$(2, -3)$ $(8, 0)$
	$(10, 1)$ $(4, -2)$

Work Quick Check 2 at the side.

Example 5 Complete the ordered pairs for the equation $x = 5$.

Equation	Ordered pairs
$x = 5$	$(\ , -2)$ $(\ , 6)$ $(\ , 3)$

The equation we are given here is $x = 5$. Therefore, no matter which value of y we might choose, we always have the same value of x, 5. Therefore, each ordered pair can be completed by placing 5 in the first position.

Equation	Ordered pairs
$x = 5$	$(5, -2)$ $(5, 6)$ $(5, 3)$

When an equation such as $x = 5$ is discussed along with equations of two variables, it is customary to think of $x = 5$ as an

Quick Check Answers
2. (a) $(0, -4)$ (b) $(6, 0)$ (c) $(3, -2)$ (d) $(-6, -8)$ (e) $\left(\frac{3}{2}, -3\right)$ (f) $\left(4, -\frac{4}{3}\right)$

equation in two variables by rewriting $x = 5$ as $x + 0 \cdot y = 5$. This form shows that for any value of y, x always equals 5.

Work Quick Check 3 at the side.

The next example shows how to decide if an ordered pair gives a solution of an equation.

Example 6 Decide whether or not the given ordered pair is a solution of the given equation.

(a) $(3, 2)$; $2x + 3y = 12$

To see whether or not the ordered pair $(3, 2)$ is a solution of the equation $2x + 3y = 12$, substitute 3 for x and 2 for y in the given equation.

$$\begin{aligned} 2x + 3y &= 12 \\ 2(3) + 3(2) &= 12 \qquad \text{Let } x = 3; \text{let } y = 2 \\ 6 + 6 &= 12 \\ 12 &= 12 \qquad \text{True} \end{aligned}$$

This result is true, so $(3, 2)$ is a solution of $2x + 3y = 12$.

(b) $(-2, -7)$; $2x + 3y = 12$

$$\begin{aligned} 2(-2) + 3(-7) &= 12 \qquad \text{Let } x = -2; \text{let } y = -7 \\ -4 + (-21) &= 12 \\ -25 &= 12 \qquad \text{False} \end{aligned}$$

This result is false, so $(-2, -7)$ is *not* a solution of $2x + 3y = 12$.

Work Quick Check 4 at the side.

The equations we worked with in this section all fit the pattern

$$ax + by = c,$$

where a, b, and c are real numbers and a and b cannot both equal 0. Such an equation is called a **linear equation in two variables.** Examples of linear equations in two variables include the following.

$$2x + 3y = 12 \qquad y = 3x - 5 \qquad x = 5$$

3. Complete the ordered pairs for each equation.

(a) $x = 3$;
(, 2), (, −4), (, 0)

(b) $y = -4$;
(2,), (6,), (−5,)

4. Decide whether or not the ordered pairs are solutions of the equation $5x + 2y = 20$.

(a) (0, 10)

(b) (2, −5)

(c) (3, 2)

(d) (−4, 20)

(e) (8, 10)

Quick Check Answers
3. (a) (3, 2), (3, −4), (3, 0)
(b) (2, −4), (6, −4), (−5, −4)
4. (a) yes (b) no (c) no
(d) yes (e) no

name date hour

7.1 EXERCISES

Complete the given ordered pairs for the equation $y = 3x + 5$. See Example 2.

1. (2,)
2. (5,)
3. (8,)
4. (0,)
5. (−3,)
6. (−4,)
7. (, 14)
8. (, −10)

Complete the given ordered pairs for the equation $y = -4x + 8$. See Example 2.

9. (0,)
10. (2,)
11. (, 16)
12. (, 24)
13. (, −4)
14. (, −8)

Complete the ordered pairs using the given equations. See Examples 2–4.

	Equation	Ordered pairs
15.	$y = 2x + 1$	(3,) (0,) (−1,)
16.	$y = 3x - 5$	(2,) (0,) (−3,)
17.	$y = 8 - 3x$	(2,) (0,) (−3,)
18.	$y = -2 - 5x$	(4,) (0,) (−4,)
19.	$2x + y = 9$	(0,) (3,) (12,)
20.	$-3x + y = 4$	(1,) (0,) (−2,)
21.	$2x + 3y = 6$	(0,) (, 0) (, 4)
22.	$4x + 3y = 12$	(0,) (, 0) (, 8)
23.	$3x - 5y = 15$	(0,) (, 0) (, −6)

24. $4x - 9y = 36$ $\quad$ (, 0) (0,) (, 4)

25. $4x + 5y = 10$ $\quad$ (0,) (, 0) (, 3)

26. $2x - 3y = 4$ $\quad$ (, 0) (0,) (, 3)

27. $6x - 4y = 5$ $\quad$ (0,) (, 0) (2,)

28. $4x - 3y = 7$ $\quad$ (, 0) (2,) (, −1)

Complete the ordered pairs using the given equations. See Example 5.

Equation	Ordered pairs
29. $x = -4$	(, 6) (, 2) (, −3)
30. $x = 8$	(, 3) (, 8) (, 0)
31. $y = 3$	(8,) (4,) (−2,)
32. $y = -8$	(4,) (0,) (−4,)
33. $x + 9 = 0$	(, 8) (, 3) (, 0)
34. $y + 4 = 0$	(9,) (2,) (0,)

Decide whether or not the given ordered pair is a solution of the given equation. See Example 6.

35. $x + y = 9$; (2, 7) ____________

36. $3x + y = 8$; (0, 8) ____________

37. $2x - y = 6$; (2, −2) ____________

38. $2x + y = 5$; (2, 1) ____________

39. $4x - 3y = 6$; (1, 2) ____________

40. $5x - 3y = 1$; (0, 1) ____________

41. $y = 3x$; (1, 3) ____________

42. $x = -4y$; (8, −2) ____________

43. $x = -6$; (−6, 8) ____________

44. $y = 2$; (9, 2) ____________

45. $x + 4 = 0$; (−5, 1) ____________

46. $x - 6 = 0$; (5, −1) ____________

OBJECTIVES

- Graph ordered pairs

In Section 3.7 we used a number line to graph the solution of an equation in one variable. Now we want to graph the solutions of an equation of *two* variables. Since the solutions of such an equation are ordered pairs of numbers in the form (x, y), we need two number lines to do this. (One is for x and one for y.) These two number lines are drawn as shown in Figure 1. The horizontal number line is called the **x-axis.** The vertical line is called the **y-axis.** Together, the x-axis and y-axis form a **coordinate system.**

A coordinate system is divided into four regions, called **quadrants.** These quadrants are numbered counterclockwise as shown in Figure 1. The point where the x-axis and y-axis meet is called the **origin.**

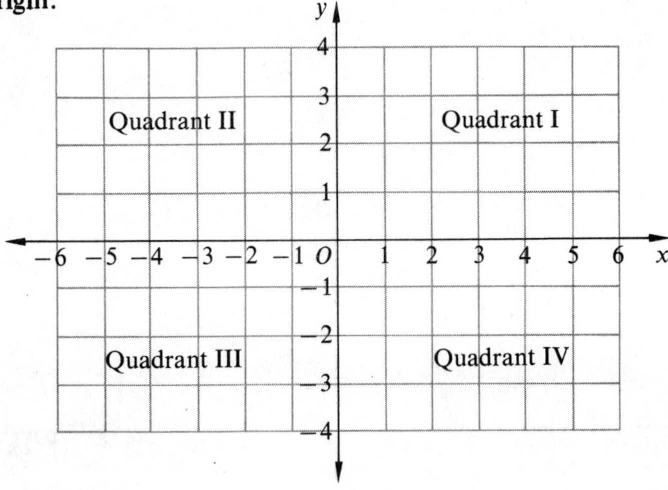

Figure 1

We can graph any ordered pair (x, y) on a coordinate system by using the x- and y- coordinates of the point. To graph the point $(2, 3)$, start at the origin. Since the x-coordinate is 2, go 2 units to the right along the x-axis. Then since the y-coordinate is 3, turn and go up 3 units on a line parallel to the y-axis. This is called **plotting** the point $(2, 3)$. (See Figure 2.)

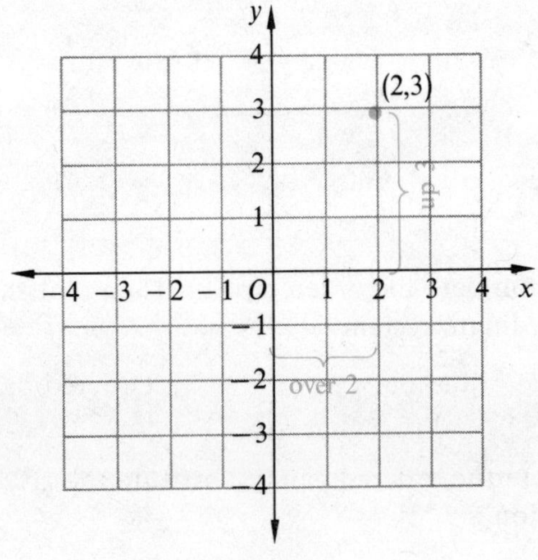

Figure 2

QUICK CHECKS

1. Plot the ordered pairs on a coordinate system.

(a) $(3, 5)$

(b) $(-2, 6)$

(c) $(-4, 0)$

(d) $(-5, -2)$

(e) $(5, -2)$

(f) $(0, -6)$

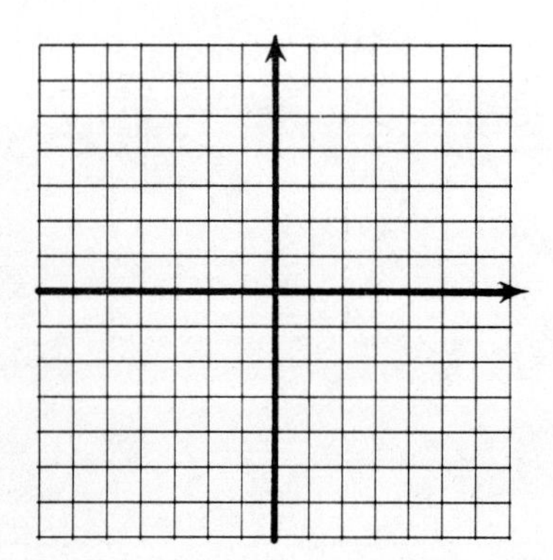

Example 1 Plot the ordered pairs on a coordinate system.

(a) $(1, 5)$

(b) $(-2, 3)$

(c) $(-1, -4)$

(d) $(7, -2)$

(e) $(3/2, 2)$

To locate the point $(-1, -4)$, for example, go 1 unit to the left along the x-axis. Then turn and go 4 units down, parallel to the y-axis. To plot the point $(3/2, 2)$, go $3/2$ (or $1\ 1/2$) units to the right along the x-axis. Then turn and go 2 units up parallel to the y-axis. Figure 3 shows the graphs of the points in this example.

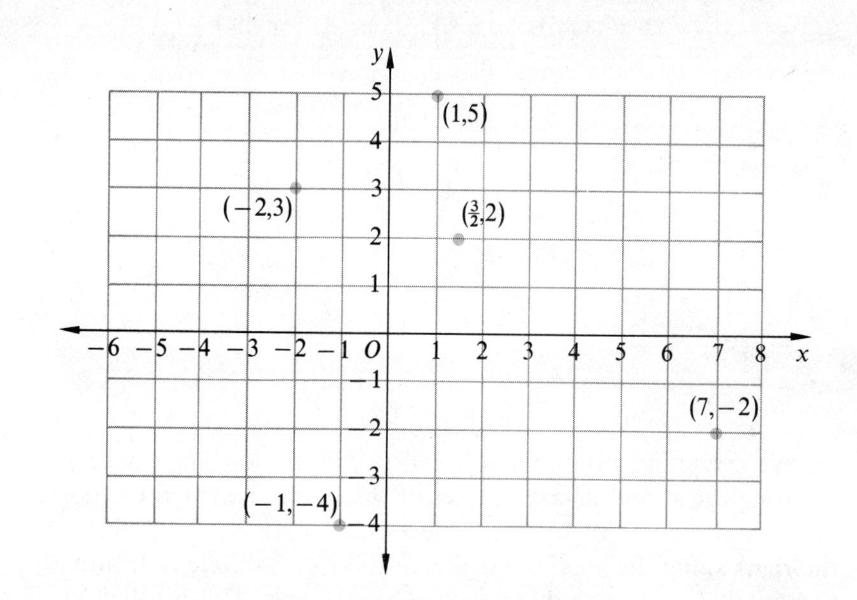

Figure 3

Work Quick Check 1 at the side.

Example 2 Complete the ordered pairs. Then plot the ordered pairs on a coordinate system.

Equation	Ordered pairs
$x + 2y = 7$	$(1,\ \ )$ $(-3,\ \ )$ $(3,\ \ )$ $(7,\ \ )$

To complete the ordered pairs, substitute the given x-values into the equation $x + 2y = 7$.

Quick Check Answers

1.

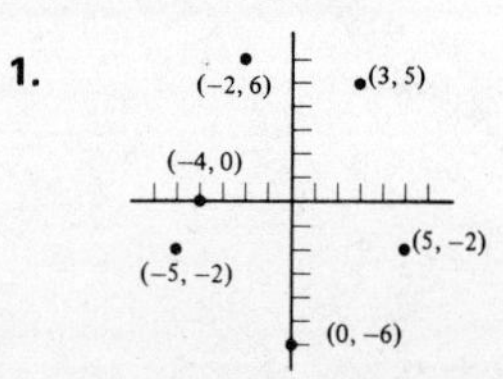

When $x = 1$,	When $x = -3$,
$x + 2y = 7$	$x + 2y = 7$
$1 + 2y = 7$	$-3 + 2y = 7$
$2y = 6$	$2y = 10$
$y = 3$	$y = 5$

This gives the ordered pairs (1, 3) and (−3, 5).

In the same way, if $x = 3$, then $y = 2$, giving (3, 2). Finally, if $x = 7$, then $y = 0$, giving (7, 0).

The completed ordered pairs are as follows.

Equation	Ordered pairs
$x + 2y = 7$	(1, 3) (−3, 5) (3, 2) (7, 0)

The graph of these ordered pairs is shown in Figure 4.

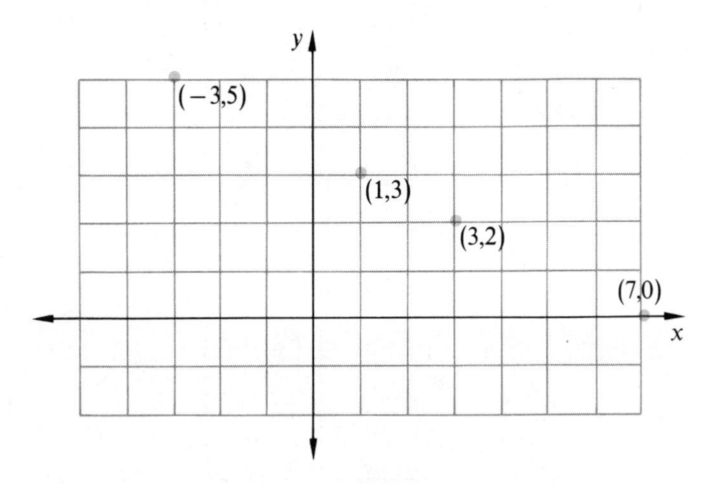

Figure 4

Work Quick Check 2 at the side.

Example 3 In a certain city, the cost of a taxi ride of x miles is given by the formula

$$y = 25x + 50,$$

where y represents the cost in cents. Complete the ordered pairs for this equation.

$$(1, \quad) \quad (2, \quad) \quad (3, \quad)$$

To complete the ordered pair (1,), we let $x = 1$.

$$y = 25x + 50$$
$$y = 25(1) + 50 \qquad \text{Let } x = 1$$
$$y = 25 + 50$$
$$y = 75$$

This gives the ordered pair (1, 75), which tells us that a taxi ride of 1 mile costs 75¢.

2. Complete the ordered pairs for the equation $2x + y = 6$ and graph your results.

(a) (0,)

(b) (2,)

(c) (4,)

(d) (, 1)

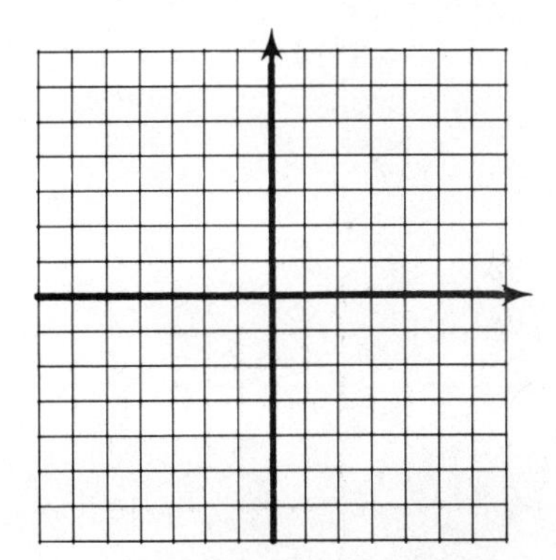

Quick Check Answers

2. (a) (0, 6) (b) (2, 2) (c) (4, −2) (d) $\left(\frac{5}{2}, 1\right)$

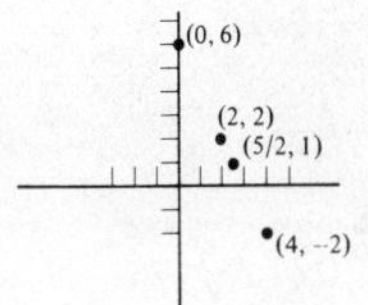

3. Complete the ordered pairs using the equation in Example 3.

(a) (4,)

(b) (5,)

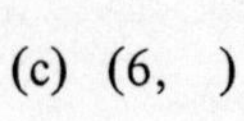

(c) (6,)

(d) (7,)

(e) (8,)

(f) (9,)

We complete (2,) and (3,) as follows.

$y = 25x + 50$		$y = 25x + 50$	
$y = 25(2) + 50$	Let $x = 2$	$y = 25(3) + 50$	Let $x = 3$
$y = 50 + 50$		$y = 75 + 50$	
$y = 100$		$y = 125$	

This gives the ordered pairs (2, 100) and (3, 125).

Work Quick Check 3 at the side.

The ordered pairs obtained above and in Quick Check 3 are plotted in Figure 5.

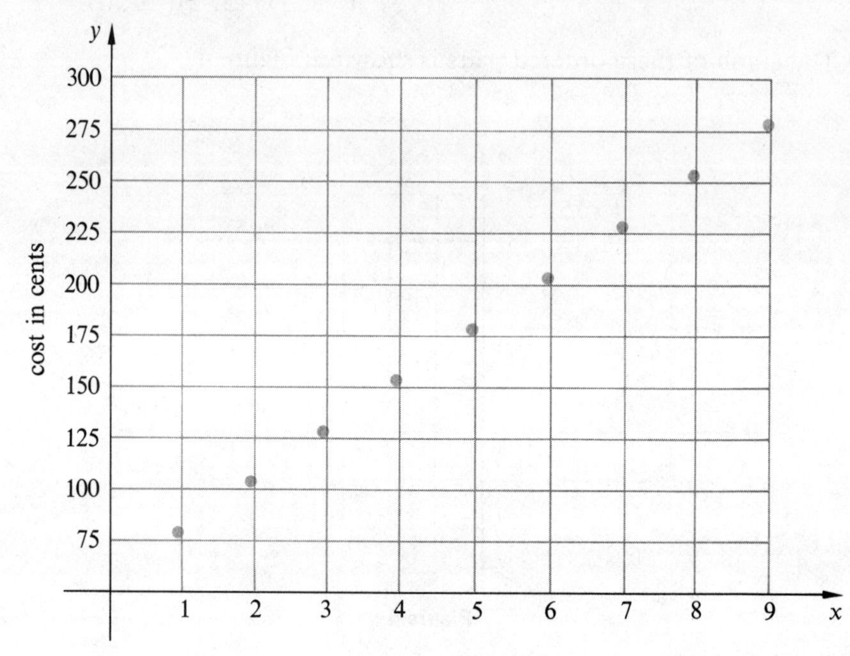

Figure 5

Quick Check Answers
3. (a) (4, 150) (b) (5, 175)
(c) (6, 200) (d) (7, 225)
(e) (8, 250) (f) (9, 275)

name date hour

7.2 EXERCISES

Write the x- and y- coordinates of the points labeled A through F in the figure.

1. *A* ______ 2. *B* ______ 3. *C* ______

4. *D* ______ 5. *E* ______ 6. *F* ______

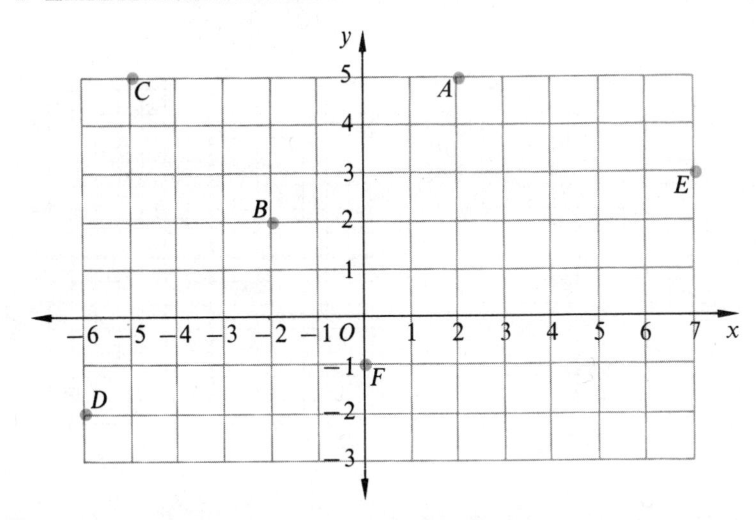

Plot the ordered pairs on the coordinate system provided. See Example 1.

7. (6, 1) 8. (4, −2) 9. (3, 5)

10. (−4, −5) 11. (−2, 4) 12. (−5, −1)

13. (−3, 5) 14. (3, −5) 15. (4, 0)

16. (1, 0) 17. (−2, 0) 18. (−5, 0)

19. (0, 3) 20. (0, 6) 21. (0, −5)

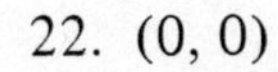

22. (0, 0)

Without plotting the given point, state the quadrant in which each point lies.

23. (2, 3) ______ 24. (2, −3) ______ 25. (−2, 3) ______

26. (−2, −3) ______ 27. (−1, −1) ______ 28. (4, 7) ______

29. $(-3, 6)$ ____________ 30. $(1, 5)$ ____________ 31. $(5, -4)$ ____________

32. $(9, -1)$ ____________ 33. $(0, 0)$ ____________ 34. $(-2, 0)$ ____________

Complete the ordered pairs using the given equation. Then plot the ordered pairs. See Example 2.

35. $x + 2y = 0$
$(0, \quad)$ $(\quad, 3)$
$(4, \quad)$ $(\quad, -1)$

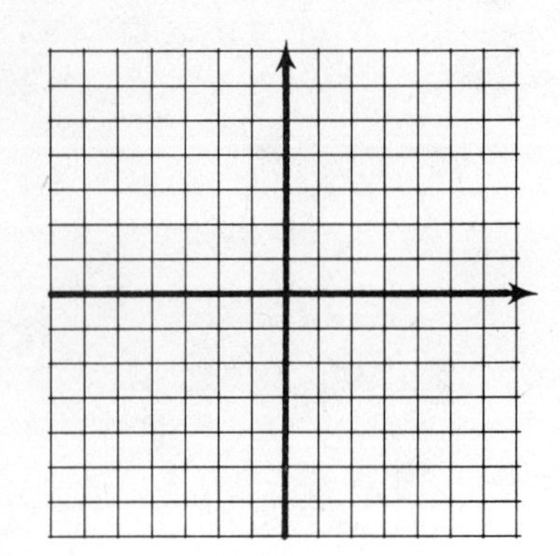

36. $x = 3$
$(\quad, 2)$ $(\quad, 5)$
$(\quad, 0)$ $(\quad, -3)$

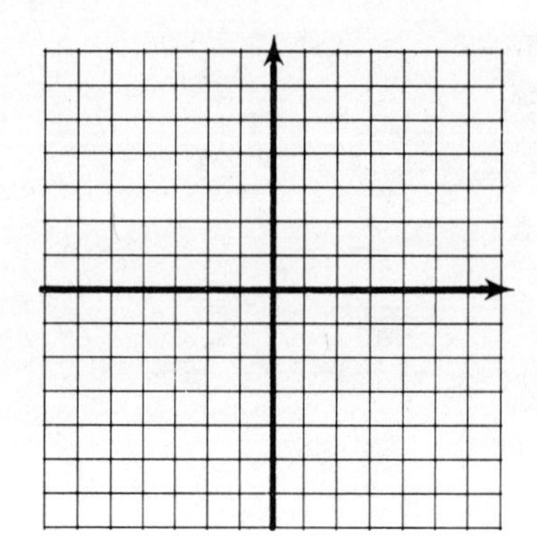

37. $y = -5$
$(2, \quad)$ $(-3, \quad)$
$(0, \quad)$ $(-1, \quad)$

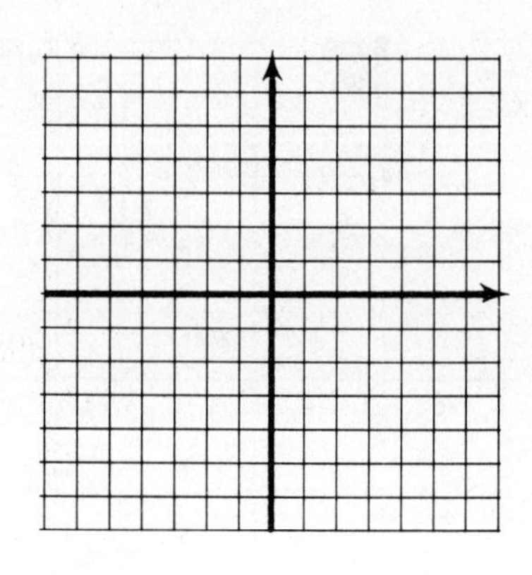

38. $y + 2 = 0$
$(5, \quad)$ $(0, \quad)$
$(-3, \quad)$ $(-2, \quad)$

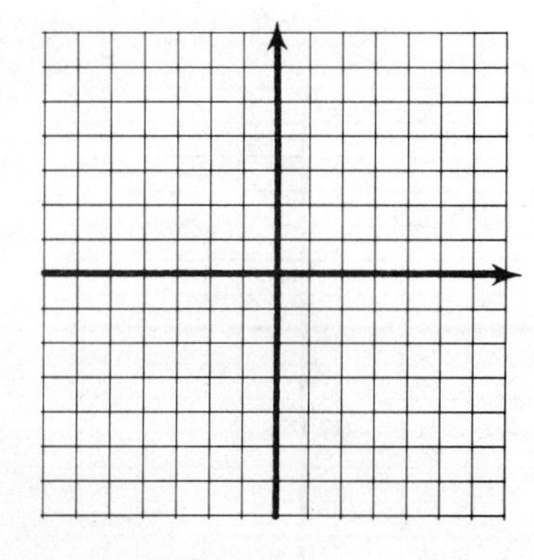

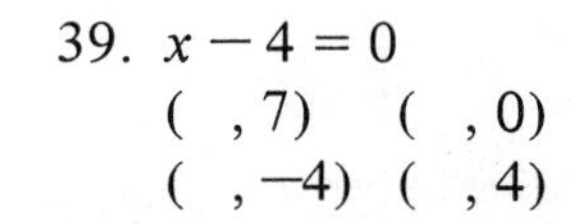

39. $x - 4 = 0$
$(\quad, 7)$ $(\quad, 0)$
$(\quad, -4)$ $(\quad, 4)$

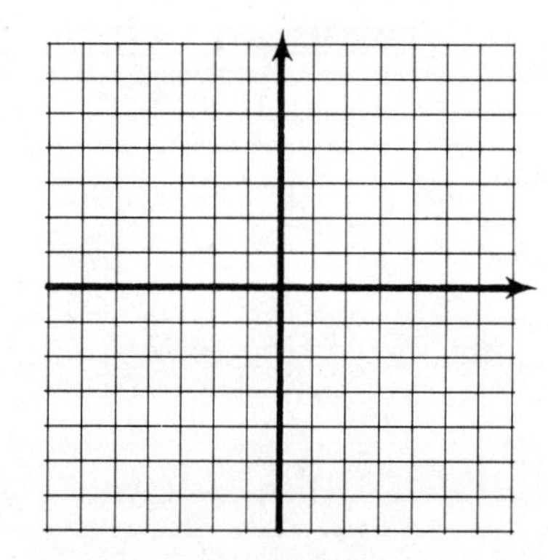

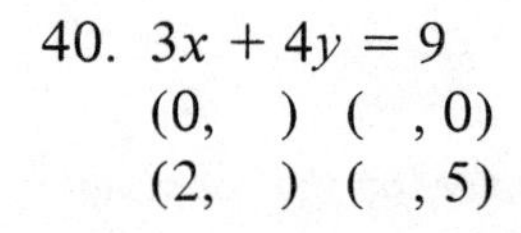

40. $3x + 4y = 9$
$(0, \quad)$ $(\quad, 0)$
$(2, \quad)$ $(\quad, 5)$

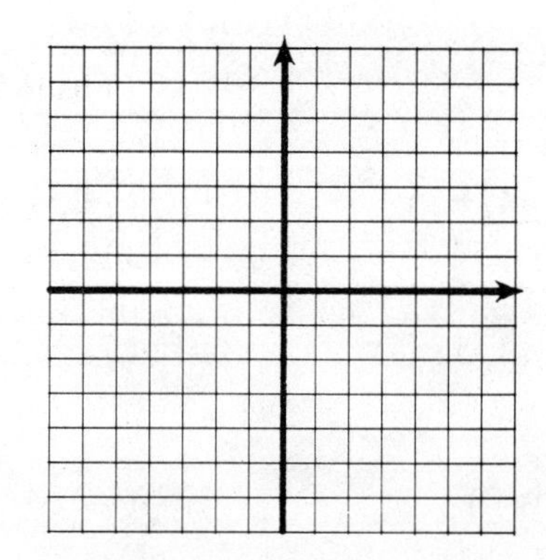

OBJECTIVES

- Graph linea. equations by completing ordered pairs
- Graph linear equations of the form y = a number
- Graph linear equations of the form x = a number

An infinite number of ordered pairs can make an equation in two variables true. For example, we can find as many ordered pairs as we want that are solutions of $x + 2y = 7$. This is done by choosing as many values of x (or y) as we want.

For example, if we choose $x = 1$, then $y = 3$, so that the ordered pair (1, 3) is a solution of the equation $x + 2y = 7$.

Work Quick Check 1 at the side.

All the ordered pairs that we have found for $x + 2y = 7$ have been graphed in Figure 6.

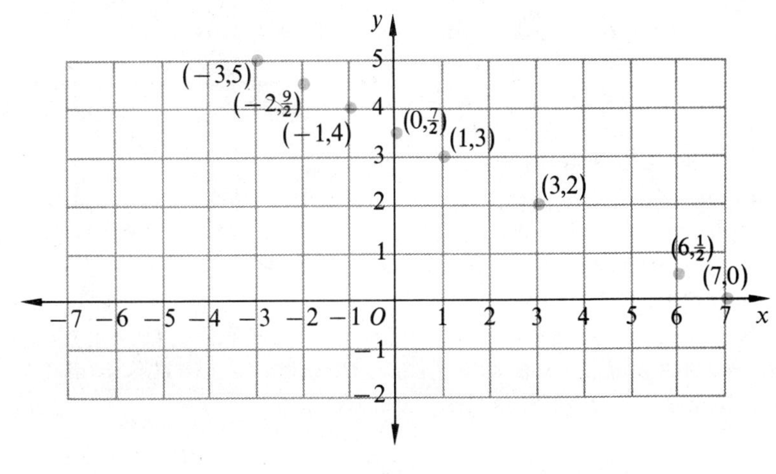

Figure 6

Notice that the points plotted in this figure all lie on a straight line. The line that goes through these points is shown in Figure 7. In fact, any point satisfying the equation $x + 2y = 7$ will lie on this straight line. (We only show a portion of this line here, but this line goes on forever in both directions.)

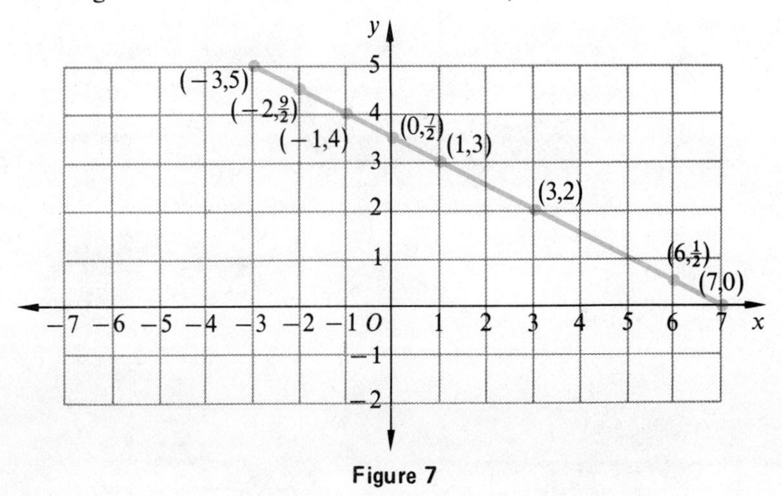

Figure 7

QUICK CHECKS

1. Complete the ordered pairs for the equation $x + 2y = 7$.

(a) (−3,)

(b) (3,)

(c) (0,)

(d) (−2,)

(e) (−1,)

(f) (6,)

(g) (7,)

Quick Check Answers

1. (a) (−3, 5) (b) (3, 2) (c) $\left(0, \frac{7}{2}\right)$ (d) $\left(-2, \frac{9}{2}\right)$ (e) (−1, 4) (f) $\left(6, \frac{1}{2}\right)$ (g) (7, 0)

2. Complete the ordered pairs and then graph the line.

$x + y = 6$
(0,)
(, 0)
(2,)

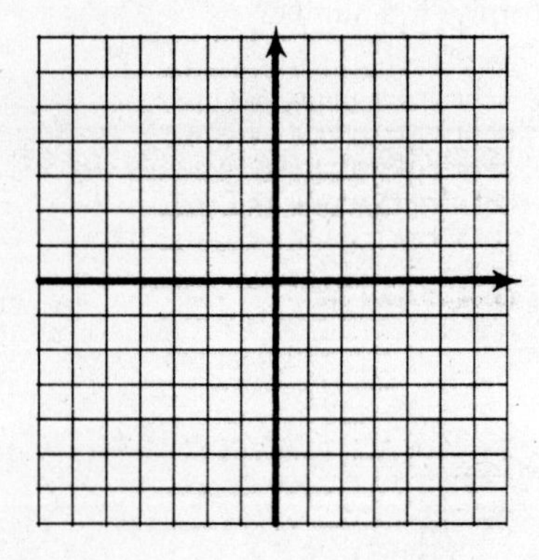

Quick Check Answers

2. (0, 6), (6, 0), (2, 4)

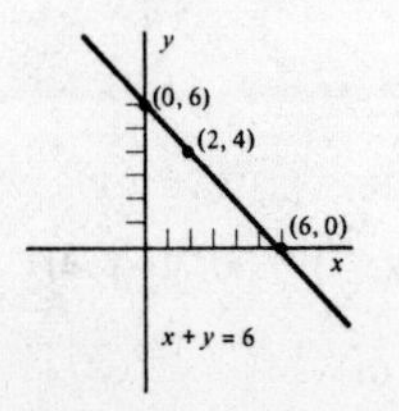

In general, the graph of any linear equation in two variables is a straight line. (Notice that the word *line* appears in the name "*linear* equation.") We can graph a straight line if we know any two points of the line. However, it is a good idea to plot a third point as a check. Knowing this, we can graph any linear equation by plotting only two or three ordered pairs which are solutions of the equation, and then drawing a straight line through them. Example 1 shows this procedure.

Example 1 Graph the linear equation $3x + 2y = 6$.

We need at least two points to draw the graph, so we first let $x = 0$ and then let $y = 0$.

$3x + 2y = 6$		$3x + 2y = 6$	
$3(0) + 2y = 6$	Let $x = 0$	$3x + 2(0) = 6$	Let $y = 0$
$0 + 2y = 6$		$3x + 0 = 6$	
$2y = 6$		$3x = 6$	
$y = 3$		$x = 2$	

This gives the ordered pairs (0, 3) and (2, 0). To get a third point (as a check), let x or y equal some number other than zero. For example, let $x = -2$. (We could have used any other number here as well.) Replace x with -2; you should find that $y = 6$, giving the ordered pair $(-2, 6)$.

Now plot the three ordered pairs we have found, (0, 3), (2, 0), and $(-2, 6)$, and draw a straight line through them. This straight line, shown in Figure 8, is the graph we want.

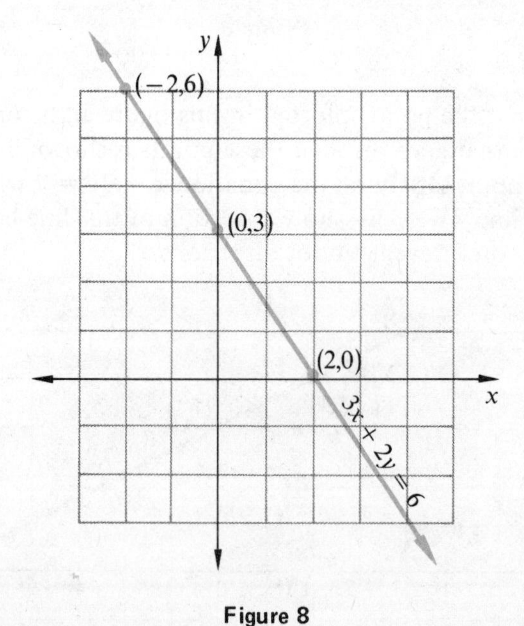

Figure 8

Work Quick Check 2 at the side.

Example 2 Graph the linear equation $4x - 5y = 20$.

Again, we need at least two points to draw the graph. By first letting $x = 0$ and then letting $y = 0$, we can complete two ordered pairs that will give us these points. This method will work for most linear equations.

$4x - 5y = 20$		$4x - 5y = 20$	
$4(0) - 5y = 20$	Let $x = 0$	$4x - 5(0) = 20$	Let $y = 0$
$-5y = 20$		$4x = 20$	
$y = -4$		$x = 5$	

This gives the ordered pairs $(0, -4)$ and $(5, 0)$.

To get a third ordered pair (as a check), choose some number other than zero for x or y. This time, let us choose $y = 2$.

$$4x - 5y = 20$$
$$4x - 5(2) = 20 \qquad \text{Let } y = 2$$
$$4x - 10 = 20$$
$$4x = 30$$
$$x = 15/2$$

This gives the ordered pair $(15/2, 2)$.

Plot the three ordered pairs we have found, $(0, -4)$, $(5, 0)$, and $(15/2, 2)$, and draw a straight line through them. This straight line, shown in Figure 9, is the graph we want.

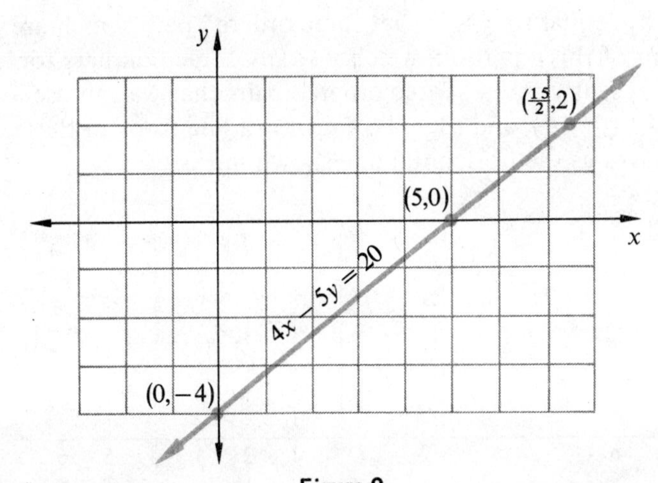

Figure 9

Work Quick Check 3 at the side.

3. Complete three ordered pairs and graph each line.

(a) $2x - 3y = 6$

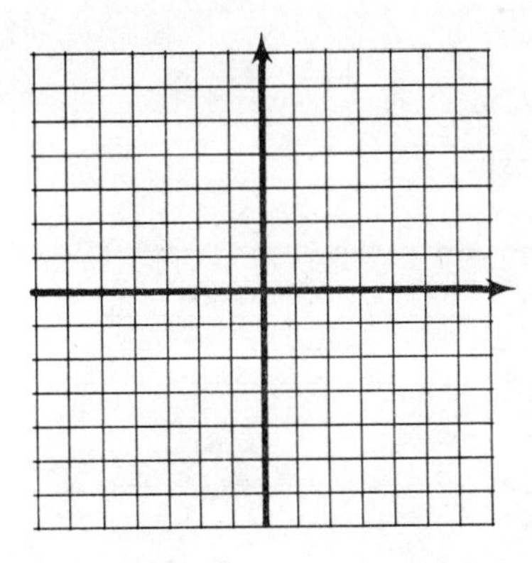

(b) $3y - 2x = 6$

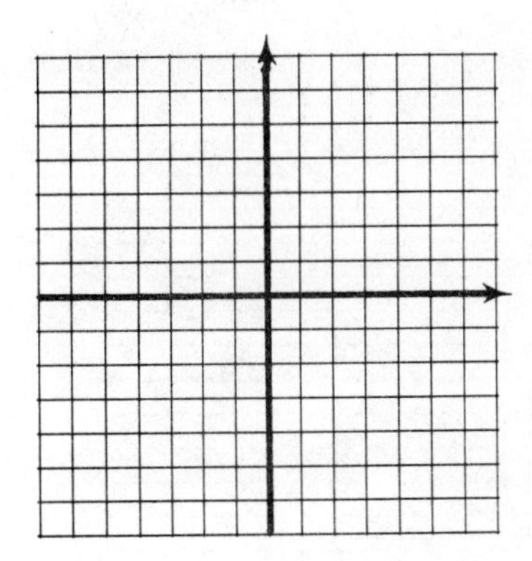

Example 3 Graph the linear equation $x = 3y$.

If we let $x = 0$, we get $y = 0$, giving $(0, 0)$. If $y = 0$, we also get the same ordered pair, $(0, 0)$. This is the same point, so we want to find two more points satisfying $x = 3y$. If we choose $y = 2$, we get $x = 6$, giving the ordered pair $(6, 2)$. Also, if we choose $x = -6$, we get $y = -2$, giving the ordered pair $(-6, -2)$. These three ordered pairs were used to get the graph shown in Figure 10.

Quick Check Answers

3. (a) (b)

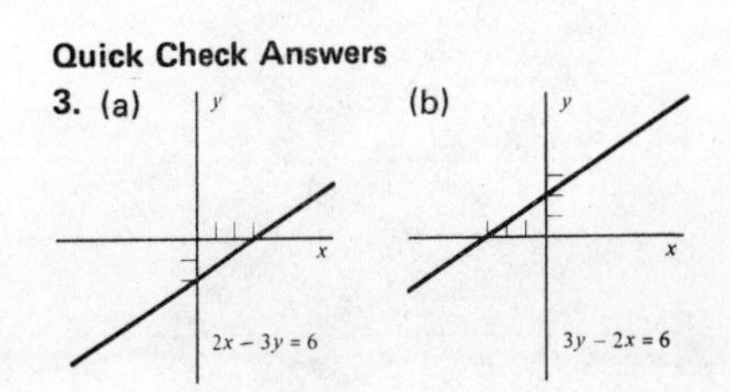

4. Graph each equation.

(a) $2x = y$

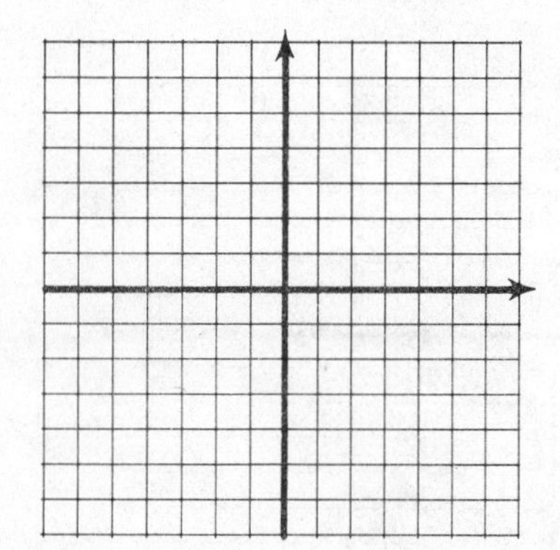

Figure 10

Work Quick Check 4 at the side.

Some equations lead to vertical or horizontal straight lines, as the next two examples show.

Example 4 Graph the linear equation $y = -4$.

This equation tells us that for any value of x we might choose, y is always equal to -4. To get three ordered pairs which are solutions of this equation, we choose any three numbers for x, and always let $y = -4$. Three ordered pairs that we can use are $(-2, -4)$, $(0, -4)$, and $(3, -4)$. Drawing a line through these points gives us the horizontal line shown in Figure 11.

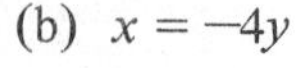

(b) $x = -4y$

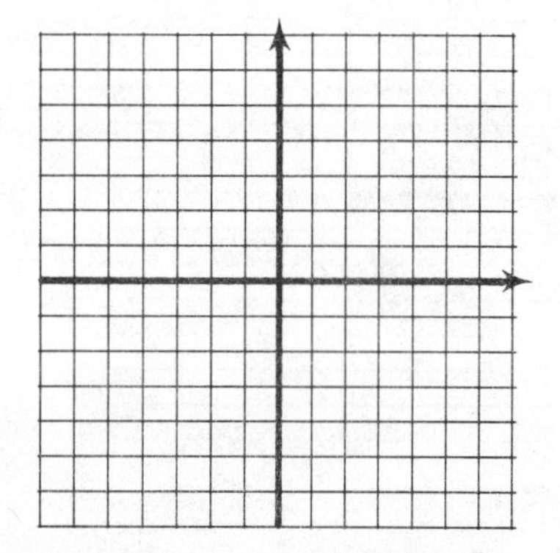

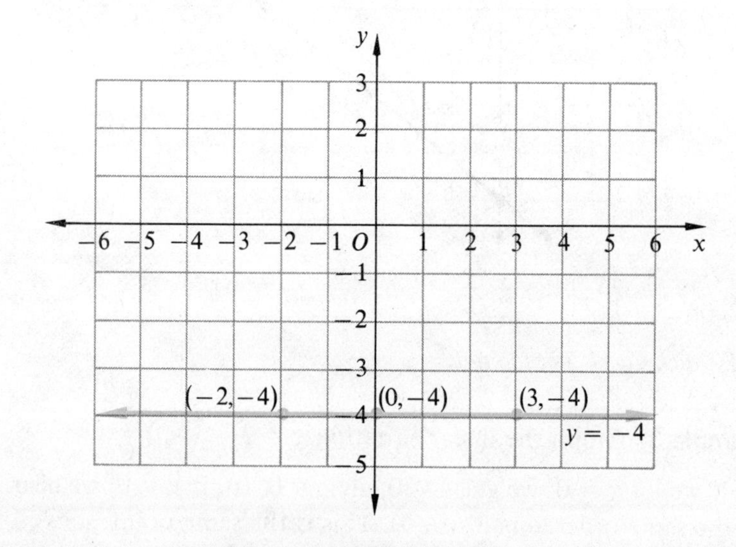

Figure 11

In general, the graph of a linear equation $y = k$, where k is a real number, is a horizontal line going through the point $(0, k)$.

Quick Check Answers

4. (a) (b)

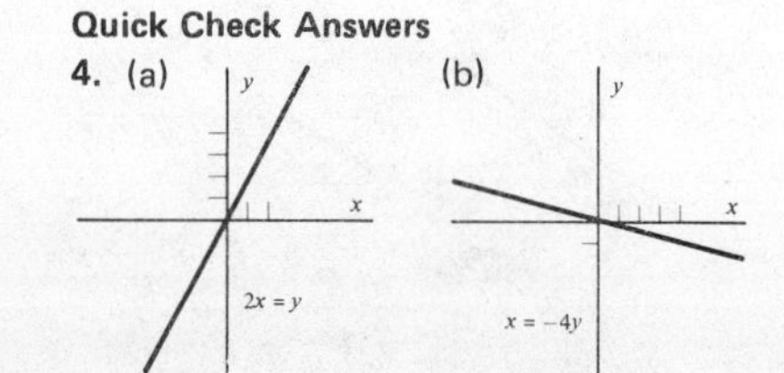

Example 5 Graph the linear equation $x = 3$.

This equation tells us that all ordered pairs that are solutions of this equation have an x-value of 3. We can use any number for y. Three ordered pairs that work are (3, 3), (3, 0), and (3, −2). Drawing a line through these points gives us the vertical line shown in Figure 12.

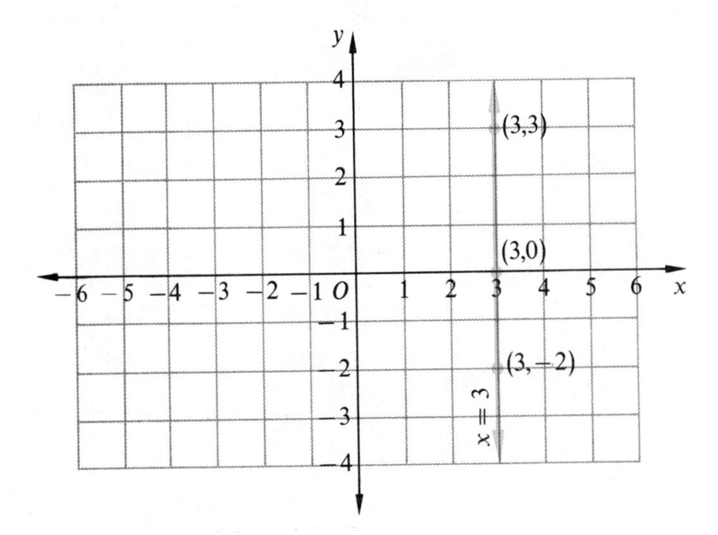

Figure 12

In general, the graph of a linear equation $x = k$, where k is a real number, is a vertical line going through the point $(k, 0)$.

5. Graph each equation.

(a) $y = -5$

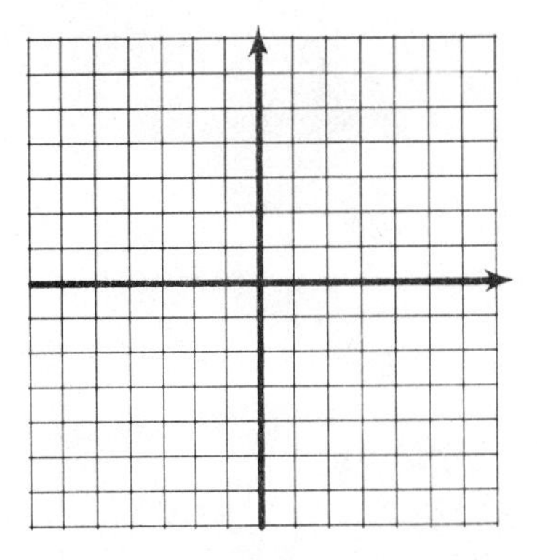

(b) $x = 2$

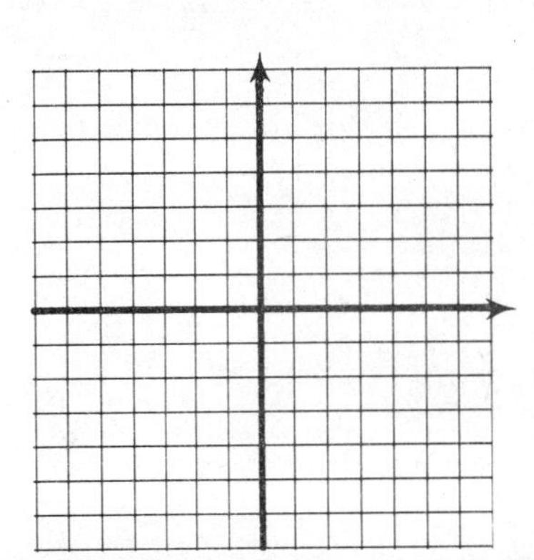

Quick Check Answers

5. (a) (b)

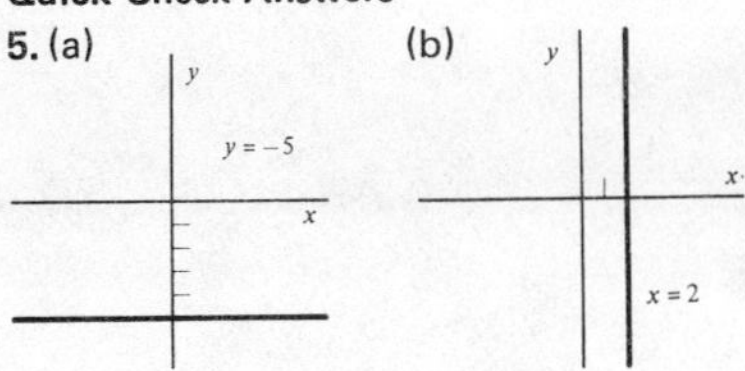

WORK SPACE

name date hour

7.3 EXERCISES

Complete the ordered pairs using the given equation. Then graph the equation by plotting the points and drawing a line through them. See Examples 1–5.

1. $x + y = 5$
   (0,) (, 0)
   (2,)

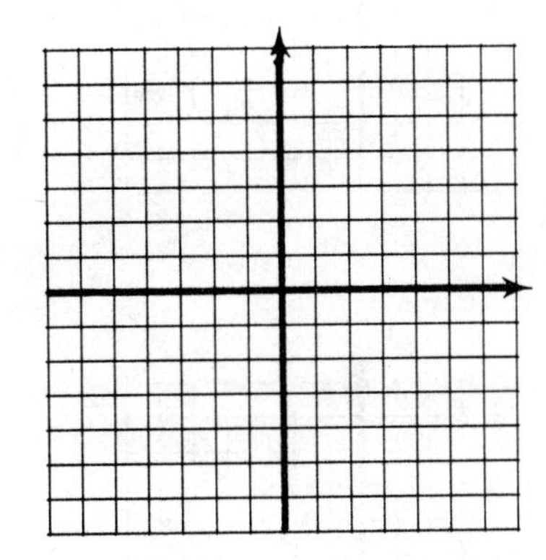

2. $y = x - 3$
   (0,) (, 0)
   (5,)

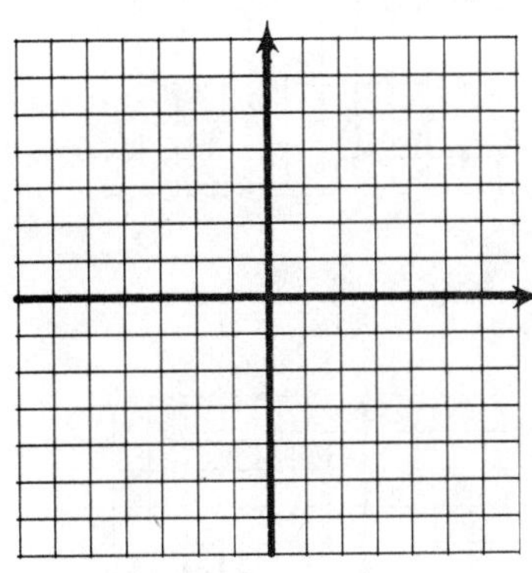

3. $y = x + 4$
   (0,) (, 0)
   (−2,)

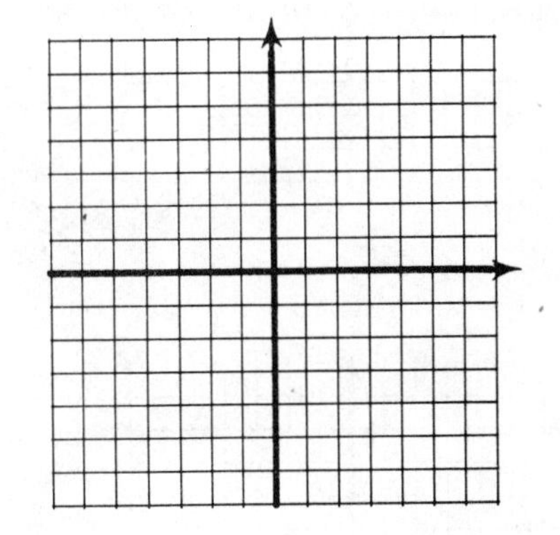

4. $y + 5 = x$
   (0,) (, 0)
   (6,)

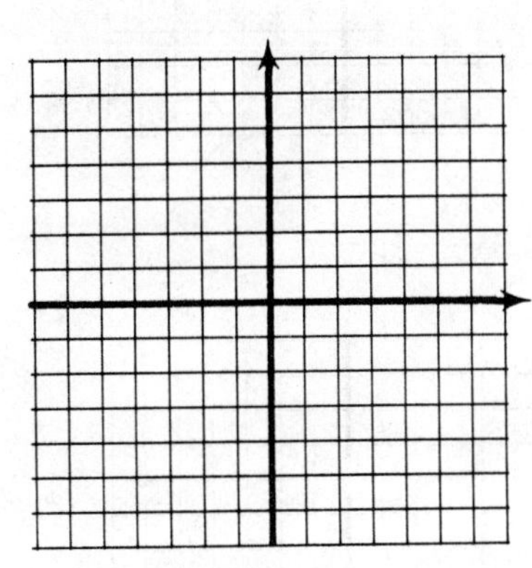

5. $y = 3x - 6$
   (0,) (, 0)
   (3,)

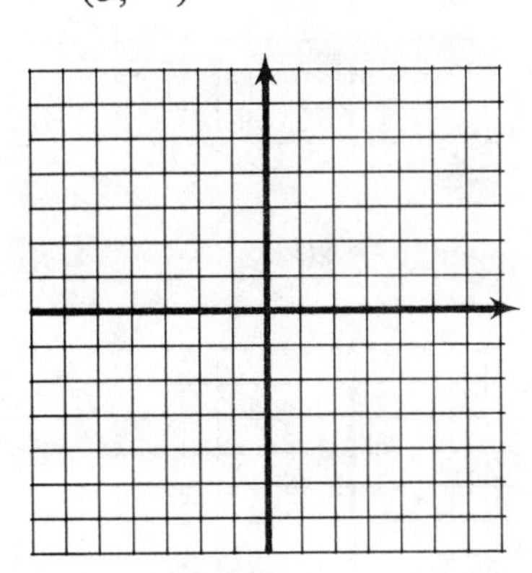

6. $x = -5$
   (, 2) (, 0)
   (, −3)

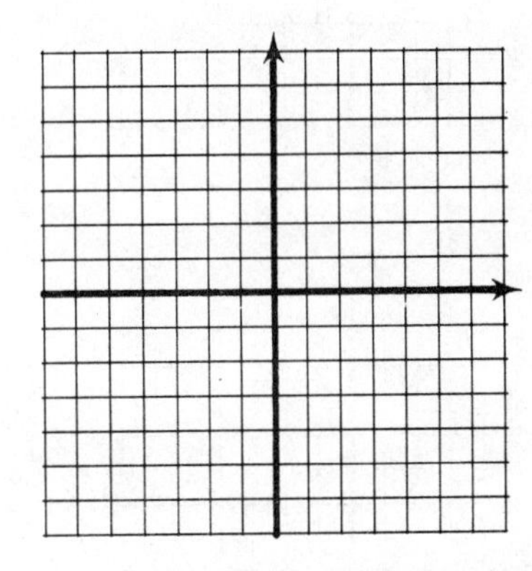

7. $y = 4$
   (3,) (0,)
   (−2,)

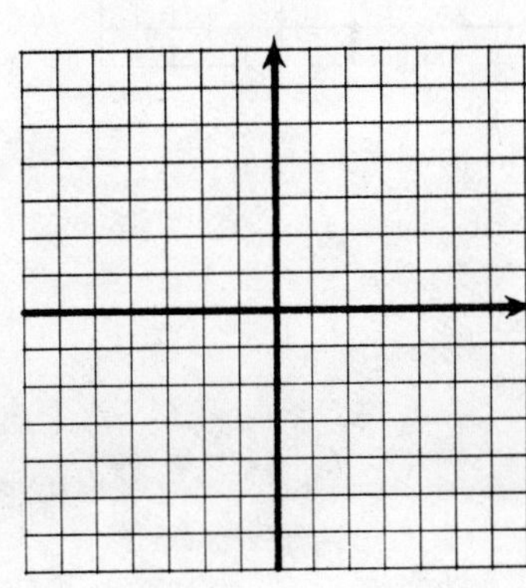

8. $y + 3 = 0$
   (4,) (5,)
   (−2,)

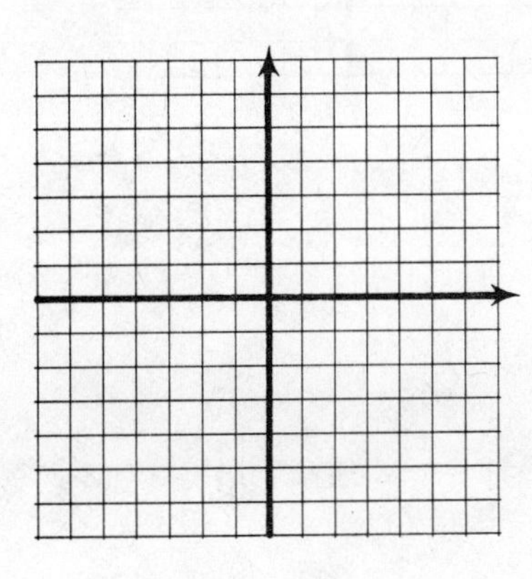

9. $x - 2 = 0$
   (, 3) (, 5)
   (, −1)

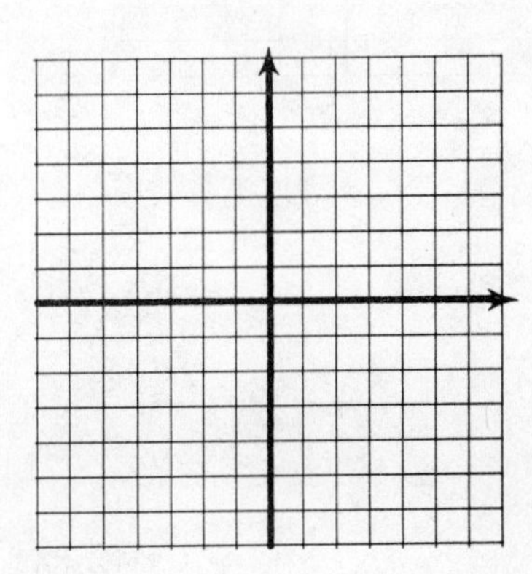

Graph each linear equation.

10. $x - y = 2$

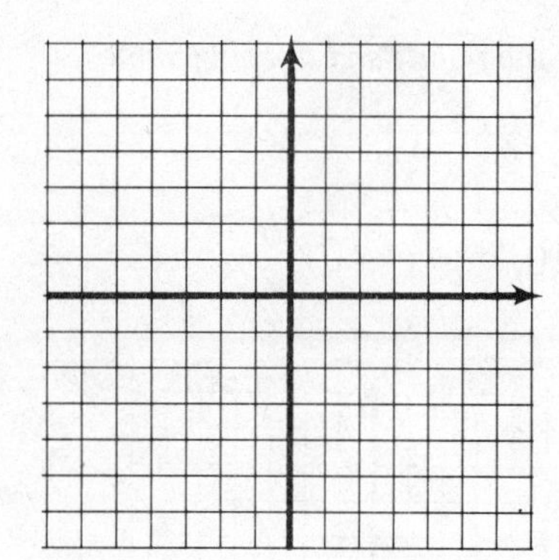

11. $x + y = 6$

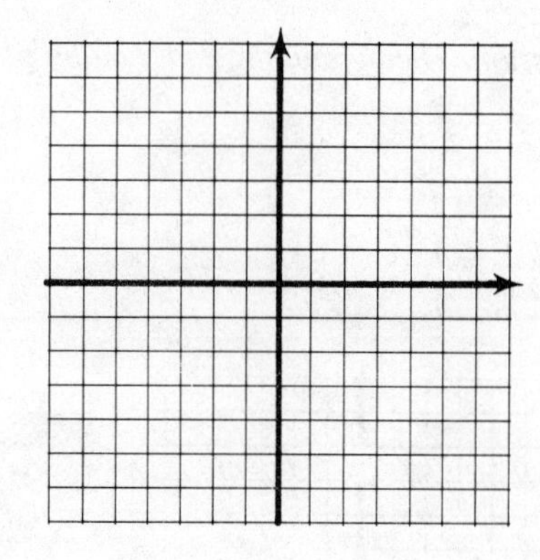

12. $y = x + 2$

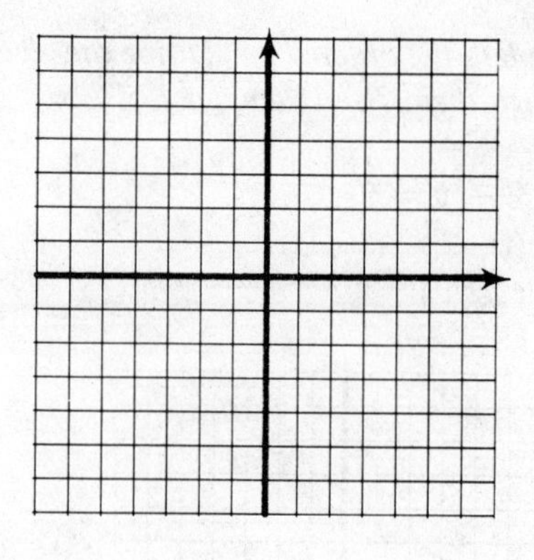

13. $y = x - 1$

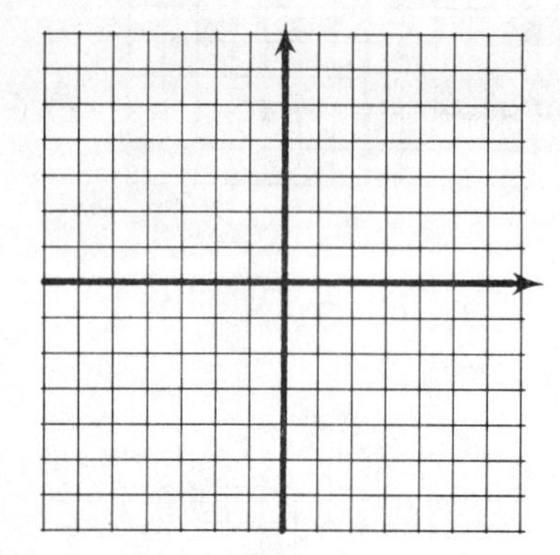

14. $y = 2x - 4$

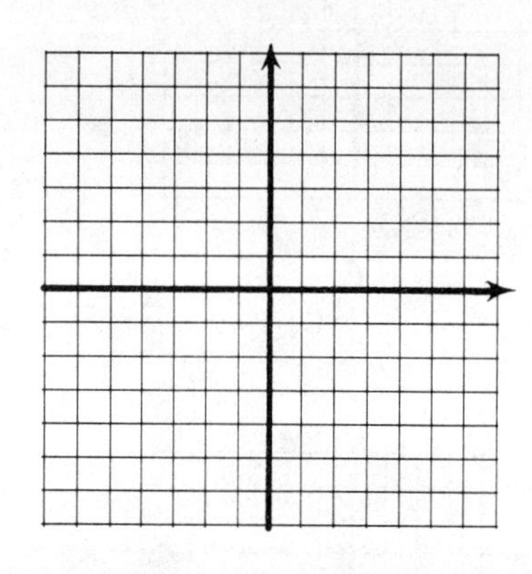

15. $y = 3x + 9$

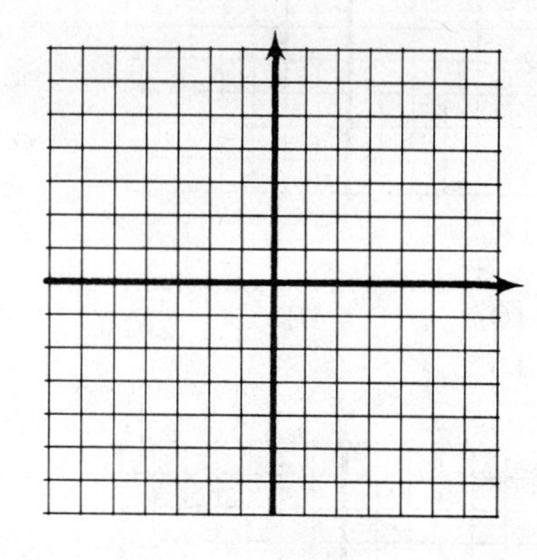

16. $x = 3y - 12$

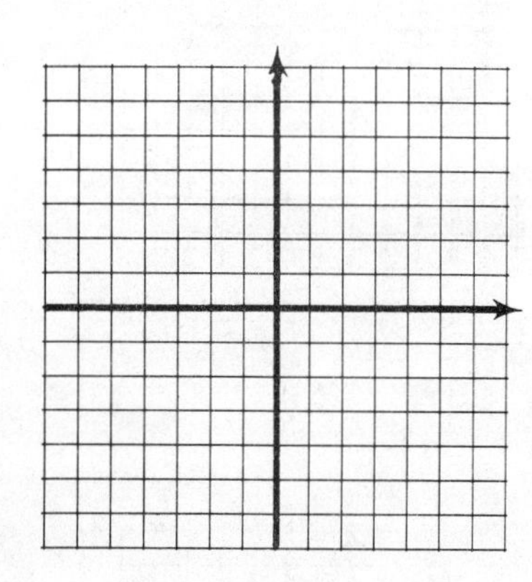

17. $x = 2y - 10$

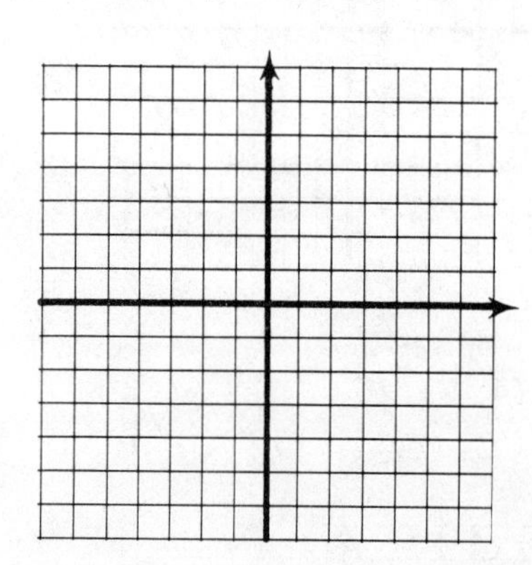

18. $3x - 2y = 6$

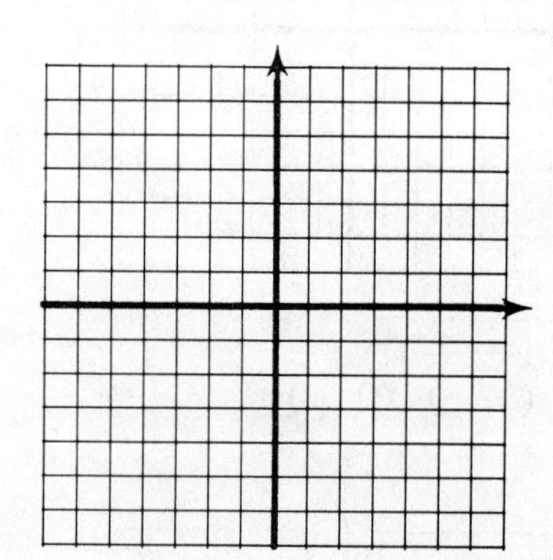

name date hour

19. $2x + 3y = 12$

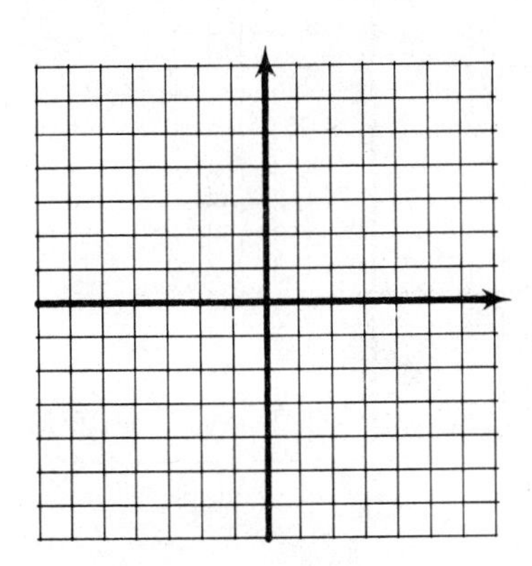

20. $2x - 7y = 14$

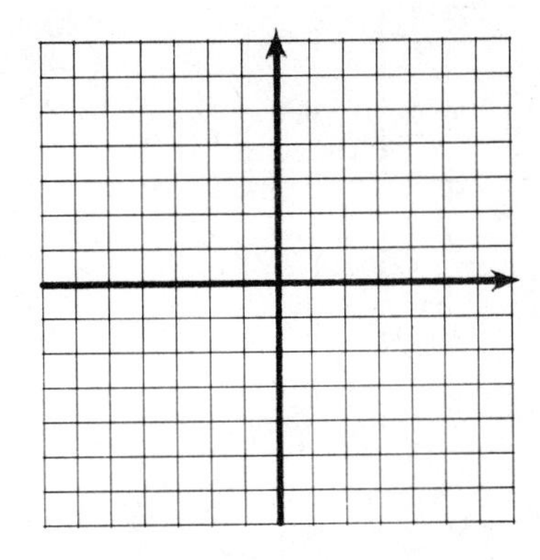

21. $3x + 5y = 15$

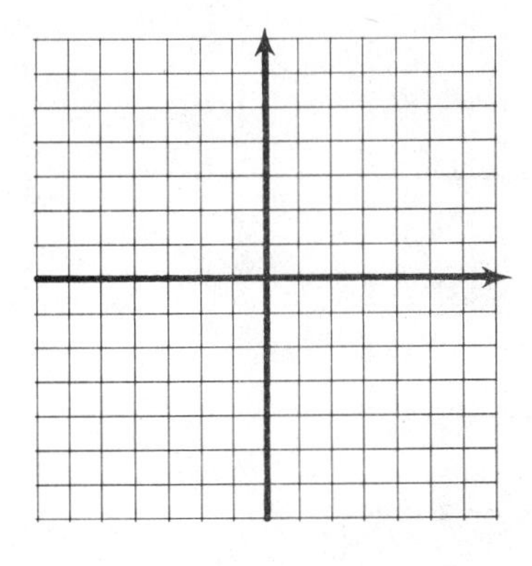

22. $3x + 7y = 21$

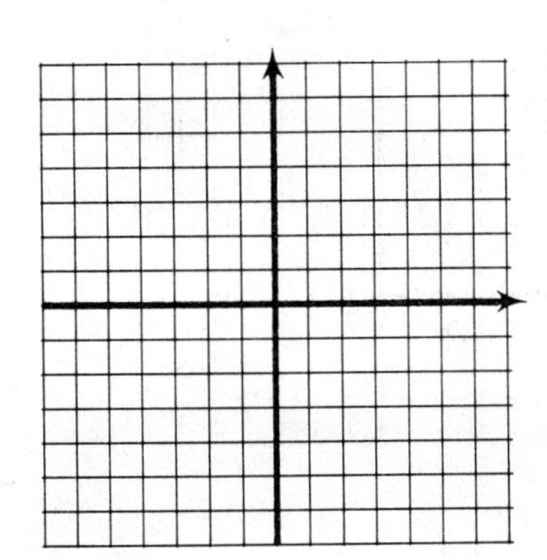

23. $6x - 5y = 30$

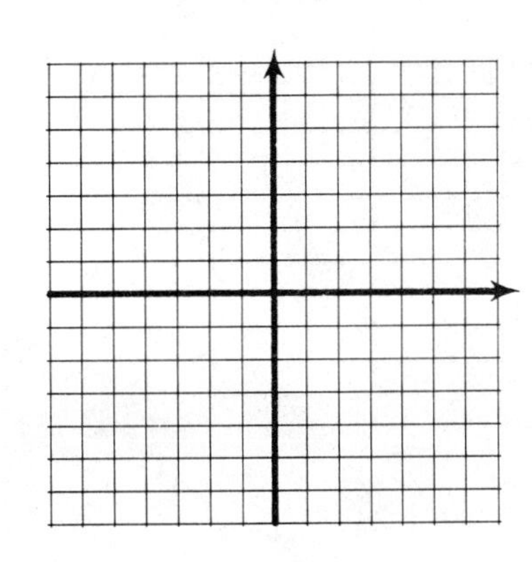

24. $y = 2x$

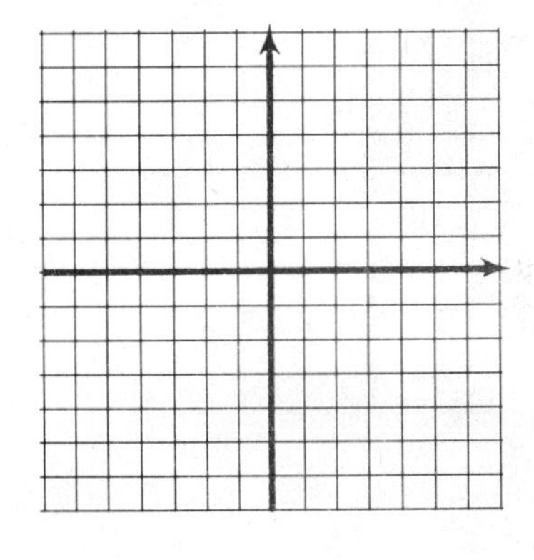

25. $y = -3x$

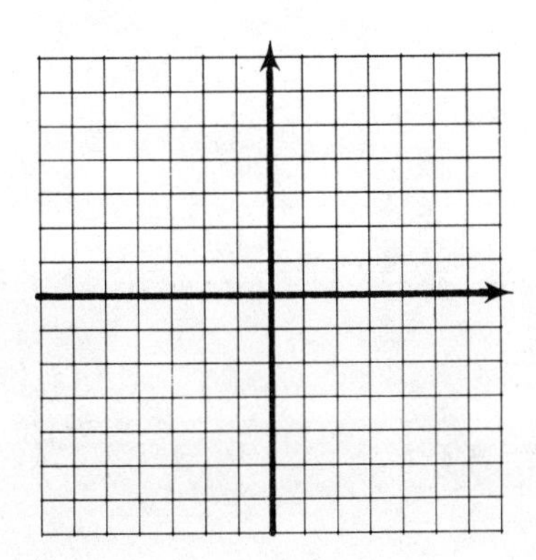

26. $y + 6x = 0$

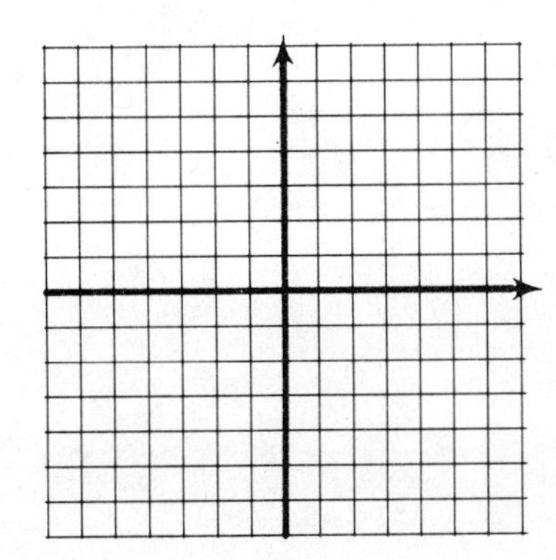

27. $y - 4x = 0$

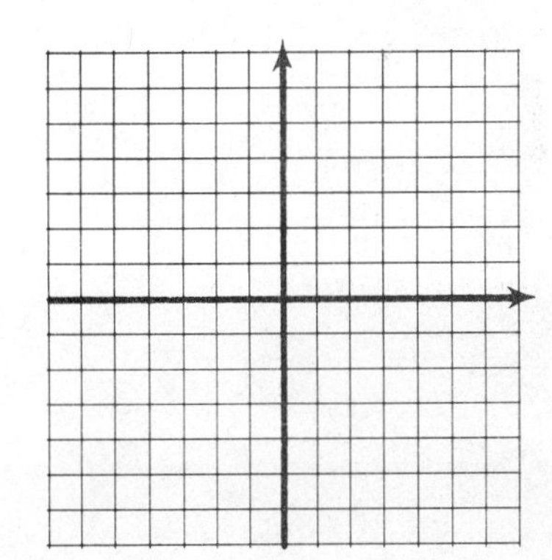

28. $2x + 3y = 0$

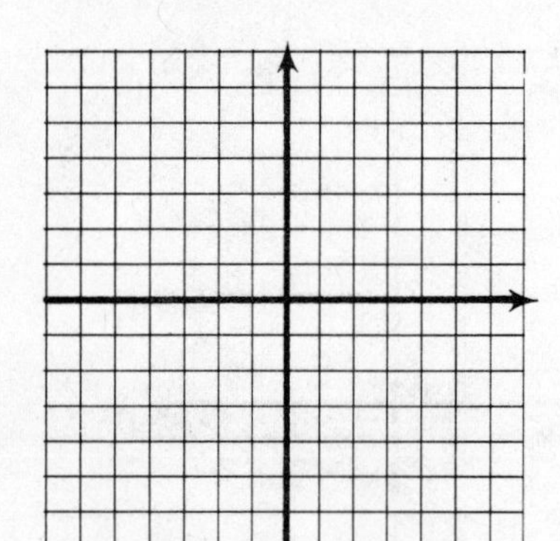

29. $3x - 4y = 0$

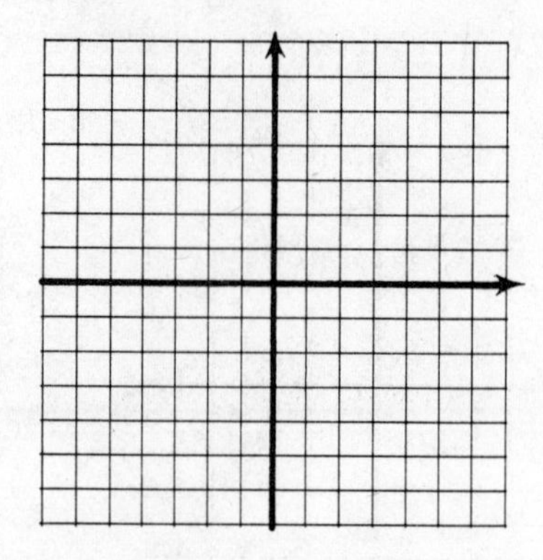

30. $x + 2 = 0$

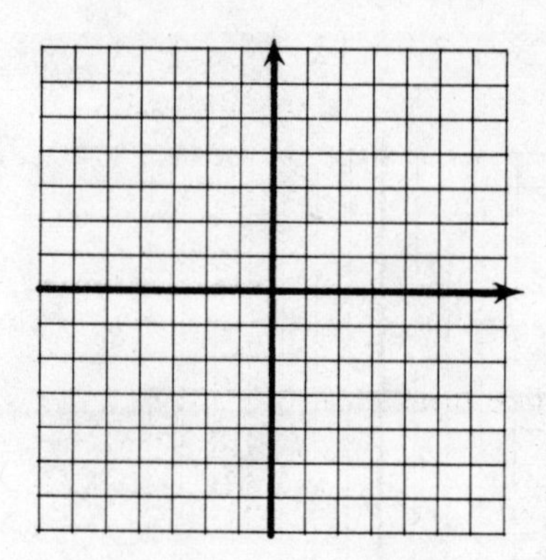

31. $y - 3 = 0$

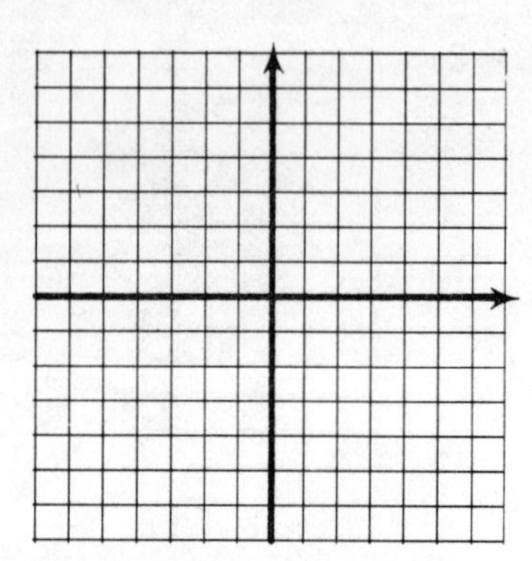

32. $y = 6$

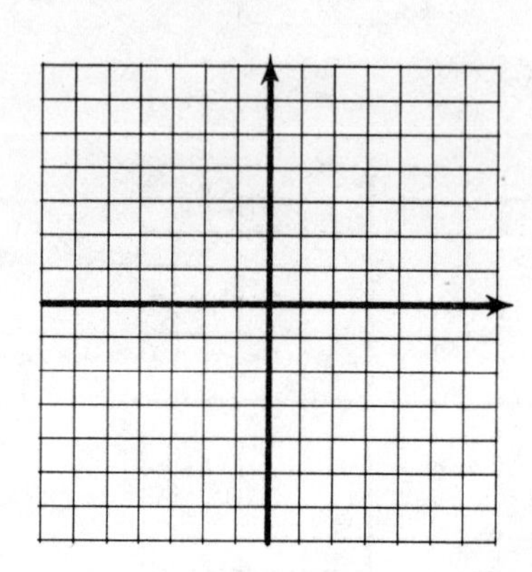

33. $x = 2$

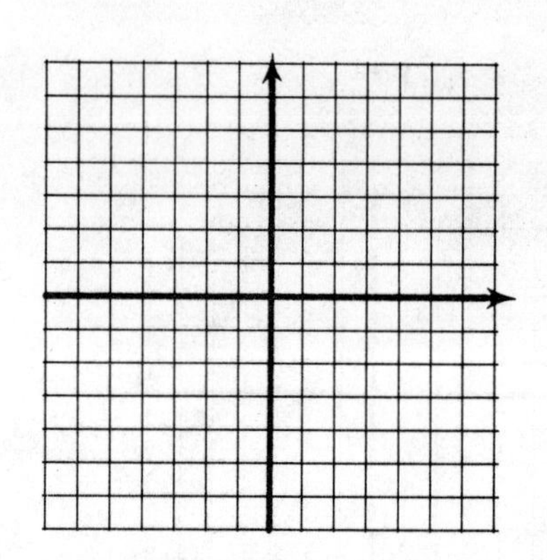

OBJECTIVES

- Graph linear inequalities

In Section 7.3 we discussed methods for graphing linear equations, such as $2x + 3y = 6$. In this section we extend this discussion to linear inequalities, such as

$$2x + 3y \leq 6. \quad \text{(Recall that } \leq \text{ is read "less than or equal to.")}$$

The points on the line $2x + 3y = 6$ are solutions of $2x + 3y \leq 6$ but are only part of the final graph. To complete the graph of the linear inequality $2x + 3y \leq 6$, go through the following steps.

Step 1 Graph the equation $2x + 3y = 6$. This graph is a line which divides the plane into three regions: one region below the line, one region above the line, and the points on the line itself.

Step 2 The graph of $2x + 3y \leq 6$ includes either all the points below the line or all the points above the line. To decide which, choose any point not on the line as a test point. The origin, (0, 0), is a good choice, and we will use it here. Substitute 0 for x and 0 for y in the original inequality $2x + 3y \leq 6$ to see if the resulting statement is true or false. We have

$2x + 3y \leq 6$	Original inequality
$2(0) + 3(0) \leq 6$	Let $x = 0$ and $y = 0$
$0 + 0 \leq 6$	
$0 \leq 6$	True

Step 3 Since this statement is true, the graph of the inequality includes the region containing (0, 0). Shade this region, as shown in Figure 13. The shaded region, along with the original line, is the graph we want.

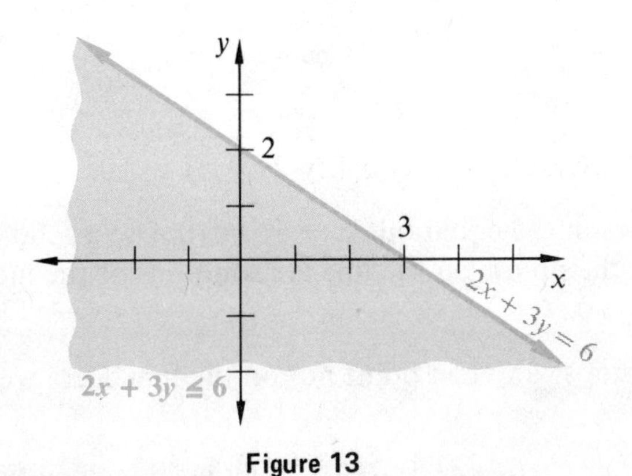

Figure 13

Work Quick Check 1 at the side.

Example 1 Graph the inequality $x - y > 5$.

This inequality is restricted to "greater than." Therefore, the points on the line $x - y = 5$ do not belong to the graph. However, the line does serve as a boundary for two regions, one of which

QUICK CHECKS

1. Graph each linear inequality.

(a) $x + 2y \geq 6$

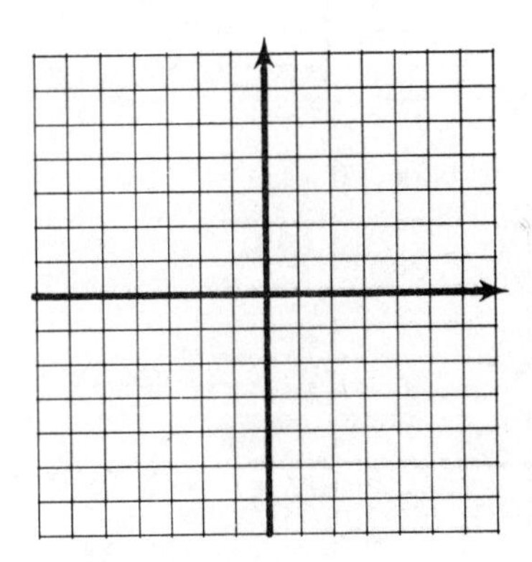

(b) $3x + 4y \leq 12$

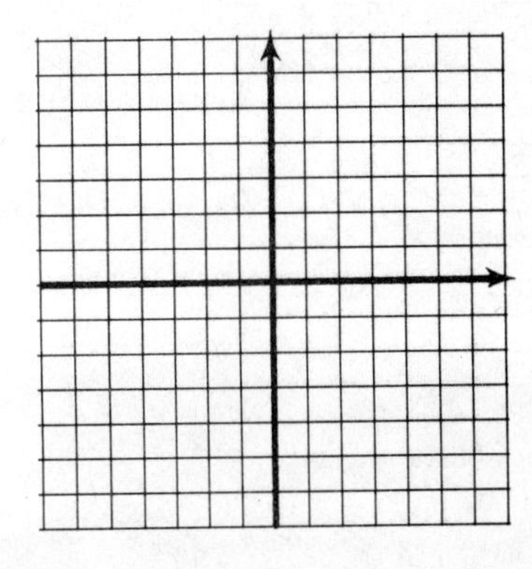

Quick Check Answers

1. (a) (b)

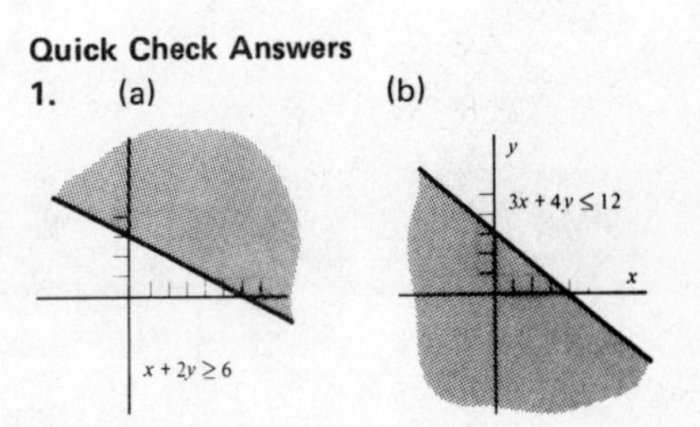

2. Graph each linear inequality.

(a) $2x - y \geqslant -4$

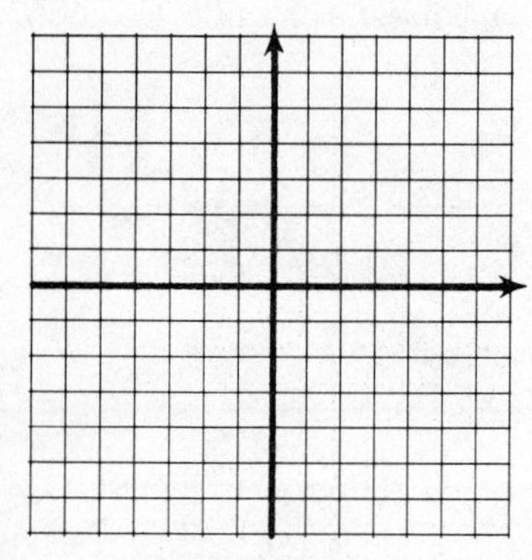

(b) $x + 3y > 6$

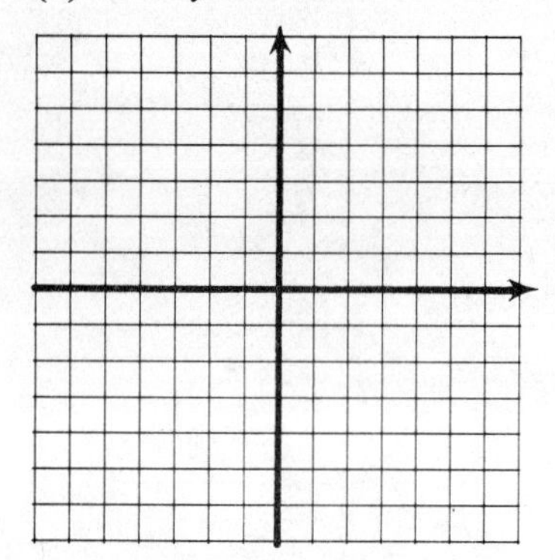

(c) $2x - y < 5$

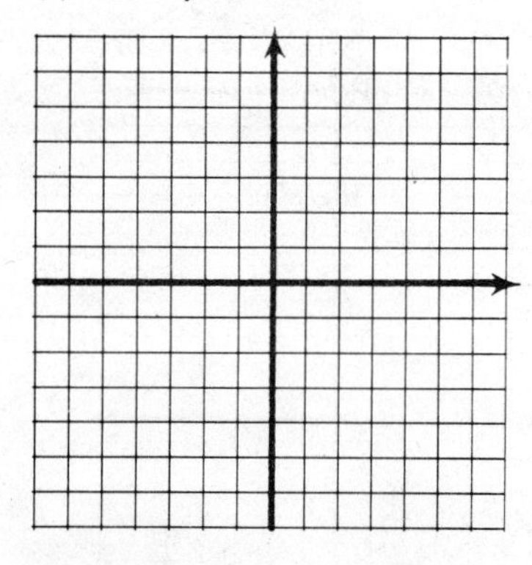

Quick Check Answers

2. (a) (b)

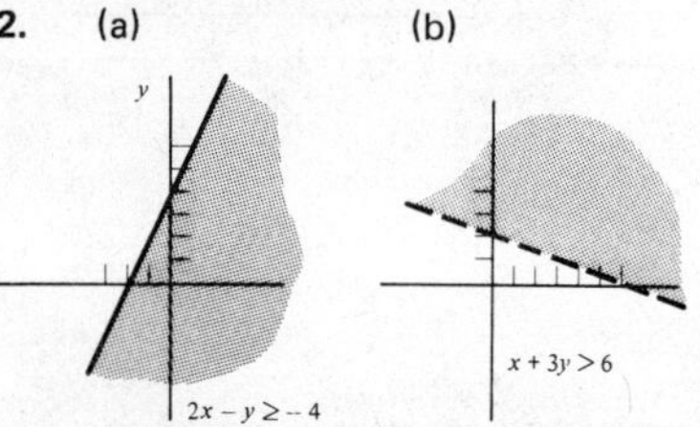

(c)

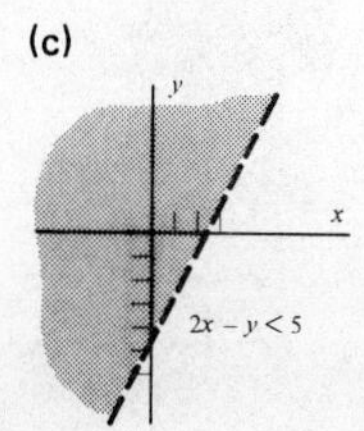

satisfies the inequality. To graph the inequality, go through the steps mentioned above.

Step 1 Graph the equation $x - y = 5$. Use a dashed line to show that the points on the line are not solutions of the inequality $x - y > 5$.

Step 2 Choose a test point to see which side of the line satisfies the inequality. Let us choose (0, 0).

$x - y > 5$	Original inequality
$0 - 0 > 5$	Let $x = 0$ and $y = 0$
$0 > 5$	False

Step 3 Since this statement is false, the graph of the inequality includes the region which does not contain (0, 0). Shade this region, as shown in Figure 14. This shaded region is the graph we want.

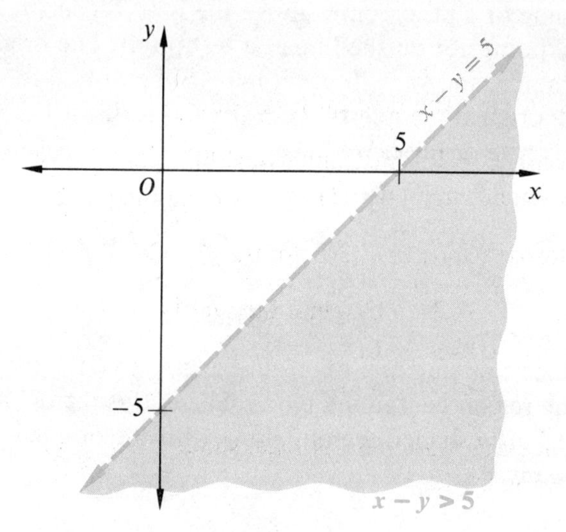

Figure 14

Example 2 Graph the inequality $2x - 5y \geqslant 10$.

Step 1 Graph the equation $2x - 5y = 10$. Use a solid line to show that the points on the line are solutions of the inequality $2x - 5y \geqslant 10$.

Step 2 Choose any test point not on the line. Here we choose (0, 0).

$2x - 5y \geqslant 10$	Original inequality
$2(0) - 5(0) \geqslant 10$	Let $x = 0$ and $y = 0$
$0 - 0 \geqslant 10$	
$0 \geqslant 10$	False

Step 3 Since this statement is false, shade the region *not* containing (0, 0). (See Figure 15)

Work Quick Check 2 at the side.

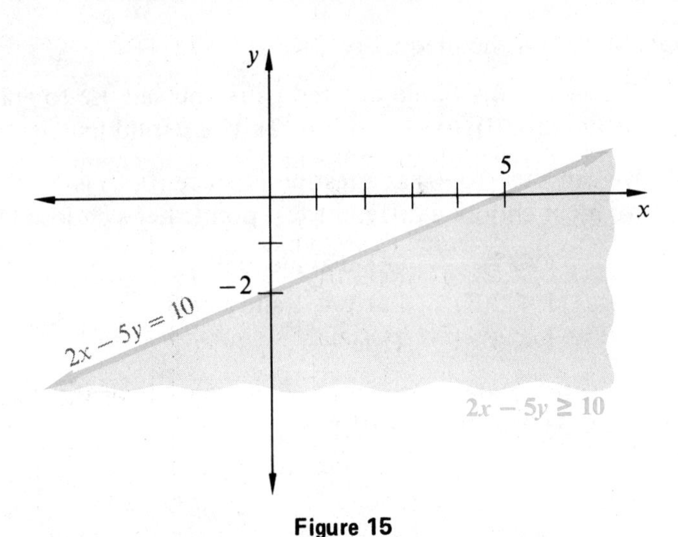

Figure 15

Example 3 Graph the inequality $x \leqslant 3$.

Step 1 Graph $x = 3$. This is a vertical line going through the point (3, 0). Use a solid line. (Why?)

Step 2 Choose (0, 0) as a test point.

$x \leqslant 3$	Original inequality
$0 \leqslant 3$	Let $x = 0$
$0 \leqslant 3$	True

Step 3 Since this statement is true, shade the region containing (0, 0). (See Figure 16.)

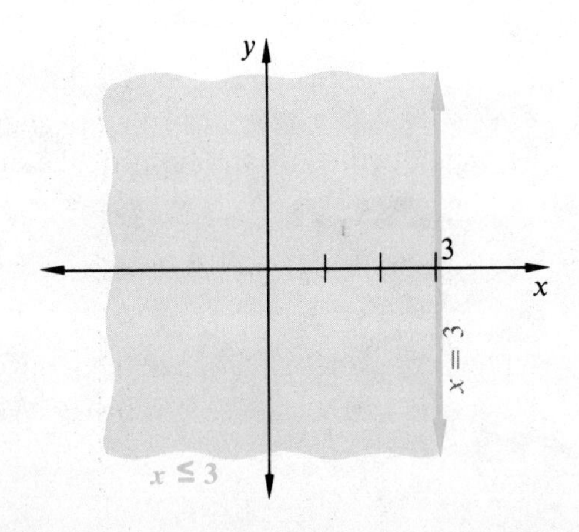

Figure 16

Work Quick Check 3 at the side.

3. Graph each linear inequality.

(a) $y \leqslant 4$

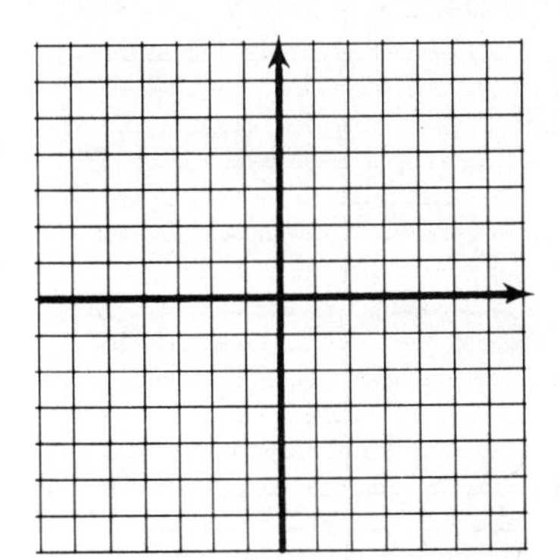

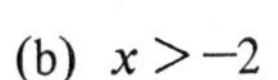

(b) $x > -2$

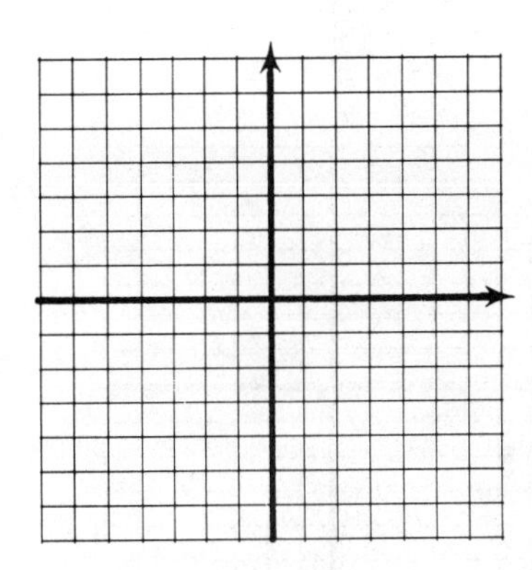

Quick Check Answers

3. (a) (b)

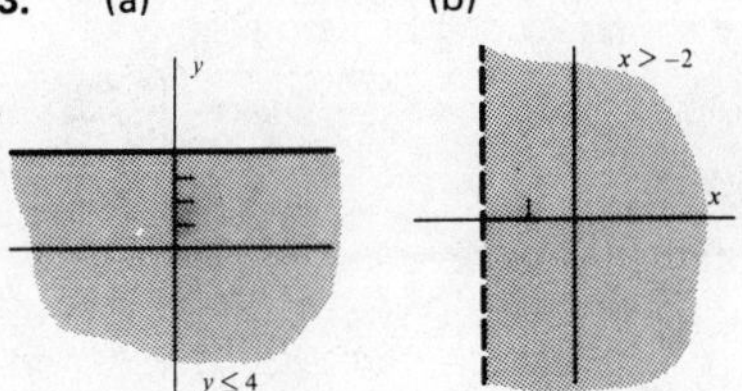

4. Graph each linear inequality.

(a) $x \geqslant -3y$

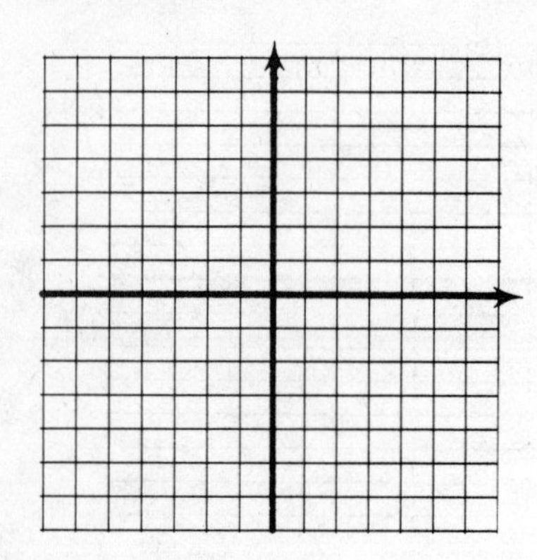

(b) $3x < y$

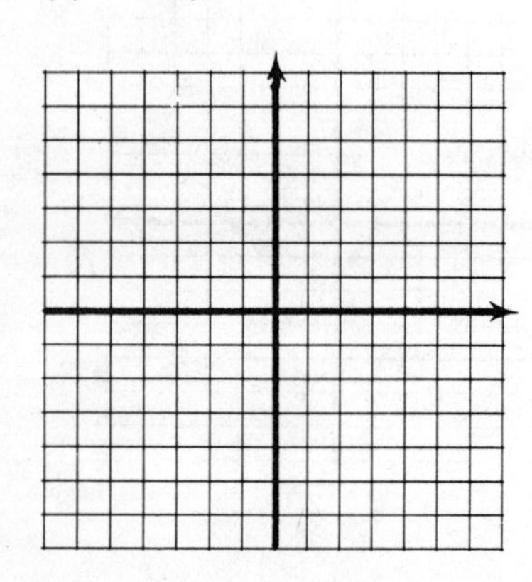

Quick Check Answers

4. (a) (b)

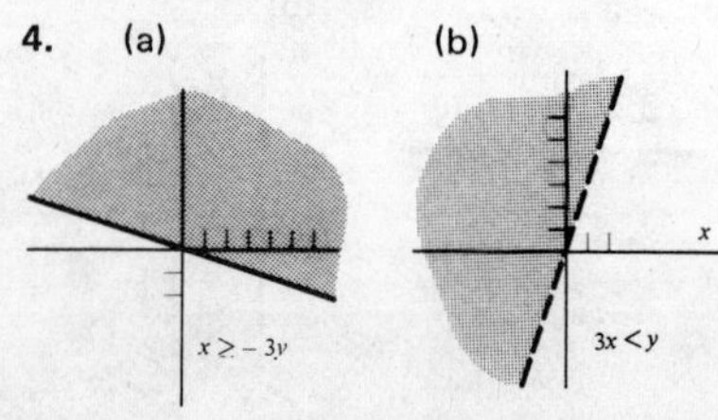

Example 4 Graph the inequality $x \leqslant 2y$.

Step 1 Graph $x = 2y$. Some ordered pairs you can use to graph this line include (0, 0), (6, 3), and (4, 2). Use a solid line.

Step 2 We can't use (0, 0) as a test point since (0, 0) is on the line $x = 2y$. We must choose a different test point. Let's choose (1, 3).

$x \leqslant 2y$	Original inequality
$1 \leqslant 2(3)$	Let $x = 1$ and $y = 3$
$1 \leqslant 6$	True

Step 3 Since this statement is true, shade the side of the graph containing (1, 3). (See Figure 17.)

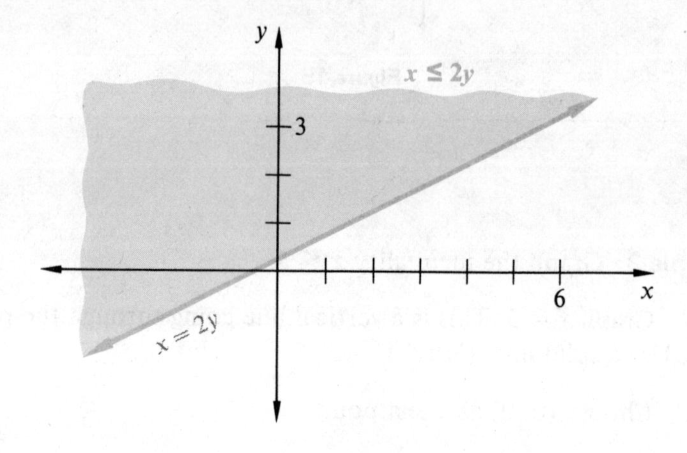

Work Quick Check 4 at the side.

name date hour

7.4 EXERCISES

In Exercises 1-12, the straight line for each inequality has been drawn. Complete each graph by shading the correct region. See Example 1, Steps 2 and 3.

1. $x + y \leqslant 4$

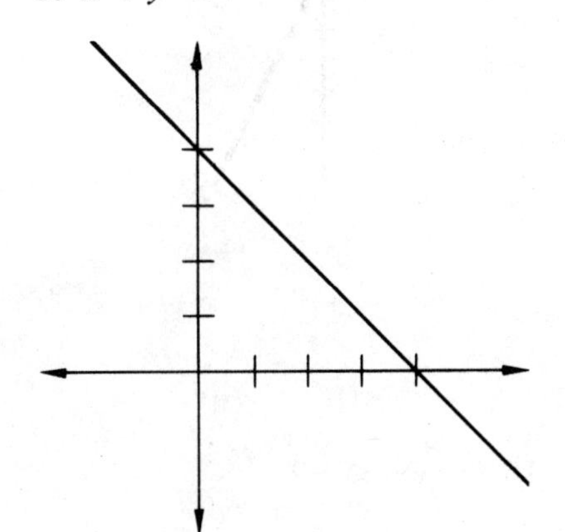

2. $x + y \geqslant 2$

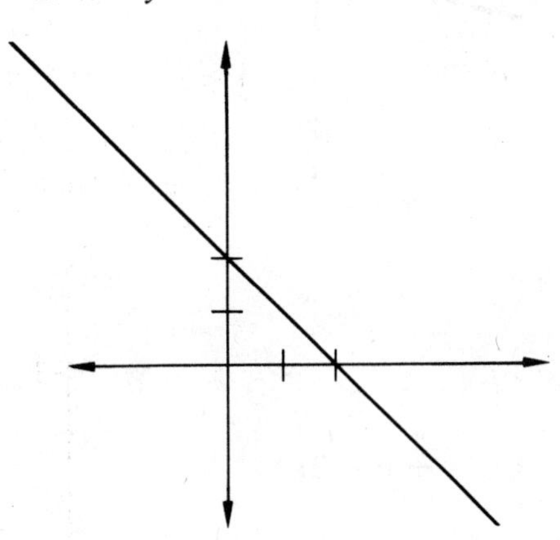

3. $x + 2y \leqslant 7$

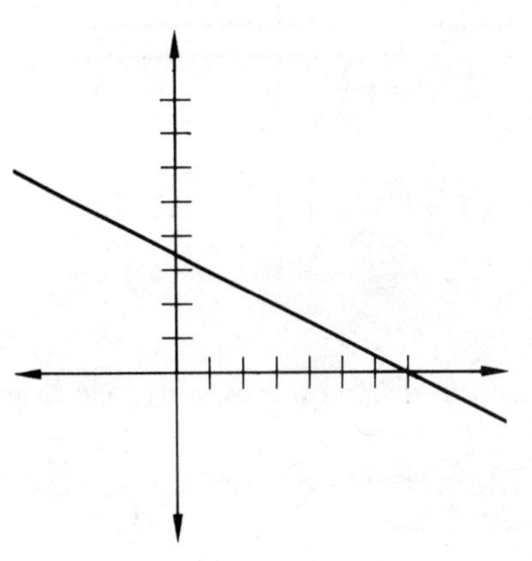

4. $2x + y \leqslant 5$

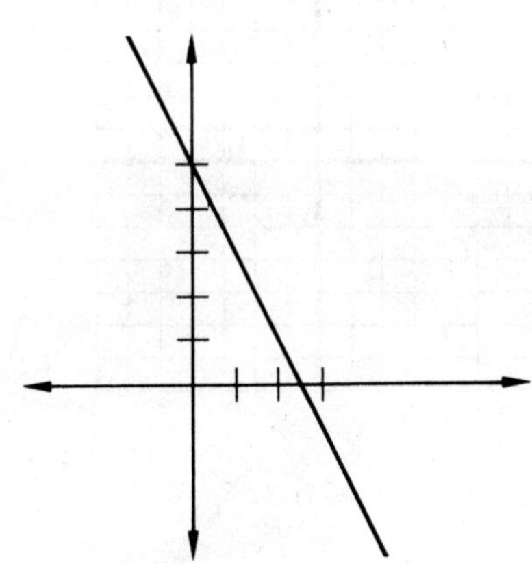

5. $-3x + 4y < 12$

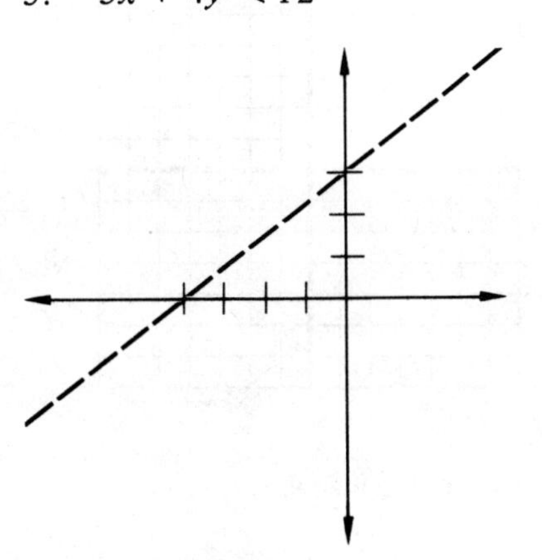

6. $4x - 5y > 20$

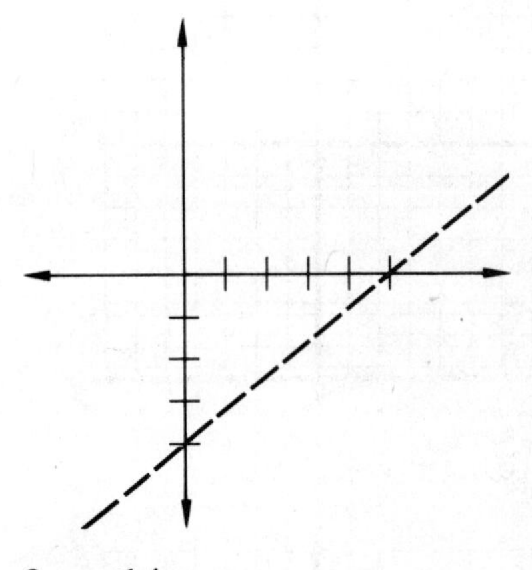

7. $5x + 3y > 15$

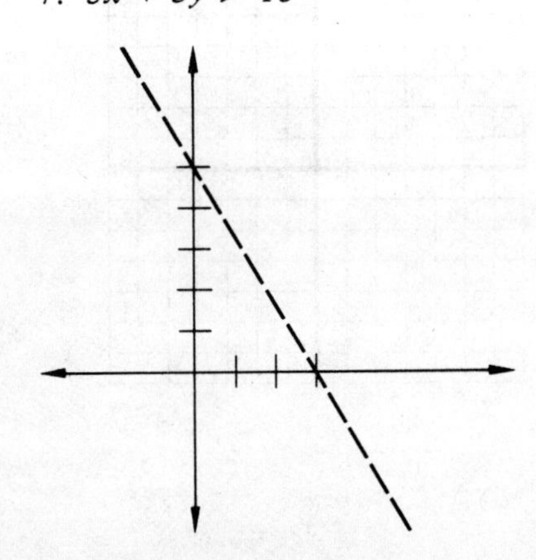

8. $6x - 5y < 30$

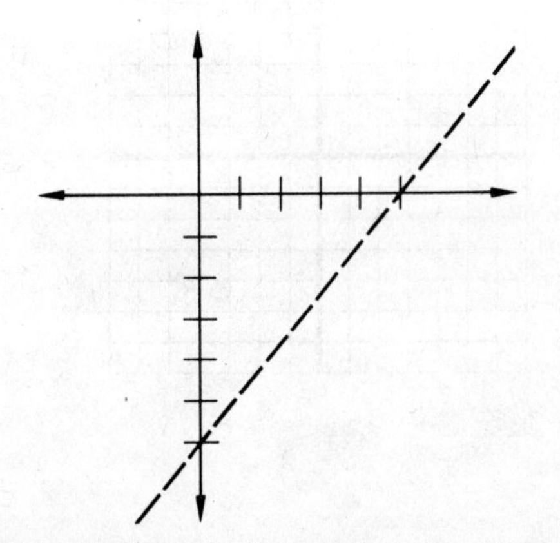

9. $x < 4$

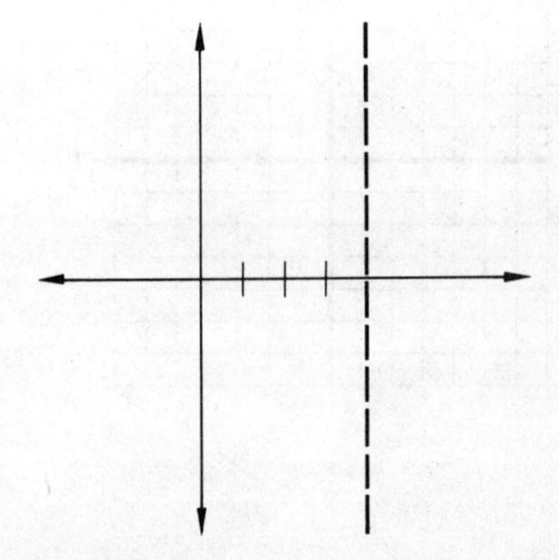

10. $y > -1$

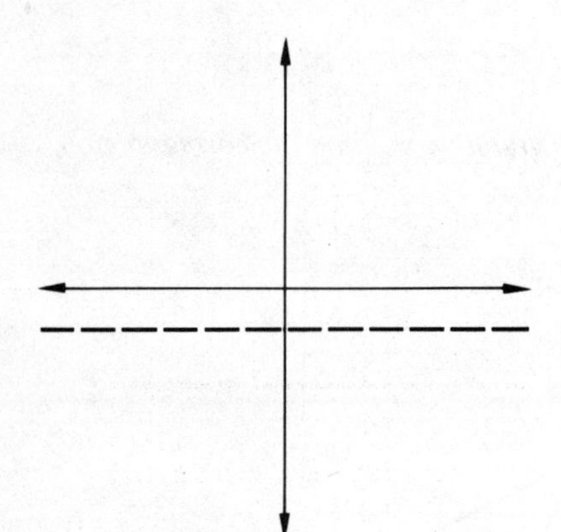

11. $x \leqslant 4y$

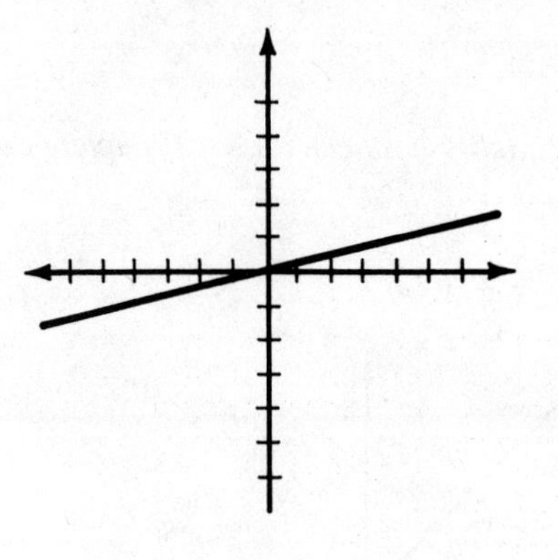

12. $-2x > y$

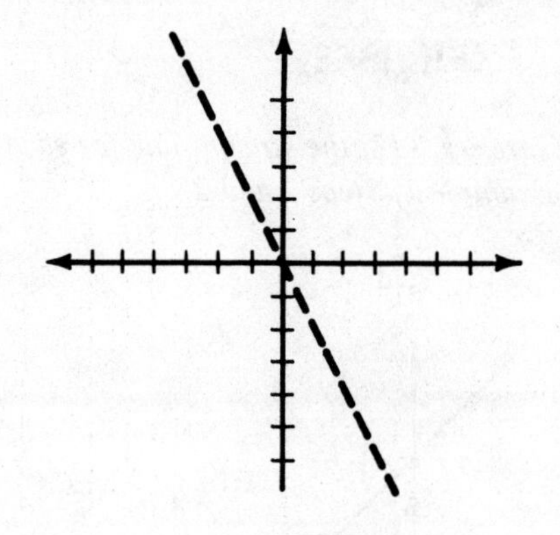

Graph each linear inequality. See Examples 1–4.

13. $x + y \leqslant 8$

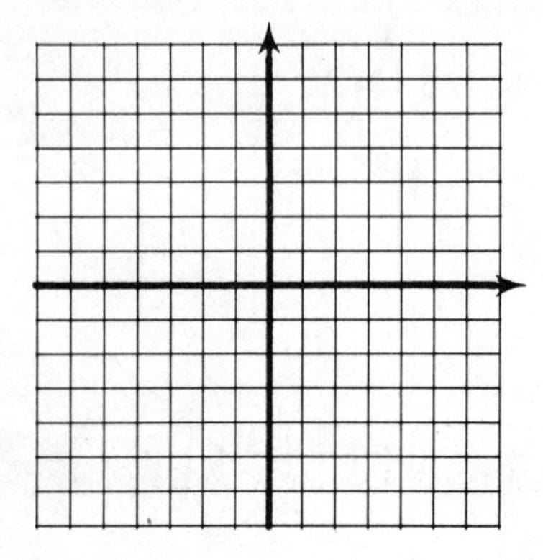

14. $x + y \geqslant 2$

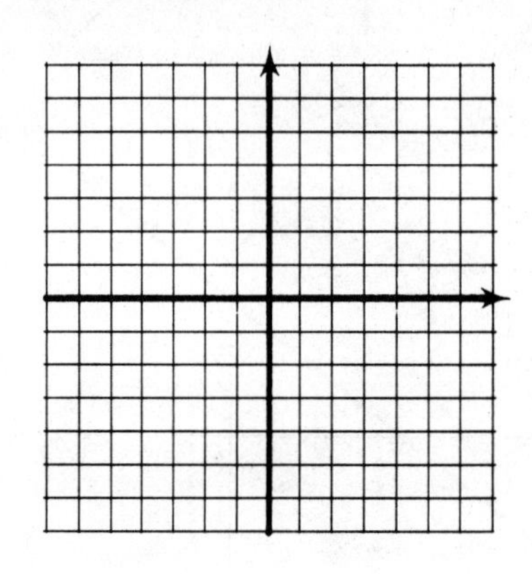

15. $x - y \leqslant -2$

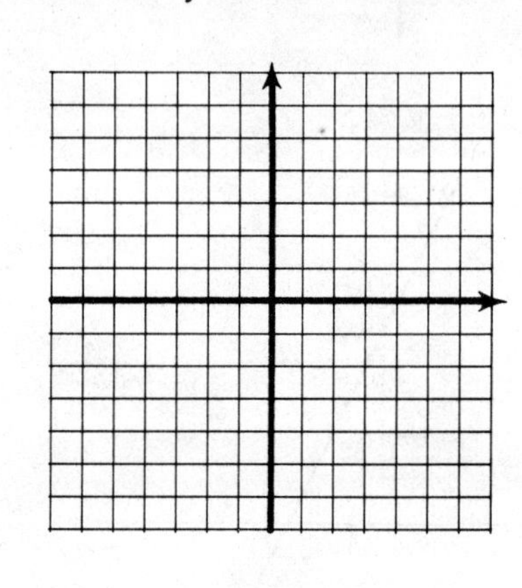

16. $x - y \leqslant 3$

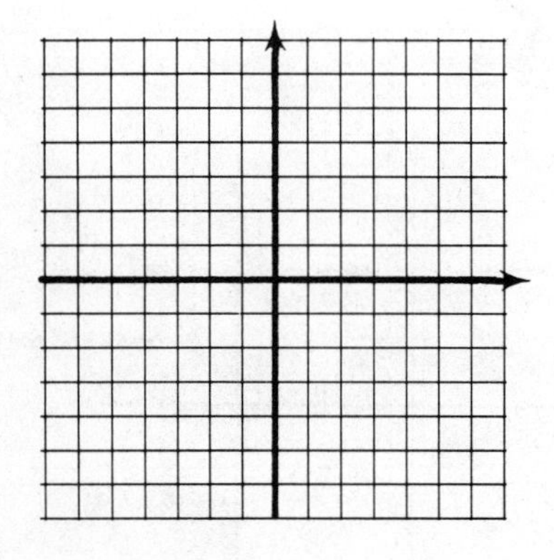

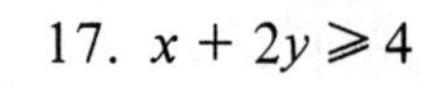

17. $x + 2y \geqslant 4$

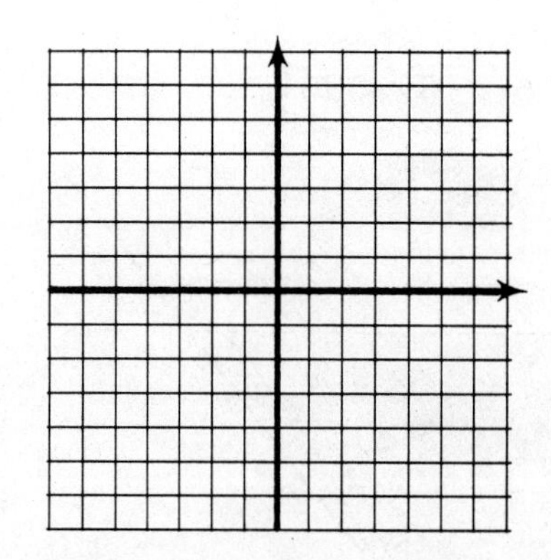

18. $x + 3y \leqslant 6$

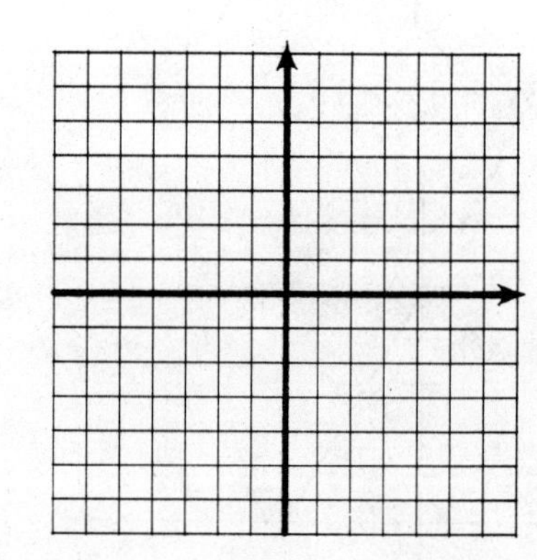

name date hour

19. $2x + 3y > 6$

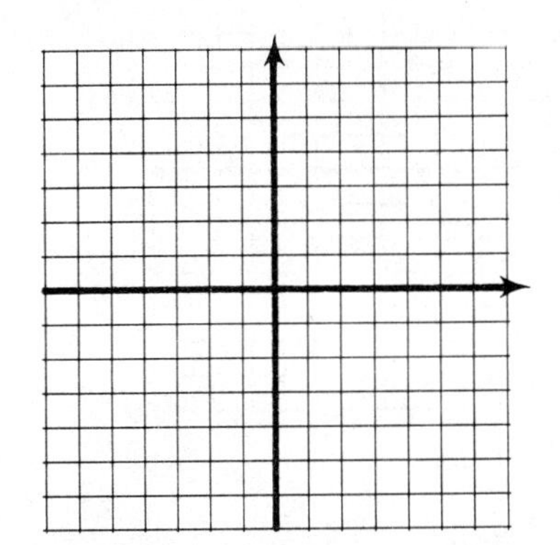

20. $3x + 4y > 12$

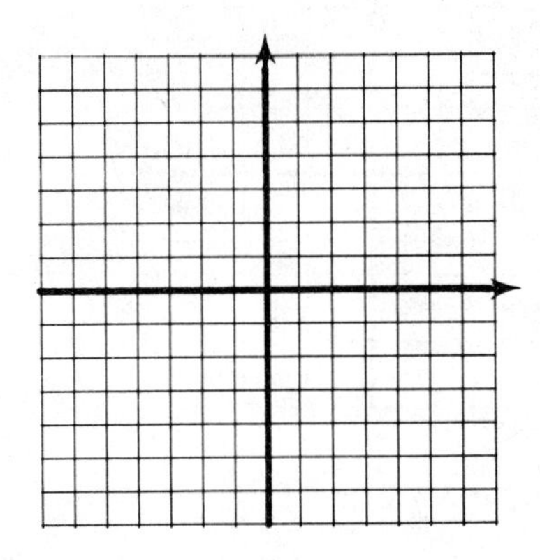

21. $3x - 4y < 12$

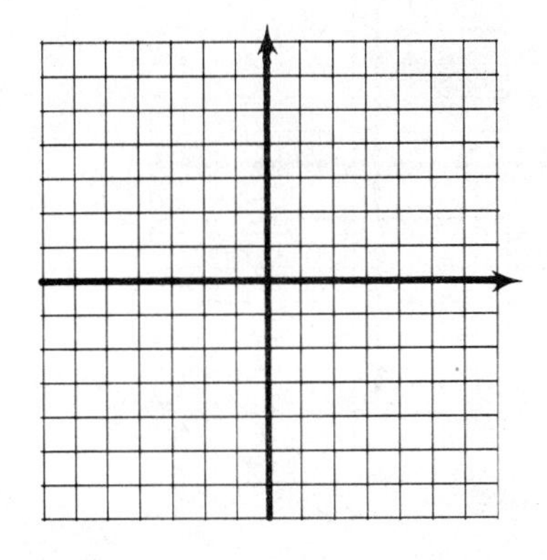

22. $2x - 3y < -6$

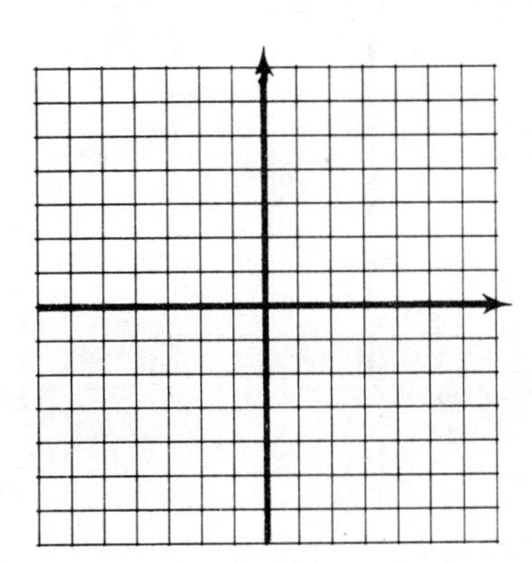

23. $3x + 7y \geqslant 21$

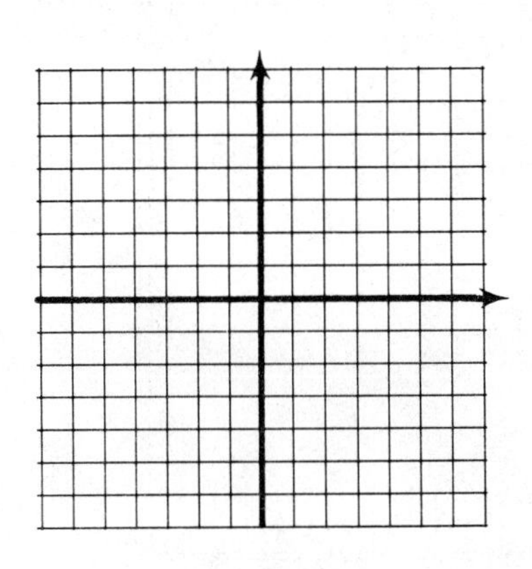

24. $2x + 5y \geqslant 10$

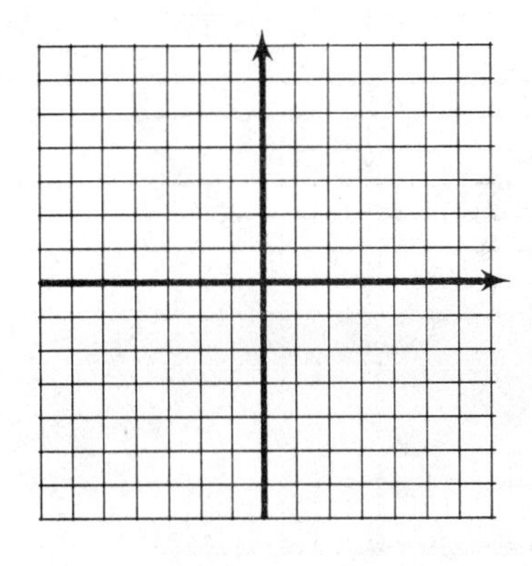

25. $4x - 5y \geqslant 20$

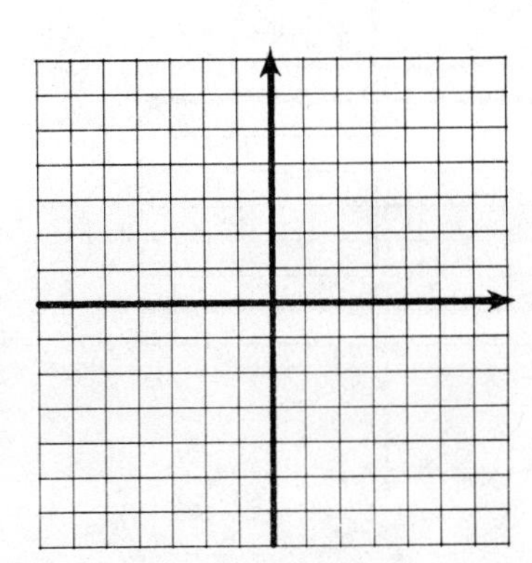

26. $3x + 5y \leqslant 15$

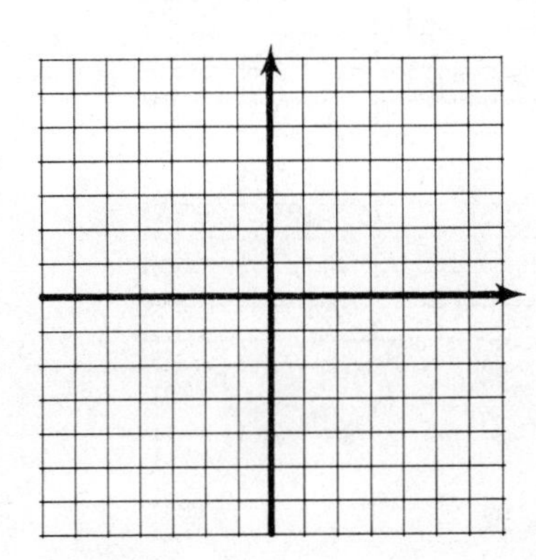

27. $4x + 7y \leqslant 14$

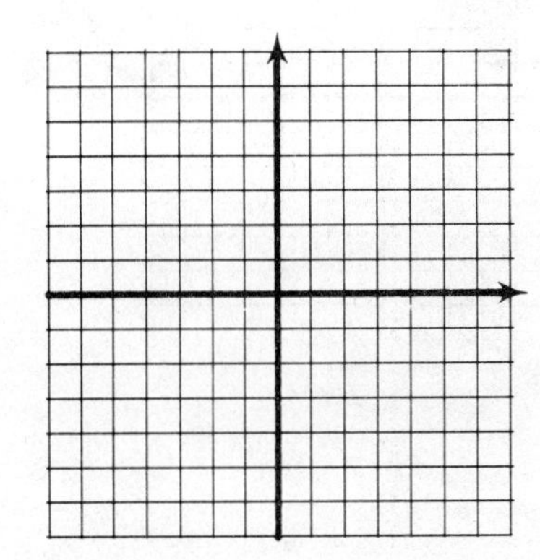

28. $x - y \leqslant 2$

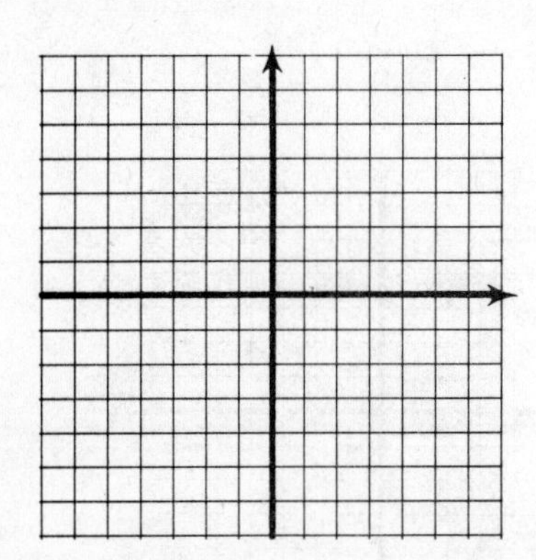

29. $x < 4$

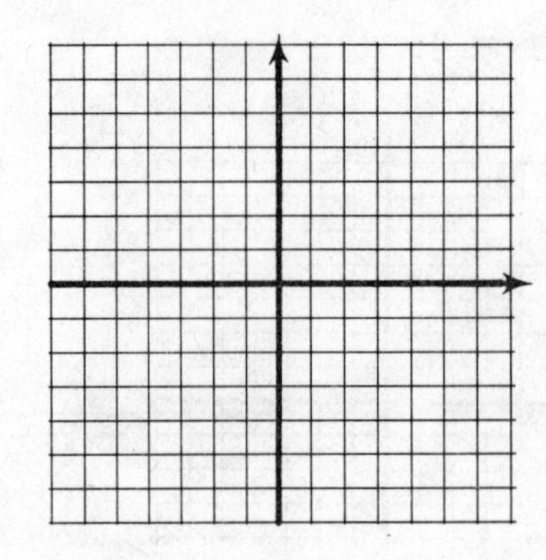

30. $x < -2$

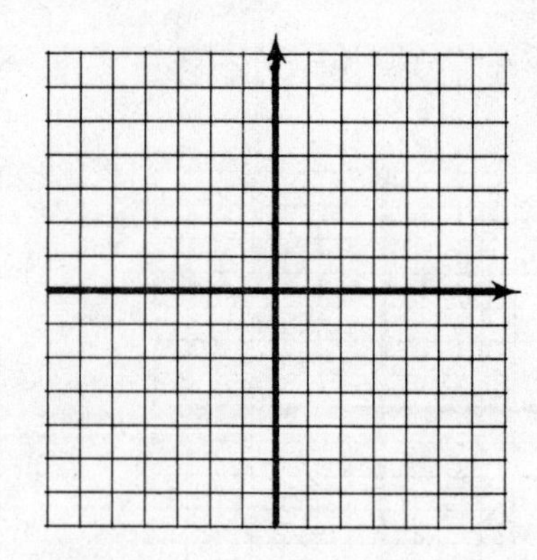

31. $y \leqslant 2$

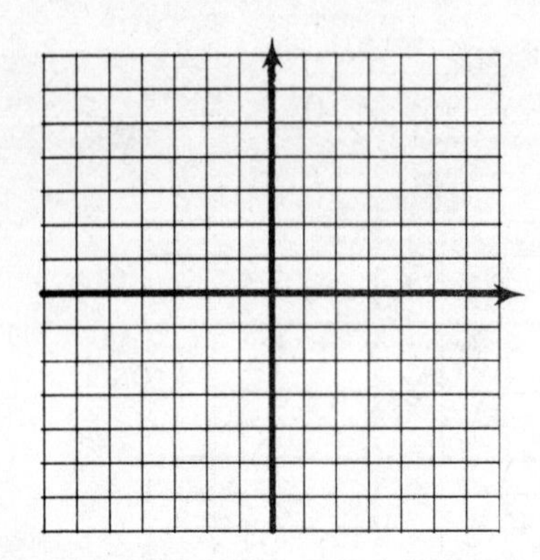

32. $y \leqslant -3$

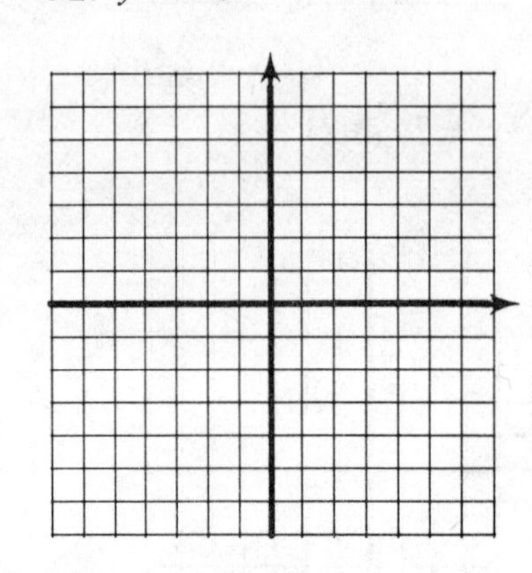

33. $x \geqslant -2$

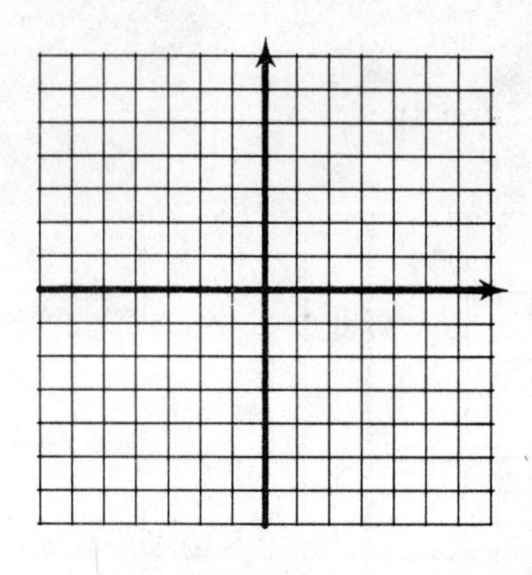

34. $x \leqslant 3y$

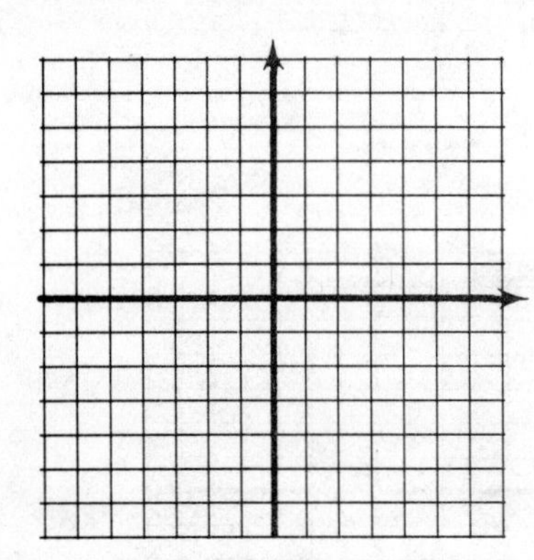

35. $x \leqslant 5y$

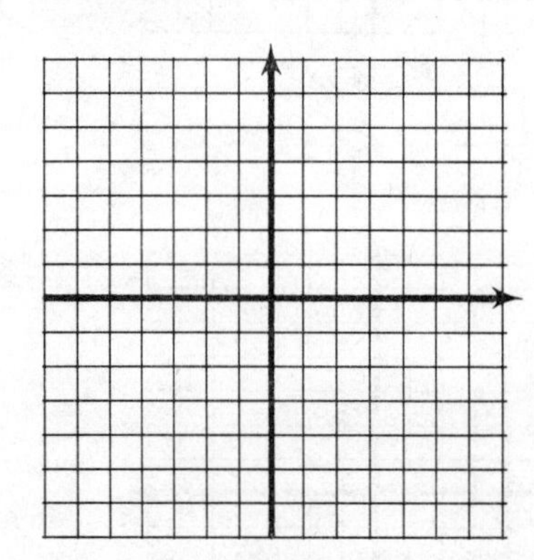

36. $x \geqslant -2y$

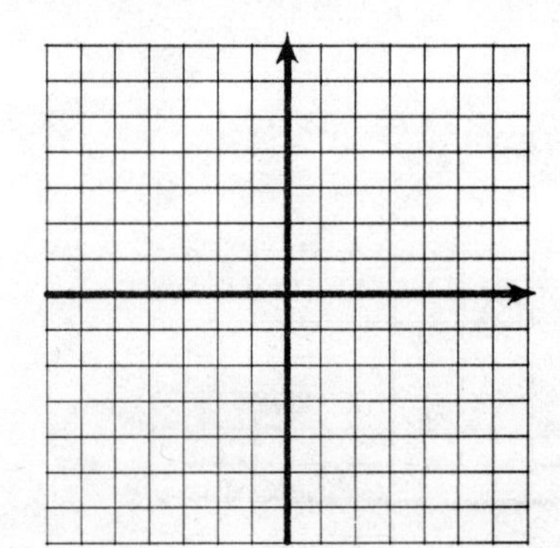

name date hour

CHAPTER 7 TEST

Complete the ordered pairs using the given equations.

Equation	Ordered pairs
1. $y = 5x - 6$	(0,) (−2,) (, 14)
2. $3x - 5y = 30$	(0,) (, 0) (5,)
3. $2x + 7y = 21$	(0,) (, 0) (3,) (, 2)
4. $x = 3y$	(0,) (, 2) (8,) (−12,)
5. $x + 4 = 0$	(, 2) (, 0) (, −3)
6. $y - 2 = 0$	(5,) (4,) (0,) (−3,)

Graph each linear equation.

7. $x + y = 9$

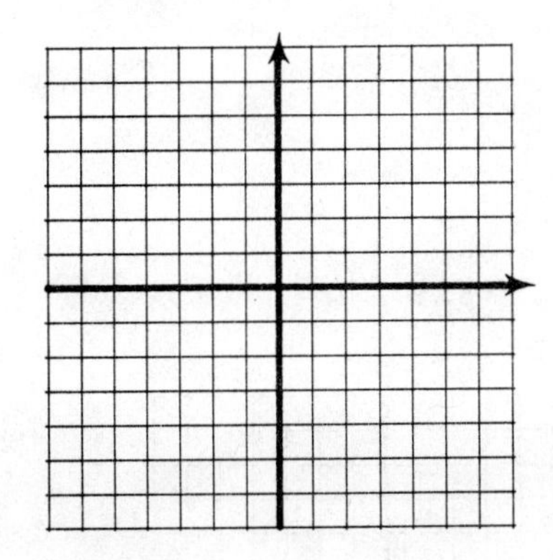

8. $2x + y = 6$

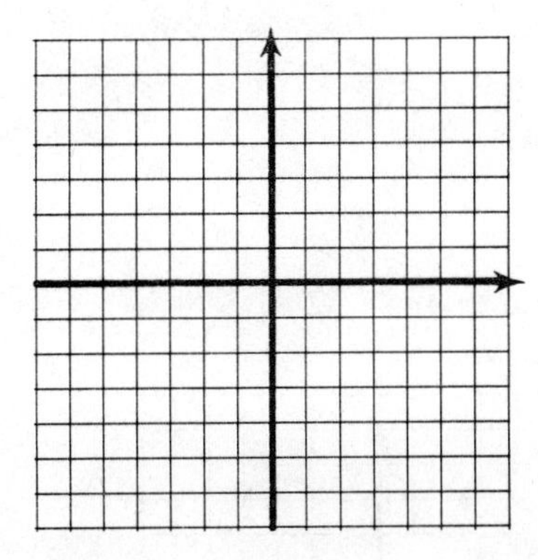

9. $3x - 2y = 18$

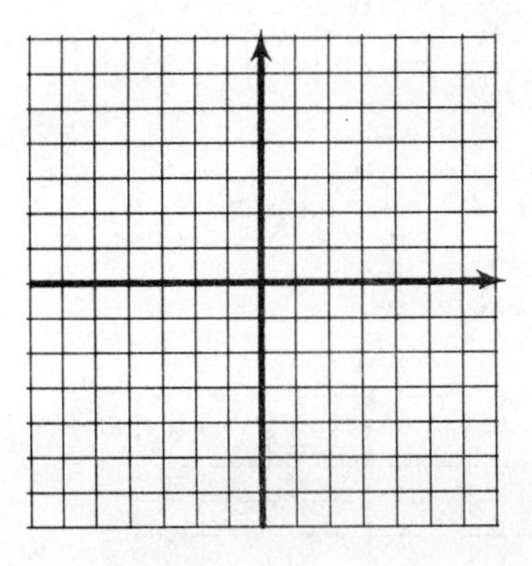

10. $4x + 5y = 10$

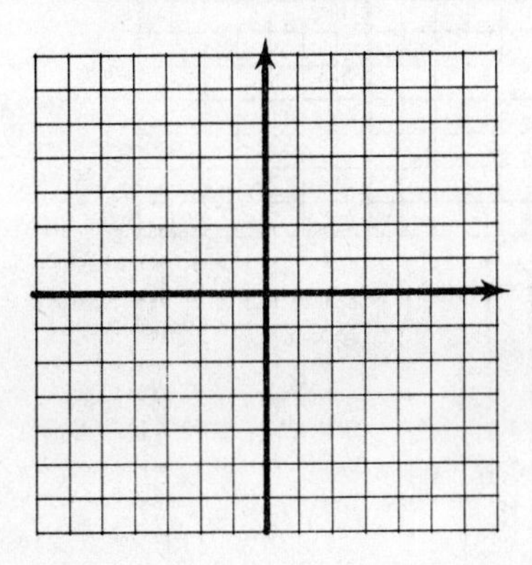

11. $2x - 3y = 7$

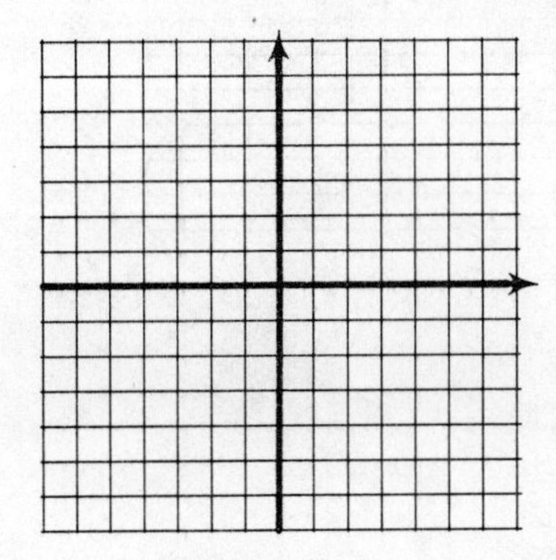

12. $x = 4y$

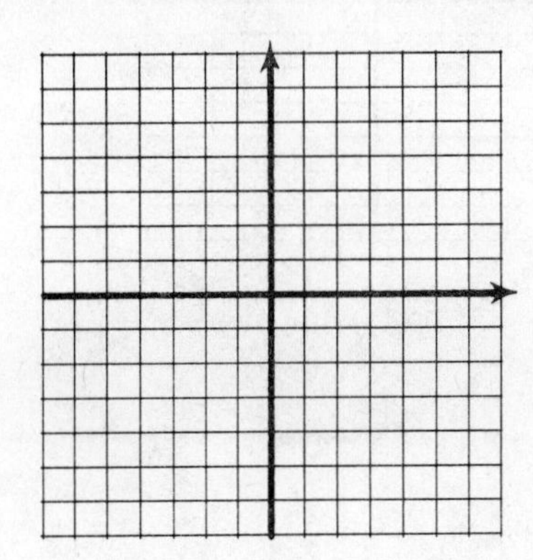

13. $2x + y = 0$

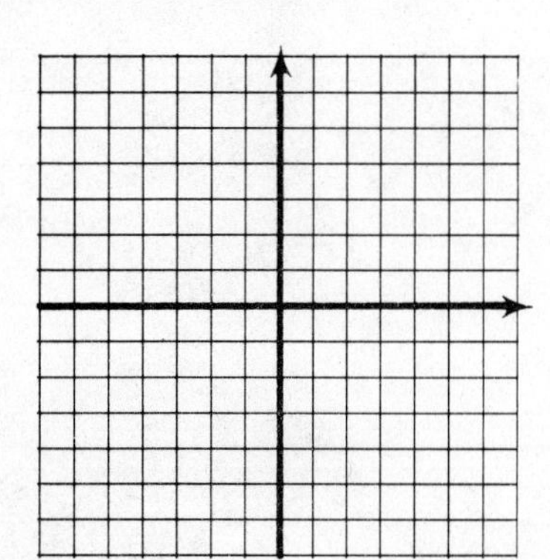

14. $x + 5 = 0$

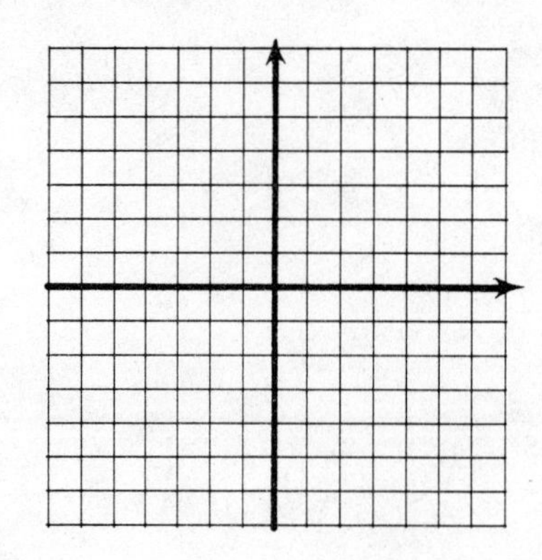

15. $y = 2$

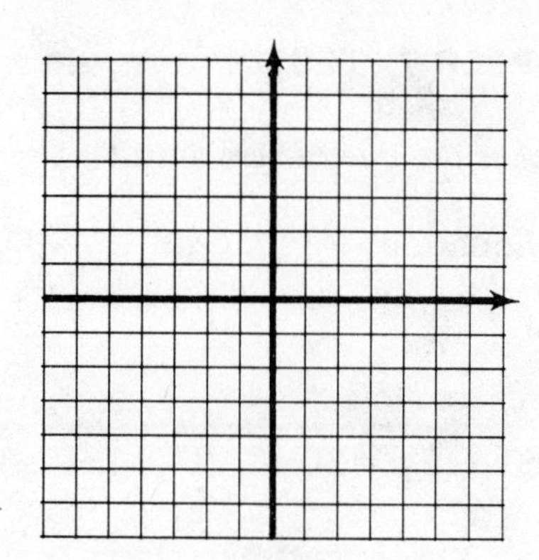

Complete the graph of each linear inequality.

16. $2x + y \leqslant 8$

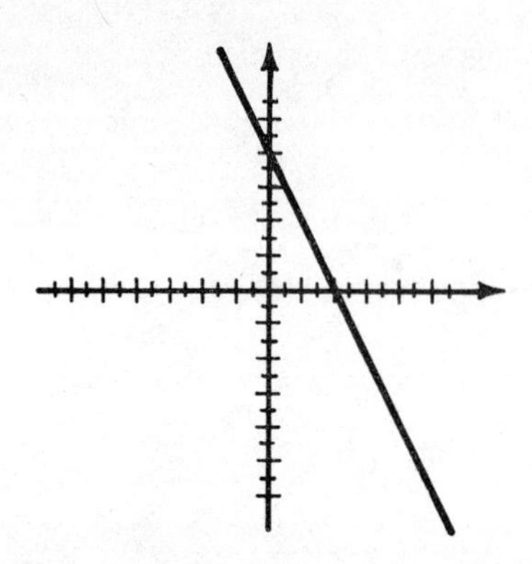

17. $2x - 3y > 6$

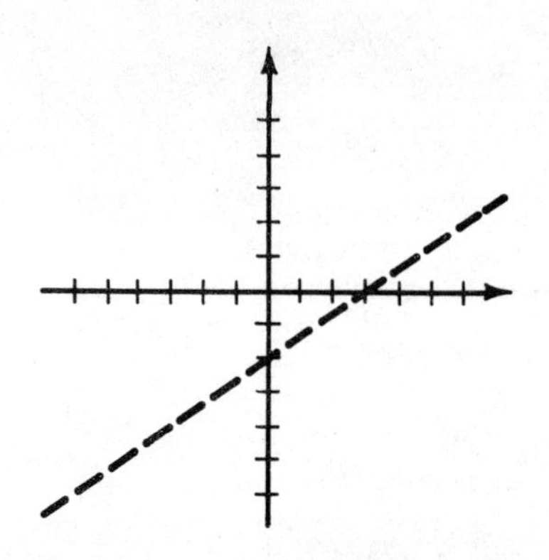

18. $y < -3$

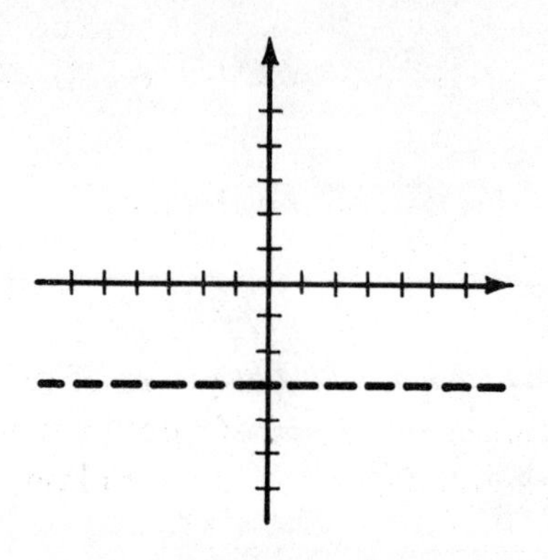

Graph each linear inequality.

19. $x + y \leqslant 6$

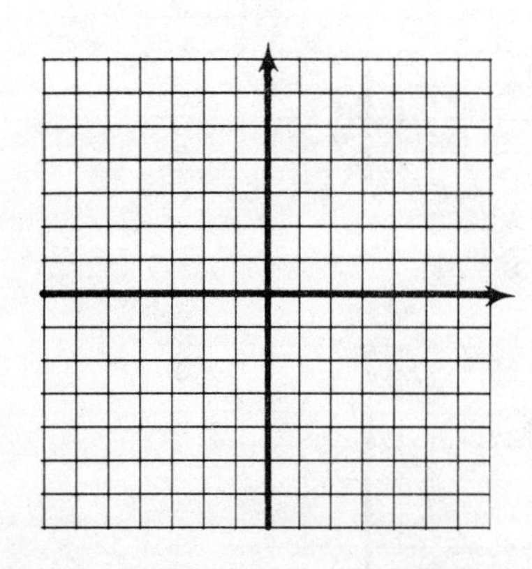

20. $3x - 4y > 12$

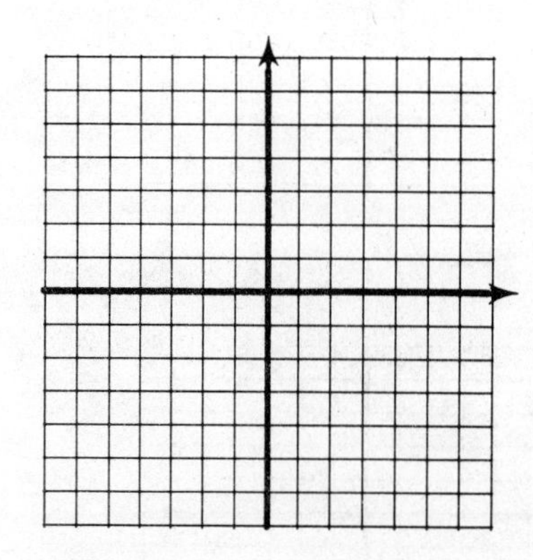

21. $x < 3y$

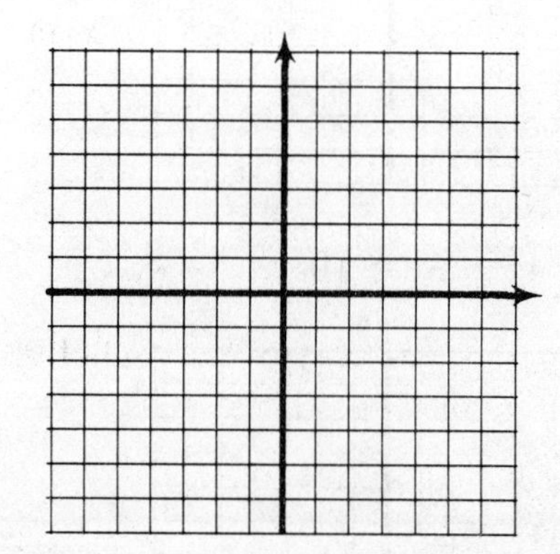

8 Linear Systems

8.1 SOLVING SYSTEMS OF LINEAR EQUATIONS BY GRAPHING

OBJECTIVES

- Decide whether a given ordered pair is a simultaneous solution of a system
- Solve linear systems by graphing

A **system** of linear equations is two or more linear equations which have the same variables. Some examples of systems of two linear equations are

System A	*System B*	*System C*
$2x + 3y = 4$	$x + 3y = 1$	$x - y = 1$
$3x - y = -5$	$-y = 4 - 2x$	$y = 3.$

In System C, you can think of $y = 3$ as an equation in two variables by writing it as

$$0x + y = 3.$$

The solutions of a system of two linear equations are all the ordered pairs that satisfy both equations at the same time. Such an ordered pair is called a **simultaneous solution** of the system.

Example 1 Is $(4, -3)$ a simultaneous solution of the following systems?

(a) $x + 4y = -8$
$3x + 2y = 6$

To decide whether or not $(4, -3)$ is a simultaneous solution of the system, substitute 4 for x and -3 for y in each equation.

$$\begin{aligned} x + 4y &= -8 \\ 4 + 4(-3) &= -8 \\ 4 + (-12) &= -8 \\ -8 &= -8 \quad \text{True} \end{aligned} \qquad \begin{aligned} 3x + 2y &= 6 \\ 3(4) + 2(-3) &= 6 \\ 12 + (-6) &= 6 \\ 6 &= 6 \quad \text{True} \end{aligned}$$

Since $(4, -3)$ satisfies both equations, it is a simultaneous solution of the system.

(b) $2x + 5y = -7$
$3x + 4y = 2$

Again, substitute 4 for x and -3 for y in both equations.

$$\begin{aligned} 2x + 5y &= -7 \\ 2(4) + 5(-3) &= -7 \\ 8 + (-15) &= -7 \\ -7 &= -7 \quad \text{True} \end{aligned} \qquad \begin{aligned} 3x + 4y &= 2 \\ 3(4) + 4(-3) &= 2 \\ 12 + (-12) &= 2 \\ 0 &= 2 \quad \text{False} \end{aligned}$$

QUICK CHECKS

1. Decide whether or not the given ordered pair is a simultaneous solution of the system.

(a) $(2, 5)$
$3x - 2y = -4$
$5x + y = 15$

(b) $(1, -2)$
$x - 3y = 7$
$4x + y = 5$

(c) $(-3, 3)$
$4x + 6y = 5$
$10x + 3y = -11$

(d) $(4, -1)$
$5x + 6y = 14$
$2x + 5y = 3$

Quick Check Answers

1. (a) yes (b) no (c) no (d) yes

Here, $(4, -3)$ is not a simultaneous solution since it does not satisfy the second equation.

Work Quick Check 1 at the side.

One way to find the simultaneous solution of a system of two equations is to graph both equations on the same axes. The coordinates of any point where the lines cross give the simultaneous solution of the system. Since two different straight lines can cross at no more than one point, there can never be more than one solution for such a system.

Example 2 Solve each system of equations by graphing both equations on the same axes.

(a) $2x + 3y = 4$
$3x - y = -5$

Figure 1 shows the graphs of $2x + 3y = 4$ and $3x - y = -5$ on the same axes. The graphs cross at the point $(-1, 2)$ which is the simultaneous solution of the system.

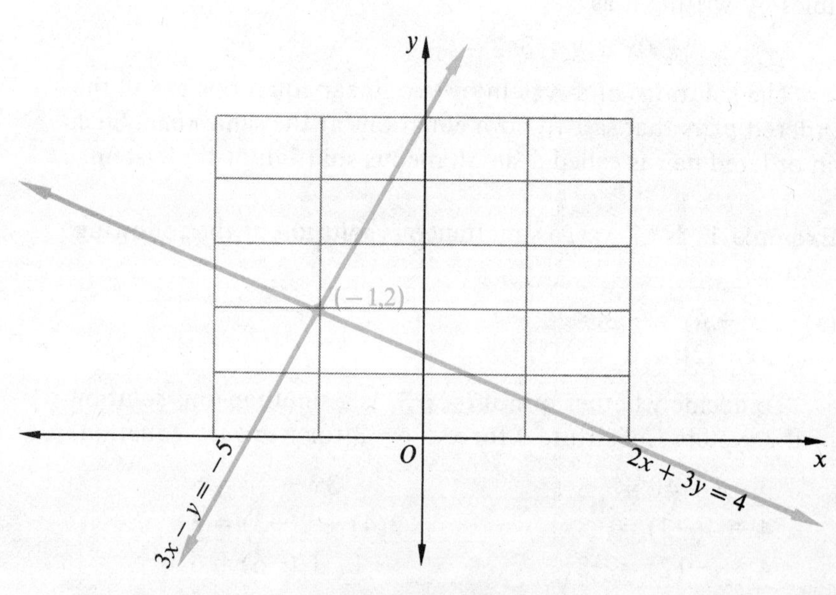

Figure 1

(b) $y = -2$
$x = 1$

To find the simultaneous solution of the system, graph the two lines on the same axes. The graph of $y = -2$ is a horizontal line. The graph of $x = 1$ is a vertical line. As shown in Figure 2, the simultaneous solution is $(1, -2)$, which is the point where the two graphs cross.

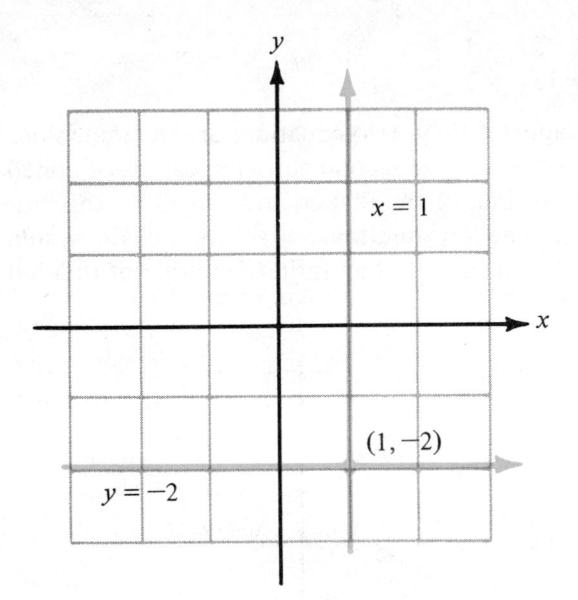

Figure 2

Work Quick Check 2 at the side.

Example 3 Solve each system by graphing.

(a) $2x + y = 2$
$2x + y = 8$

The graphs of these lines are shown in Figure 3. The two lines are parallel and therefore have no points in common. Thus, there is no simultaneous solution for this system.

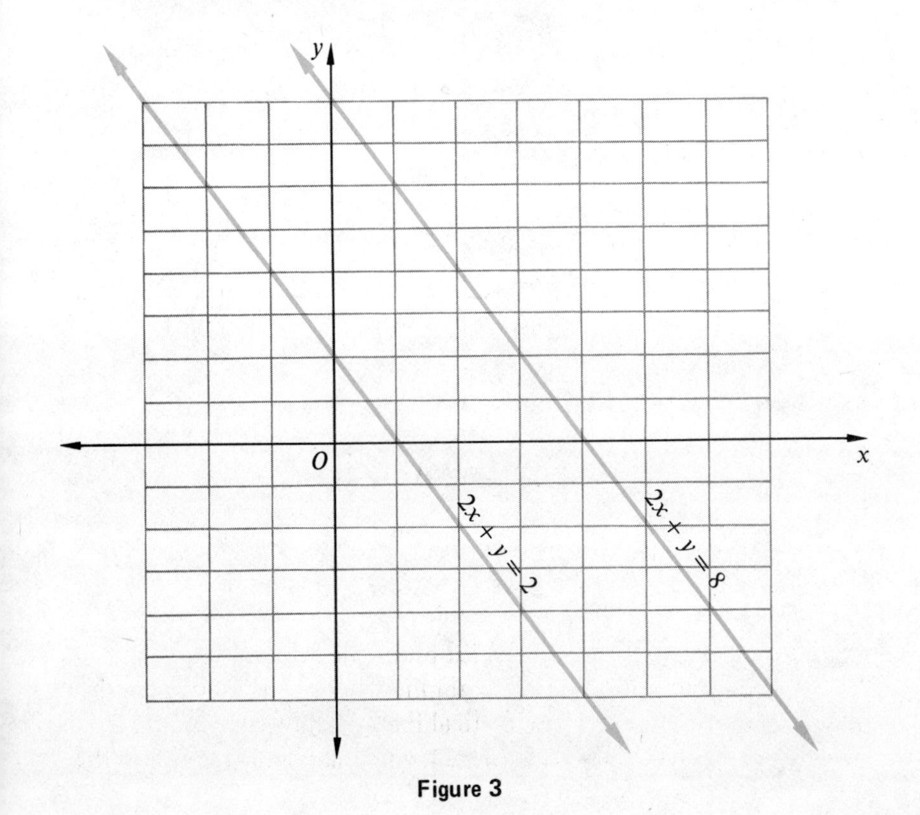

Figure 3

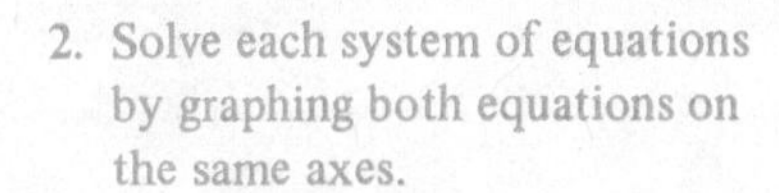

2. Solve each system of equations by graphing both equations on the same axes.

(a) $x + 3y = 9$
$4x - 2y = 8$

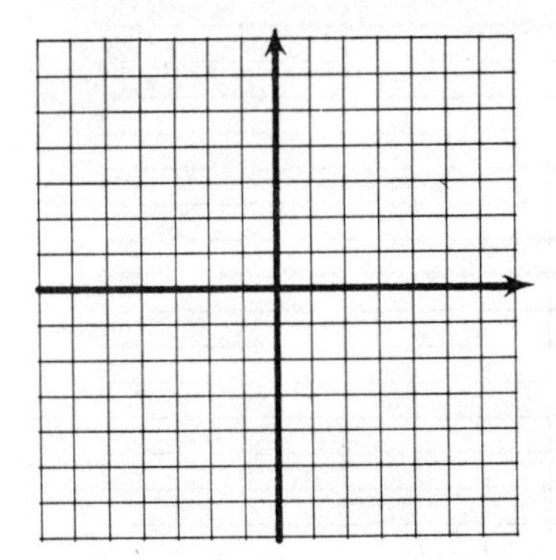

(b) $x + y = 4$
$2x - y = -1$

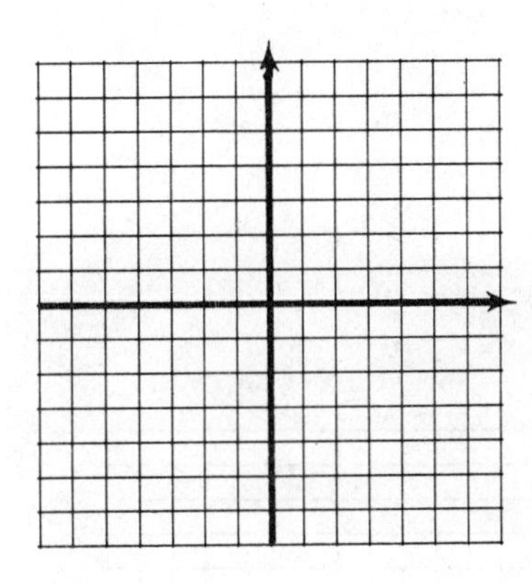

(c) $y = 1$
$x = 2$

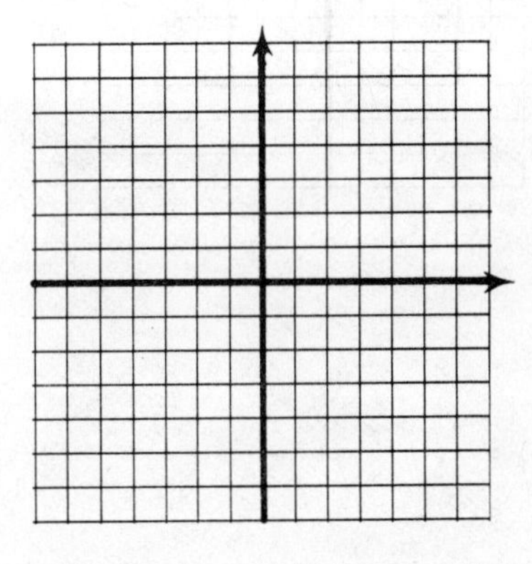

Quick Check Answers

2. (a) (3, 2) (b) (1, 3)
(c) (2, 1)

3. Solve each system of equations by graphing both equations on the same axes.

(a) $3x - y = 4$
$6x - 2y = 12$

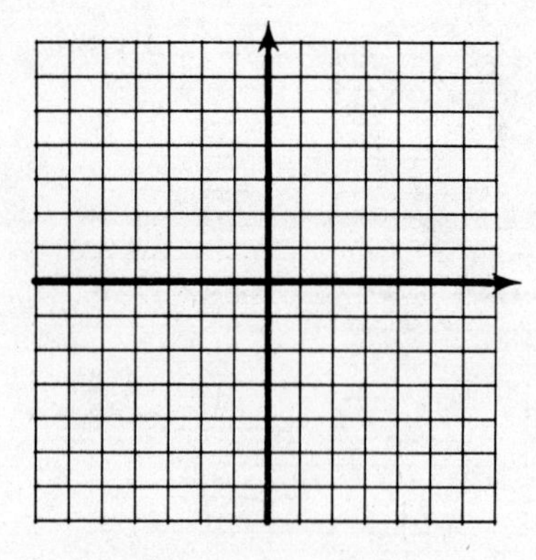

(b) $-x + 3y = 2$
$2x - 6y = -4$

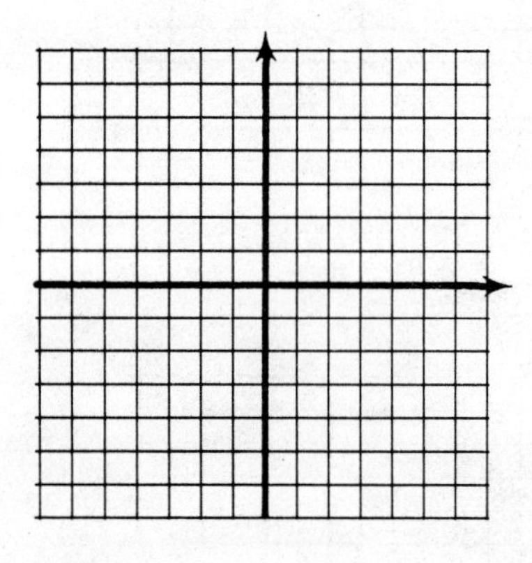

Quick Check Answers
3. (a) no solution (b) same line

(b) $2x + 5y = 1$
$6x + 15y = 3$

The graphs of these two equations are the same line. See Figure 4. Note that the second equation can be obtained by multiplying both sides of the first equation by 3. In this case, every point on the line is a simultaneous solution of the system. Thus, the solution is made up of an infinite number of ordered pairs.

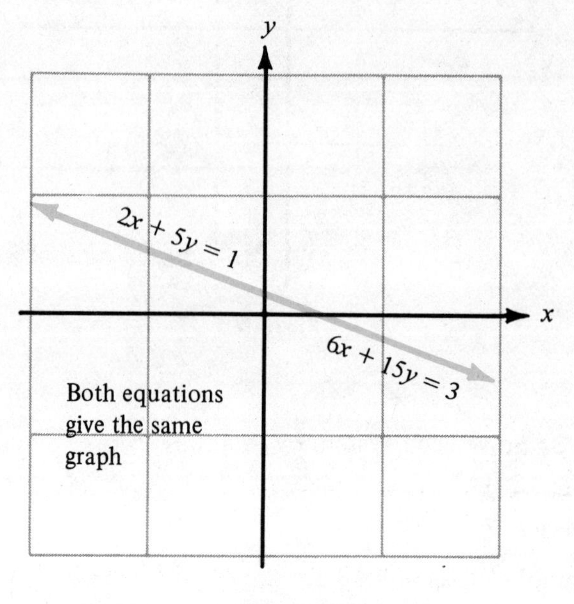

Figure 4

Work Quick Check 3 at the side.

name date hour

8.1 EXERCISES

Decide whether or not the given ordered pair is the simultaneous solution of the given system. See Example 1.

1. $(2, -5)$ ____________
$3x + y = 1$
$2x + 3y = -11$

2. $(-1, 6)$ ____________
$2x + y = 4$
$3x + 2y = 9$

3. $(4, -2)$ ____________
$x + y = 2$
$2x + 5y = 2$

4. $(-6, 3)$ ____________
$x + 2y = 0$
$3x + 5y = 3$

5. $(2, 0)$ ____________
$3x + 5y = 6$
$4x + 2y = 5$

6. $(0, -4)$ ____________
$2x - 5y = 20$
$3x + 6y = -20$

7. $(5, 2)$ ____________
$4x + 3y = 26$
$3x + 7y = 29$

8. $(9, 1)$ ____________
$2x + 5y = 23$
$3x + 2y = 29$

9. $(6, -8)$ ____________
$x + 2y + 10 = 0$
$2x - 3y + 30 = 0$

10. $(-5, 2)$ ____________
$3x - 5y + 20 = 0$
$2x + 3y + 4 = 0$

11. $(5, -2)$ ____________
$x - 5 = 0$
$y + 2 = 0$

12. $(-8, 3)$ ____________
$x = 8$
$y = 3$

Solve each system by graphing both equations on the same axes. See Example 2.

13. $x + y = 8$
$x - y = 2$

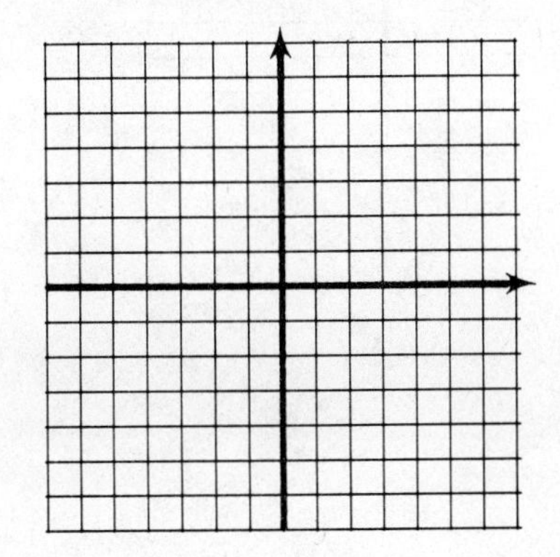

14. $x + y = -1$
$x - y = 3$

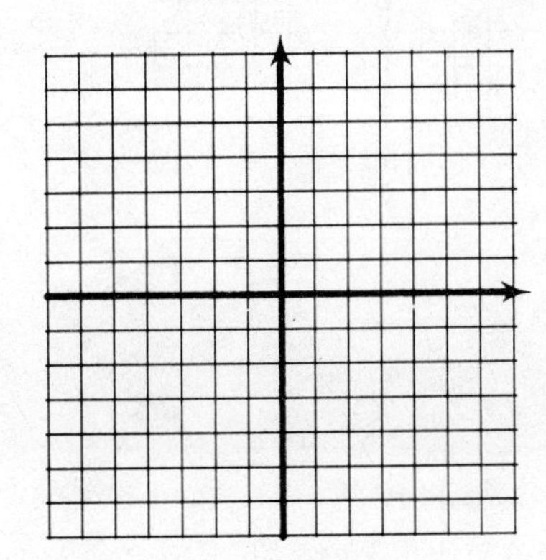

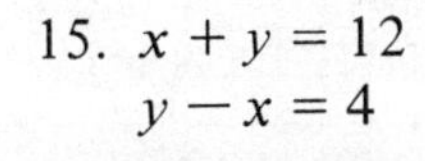

15. $x + y = 12$
$y - x = 4$

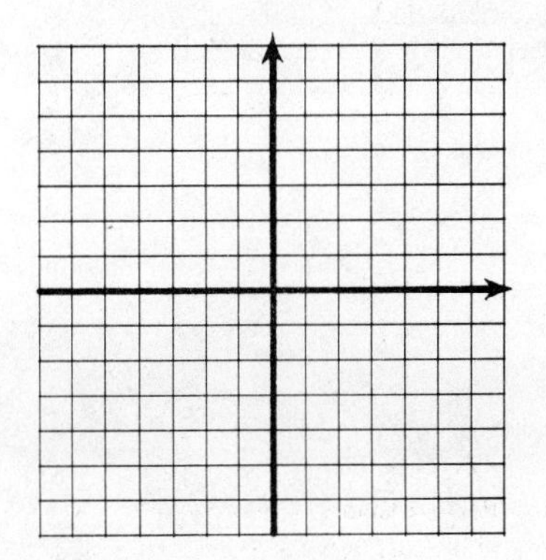

16. $y - x = -5$
$x + y = 1$

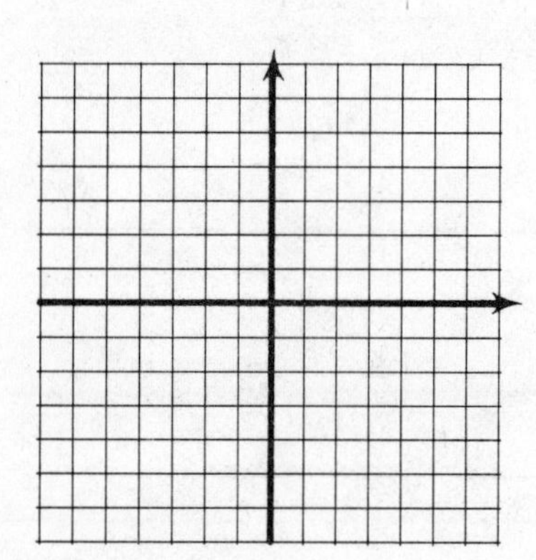

17. $2x + y = 6$
$x - 3y = 10$

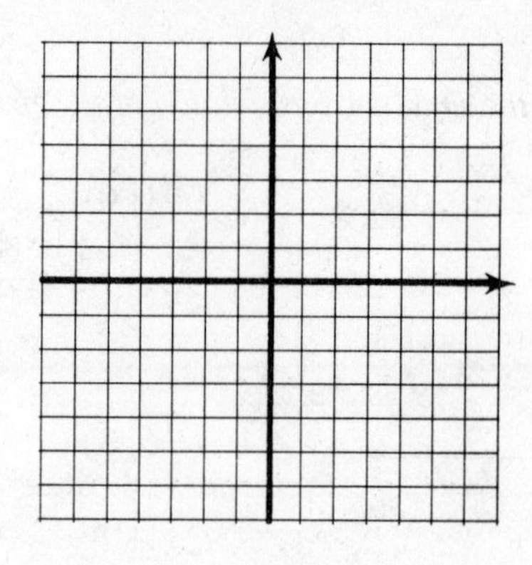

18. $4x + y = 10$
$2x - y = 8$

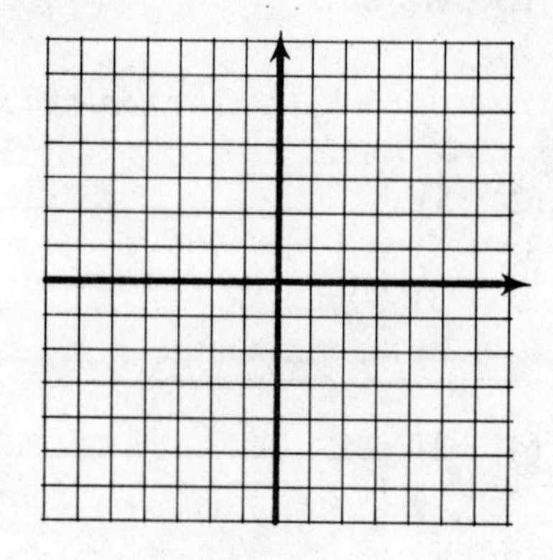

19. $2x - 5y = 17$
$3x + y = 0$

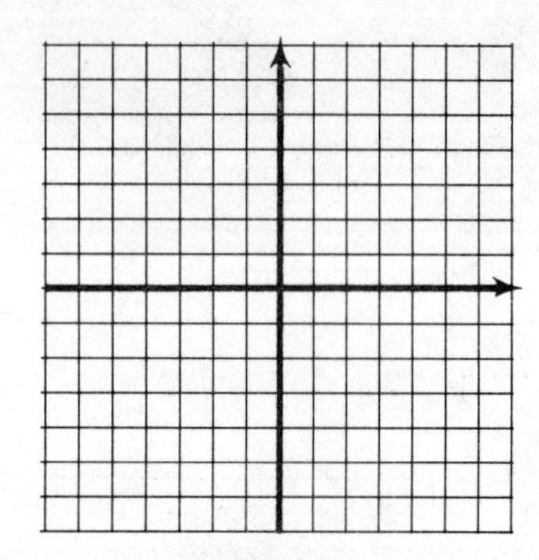

20. $2x + 3y = 11$
$3x - y = 11$

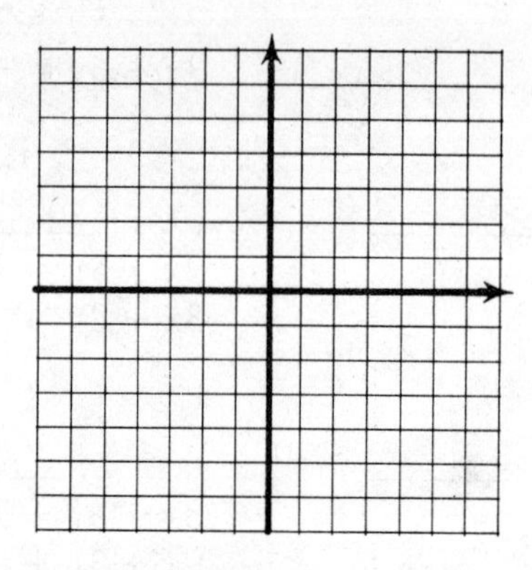

21. $5x + 4y = 7$
$2x - 3y = 12$

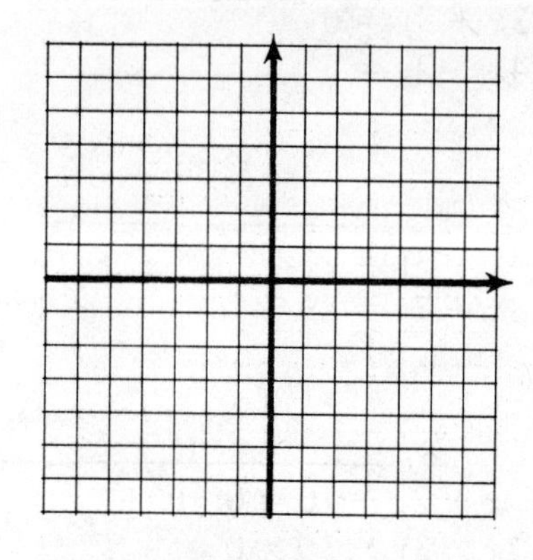

22. $2x + 5y = 17$
$3x - 4y = -9$

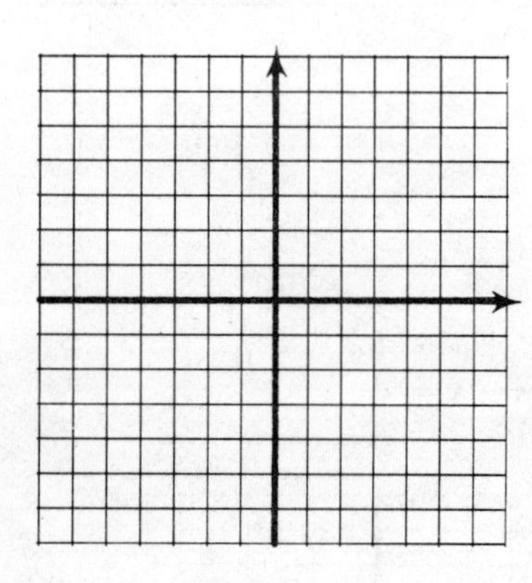

23. $4x + 5y = 3$
$2x - 5y = 9$

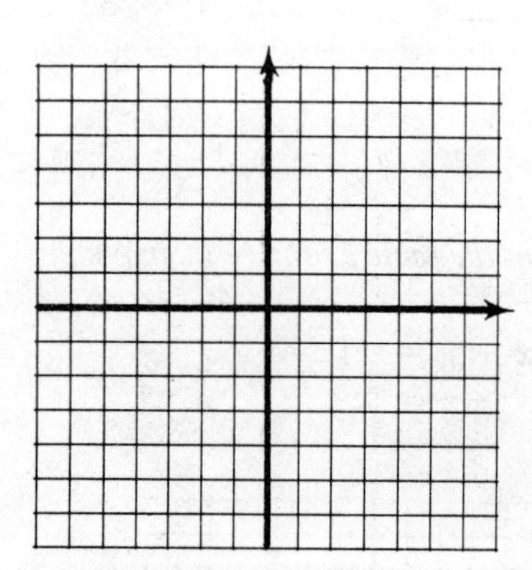

24. $2x + y = 1$
$3x - 4y = 29$

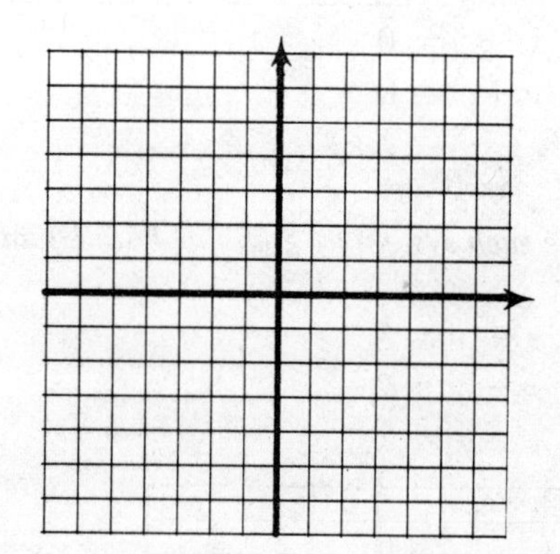

name date hour

25. $3x + 2y = -12$
$x - 2y = -20$

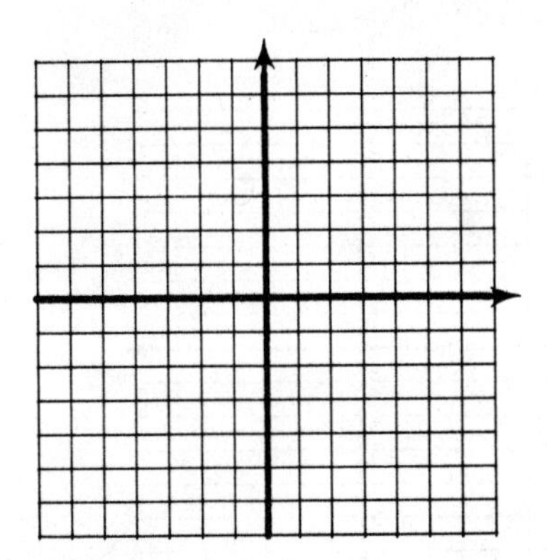

26. $4x + y = -14$
$3x - 2y = -5$

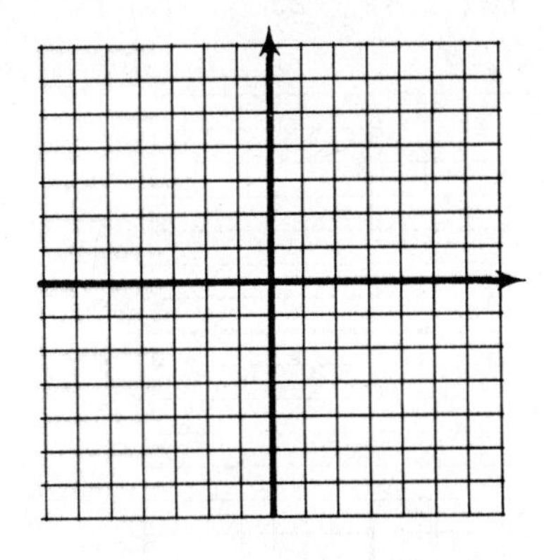

27. $3x - 4y = -8$
$5x + 2y = -22$

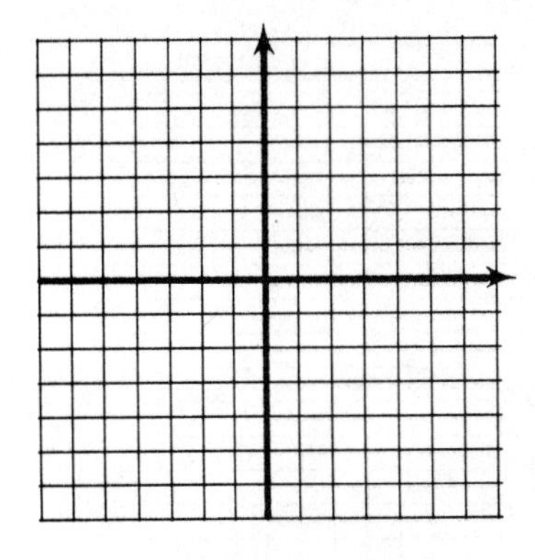

28. $x - 2y = -3$
$4x + 3y = -34$

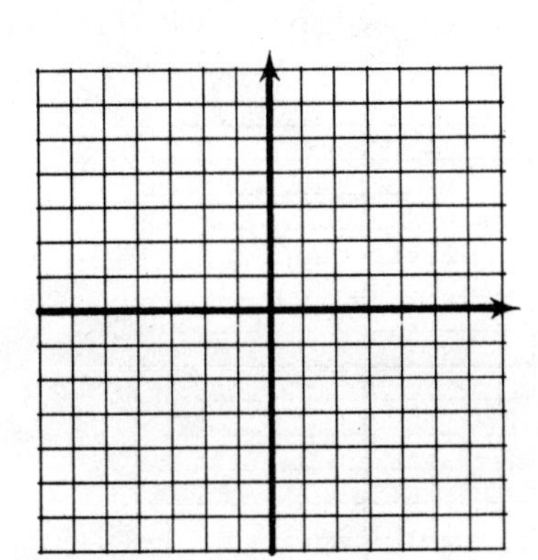

29. $3x - 2y = 15$
$4x + 3y = 20$

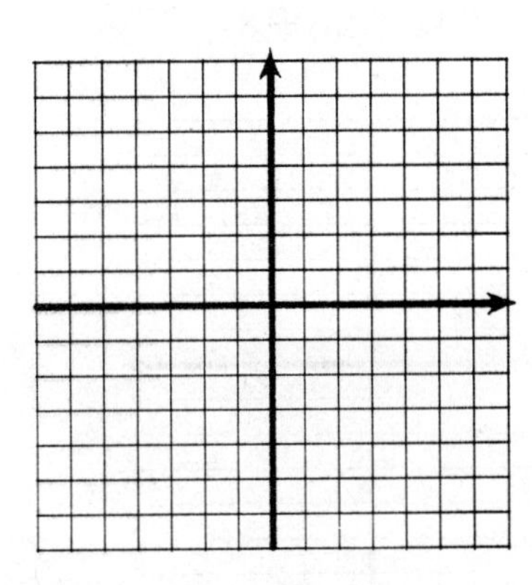

30. $-4x + 3y = 16$
$2x - 3y = -8$

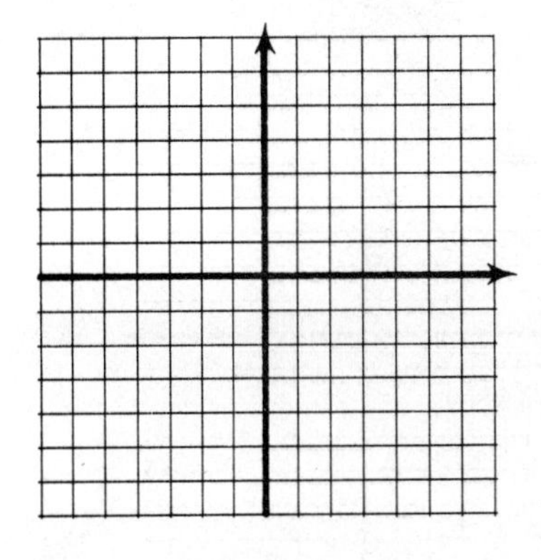

31. $3x - 4y = 8$
$4x + 5y = -10$

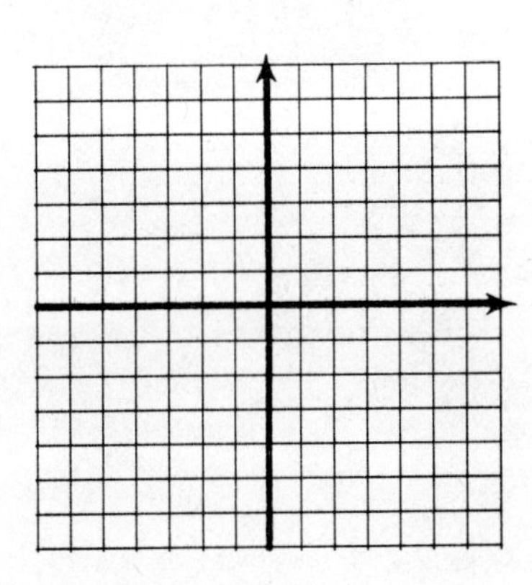

32. $3x + 2y = 10$
$4x - 3y = -15$

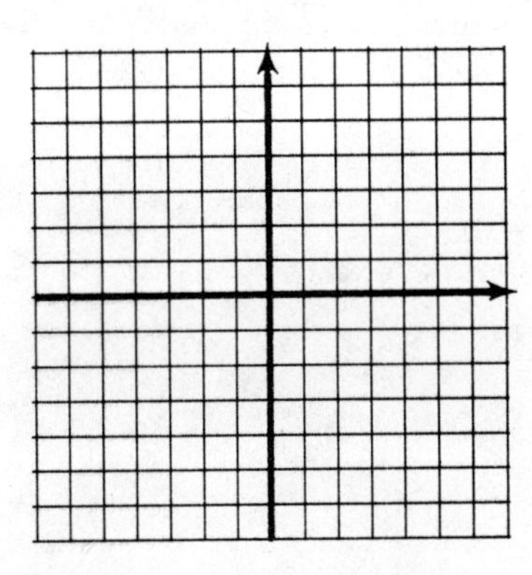

33. $2x + 5y = 20$
$x - 2y = 1$

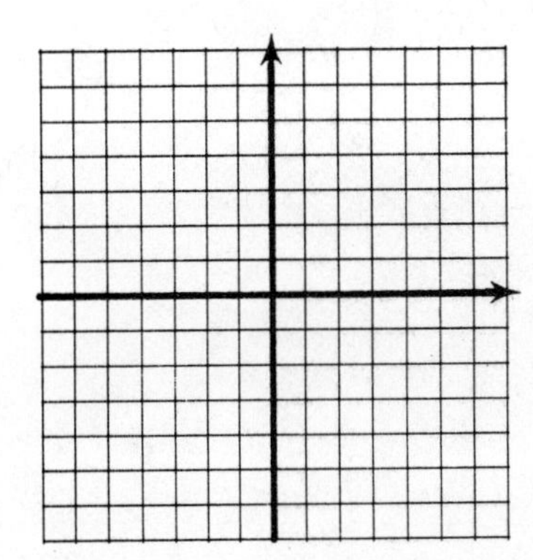

Solve each system by graphing. If the two equations produce parallel lines, write "no solution." If the two equations produce the same line, write "same line." See Example 3.

34. $2x + 3y = 5$
$4x + 6y = 9$

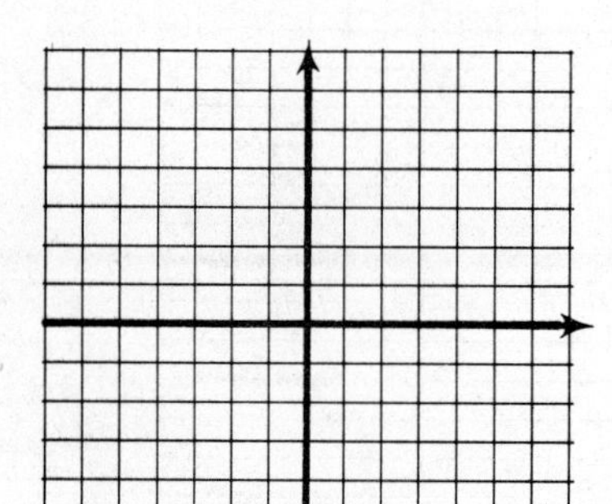

35. $5x - 4y = 5$
$10x - 8y = 23$

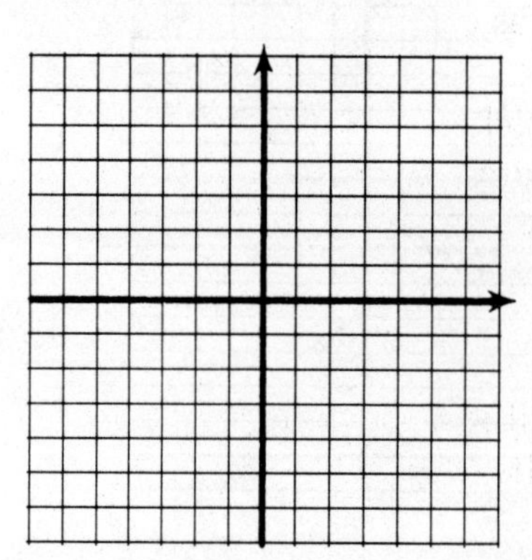

36. $3x = y + 5$
$6x - 2y = 5$

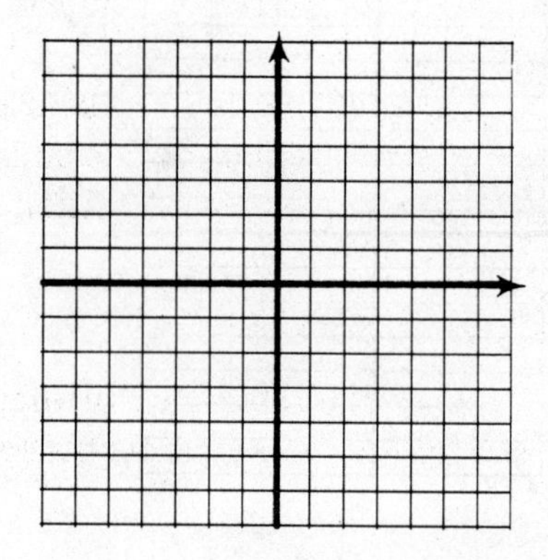

37. $4y + 1 = x$
$2x - 3 = 8y$

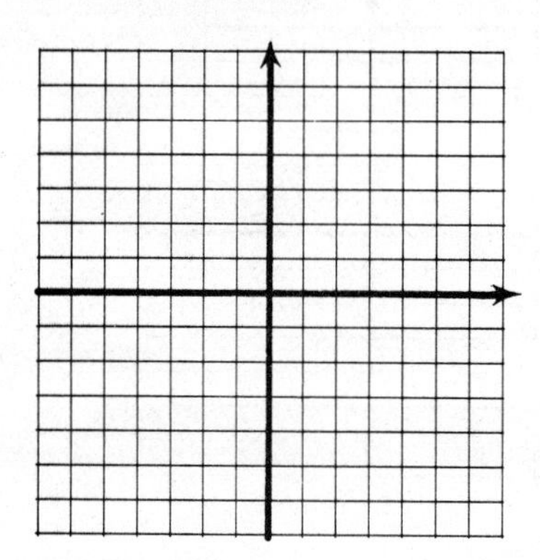

38. $2x - y = 4$
$4x = 2y + 8$

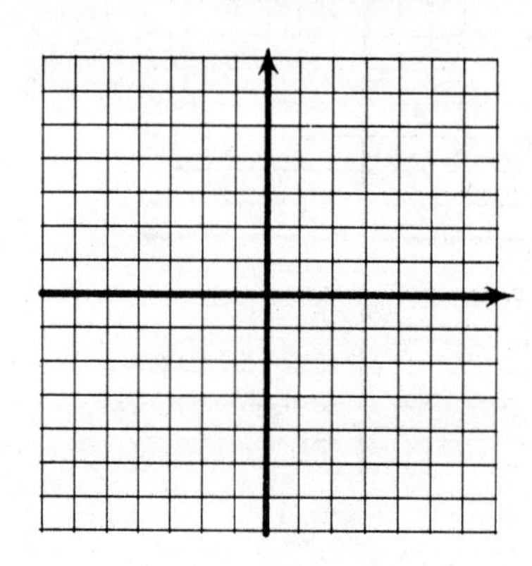

39. $3x = 5 - y$
$6x + 2y = 10$

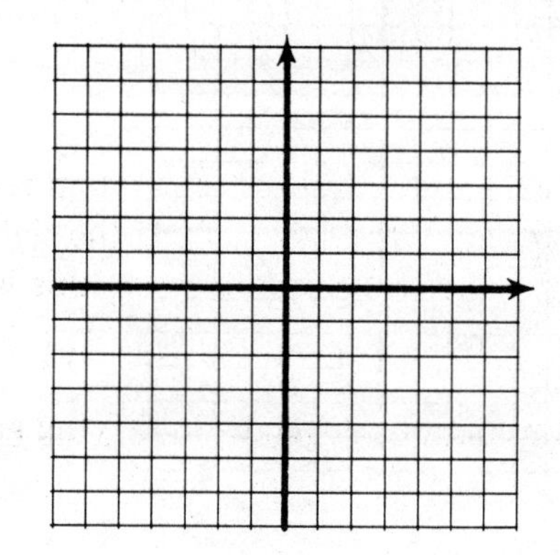

8.2 SOLVING SYSTEMS OF LINEAR EQUATIONS BY ADDITION

OBJECTIVES

- Solve linear systems by addition or elimination

Graphing to solve a system of equations has a serious drawback: it is difficult to estimate a solution such as $(1/3, -5/6)$ accurately from a graph.

An algebraic method can be used which depends on the addition property of equality. The addition property says that if two quantities are equal, then addition of the same quantity to each results in equal sums.

$$\text{If } A = B, \text{ then } A + C = B + C.$$

We now take this a step further. Addition of *equal* quantities, rather than the *same* quantity, also results in equal sums.

$$\text{If } A = B \text{ and } C = D, \text{ then } A + C = B + D.$$

Example 1 Use the addition property of equality to find the simultaneous solution of the system

$$x + y = 5$$
$$x - y = 3.$$

Each equation of this system is a statement of equality, so, as discussed above, the sum of the right-hand sides equals the sum of the left-hand sides. Adding in this way, we have

$$(x + y) + (x - y) = 5 + 3.$$

Combining terms and simplifying gives

$$2x = 8$$
$$x = 4.$$

Thus, $x = 4$ gives the x-value of the simultaneous solution of the given system. To find the y-value of the solution, substitute 4 for x in either of the two equations of the system. Choose the first equation, $x + y = 5$.

$$x + y = 5$$
$$4 + y = 5 \qquad \text{Let } x = 4$$
$$y = 1$$

Then $y = 1$ gives the y-value of the simultaneous solution. The solution is given by the ordered pair $(4, 1)$.

This can be checked by substituting the values we found into the other equation of the given system, $x - y = 3$.

$$x - y = 3$$
$$4 - 1 = 3 \qquad \text{Let } x = 4; \text{let } y = 1$$
$$3 = 3 \qquad \text{True}$$

This result is true, so that the simultaneous solution of the given system is $(4, 1)$.

QUICK CHECKS

1. Solve each system by addition.

(a) $2x - y = 2$
$4x + y = 10$

(b) $x + 3y = -2$
$-x - 2y = 0$

Example 2 Solve the system

$$-2x + y = -11$$
$$5x - y = 26.$$

We need to add left-hand sides and right-hand sides. The work is easier if we draw a line under the second equation and add vertically. To do this, the like terms must be lined up in columns.

$$\begin{array}{r} -2x + y = -11 \\ 5x - y = 26 \\ \hline 3x = 15 \end{array}$$

$$x = 5$$

Substitute 5 for x in either of the original equations. We choose the first.

$$-2x + y = -11$$
$$-2(5) + y = -11 \qquad \text{Let } x = 5$$
$$-10 + y = -11$$
$$y = -1$$

The simultaneous solution is $(5, -1)$. The solution can be checked by substitution into both of the original equations.

Work Quick Check 1 at the side.

Sometimes it is necessary to multiply one or both equations in a system by some number before we can use the addition method.

Example 3 Solve the system

$$x + 3y = 7 \qquad (1)$$
$$2x + 5y = 12. \qquad (2)$$

If we add the two equations, we get $3x + 8y = 19$, which does not help us find the solution. However, if we multiply both sides of equation (1) by -2, the terms with the variable x would drop out when we add.

$$-2(x + 3y) = -2(7)$$
$$-2x - 6y = -14 \qquad (3)$$

Now add equations (3) and (2).

$$\begin{array}{rr} -2x - 6y = -14 & (3) \\ 2x + 5y = 12 & (2) \\ \hline -y = -2 & \end{array}$$

From this result, we get $y = 2$. Substituting back into equation (1) gives

$$x + 3y = 7$$
$$x + 3(2) = 7 \qquad \text{Let } y = 2$$
$$x + 6 = 7$$
$$x = 1.$$

Quick Check Answers

1. (a) (2, 2) (b) (4, −2)

The solution of this system is (1, 2).

Work Quick Check 2 at the side.

Example 4 Solve the system

$$2x + 3y = -15 \qquad (4)$$
$$5x + 2y = 1. \qquad (5)$$

Here we use the multiplication property of equality with both equations instead of just one, as in Example 3. Multiply by numbers that will cause the coefficients of x (or of y) in the two equations to be negatives of each other. For example, multiply both sides of equation (4) by 5, and both sides of equation (5) by -2.

$$\begin{array}{r} 10x + 15y = -75 \\ -10x - 4y = -2 \\ \hline 11y = -77 \end{array}$$

This gives $y = -7$. By substituting -7 for y in equation (4) or (5), we get $x = 3$. The solution of the system is $(3, -7)$.

We could have obtained the same result by multiplying equation (4) by 2 and equation (5) by -3. (Check this.)

Work Quick Check 3 at the side.

The method of solution discussed in this section is called the **addition method** or **elimination method.** For most systems, this method is more efficient than graphing. The solution of a linear system of equations having exactly one solution can be found using the addition method summarized in Steps 1–7.

1. Write both equations of the system in the form $ax + by = c$.
2. Multiply one or both equations by appropriate numbers so that the coefficients of x (or y) are additive inverses of each other.
3. Add the two equations to get an equation with only one variable.
4. Solve the equation from Step 3.
5. Substitute the solution from Step 4 into either of the original equations.
6. Solve the resulting equation from Step 5 for the remaining variable.
7. Check the answer.

2. Solve each system by addition.

(a) $x - 3y = -7$
$3x + 2y = 23$

(b) $8x + 2y = 2$
$3x - y = 6$

3. Solve each system of equations.

(a) $4x - 5y = -18$
$3x + 2y = -2$

(b) $6x + 7y = 4$
$5x + 8y = -1$

Quick Check Answers
2. (a) (5, 4) (b) (1, −3)
3. (a) (−2, 2) (b) (3, −2)

WORK SPACE

name date hour

8.2 EXERCISES

Solve each system by the addition method. See Examples 1 and 2.

1. $x - y = 3$ \
$x + y = -1$ ______

2. $x + y = 7$ \
$x - y = -3$ ______

3. $x + y = 2$ \
$2x - y = 4$ ______

4. $3x - y = 8$ \
$x + y = 4$ ______

5. $2x + y = 14$ \
$x - y = 4$ ______

6. $2x + y = 2$ \
$-x - y = 1$ ______

7. $3x + 2y = 6$ \
$-3x - y = 0$ ______

8. $5x - y = 9$ \
$-5x + 2y = -8$ ______

9. $6x - y = 1$ \
$-6x + 5y = 7$ ______

10. $6x + y = -2$ \
$-6x + 3y = -14$ ______

11. $2x - y = 5$ \
$4x + y = 4$ ______

12. $x - 4y = 13$ \
$-x + 6y = -18$ ______

13. $5x - y = 15$ \
$7x + y = 21$ ______

14. $x - 4y = 12$ \
$-x + 6y = -18$ ______

Solve each system by the addition method. See Example 3.

15. $2x - y = 7$ \
$3x + 2y = 0$ ______

16. $x + y = 7$ \
$-3x + 3y = -9$ ______

17. $x + 3y = 16$ \
$2x - y = 4$ ______

18. $4x - 3y = 8$ \
$2x + y = 14$ ______

19. $x + 4y = -18$ \
$3x + 5y = -19$ ______

20. $2x + y = 3$ \
$5x - 2y = -15$ ______

21. $3x - 2y = -6$ \
$-5x + 4y = 16$ ______

22. $-4x + 3y = 0$ \
$5x - 6y = 9$ ______

23. $2x - y = -8$ ______
$5x + 2y = -20$

24. $5x + 3y = -9$ ______
$7x + y = -3$

25. $2x + y = 5$ ______
$5x + 3y = 11$

26. $2x + 7y = -53$ ______
$4x + 3y = -7$

Solve each system by the addition method. See Example 4.

27. $5x - 4y = -1$ ______
$-7x + 5y = 8$

28. $3x + 2y = 12$ ______
$5x - 3y = 1$

29. $3x + 5y = 33$ ______
$4x - 3y = 15$

30. $2x + 5y = 3$ ______
$5x - 3y = 23$

31. $3x + 5y = -7$ ______
$5x + 4y = 10$

32. $2x + 3y = -11$ ______
$5x + 2y = 22$

33. $2x + 3y = -12$ ______
$5x - 7y = -30$

34. $2x + 9y = 16$ ______
$5x - 6y = 40$

35. $4x - 3y = 0$ ______
$6x + 6y = 7$

36. $8x + 3y = 9$ ______
$12x + 6y = 13$

37. $8x + 12y = 13$ ______
$16x - 18y = -9$

38. $9x + 6y = -9$ ______
$6x + 8y = -16$

39. $3x - 2y = 3$ ______
$3x + 3y = 78$

40. $3x - 2y = 27$ ______
$2x - 7y = -50$

41. $5x - 7y = 6$ ______
$3x - 6y = 2$

42. $3x + 7y = -12$ ______
$-4x + 3y = 16$

8.3 TWO SPECIAL CASES

OBJECTIVES

- Solve linear systems having parallel lines or the same line as their graphs

In Section 8.1 we saw that the graphs of a linear system are sometimes two parallel lines and sometimes the same line. In this section, we use the addition method to solve such systems.

Example 1 Solve by the addition method.

$$2x + 4y = 5$$
$$4x + 8y = -9$$

If we multiply both sides of $2x + 4y = 5$ by -2, and then add $4x + 8y = -9$, we get

$$\begin{aligned} -4x - 8y &= -10 \\ 4x + 8y &= -9 \\ \hline 0 &= -19. \quad \text{False} \end{aligned}$$

The false statement, $0 = -19$, shows that the given system is self-contradictory. *It has no simultaneous solution.* This means that the graphs of the equations of this system are parallel lines, as shown in Figure 5.

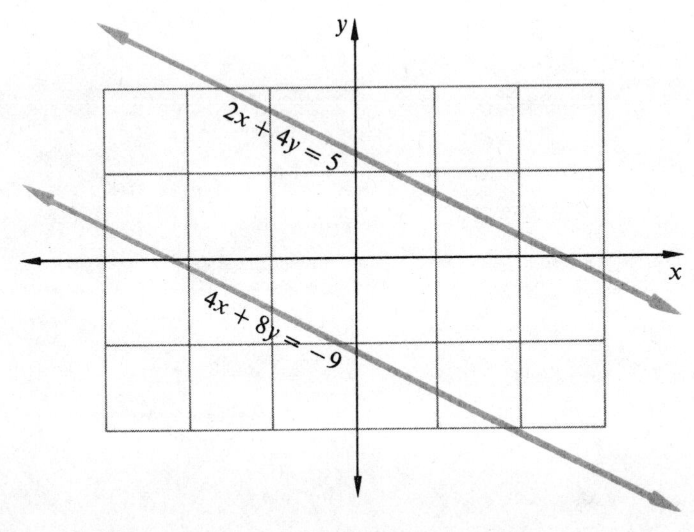

Figure 5

Work Quick Check 1 at the side.

Example 2 Solve by the addition method.

$$3x - y = 4$$
$$-9x + 3y = -12$$

If we multiply both sides of the first equation by 3 and add, we get

$$\begin{aligned} 9x - 3y &= 12 \\ -9x + 3y &= -12 \\ \hline 0 &= 0. \quad \text{True} \end{aligned}$$

QUICK CHECKS

1. Solve each system by the addition method.

(a) $4x + 3y = 10$
$2x + \frac{3}{2}y = 12$

(b) $-2x - 4y = -1$
$5x + 10y = 15$

Quick Check Answers

1. (a) no solution (b) no solution

2. Solve each system by the addition method.

(a) $6x + 3y = 9$
$-8x - 4y = -12$

(b) $4x - 6y = 10$
$-10x + 15y = -25$

This result, $0 = 0$, is true for any ordered pair (x, y) which satisfies either equation. Thus, any ordered pair which satisfies either equation is a simultaneous solution, and there is an infinite number of solutions. The solutions can be written as the solution set $\{(x, y) | 3x - y = 4\}$*, or as the solution set $\{(x, y) | -9x + 3y = -12\}$. However, in the answers at the back of this book, such a solution is indicated by the words "same line."

Work Quick Check 2 at the side.

There are three possibilities when using the addition method to solve a linear system of equations:

1. The result of the addition step is a statement such as $x = 2$ or $y = -3$. The solution will be exactly one ordered pair. The graphs of the equations of the system will cross at exactly one point.

2. The result of the addition step is a false statement, such as $0 = 4$. In this case, the graphs are parallel lines, and there is no simultaneous solution for the system.

3. The result of the addition step is a true statement, such as $0 = 0$. The graphs of the equations of the system are the same line, and there is an infinite number of ordered pairs which are solutions.

Quick Check Answers

2. (a) same line (b) same line

*Read "the set of all ordered pairs (x, y) such that $3x - y = 4$.

name date hour

8.3 EXERCISES

Use the addition method to solve each system. See Examples 1 and 2.

1. $x + y = 4$ ______
   $x + y = -2$

2. $2x - y = 1$ ______
   $2x - y = 4$

3. $5x - 2y = 6$ ______
   $10x - 4y = 10$

4. $3x - 5y = 2$ ______
   $6x - 10y = 8$

5. $x + 3y = 5$ ______
   $2x + 6y = 10$

6. $6x - 2y = 12$ ______
   $-3x + y = -6$

7. $2x + 3y = 8$ ______
   $4x + 6y = 12$

8. $4x + y = 6$ ______
   $-8x - 2y = 21$

9. $5x = y + 4$ ______
   $5x = y - 4$

10. $4y = 3x - 2$ ______
    $4y = 3x + 5$

11. $6x + 3y = 0$ ______
    $-12x - 6y = 0$

12. $3x - 5y = 0$ ______
    $6x - 10y = 0$

13. $2x - 3y = 0$ ______
$4x + 5y = 0$

14. $3x - 5y = 0$ ______
$6x + 10y = 0$

15. $3x + 5y = 19$ ______
$4x - 3y = 6$

16. $2x + 5y = 17$ ______
$4x + 3y = -1$

17. $4x - 2y = 1$ ______
$8x - 4y = 1$

18. $-2x + 3y = 5$ ______
$4x - 6y = 5$

19. $3x - 2y = 8$ ______
$-3x + 2y = -8$

20. $4x + y = 4$ ______
$-8x - 2y = -8$

21. $4x - y = 3$ ______
$-2x + \frac{1}{2}y = -\frac{3}{2}$

22. $5x - 2y = 8$ ______
$-\frac{5}{2}x + y = -4$

8.4 SOLVING SYSTEMS OF LINEAR EQUATIONS BY SUBSTITUTION

OBJECTIVES

- Solve linear systems by substitution

We have looked at the graphical method and the addition method for solving systems of linear equations. A third method is the **substitution method,** which is useful for solving systems where one equation is solved for one of the variables. However, the substitution method can be used to solve any system of linear equations.

Example 1 Solve the system

$$\begin{aligned} 3x + 5y &= 26 \\ y &= 2x. \end{aligned}$$

From the second of these two equations, we observe that $y = 2x$. Using this fact, we can substitute $2x$ for y in the first equation.

$$\begin{aligned} 3x + 5y &= 26 \\ 3x + 5(2x) &= 26 \qquad \text{Let } y = 2x \\ 3x + 10x &= 26 \\ 13x &= 26 \\ x &= 2 \end{aligned}$$

Since $y = 2x$, we have $y = 2(2) = 4$. The solution of the given system is $(2, 4)$.

Example 2 Use substitution to solve the system

$$\begin{aligned} 2x + 5y &= 7 \\ x &= -1 - y. \end{aligned}$$

The second equation gives x in terms of y. Substitute $-1 - y$ for x in the first equation.

$$\begin{aligned} 2x + 5y &= 7 \\ 2(-1 - y) + 5y &= 7 \qquad \text{Let } x = -1 - y \\ -2 - 2y + 5y &= 7 \\ -2 + 3y &= 7 \\ 3y &= 9 \\ y &= 3 \end{aligned}$$

Since $x = -1 - y$, we have $x = -1 - 3$, or $x = -4$. The solution of the given system is $(-4, 3)$.

Example 3 Use substitution to solve the system

$$\begin{aligned} x &= 5 - 2y \\ 2x + 4y &= 6. \end{aligned}$$

Substitute $5 - 2y$ for x in the second equation.

QUICK CHECKS

1. Solve each system by substitution.

(a) $3x - 4y = -11$
$x = y - 2$

$$2x + 4y = 6$$
$$2(5 - 2y) + 4y = 6 \qquad \text{Let } x = 5 - 2y$$
$$10 - 4y + 4y = 6$$
$$10 = 6 \qquad \text{False}$$

The false result means that the equations of the system have graphs that are parallel lines. Thus there is no simultaneous solution for this system.

Work Quick Check 1 at the side.

(b) $5x + 2y = -2$
$y = -2x$

Example 4 Use substitution to solve the system

$$2x + 3y = 8$$
$$-4x - 2y = 0.$$

To use the substitution method, we need an equation giving x in terms of y (or y in terms of x). We can choose the first equation of the system, which is $2x + 3y = 8$, and solve the equation for x. This means that we need to have x alone on one side of the equation. To get this, we need to first add $-3y$ to both sides.

$$2x + 3y = 8$$
$$2x = 8 - 3y$$

Now divide both sides of this equation by 2.

$$x = \frac{8 - 3y}{2}$$

2. Solve each system by substitution.

(a) $x + 4y = -1$
$2x - 5y = 11$

Finally, substitute this result for x in the second equation of the system.

$$-4x - 2y = 0$$
$$-4\left(\frac{8 - 3y}{2}\right) - 2y = 0 \qquad \text{Let } x = \frac{8 - 3y}{2}$$
$$-2(8 - 3y) - 2y = 0$$
$$-16 + 6y - 2y = 0$$
$$-16 + 4y = 0$$
$$4y = 16$$
$$y = 4$$

(b) $2x + 5y = 4$
$6x + 7y = -4$

Let $y = 4$ in $x = (8 - 3y)/2$.

$$x = \frac{8 - 3 \cdot 4}{2}$$
$$x = \frac{8 - 12}{2}$$
$$x = \frac{-4}{2}$$
$$x = -2$$

The solution of the given system is $(-2, 4)$.

Work Quick Check 2 at the side.

Quick Check Answers
1. (a) $(3, 5)$ (b) $(-2, 4)$
2. (a) $(3, -1)$ (b) $(-3, 2)$

Example 5 Use substitution to solve the system

$$2x = 3 - 2y \quad (1)$$
$$6 + 3y + x = 10 - x. \quad (2)$$

To begin, simplify the second equation by adding x and -6 to both sides. This gives the simplified system

$$2x = 3 - 2y \quad (1)$$
$$2x + 3y = 4. \quad (3)$$

We can use the substitution method if we solve one of the equations for either x or y. We can solve equation (1) for x if we multiply both sides by 1/2. This gives

$$2x = 3 - 2y \quad (1)$$
$$\frac{1}{2}(2x) = \frac{1}{2}(3 - 2y)$$
$$x = \frac{3}{2} - y.$$

Substitute $\frac{3}{2} - y$ for x in equation (3) from above.

$$2x + 3y = 4 \quad (3)$$
$$2\left(\frac{3}{2} - y\right) + 3y = 4$$
$$3 - 2y + 3y = 4$$
$$3 + y = 4$$
$$y = 1$$

Since $x = \frac{3}{2} - y$, and $y = 1$, we have

$$x = \frac{3}{2} - y = \frac{3}{2} - 1 = \frac{1}{2},$$

so the solution is $\left(\frac{1}{2}, 1\right)$.

Work Quick Check 3 at the side.

Example 6 Solve the system

$$3x + \frac{1}{4}y = 2 \quad (4)$$
$$\frac{1}{2}x + \frac{3}{4}y = \frac{-5}{2} \quad (5)$$

by any method.

Begin by clearing both equations of fractions. To clear equation (4) of fractions, multiply both sides by 4.

$$4\left(3x + \frac{1}{4}y\right) = 4(2)$$
$$4(3x) + 4\left(\frac{1}{4}y\right) = 4(2)$$
$$12x + y = 8 \quad (6)$$

3. Solve each system by substitution. First simplify where necessary.

(a) $x = 5 - 3y$
$2x + 3 = 5x - 4y + 14$

(b) $4x + 2x - y + 1 = 30$
$y = -4 + x$

Quick Check Answers

3. (a) $(-1, 2)$ (b) $(5, 1)$

4. Solve each system by any method. First clear all fractions.

(a) $\frac{2}{3}x + \frac{1}{2}y = 6$
$\frac{1}{2}x - \frac{3}{4}y = 0$

(b) $\frac{3}{5}x + \frac{1}{2}y = 7$
$\frac{7}{10}x - \frac{1}{5}y = 16$

Now clear equation (5) of fractions by multiplying both sides by the common denominator 4.

$$4\left(\frac{1}{2}x + \frac{3}{4}y\right) = 4\left(\frac{-5}{2}\right)$$
$$4\left(\frac{1}{2}x\right) + 4\left(\frac{3}{4}y\right) = 4\left(\frac{-5}{2}\right)$$
$$2x + 3y = -10 \qquad (7)$$

Now solve the system of equations (6) and (7).

$$12x + y = 8 \qquad (6)$$
$$2x + 3y = -10 \qquad (7)$$

If we choose the substitution method, we must solve one equation for either x or y. Let us solve equation (6) for y.

$$12x + y = 8$$
$$y = -12x + 8$$

Now substitute the result for y in equation (7).

$$2x + 3(-12x + 8) = -10$$
$$2x - 36x + 24 = -10$$
$$-34x = -34$$
$$x = 1$$

From $y = -12x + 8$ with $x = 1$, we get $y = -4$. The solution is $(1, -4)$. Check by substituting 1 for x and -4 for y in both of the original equations.

Work Quick Check 4 at the side.

Quick Check Answers
4. (a) (6, 4) (b) (20, −10)

name date hour

8.4 EXERCISES

Solve each system by the substitution method. See Examples 1–4.

1. $x + y = 6$ ______
   $y = 2x$

2. $x + 3y = -11$ ______
   $y = -4x$

3. $3x + 2y = 26$ ______
   $x = y + 2$

4. $4x + 3y = -14$ ______
   $x = y - 7$

5. $x + 5y = 3$ ______
   $x = 2y + 10$

6. $5x + 2y = 14$ ______
   $y = 2x - 11$

7. $5x + 7y = 40$ ______
   $x = 2y - 9$

8. $4x + 9y = -7$ ______
   $y = 2x - 13$

9. $3x - 2y = 14$ ______
   $2x + y = 0$

10. $2x - 5 = -y$ ______
    $x + 3y = 0$

11. $x + y = 6$ ______
    $x - y = 4$

12. $3x - 2y = 13$ ______
    $x + y = 6$

13. $3x - y = 6$ ______
    $y = 3x - 5$

14. $4x - y = 4$ ______
    $y = 4x + 3$

15. $6x - 8y = 4$ ______
    $3x = 4y + 2$

16. $12x + 18y = 12$ ______
    $2x = 2 - 3y$

17. $4x + 5y = 5$ ______
    $2x + 3y = 1$

18. $3x + 4y = 10$ ______
    $4x + 5y = 14$

19. $2x + 3y = 11$ ______
    $y = 1$

20. $3x + 4y = -10$ ______
    $x = -6$

21. $4x + y = 5$ ______
    $x - 2 = 0$

22. $5x + 2y = -19$ ______
    $y - 3 = 0$

Solve each system by either the addition method or the substitution method. First simplify any equations where necessary. See Example 5.

23. $x + 4y = 34$ __________
$y = 4x$

24. $3x - y = -14$ __________
$x = -2y$

25. $4 + 4x - 3y = 34 + x$ __________
$4x = -y - 2 + 3x$

26. $5x - 4y = 42 - 8y - 2$ __________
$2x + y = x + 1$

27. $4x - 2y + 8 = 3x + 4y - 1$ __________
$3x + y = x + 8$

28. $5x - 4y - 8x - 2 = 6x + 3y - 3$ __________
$4x - y = -2y - 8$

29. $2x - 8y + 3y + 2 = 5y + 16$ __________
$8x - 2y = 4x + 28$

30. $7x - 9 + 2y - 8 = -3y + 4x + 13$ __________
$4y - 8x = -8 + 9x + 32$

31. $2x + 3y = 10$ __________
$4x + 5y = 10 - y$

32. $10x + 21y = 90$ __________
$5x + 11y = 10 - 5x - 10y$

33. $-2x + 3y = 12 + 2y$ __________
$2x - 5y + 4 = -8 - 4y$

34. $2x + 5y = 7 + 4y - x$ __________
$5x + 3y + 8 = 22 - x + y$

35. $y + 9 = 3x - 2y + 6$ __________
$5 - 3x + 24 = -2x + 4y + 3$

36. $5x - 2y = 16 + 4x - 10$ __________
$4x + 3y = 60 + 2x + y$

Solve each system by either the addition method or the substitution method. First clear all fractions. See Example 6.

37. $x + \frac{1}{3}y = y - 2$ __________
$\frac{1}{4}x - y = x - y$

38. $\frac{5}{3}x + 2y = \frac{1}{3} + y$ __________
$2x - 3 + \frac{y}{3} = -2 + x$

39. $\frac{x}{6} + \frac{y}{6} = 1$ __________
$-\frac{1}{2}x - \frac{1}{3}y = -5$

40. $\frac{x}{2} - \frac{y}{3} = \frac{5}{6}$ __________
$\frac{x}{5} - \frac{y}{4} = \frac{1}{10}$

41. $\frac{x}{3} - \frac{3y}{4} = -\frac{1}{2}$ __________
$\frac{2x}{3} + \frac{y}{2} = 3$

42. $\frac{x}{5} + 2y = \frac{8}{5}$ __________
$\frac{3x}{5} + \frac{y}{2} = \frac{-7}{10}$

43. $\frac{x}{2} + \frac{y}{3} = \frac{7}{6}$ __________
$\frac{x}{4} - \frac{3y}{2} = \frac{9}{4}$

44. $\frac{5x}{2} - \frac{y}{3} = \frac{5}{6}$ __________
$\frac{4x}{3} + y = \frac{19}{3}$

OBJECTIVES

- Use linear systems to solve word problems

Many practical problems are more easily translated into equations if two variables are used. With two variables, we need two equations to find the desired solution. The examples in this section illustrate the method of solving word problems using two equations and two variables.

Example 1 The sum of two numbers is 63. Their difference is 19. Find the two numbers.

Let x represent one number and y the other. From the information given in the problem, we set up a system of equations.

$$x + y = 63$$
$$x - y = 19$$

This system can be solved by the addition method. We have

$$\begin{aligned} x + y &= 63 \\ x - y &= 19 \\ \hline 2x \quad &= 82. \end{aligned}$$

From this last equation, $x = 41$. Substitute 41 for x in the first equation and check that $y = 22$. The numbers required in the problem are 41 and 22.

Work Quick Check 1 at the side.

1. (a) The sum of two numbers is 97. Their difference is 41. What are the numbers?

(b) The sum of two numbers is 38. If twice the first is added to three times the second, the result is 99. Find the numbers.

QUICK CHECKS

Example 2 Admission prices at a football game were \$1.25 for adults and \$.50 for children. The total receipts from the game were \$530.75. Tickets were sold to 454 people. How many adults and how many children attended the game?

Let a represent the number of adult tickets that were sold, and let c represent the number of children's tickets. The information given in the problem is summarized in the table.

Kind of ticket	Number sold	Cost of each (in dollars)	Receipts (in dollars)
Adult	a	1.25	$1.25a$
Child	c	.50	$.50c$

The total number of tickets sold was 454, so that

$$a + c = 454.$$

The receipts from the sale of a adult tickets at \$1.25 each are \$$1.25a$, while the receipts from the sale of c children's tickets at \$.50 each are \$$.50c$. Since the total receipts were \$530.75,

$$1.25a + .50c = 530.75.$$

Quick Check Answers.
1. (a) 69, 28 (b) 15, 23

2. The receipts from a concert were \$1850. The price for a regular ticket was \$5.00, while student tickets were \$3.50. A total of 400 tickets were sold. How many of each type were sold?

We have used the information in the problem to set up the system of equations

$$a + c = 454 \quad (1)$$
$$1.25a + .50c = 530.75. \quad (2)$$

Equation (2) can be simplified if we multiply both sides by 100 to clear the decimals.

$$100(1.25a + .50c) = 100(530.75)$$
$$125a + 50c = 53{,}075 \quad (3)$$

To solve the system of equations, multiply equation (1) on both sides by −50, then add to equation (3).

$$-50a - 50c = -50(454)$$
$$-50a - 50c = -22{,}700$$

Now we add.

$$\begin{aligned} -50a - 50c &= -22{,}700 \\ 125a + 50c &= 53{,}075 \\ \hline 75a &= 30{,}375 \end{aligned}$$

From the equation $75a = 30375$, we get $a = 405$. We know that $a + c = 454$; therefore, $c = 49$. (Check this.) There were 405 adults and 49 children at the game.

Work Quick Check 2 at the side.

Example 3 A pharmacist needs 100 liters of 50% alcohol solution. She has on hand 30% alcohol solution and 80% alcohol solution, which she can mix. How many liters of each will be required to make the 100 liters of 50% alcohol solution?

Let x represent the number of liters of 30% alcohol needed, and let y represent the number of liters of 80% alcohol. The information of the problem is summarized in the table.

Liters of solution	Percent	Liters of pure alcohol
x	30	$.30x$
y	80	$.80y$
100	50	$.50(100)$

She will have $.30x$ liters of alcohol from the x liters of 30% solution and $.80y$ liters of alcohol from the y liters of 80% solution. The total is $.30x + .80y$ liters of pure alcohol. In the mixture, she wants 100 liters of 50% solution. This 100 liters would contain $.50(100) = 50$ liters of pure alcohol. Since the amounts of pure alcohol must be equal,

$$.30x + .80y = 50.$$

Quick Check Answers
2. 300 regular, 100 student

We also know that the total number of liters is 100, or

$$x + y = 100.$$

These two equations give the system

$$.30x + .80y = 50$$
$$x + y = 100.$$

Let us solve this system by the substitution method. From the second equation of the system, we have $x = 100 - y$. If we substitute $100 - y$ for x in the first equation, we get

$$.30(100 - y) + .80y = 50 \qquad \text{Let } x = 100 - y$$
$$30 - .30y + .80y = 50$$
$$.50y = 20$$
$$y = 40.$$

Since $x + y = 100$, then $x = 60$. The pharmacist should use 60 liters of the 30% solution and 40 liters of the 80% solution.

Work Quick Check 3 at the side.

3. Joe needs 100 cc (cubic centimeters) of 20% salt solution for a chemistry experiment. The lab has on hand only 10% and 25% salt solutions. How much of each should he mix together to get the amount he needs of 20% solution?

Example 4 Two cars start from positions 400 miles apart and travel toward each other. They meet after four hours. Find the average speed of each car if one car travels 20 miles per hour faster than the other.

We need the formula that relates distance, rate, and time. As we learned earlier, this formula is $d = rt$. Let x be the average speed of the first car, and y the average speed of the second car. This information is shown in the chart.

	r	t	d
First car	x	4	$4x$
Second car	y	4	$4y$

Since each car travels for four hours, t for each car is 4. The distance is found by using the formula $d = rt$ and the amounts already entered in the chart. Since the total distance traveled by both cars is 400 miles,

$$4x + 4y = 400.$$

One car traveled 20 miles per hour faster than the other. Assume that the first car was faster. Then

$$x = 20 + y.$$

We now have the system of equations.

$$4x + 4y = 400$$
$$x = 20 + y$$

Quick Check Answers

3. $33\frac{1}{3}$ cc of 10%, $66\frac{2}{3}$ cc of 25%

4. Two cars which were 450 miles apart traveled toward each other. They met after 5 hours. If one car traveled twice as fast as the other, what were their speeds?

This system can be solved by substitution. Replace x with $20 + y$ in the first equation of the system.

$$4(20 + y) + 4y = 400$$
$$80 + 4y + 4y = 400$$
$$80 + 8y = 400$$
$$8y = 320$$
$$y = 40$$

Since $x = 20 + y$, we have $x = 60$. Thus, the speeds of the two cars were 40 miles per hour and 60 miles per hour.

Work Quick Check 4 at the side.

Quick Check Answers

4. 30 mph, 60 mph

name date hour

8.5 EXERCISES

Write a system of equations for each problem. Then solve the system. Formulas are in the back of the book. See Example 1.

1. The sum of two numbers is 52, and their difference is 34. Find the numbers. __________

2. Find two numbers whose sum is 56 and whose difference is 18. __________

3. A certain number is three times as large as a second number. Their sum is 96. What are the two numbers? __________

4. One number is five times as large as another. The difference of the numbers is 48. Find the numbers. __________

5. A rectangle is twice as long as it is wide. Its perimeter is 60 inches. Find the dimensions of the rectangle. __________

6. The perimeter of a triangle is 21 inches. If two sides are of equal length, and the third side is 3 inches longer than one of the equal sides, find the length of the three sides. __________

See Example 2.

7. The cashier at the Evergreen Ranch has some \$10 bills and some \$20 bills. The total value of the money is \$1480. If there is a total of 85 bills, how many of each type are there? __________

8. A bank teller has 154 bills of \$1 and \$5 denominations. How many of each type of bill does he have if the total value of the money is \$466? __________

9. A club secretary bought 8¢ and 10¢ pieces of candy to give to the members. She spent a total of \$15.52. If she bought 170 pieces of candy, how many of each kind did she buy? __________

10. There were 311 tickets sold for a basketball game, some for students and some for non-students. Student tickets cost 25¢ each and non-student tickets cost 75¢ each. The total receipts were \$108.75. How many of each type of ticket were sold? __________

11. Ms. Sullivan has \$10,000 to invest, part at 5% and part at 7%. She wants the income from simple interest on the two investments to total \$550 yearly. How much should she invest at each rate? ______

12. Mr. Emerson has twice as much money invested at 7% as he has at 8%. If his yearly income from investments is \$440, how much does he have invested at each rate? ______

See Example 3.

13. A 90% antifreeze solution is to be mixed with a 75% solution to make 20 liters of a 78% solution. How many liters of 90% and 75% solutions should be used? ______

14. A grocer wishes to blend candy selling for 60¢ a pound with candy selling for 90¢ a pound to get a mixture which will be sold for 70¢ a pound. How many pounds of the 60¢ and the 90¢ candy should be used to get 30 pounds of the mixture? ______

15. How many barrels of olives worth \$40 per barrel must be mixed with olives worth \$60 per barrel to get 50 barrels of a mixture worth \$48 per barrel? ______

16. A glue merchant wishes to mix some glue worth \$70 per barrel with some glue worth \$90 per barrel, to get 80 barrels of a mixture worth \$77.50 per barrel. How many barrels of each type should be used? ______

See Example 4.

17. If a plane can travel 400 miles per hour into the wind and 540 miles per hour with the wind, find the speed of the wind, and the speed of the plane in still air. ______

18. It takes a boat 1 1/2 hours to go 12 miles downstream, and 6 hours to return. Find the speed of the current and the speed of the boat in still water. ______

19. At the beginning of a walk for charity, John and Harriet are 30 miles apart. If they leave at the same time and walk in the same direction, John overtakes Harriet in 60 hours. If they walk toward each other, they meet in 5 hours. What are their speeds? ______

20. Mr. Anderson left Farmersville in a plane at noon to travel to Exeter. Mr. Bentley left Exeter in his automobile at 2 P.M. to travel to Farmersville. It is 400 miles from Exeter to Farmersville. If the sum of their speeds is 120 miles per hour, and if they met at 4 P.M., find the speed of each. ______________

The next problems are "brain-busters."

21. The Smith family is coming to visit, and no one knows how many children they have. Janet, one of the girls, says she has as many brothers as sisters; her brother Steve says he has twice as many sisters as brothers. How many boys and how many girls are in the family? ______________

22. In the Lopez family, the number of boys is one more than half the number of girls. One of the Lopez boys, Rico, says that he has one more sister than brothers. How many boys and girls are in the family? ______________

WORK SPACE

OBJECTIVES

- Solve systems of linear inequalities by graphing

In Section 7.4, we saw how to graph the solution of a linear inequality. Let us review the method. To graph the solution of $x + 3y > 12$, for example, first graph the line $x + 3y = 12$ by finding a few ordered pairs that satisfy the equation. Because the points on the line do not satisfy the inequality, make the line dashed. Choose a test point not on the line, say (0, 0). Substitute 0 for x and 0 for y in the given inequality.

$$x + 3y > 12$$
$$0 + 0 > 12$$
$$0 > 12 \quad \text{False}$$

Since the test point does not satisfy the inequality, shade the region on the side of the line that does not include (0, 0), as in Figure 6.

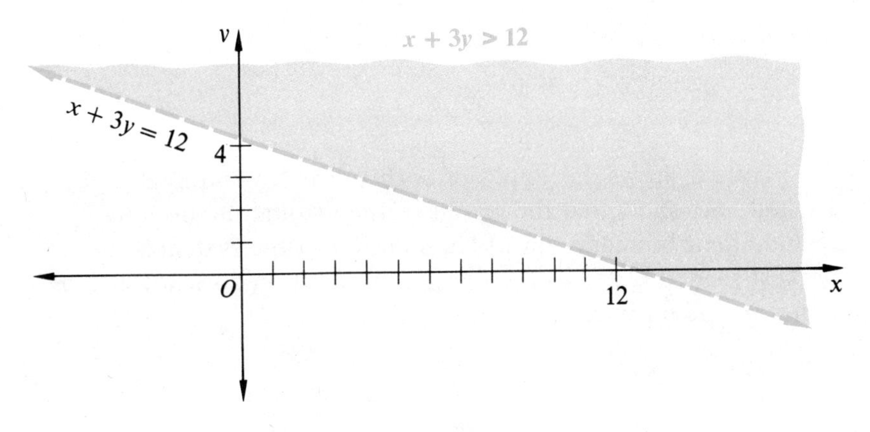

Figure 6

The same procedure is used to determine the solution of a system of two linear inequalities, as shown in Examples 1–3.

Example 1 Graph the solution of the system

$$3x + 2y \leqslant 6$$
$$2x - 5y \geqslant 10.$$

First graph the inequality $3x + 2y \leqslant 6$, using the steps described above. Then, on the same axes, graph the second inequality, $2x - 5y \geqslant 10$. The solution of the system is given by the overlap of the regions of the two graphs. This solution is the darkest shaded region in Figure 7 and includes portions of the two boundary lines.

QUICK CHECKS

1. Graph the solution of the system

$$x - 2y \leq 8$$
$$3x + y \geq 6.$$

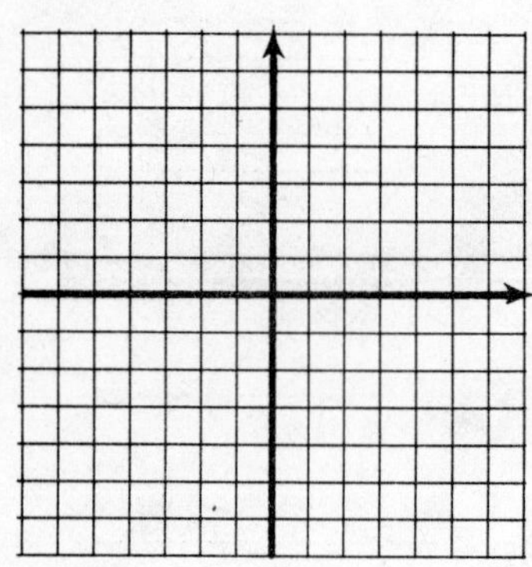

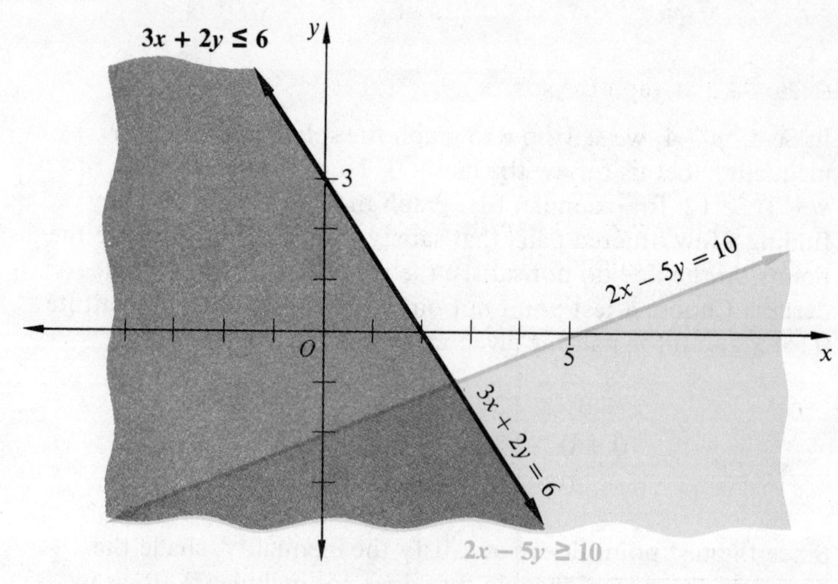

Figure 7

Work Quick Check 1 at the side.

Example 2 Graph the solution of the system

$$x - y > 5$$
$$2x + y < 2.$$

Figure 8 shows the graphs of both $x - y > 5$ and $2x + y < 2$. Dashed lines show that the graphs of the inequalities do not include their boundary lines. The solution of the system is the darkest shaded region in the figure. The solution does not include either boundary line.

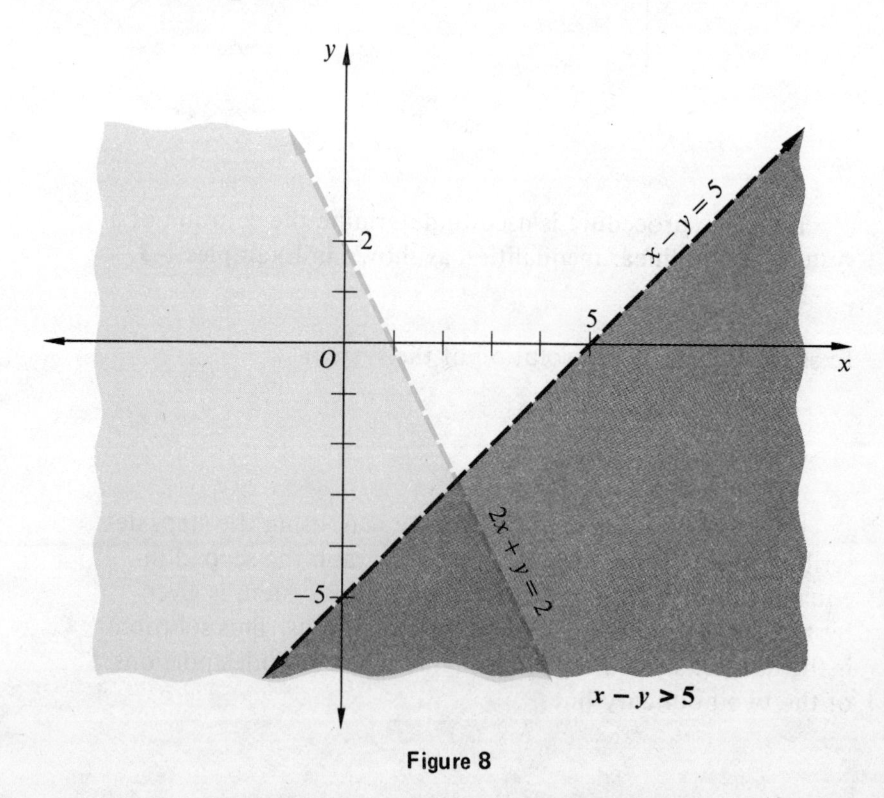

Figure 8

Quick Check Answers

1.

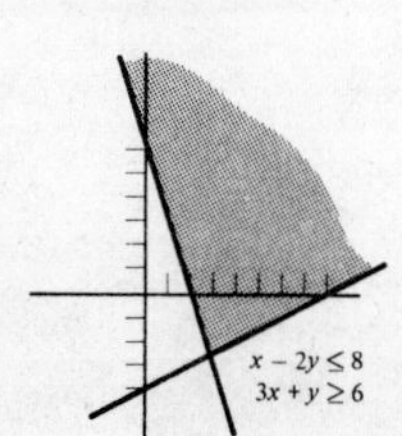

Work Quick Check 2 at the side.

Example 3 Graph the solution of the system

$$4x - 3y \leqslant 8$$
$$x \geqslant 2.$$

Recall that $x = 2$ is a vertical line through the point (2, 0). The graph of the solution is the darkest shaded region in Figure 9.

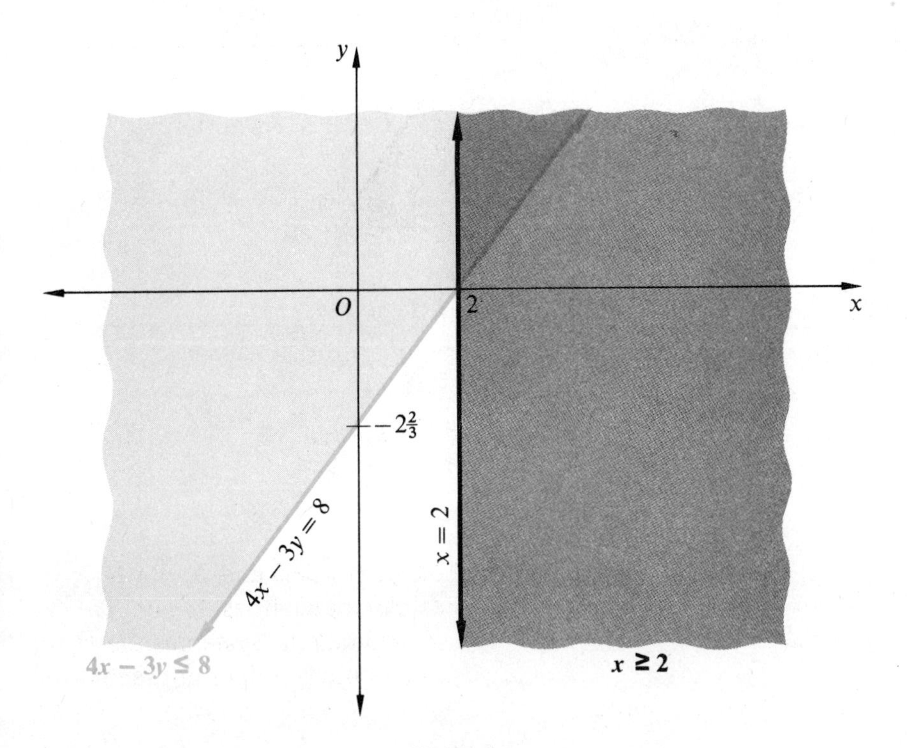

Figure 9

Work Quick Check 3 at the side.

2. Graph the solution of the system

$$x + 2y < 0$$
$$3x - 4y < 12.$$

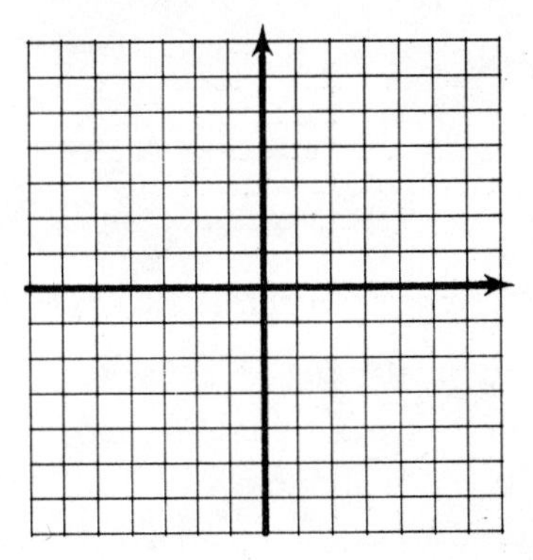

3. Graph the solution of the system

$$3x - y \geqslant 6$$
$$y \leqslant 2.$$

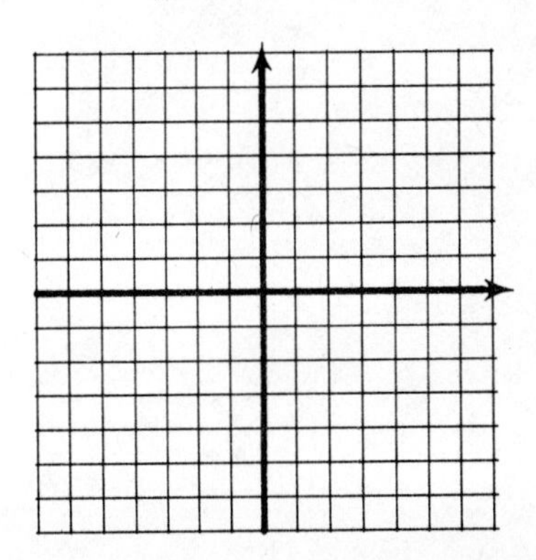

Quick Check Answers

2. $x + 2y < 0$, $3x - 4y < 12$

3. $3x - y \geq 6$, $y \leq 2$

WORK SPACE

name date hour

8.6 EXERCISES

Graph each system of linear inequalities. See Examples 1–3.

1. $x + y \leqslant 6$
   $x - y \leqslant 1$

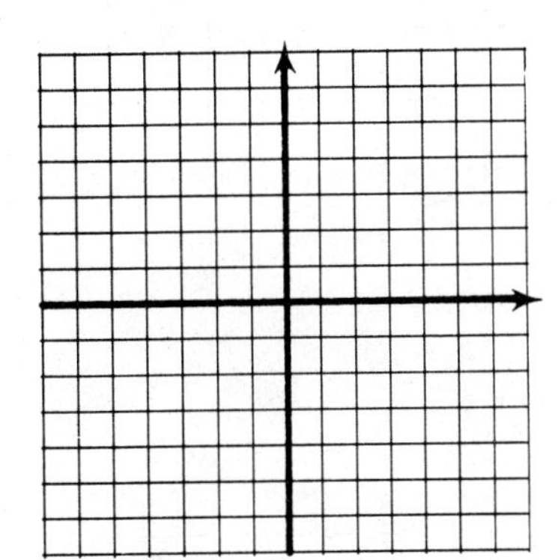

2. $x + y \geqslant 2$
   $x - y \leqslant 3$

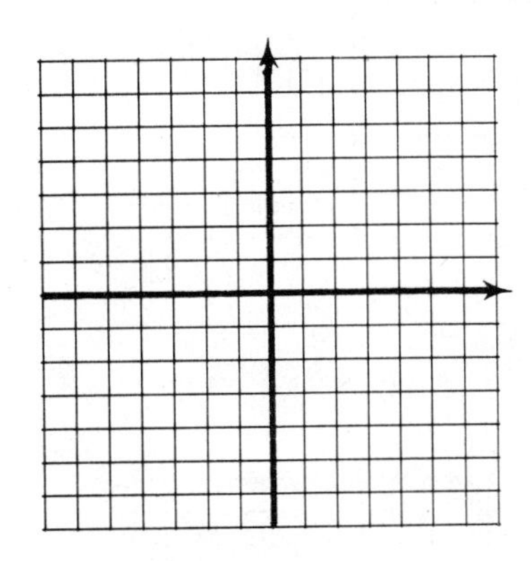

3. $2x - 3y \leqslant 6$
   $x + y \geqslant -1$

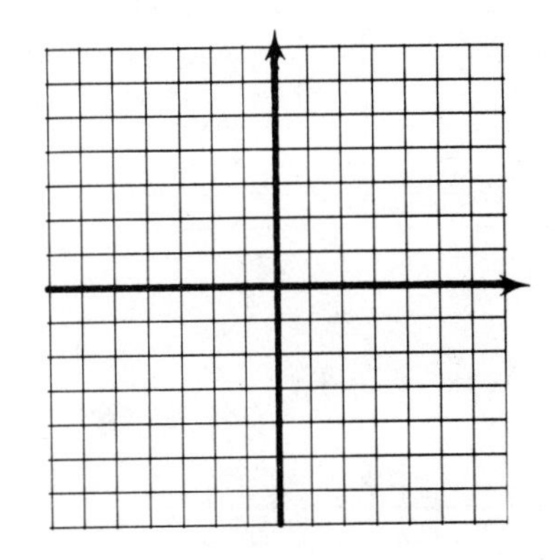

4. $4x + 5y \leqslant 20$
   $x - y \leqslant 3$

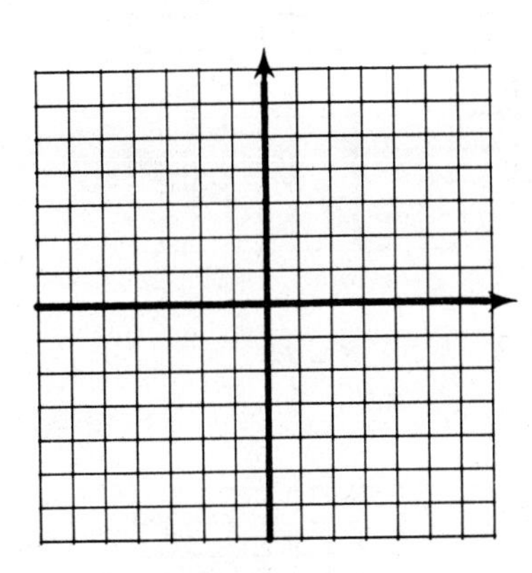

5. $x + 4y \leqslant 8$
   $2x - y \leqslant 4$

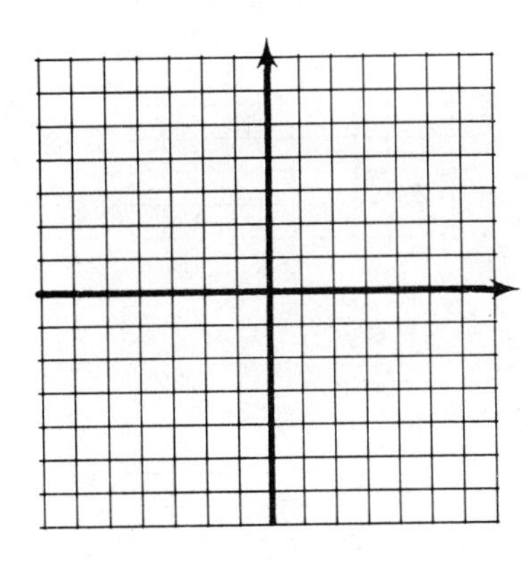

6. $3x + y \leqslant 6$
   $2x - y \leqslant 8$

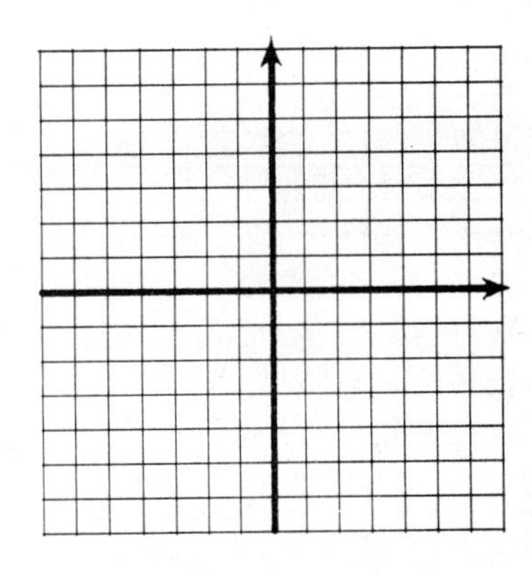

7. $x - 4y \leqslant 3$
   $x \geqslant 2y$

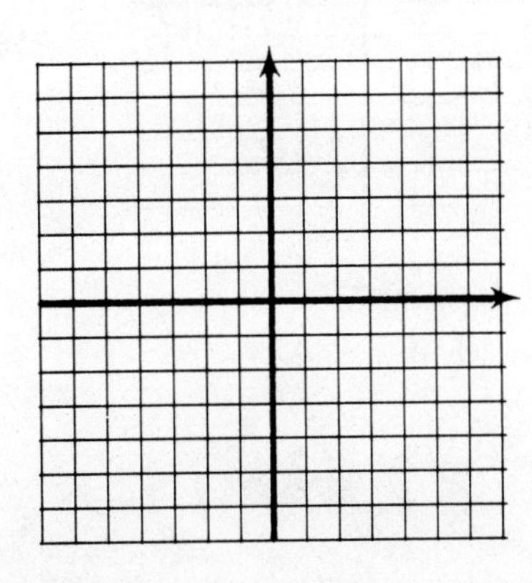

8. $2x + 3y \leqslant 6$
   $x - y \geqslant 5$

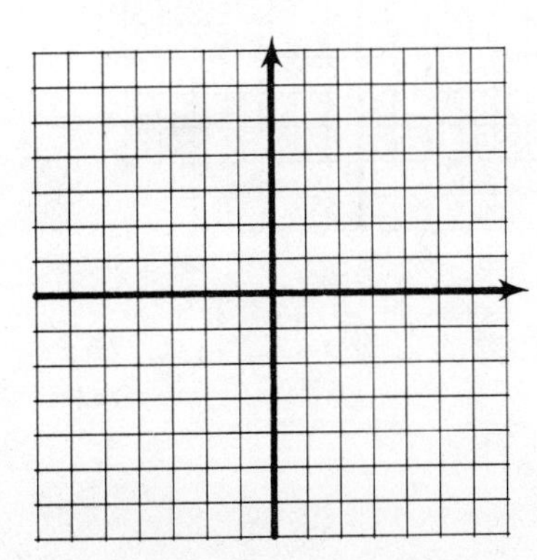

9. $x + 2y \leqslant 4$
   $x + 1 \geqslant y$

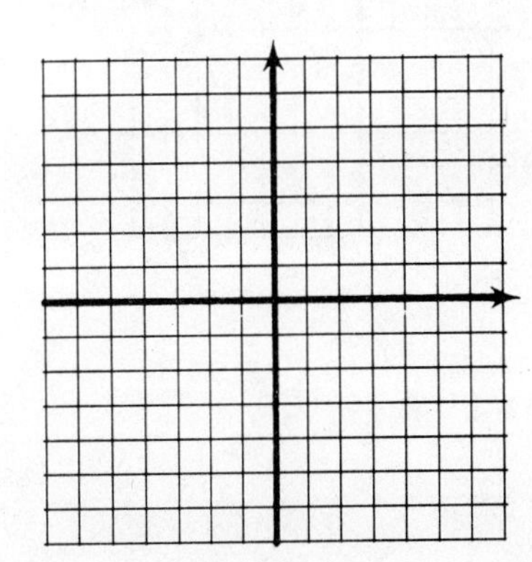

10. $y \leqslant 2x - 5$
$x - 3y \leqslant 2$

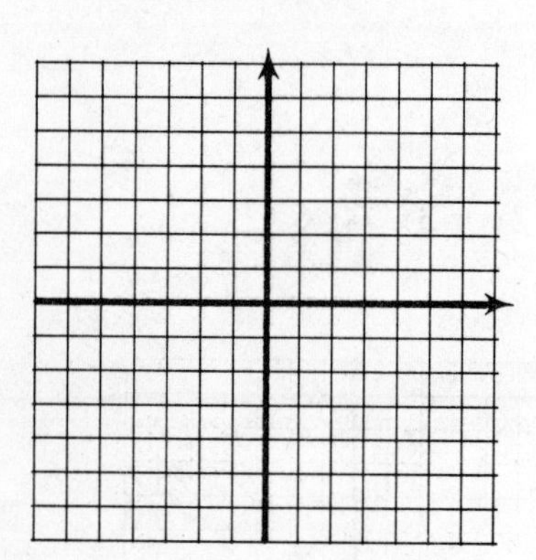

11. $4x + 3y \leqslant 6$
$x - 2y \geqslant 4$

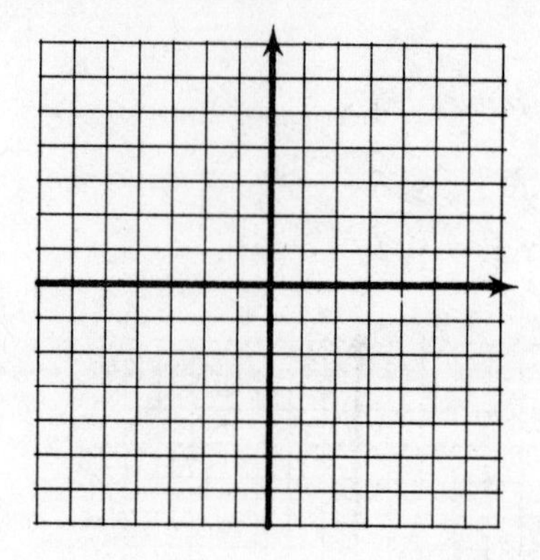

12. $3x - y \leqslant 4$
$-6x + 2y \leqslant -10$

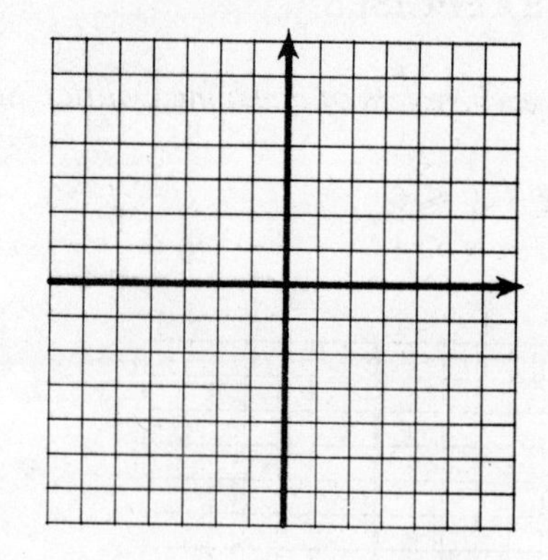

13. $x - 2y > 6$
$2x + y > 4$

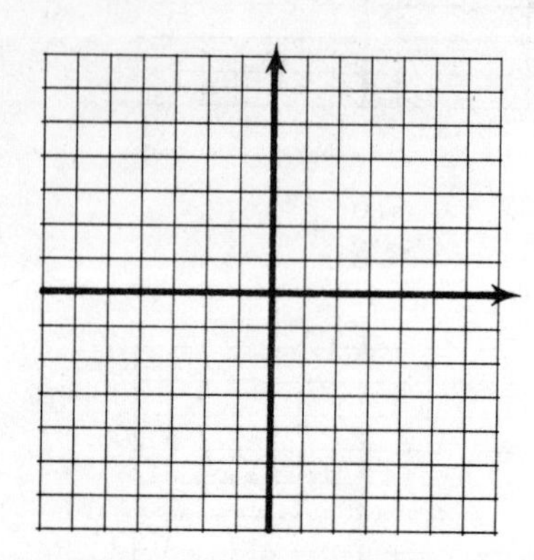

14. $3x + y < 4$
$x + 2y > 2$

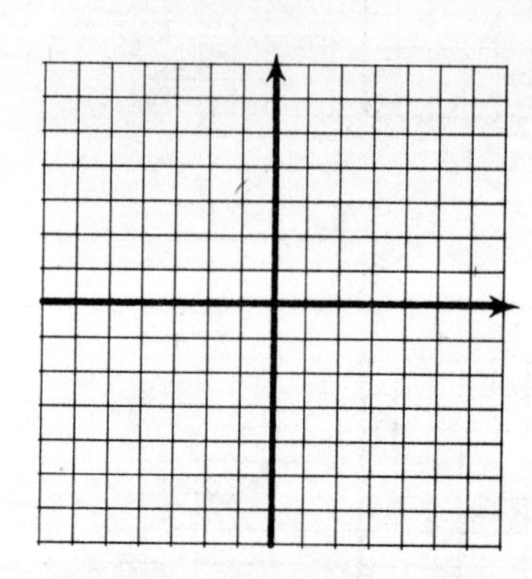

15. $x < 2y + 3$
$x + y > 0$

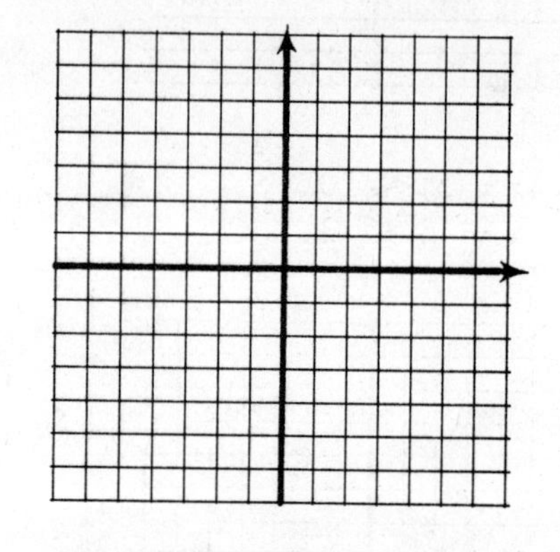

16. $2x + 3y < 6$
$4x + 6y > 18$

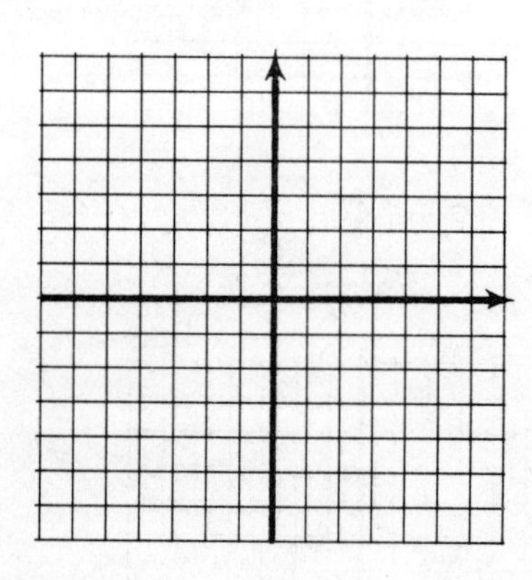

17. $x - 3y \leqslant 6$
$x \geqslant -1$

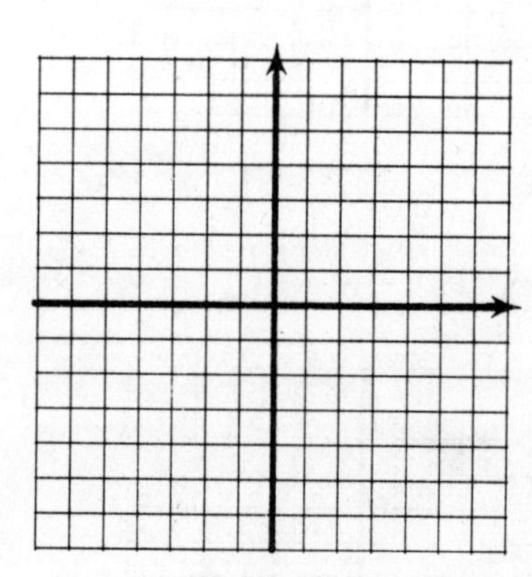

18. $2x + 5y \geqslant 20$
$x \leqslant 4$

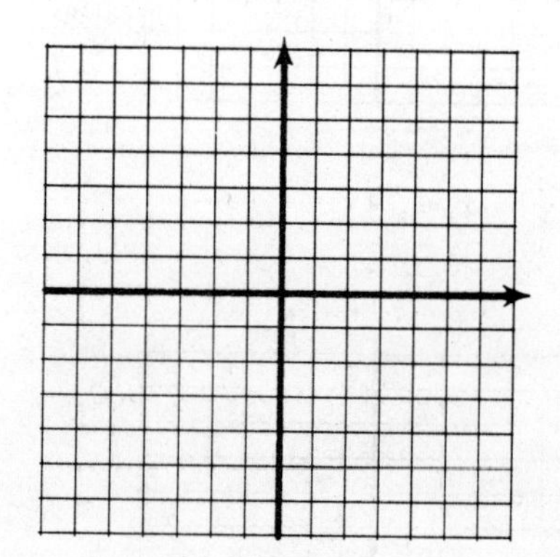

name date hour

CHAPTER 8 TEST

Solve each system by graphing.

1. $2x + y = 5$
   $3x - y = 15$

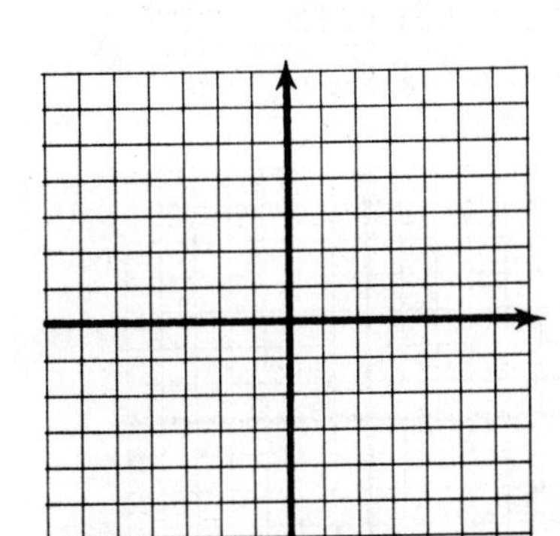

2. $3x + 2y = 8$
   $5x + 4y = 10$

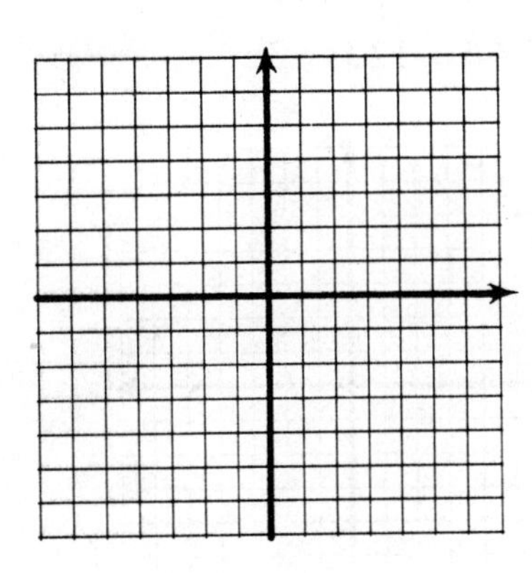

3. $x + 2y = 6$
   $2x - y = 7$

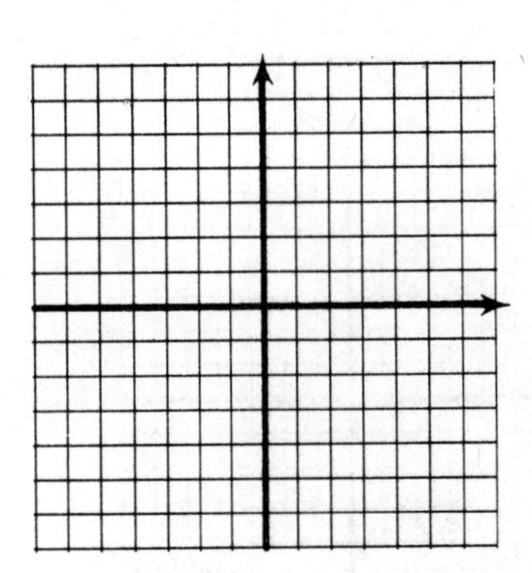

Solve each system by the addition method.

4. $2x - 5y = -13$
   $3x + 5y = 43$ __________

5. $4x + 3y = 26$
   $5x + 4y = 32$ __________

6. $6x + 5y = -13$
   $3x + 2y = -4$ __________

7. $4x + 5y = 8$
   $-8x - 10y = -6$ __________

8. $4x - 3y = 6$
   $x + 2y = -4$ __________

9. $6x - 5y = 2$
   $-2x + 3y = 2$ __________

10. $2x - y = 5$
    $4x + 3y = 0$ __________

11. $3x + 2y = 16$
    $9x - 3y = -6$ __________

Solve each system by substitution.

12. $2x + y = 1$
    $x = 8 + y$ __________

13. $4x + 3y = 0$
    $x = 2 - y$ __________

Solve each word problem by any method.

14. The local record shop is having a sale. Some records cost \$2.50 and some cost \$3.75. Joe has exactly \$20 to spend and wants to buy six records. How many can he buy at each price? __________

15. Two cars leave from the same place and travel in the same direction. One car travels one and one-third times as fast as the other. After three hours, they are 45 miles apart. What is the speed of each car? ________________

Graph the solution of each system of inequalities.

16. $2x + 7y \leqslant 14$
    $x - y \geqslant 1$

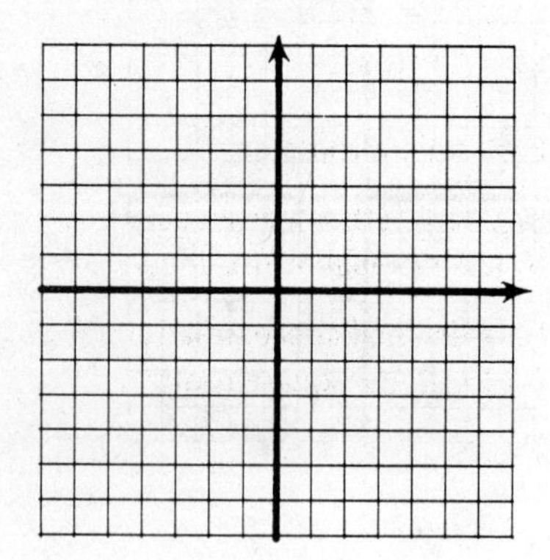

17. $2x - y \leqslant 6$
    $4y + 12 \geqslant -3x$

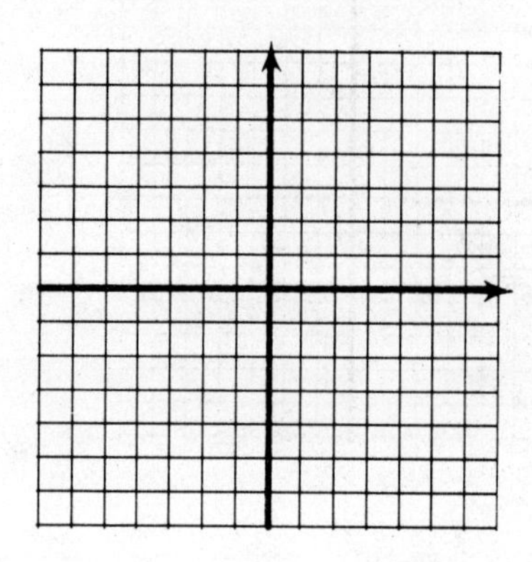

18. $3x - 5y < 15$
    $y < 2$

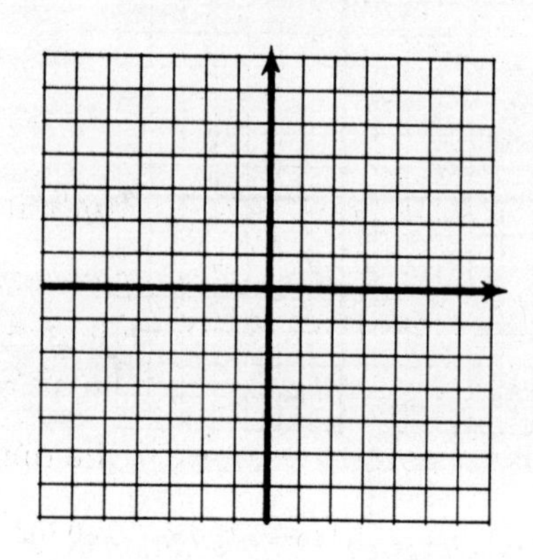

9
Roots and Radicals

9.1 SQUARE ROOTS

OBJECTIVES

- Find square roots
- Decide whether a given square root is rational or irrational
- Find decimal approximations for irrational square roots

To square a number, we multiply it by itself.

If $a = 7$, then $a^2 = 7 \cdot 7 = 49$.
If $a = 10$, then $a^2 = 10 \cdot 10 = 100$.
If $a = -5$, then $a^2 = (-5) \cdot (-5) = 25$.

In this chapter, we consider the opposite situation.

If $a^2 = 49$, then $a = ?$
If $a^2 = 100$, then $a = ?$
If $a^2 = 25$, then $a = ?$

To answer these questions, we must find a number which, when multiplied by itself, will result in the given number. The number that we find is called the **square root** of the given number.

Example 1 One square root of 49 is 7, since $7^2 = 49$. Another square root of 49 is -7, since $(-7)^2 = 49$. Therefore, 49 has two square roots, 7 and -7. One of these is positive and one is negative.

Work Quick Check 1 at the side.

QUICK CHECKS

1. Find all square roots.

(a) 1

(b) 9

(c) 36

(d) 64

All numbers which have integer square roots are called **perfect squares**. The first 100 perfect squares are listed in the table in the back of the book.

The positive square root of a number is written with the symbol $\sqrt{}$. For example, the positive square root of 121 is 11, written

$$\sqrt{121} = 11.$$

The symbol $-\sqrt{}$ is used for the negative square root of a number. For example, the negative square root of 121 is -11, written

$$-\sqrt{121} = -11.$$

The symbol $\sqrt{}$ is called a **radical sign** and, used alone, always represents the positive square root (except that $\sqrt{0} = 0$). The number inside the radical sign is called the **radicand** and the entire expression, radical sign and radicand, is called a **radical.**

Quick Check Answers
1. (a) 1, −1 (b) 3, −3 (c) 6, −6 (d) 8, −8

2. Find each square root.

(a) $\sqrt{16}$

(b) $-\sqrt{169}$

(c) $-\sqrt{225}$

(d) $\sqrt{729}$

(e) $\sqrt{2500}$

3. Is each square root rational or irrational?

(a) $\sqrt{9}$

(b) $\sqrt{7}$

(c) $\sqrt{16}$

(d) $\sqrt{72}$

Quick Check Answers

2. (a) 4 (b) −13 (c) −15 (d) 27 (e) 50

3. (a) rational (3) (b) irrational (c) rational (4) (d) irrational

Example 2 Find each square root.

(a) $\sqrt{144}$

The radical $\sqrt{144}$ represents the positive square root of 144, which is 12 (see the table of perfect squares). Thus,

$$\sqrt{144} = 12.$$

(b) $-\sqrt{1024}$

This symbol represents the negative square root of 1024, so

$$-\sqrt{1024} = -32.$$

(c) $\sqrt{256} = 16$

(d) $-\sqrt{900} = -30$

Work Quick Check 2 at the side.

Not every number has a square root. For example, there is no real number which we can square and get −36. (The square of a real number can never be negative.) Because of this $\sqrt{-36}$ is not a real number.

A real number which is not rational is called an **irrational number**. In general, it is difficult to prove that a given number is irrational. However, if a is a positive integer that is not a perfect square, then $\sqrt{a}$ is irrational.

Example 3 Tell whether each square root is rational or irrational.

(a) $\sqrt{17}$

Since 17 is not a perfect square (see the table), $\sqrt{17}$ is irrational.

(b) $\sqrt{64}$

The number 64 is a perfect square because $\sqrt{64} = 8$, a rational number.

(c) $\sqrt{85}$ is irrational.

(d) $\sqrt{81}$ is rational ($\sqrt{81} = 9$).

Work Quick Check 3 at the side.

Not all irrational numbers are roots of integers. For example, π (approximately 3.14159) is an irrational number that is not a square root of any integer.

If a number is irrational, we can still find a decimal that approximately equals the number. This can be done with calculators. For square roots, we can use the square root table, as shown in Example 4.

Example 4 Find a decimal approximation for each square root.

(a) $\sqrt{11}$

Look in the square root table. Find 11 at the left. The approximate square root is given in the column having $\sqrt{n}$ at the top. You should find that

$$\sqrt{11} \approx 3.317,$$

where $\approx$ means approximately equal to.

(b) $\sqrt{39} \approx 6.245$

(c) $\sqrt{740}$

There is no 740 in the "n" column of the square root table. However, $740 = 74 \times 10$, so that we can find $\sqrt{740}$ in the "$\sqrt{10n}$" column. You should find that

$$\sqrt{740} \approx 27.203.$$

(d) $\sqrt{180} \approx 13.416$

Look in the "n" column for 18; then read across to the "$\sqrt{10n}$" column.

(e) $\sqrt{810} \approx 28.460$

Work Quick Check 4 at the side.

4. Find decimal approximations for each square root.

(a) $\sqrt{28}$

(b) $\sqrt{63}$

(c) $\sqrt{190}$

(d) $\sqrt{270}$

(e) $\sqrt{1000}$

Quick Check Answers
4. (a) 5.292 (b) 7.937
(c) 13.784 (d) 16.432
(e) 31.623

WORK SPACE

name date hour

9.1 EXERCISES

Find all square roots of each number. See Example 1.

1. 9 ______ 2. 16 ______ 3. 121 ______

4. 196 ______ 5. 400 ______ 6. 900 ______

7. 625 ______ 8. 961 ______ 9. 1521 ______

10. 2209 ______ 11. 3969 ______ 12. 4624 ______

Find each root that exists. See Example 2.

13. $\sqrt{4}$ ______ 14. $\sqrt{9}$ ______ 15. $\sqrt{25}$ ______

16. $\sqrt{36}$ ______ 17. $-\sqrt{64}$ ______ 18. $-\sqrt{100}$ ______

19. $\sqrt{169}$ ______ 20. $\sqrt{196}$ ______ 21. $\sqrt{900}$ ______

22. $\sqrt{1600}$ ______ 23. $-\sqrt{1681}$ ______ 24. $-\sqrt{2116}$ ______

25. $\sqrt{2601}$ ______ 26. $\sqrt{3025}$ ______ 27. $-\sqrt{4900}$ ______

28. $-\sqrt{5625}$ ______ 29. $\sqrt{-9}$ ______ 30. $\sqrt{-25}$ ______

31. $-\sqrt{-49}$ ______ 32. $-\sqrt{-81}$ ______

Write "rational" or "irrational" for each number. If a number is rational, give its exact value. If a number is irrational, give a decimal approximation for the square root. See Examples 3 and 4.

33. $\sqrt{16}$ ______ 34. $\sqrt{81}$ ______

35. $\sqrt{15}$ ______ 36. $\sqrt{31}$ ______

37. $\sqrt{47}$ ______ 38. $\sqrt{53}$ ______

39. $\sqrt{68}$ ____________________ 40. $\sqrt{72}$ ____________________

41. $-\sqrt{121}$ ____________________ 42. $-\sqrt{144}$ ____________________

43. $\sqrt{110}$ ____________________ 44. $\sqrt{170}$ ____________________

45. $-\sqrt{200}$ ____________________ 46. $-\sqrt{260}$ ____________________

47. $\sqrt{400}$ ____________________ 48. $\sqrt{900}$ ____________________

49. $\sqrt{570}$ ____________________ 50. $\sqrt{690}$ ____________________

OBJECTIVES

- Multiply and divide radicals
- Simplify radicals

In this section, we develop rules for finding products and quotients of radicals. To get a rule for products, first look at these examples.

$$\sqrt{4} \cdot \sqrt{9} = 2 \cdot 3 = 6 \quad \text{and} \quad \sqrt{4 \cdot 9} = \sqrt{36} = 6$$

We conclude that

$$\sqrt{4} \cdot \sqrt{9} = \sqrt{4 \cdot 9}.$$

This example generalizes as the **product rule for radicals**:

$$\sqrt{x} \cdot \sqrt{y} = \sqrt{x \cdot y},$$

as long as x and y are not negative. That is, the product of two radicals is the radical of the product.

Example 1 Use the product rule for radicals to find each product.

(a) $\sqrt{2} \cdot \sqrt{3} = \sqrt{2 \cdot 3} = \sqrt{6}$

(b) $\sqrt{7} \cdot \sqrt{5} = \sqrt{35}$

(c) $\sqrt{11} \cdot \sqrt{a} = \sqrt{11a}$

Work Quick Check 1 at the side.

QUICK CHECKS

1. Use the product rule for radicals to find each product.

(a) $\sqrt{6} \cdot \sqrt{11}$

(b) $\sqrt{2} \cdot \sqrt{7}$

(c) $\sqrt{17} \cdot \sqrt{3}$

(d) $\sqrt{10} \cdot \sqrt{r}$

One very important use of the product rule is in simplifying radicals. To **simplify** a radical, we write it so that there is no perfect square factor under the radical sign. Examples 2–5 show how this is done.

Example 2 Simplify $\sqrt{20}$.

We can divide 20 by the perfect square 4. Therefore,

$$\begin{aligned} \sqrt{20} &= \sqrt{4 \cdot 5} && \text{4 is a perfect square} \\ &= \sqrt{4} \cdot \sqrt{5} && \text{Product rule} \\ &= 2\sqrt{5} && \sqrt{4} = 2 \end{aligned}$$

Thus, $\sqrt{20} = 2\sqrt{5}$. Since 5 is not divisible by a perfect square (other than 1), $2\sqrt{5}$ is called the **simplified form** of $\sqrt{20}$.

Example 3 Simplify $\sqrt{72}$.

Look down the list of perfect squares. Find the largest of these numbers that divides into 72. The largest is 36, so that

$$\begin{aligned} \sqrt{72} &= \sqrt{36 \cdot 2} && \text{36 is a perfect square} \\ &= \sqrt{36} \cdot \sqrt{2} \\ \sqrt{72} &= 6\sqrt{2}. \end{aligned}$$

Example 4

$$\begin{aligned} \sqrt{300} &= \sqrt{100 \cdot 3} && \text{100 is a perfect square} \\ &= \sqrt{100} \cdot \sqrt{3} \\ \sqrt{300} &= 10\sqrt{3} \end{aligned}$$

Quick Check Answers

1. (a) $\sqrt{66}$ (b) $\sqrt{14}$ (c) $\sqrt{51}$ (d) $\sqrt{10r}$

2. Simplify each radical.

(a) $\sqrt{8}$

(b) $\sqrt{27}$

(c) $\sqrt{50}$

(d) $\sqrt{60}$

(e) $\sqrt{200}$

3. Find the product and simplify.

(a) $\sqrt{3} \cdot \sqrt{15}$

(b) $\sqrt{7} \cdot \sqrt{14}$

(c) $\sqrt{10} \cdot \sqrt{50}$

(d) $\sqrt{12} \cdot \sqrt{12}$

4. Simplify each quotient.

(a) $\sqrt{\frac{81}{16}}$

(b) $\sqrt{\frac{100}{9}}$

(c) $\sqrt{\frac{10}{49}}$

(d) $\sqrt{\frac{5}{16}}$

(e) $\frac{8\sqrt{50}}{4\sqrt{5}}$

Quick Check Answers

2. (a) $2\sqrt{2}$ (b) $3\sqrt{3}$ (c) $5\sqrt{2}$ (d) $2\sqrt{15}$ (e) $10\sqrt{2}$

3. (a) $3\sqrt{5}$ (b) $7\sqrt{2}$ (c) $10\sqrt{5}$ (d) 12

4. (a) $\frac{9}{4}$ (b) $\frac{10}{3}$ (c) $\frac{\sqrt{10}}{7}$ (d) $\frac{\sqrt{5}}{4}$ (e) $2\sqrt{10}$

Example 5 The number 15 is not divisible by any perfect square. Because of this, $\sqrt{15}$ cannot be further simplified.

Work Quick Check 2 at the side.

Sometimes the product rule can be used to simplify an answer, as Examples 6 and 7 show.

Example 6

$$\begin{aligned} \sqrt{9} \cdot \sqrt{75} &= 3\sqrt{75} \qquad \sqrt{9} = 3 \\ &= 3\sqrt{25 \cdot 3} \\ &= 3\sqrt{25} \cdot \sqrt{3} \\ &= 3 \cdot 5\sqrt{3} \\ &= 15\sqrt{3} \end{aligned}$$

Example 7

$$\begin{aligned} \sqrt{8} \cdot \sqrt{12} &= \sqrt{8 \cdot 12} \\ &= \sqrt{96} \\ &= \sqrt{16 \cdot 6} \\ &= \sqrt{16} \cdot \sqrt{6} \\ &= 4\sqrt{6} \end{aligned}$$

Work Quick Check 3 at the side.

There is a **quotient rule** for radicals that is very similar to the product rule:

$$\frac{\sqrt{x}}{\sqrt{y}} = \sqrt{\frac{x}{y}},$$

as long as x and y are not negative, and y is not 0. That is, the quotient of the radicals is the radical of a quotient.

Example 8 Use the quotient rule to simplify each radical.

(a) $\sqrt{\frac{25}{9}} = \frac{\sqrt{25}}{\sqrt{9}} = \frac{5}{3}$

(b) $\sqrt{\frac{144}{49}} = \frac{\sqrt{144}}{\sqrt{49}} = \frac{12}{7}$

(c) $\sqrt{\frac{3}{4}} = \frac{\sqrt{3}}{\sqrt{4}} = \frac{\sqrt{3}}{2}$

Example 9

$$\begin{aligned} \frac{27\sqrt{15}}{9\sqrt{3}} &= \frac{27}{9} \cdot \frac{\sqrt{15}}{\sqrt{3}} \\ &= \frac{27}{9} \cdot \sqrt{\frac{15}{3}} \\ &= 3\sqrt{5} \end{aligned}$$

Work Quick Check 4 at the side.

Some problems require both the product and quotient rules, as Examples 10 and 11 show.

Example 10 Simplify $\sqrt{\frac{3}{5}} \cdot \sqrt{10}$.

Use the product and quotient rules.

$$\sqrt{\frac{3}{5}} \cdot \sqrt{10} = \frac{\sqrt{3}}{\sqrt{5}} \cdot \frac{\sqrt{10}}{1}$$

$$= \frac{\sqrt{3} \cdot \sqrt{10}}{\sqrt{5}}$$

$$= \frac{\sqrt{30}}{\sqrt{5}}$$

$$= \sqrt{\frac{30}{5}}$$

$$= \sqrt{6}$$

Finally, the properties of this section are also valid when variables appear under the radical sign, as long as all the variables represent only positive numbers.

Example 11 Simplify each radical.

(a) $\sqrt{25m^4} = \sqrt{25} \cdot \sqrt{m^4} = 5m^2$

(b) $\sqrt{64p^{10}} = 8p^5$

(c) $\sqrt{r^9} = \sqrt{r^8 \cdot r} = \sqrt{r^8} \cdot \sqrt{r} = r^4\sqrt{r}$

(d) $\sqrt{\frac{5}{x^2}} = \frac{\sqrt{5}}{\sqrt{x^2}} = \frac{\sqrt{5}}{x}$

Work Quick Check 5 at the side.

5. Multiply and then simplify each radical, as needed.

(a) $\sqrt{\frac{5}{6}} \cdot \sqrt{120}$

(b) $\sqrt{\frac{3}{8}} \cdot \sqrt{40}$

(c) $\sqrt{36y^6}$

(d) $\sqrt{100p^8}$

(e) $\sqrt{a^5}$

(f) $\sqrt{\frac{7}{p^2}}$

Quick Check Answers

5. (a) 10 (b) $\sqrt{15}$ (c) $6y^3$ (d) $10p^4$ (e) $a^2\sqrt{a}$ (f) $\frac{\sqrt{7}}{p}$

WORK SPACE

name date hour

9.2 EXERCISES

Use the product rule to simplify each expression. See Examples 1–7.

1. $\sqrt{8} \cdot \sqrt{2}$ ______

2. $\sqrt{27} \cdot \sqrt{3}$ ______

3. $\sqrt{6} \cdot \sqrt{6}$ ______

4. $\sqrt{11} \cdot \sqrt{11}$ ______

5. $\sqrt{21} \cdot \sqrt{21}$ ______

6. $\sqrt{17} \cdot \sqrt{17}$ ______

7. $\sqrt{3} \cdot \sqrt{7}$ ______

8. $\sqrt{2} \cdot \sqrt{5}$ ______

9. $\sqrt{27}$ ______

10. $\sqrt{45}$ ______

11. $\sqrt{28}$ ______

12. $\sqrt{40}$ ______

13. $\sqrt{18}$ ______

14. $\sqrt{75}$ ______

15. $\sqrt{48}$ ______

16. $\sqrt{80}$ ______

17. $\sqrt{125}$ ______

18. $\sqrt{150}$ ______

19. $\sqrt{700}$ ______

20. $\sqrt{1100}$ ______

21. $\sqrt{100} \cdot \sqrt{27}$ ______

22. $\sqrt{16} \cdot \sqrt{8}$ ______

23. $\sqrt{80} \cdot \sqrt{20}$ ______

24. $\sqrt{200} \cdot \sqrt{2}$ ______

25. $\sqrt{27} \cdot \sqrt{48}$ ______

26. $\sqrt{75} \cdot \sqrt{27}$ ______

27. $\sqrt{50} \cdot \sqrt{72}$ ______

28. $\sqrt{98} \cdot \sqrt{8}$ ______

29. $\sqrt{7} \cdot \sqrt{21}$ ______

30. $\sqrt{12} \cdot \sqrt{48}$ ______

31. $\sqrt{15} \cdot \sqrt{45}$ ______

32. $\sqrt{20} \cdot \sqrt{45}$ ______

33. $\sqrt{80} \cdot \sqrt{15}$ ____________

34. $\sqrt{60} \cdot \sqrt{12}$ ____________

35. $\sqrt{50} \cdot \sqrt{20}$ ____________

36. $\sqrt{72} \cdot \sqrt{12}$ ____________

Use the quotient rule and product rule, as necessary, to simplify each expression. See Examples 8–10.

37. $\sqrt{\frac{100}{9}}$ ____________

38. $\sqrt{\frac{225}{16}}$ ____________

39. $\sqrt{\frac{36}{49}}$ ____________

40. $\sqrt{\frac{256}{9}}$ ____________

41. $\sqrt{\frac{5}{16}}$ ____________

42. $\sqrt{\frac{11}{25}}$ ____________

43. $\sqrt{\frac{30}{49}}$ ____________

44. $\sqrt{\frac{10}{121}}$ ____________

45. $\sqrt{\frac{1}{5}} \cdot \sqrt{\frac{4}{5}}$ ____________

46. $\sqrt{\frac{2}{3}} \cdot \sqrt{\frac{2}{27}}$ ____________

47. $\sqrt{\frac{2}{5}} \cdot \sqrt{\frac{8}{125}}$ ____________

48. $\sqrt{\frac{3}{8}} \cdot \sqrt{\frac{3}{2}}$ ____________

49. $\frac{\sqrt{75}}{\sqrt{3}}$ ____________

50. $\frac{\sqrt{200}}{\sqrt{2}}$ ____________

51. $\frac{\sqrt{48}}{\sqrt{3}}$ ____________

52. $\frac{\sqrt{72}}{\sqrt{8}}$ ____________

53. $\frac{15\sqrt{10}}{5\sqrt{2}}$ ____________

54. $\frac{18\sqrt{20}}{2\sqrt{10}}$ ____________

55. $\frac{25\sqrt{50}}{5\sqrt{5}}$ ____________

56. $\frac{26\sqrt{10}}{13\sqrt{5}}$ ____________

name date hour

Simplify each expression. Assume that all variables represent positive numbers. See Example 11.

57. $\sqrt{y} \cdot \sqrt{y}$ ____________

58. $\sqrt{m} \cdot \sqrt{m}$ ____________

59. $\sqrt{x} \cdot \sqrt{z}$ ____________

60. $\sqrt{p} \cdot \sqrt{q}$ ____________

61. $\sqrt{x^2}$ ____________

62. $\sqrt{y^2}$ ____________

63. $\sqrt{x^4}$ ____________

64. $\sqrt{y^4}$ ____________

65. $\sqrt{x^2 y^4}$ ____________

66. $\sqrt{x^4 y^8}$ ____________

67. $\sqrt{x^3}$ ____________

68. $\sqrt{y^3}$ ____________

69. $\sqrt{\frac{16}{x^2}}$ ____________

70. $\sqrt{\frac{100}{m^4}}$ ____________

71. $\sqrt{\frac{11}{r^4}}$ ____________

72. $\sqrt{\frac{23}{y^6}}$ ____________

WORK SPACE

OBJECTIVES

- Add and subtract radicals

We add or subtract radicals using the distributive property. For example,

$$8\sqrt{3} + 6\sqrt{3} = (8 + 6)\sqrt{3} = 14\sqrt{3}.$$

Also, $2\sqrt{11} - 7\sqrt{11} = -5\sqrt{11}$. Only like radicals can be added or subtracted.

Example 1 Add or subtract, as indicated.

(a) $3\sqrt{6} + 5\sqrt{6} = (3 + 5)\sqrt{6} = 8\sqrt{6}$

(b) $5\sqrt{10} - 7\sqrt{10} = (5 - 7)\sqrt{10} = -2\sqrt{10}$

(c) $\sqrt{5} + \sqrt{5} = 1\sqrt{5} + 1\sqrt{5} = (1 + 1)\sqrt{5} = 2\sqrt{5}$

(d) $\sqrt{7} + 2\sqrt{7} = 1\sqrt{7} + 2\sqrt{7} = 3\sqrt{7}$

(e) $\sqrt{3} + \sqrt{7}$ cannot be further simplified.

Work Quick Check 1 at the side.

Sometimes it is necessary to first simplify each radical in a sum or difference. Then add or subtract, if possible.

Example 2

$$\begin{aligned} 3\sqrt{2} + \sqrt{8} &= 3\sqrt{2} + \sqrt{4 \cdot 2} && \text{Simplify } \sqrt{8} \\ &= 3\sqrt{2} + \sqrt{4} \cdot \sqrt{2} \\ &= 3\sqrt{2} + 2\sqrt{2} \\ &= 5\sqrt{2} \end{aligned}$$

Example 3

$$\begin{aligned} \sqrt{18} - \sqrt{27} &= \sqrt{9 \cdot 2} - \sqrt{9 \cdot 3} \\ &= \sqrt{9} \cdot \sqrt{2} - \sqrt{9} \cdot \sqrt{3} \\ &= 3\sqrt{2} - 3\sqrt{3} \end{aligned}$$

Since $\sqrt{2}$ and $\sqrt{3}$ are unlike radicals, this difference cannot be further simplified.

Example 4

$$\begin{aligned} 2\sqrt{12} + 3\sqrt{75} &= 2(\sqrt{4} \cdot \sqrt{3}) + 3(\sqrt{25} \cdot \sqrt{3}) \\ &= 2(2\sqrt{3}) + 3(5\sqrt{3}) \\ &= 4\sqrt{3} + 15\sqrt{3} \\ &= 19\sqrt{3} \end{aligned}$$

Work Quick Check 2 at the side.

QUICK CHECKS

1. Add or subtract, as indicated.

(a) $8\sqrt{5} + 2\sqrt{5}$

(b) $-4\sqrt{3} + 9\sqrt{3}$

(c) $12\sqrt{11} - 2\sqrt{11}$

(d) $\sqrt{15} + \sqrt{15}$

(e) $3\sqrt{19} + \sqrt{19}$

(f) $2\sqrt{7} + 2\sqrt{10}$

2. Simplify as much as possible.

(a) $\sqrt{8} + 4\sqrt{2}$

(b) $\sqrt{27} + \sqrt{12}$

(c) $5\sqrt{200} - 6\sqrt{18}$

(d) $8\sqrt{20} + 7\sqrt{45}$

Quick Check Answers

1. (a) $10\sqrt{5}$ (b) $5\sqrt{3}$ (c) $10\sqrt{11}$ (d) $2\sqrt{15}$ (e) $4\sqrt{19}$ (f) cannot be further simplified

2. (a) $6\sqrt{2}$ (b) $5\sqrt{3}$ (c) $32\sqrt{2}$ (d) $37\sqrt{5}$

3. Simplify as much as possible.

(a) $\sqrt{7} \cdot \sqrt{21} + 2\sqrt{27}$

(b) $\sqrt{3} \cdot \sqrt{48} + 5\sqrt{3}$

(c) $\sqrt{18r} + \sqrt{8r}$

(d) $6\sqrt{5m} - 9\sqrt{125m}$

Example 5
$$\begin{aligned}\sqrt{5} \cdot \sqrt{15} + 4\sqrt{3} &= \sqrt{5 \cdot 15} + 4\sqrt{3} \\ &= \sqrt{75} + 4\sqrt{3} \\ &= \sqrt{25 \cdot 3} + 4\sqrt{3} \\ &= \sqrt{25} \cdot \sqrt{3} + 4\sqrt{3} \\ &= 5\sqrt{3} + 4\sqrt{3} \\ &= 9\sqrt{3}\end{aligned}$$

Example 6
$$\begin{aligned}\sqrt{12k} + \sqrt{27k} &= \sqrt{4 \cdot 3k} + \sqrt{9 \cdot 3k} \\ &= \sqrt{4} \cdot \sqrt{3k} + \sqrt{9} \cdot \sqrt{3k} \\ &= 2\sqrt{3k} + 3\sqrt{3k} \\ &= 5\sqrt{3k}\end{aligned}$$

Here, we must assume that k is not negative.

Work Quick Check 3 at the side.

We emphasize that

a sum or difference of radicals can be simplified only if the radicals are **like** radicals.

Quick Check Answers

3. (a) $13\sqrt{3}$ (b) $12 + 5\sqrt{3}$
(c) $5\sqrt{2r}$ (d) $-39\sqrt{5m}$

name date hour

9.3 EXERCISES

Simplify and combine terms wherever possible. See Examples 1–5.

1. $2\sqrt{3} + 5\sqrt{3}$ ________
2. $6\sqrt{5} + 8\sqrt{5}$ ________
3. $4\sqrt{7} - 9\sqrt{7}$ ________
4. $6\sqrt{2} - 8\sqrt{2}$ ________
5. $\sqrt{6} + \sqrt{6}$ ________
6. $\sqrt{11} + \sqrt{11}$ ________
7. $\sqrt{17} + 2\sqrt{17}$ ________
8. $3\sqrt{19} + \sqrt{19}$ ________
9. $5\sqrt{7} - \sqrt{7}$ ________
10. $12\sqrt{14} - \sqrt{14}$ ________
11. $5\sqrt{8} + \sqrt{8}$ ________
12. $3\sqrt{27} - \sqrt{27}$ ________
13. $\sqrt{45} + 2\sqrt{20}$ ________
14. $\sqrt{24} + 5\sqrt{54}$ ________
15. $3\sqrt{18} + \sqrt{8}$ ________
16. $2\sqrt{27} - \sqrt{3}$ ________
17. $-\sqrt{12} + \sqrt{75}$ ________
18. $2\sqrt{27} - \sqrt{300}$ ________
19. $5\sqrt{72} - 2\sqrt{50}$ ________
20. $6\sqrt{18} - 4\sqrt{32}$ ________
21. $-5\sqrt{32} + \sqrt{98}$ ________
22. $4\sqrt{75} + 3\sqrt{12}$ ________
23. $5\sqrt{7} - 2\sqrt{28} + 6\sqrt{63}$ ________
24. $3\sqrt{11} + 5\sqrt{44} - 3\sqrt{99}$ ________
25. $6\sqrt{5} + 3\sqrt{20} - 8\sqrt{45}$ ________
26. $7\sqrt{3} + 2\sqrt{12} - 5\sqrt{27}$ ________
27. $6\sqrt{2} + 5\sqrt{27} - 4\sqrt{12}$ ________
28. $9\sqrt{24} - 2\sqrt{54} + 3\sqrt{20}$ ________
29. $2\sqrt{8} - 5\sqrt{32} + 2\sqrt{48}$ ________
30. $5\sqrt{72} - 3\sqrt{48} - 4\sqrt{128}$ ________
31. $4\sqrt{50} + 3\sqrt{12} + 5\sqrt{45}$ ________
32. $6\sqrt{18} + 2\sqrt{48} - 6\sqrt{28}$ ________

33. $\frac{1}{4}\sqrt{288}-\frac{1}{6}\sqrt{72}$ ____________

34. $\frac{2}{3}\sqrt{27}-\frac{3}{4}\sqrt{48}$ ____________

35. $\frac{3}{5}\sqrt{75}-\frac{2}{3}\sqrt{45}$ ____________

36. $\frac{5}{8}\sqrt{128}-\frac{3}{4}\sqrt{160}$ ____________

37. $\sqrt{6}\cdot\sqrt{2}+3\sqrt{3}$ ____________

38. $4\sqrt{15}\cdot\sqrt{3}-2\sqrt{5}$ ____________

39. $\sqrt{3}\cdot\sqrt{7}+2\sqrt{21}-\sqrt{7}$ ____________

40. $\sqrt{13}\cdot\sqrt{2}+3\sqrt{26}$ ____________

Simplify each expression. Assume that all variables represent nonnegative real numbers. See Example 6.

41. $\sqrt{9x}+\sqrt{49x}-\sqrt{16x}$ ____________

42. $\sqrt{4a}-\sqrt{16a}+\sqrt{9a}$ ____________

43. $\sqrt{4a}+6\sqrt{a}+\sqrt{25a}$ ____________

44. $\sqrt{6x^2}+x\sqrt{54}$ ____________

45. $\sqrt{75x^2}+x\sqrt{300}$ ____________

46. $\sqrt{20y^2}-3y\sqrt{5}$ ____________

47. $3\sqrt{8x^2}-4x\sqrt{2}$ ____________

48. $6r\sqrt{27r^2s}+3r^2\sqrt{3s}$ ____________

9.4 RATIONALIZING THE DENOMINATOR

OBJECTIVES

- Rationalize the denominator
- Simplify expressions by rationalizing the denominator

We learned decimal approximation for radicals in the first section of this chapter. It is easier to find these decimals for more complicated radicals if the denominators do not contain any radicals. For example, to find a decimal for

$$\frac{\sqrt{3}}{\sqrt{2}},$$

we can look up $\sqrt{3}$ and $\sqrt{2}$ in the square root table, and get

$$\frac{\sqrt{3}}{\sqrt{2}} \approx \frac{1.732}{1.414}.$$

We would then need to divide 1.414 into 1.732, which would be very difficult. This calculation would be easier if there were no radical in the denominator.

To get rid of this radical, multiply the numerator and the denominator by $\sqrt{2}$.

$$\frac{\sqrt{3}}{\sqrt{2}} = \frac{\sqrt{3} \cdot \sqrt{2}}{\sqrt{2} \cdot \sqrt{2}} = \frac{\sqrt{6}}{2} \qquad \sqrt{2} \cdot \sqrt{2} = \sqrt{4} = 2$$

This process of changing the denominator from a radical (irrational number) to a rational number is called **rationalizing the denominator.** The value of the number is not changed; only the form of the number is changed.

Example 1 Rationalize the denominator of $\frac{9}{\sqrt{6}}$.

Multiply both numerator and denominator by $\sqrt{6}$.

$$\frac{9}{\sqrt{6}} = \frac{9 \cdot \sqrt{6}}{\sqrt{6} \cdot \sqrt{6}} = \frac{9\sqrt{6}}{6} = \frac{3\sqrt{6}}{2} \qquad \sqrt{6} \cdot \sqrt{6} = 6$$

Example 2 Rationalize the denominator of $\frac{12}{\sqrt{8}}$.

We could rationalize the denominator here by multiplying by $\sqrt{8}$. However, we can get the answer faster if we multiply by $\sqrt{2}$. This is because $\sqrt{8} \cdot \sqrt{2} = \sqrt{16} = 4$, a rational number.

$$\frac{12}{\sqrt{8}} = \frac{12 \cdot \sqrt{2}}{\sqrt{8} \cdot \sqrt{2}} = \frac{12\sqrt{2}}{\sqrt{16}} = \frac{12\sqrt{2}}{4} = 3\sqrt{2}$$

Work Quick Check 1 at the side.

Some radicals can be simplified by rationalizing the denominator, as Examples 3–5 show.

QUICK CHECKS

1. Rationalize each of the denominators.

(a) $\frac{3}{\sqrt{5}}$

(b) $\frac{-6}{\sqrt{11}}$

(c) $\frac{\sqrt{7}}{\sqrt{2}}$

(d) $\frac{20}{\sqrt{18}}$

(e) $\frac{-12}{\sqrt{32}}$

Quick Check Answers

1. (a) $\frac{3\sqrt{5}}{5}$ (b) $\frac{-6\sqrt{11}}{11}$ (c) $\frac{\sqrt{14}}{2}$ (d) $\frac{10\sqrt{2}}{3}$ (e) $\frac{-3\sqrt{2}}{2}$

2. Simplify by rationalizing each of the denominators.

(a) $\sqrt{\frac{1}{7}}$

(b) $\sqrt{\frac{3}{2}}$

(c) $\sqrt{\frac{5}{18}}$

(d) $\sqrt{\frac{16}{11}}$

3. Simplify.

(a) $\sqrt{\frac{1}{2}} \cdot \sqrt{\frac{5}{6}}$

(b) $\sqrt{\frac{3}{5}} \cdot \sqrt{\frac{2}{3}}$

(c) $\sqrt{\frac{1}{10}} \cdot \sqrt{20}$

(d) $\sqrt{\frac{5}{8}} \cdot \sqrt{\frac{24}{10}}$

4. Rationalize each of the denominators. Assume that all variables represent positive numbers.

(a) $\frac{\sqrt{2r}}{\sqrt{a}}$

(b) $\sqrt{\frac{m^2}{k}}$

(c) $\sqrt{\frac{z^2 x}{9m}}$

(d) $\sqrt{\frac{7r^2 s^2}{m}}$

Quick Check Answers

2. (a) $\frac{\sqrt{7}}{7}$ (b) $\frac{\sqrt{6}}{2}$ (c) $\frac{\sqrt{10}}{6}$ (d) $\frac{4\sqrt{11}}{11}$

3. (a) $\frac{\sqrt{15}}{6}$ (b) $\frac{\sqrt{10}}{5}$ (c) $\sqrt{2}$ (d) $\frac{\sqrt{6}}{2}$

4. (a) $\frac{\sqrt{2ra}}{a}$ (b) $\frac{m\sqrt{k}}{k}$ (c) $\frac{z\sqrt{xm}}{3m}$ (d) $\frac{rs\sqrt{7m}}{m}$

Example 3 Simplify $\sqrt{\frac{27}{5}}$ by rationalizing the denominator.

First, use the quotient rule for radicals.

$$\sqrt{\frac{27}{5}} = \frac{\sqrt{27}}{\sqrt{5}}$$

Now multiply both numerator and denominator by $\sqrt{5}$.

$$\frac{\sqrt{27}}{\sqrt{5}} = \frac{\sqrt{27} \cdot \sqrt{5}}{\sqrt{5} \cdot \sqrt{5}}$$
$$= \frac{\sqrt{9 \cdot 3} \cdot \sqrt{5}}{5}$$
$$= \frac{\sqrt{9} \cdot \sqrt{3} \cdot \sqrt{5}}{5}$$
$$= \frac{3\sqrt{3} \cdot \sqrt{5}}{5}$$
$$= \frac{3\sqrt{15}}{5}$$

Work Quick Check 2 at the side.

Example 4 Simplify $\sqrt{\frac{5}{8}} \cdot \sqrt{\frac{1}{6}}$.

Use both the quotient rule and the product rule.

$$\sqrt{\frac{5}{8}} \cdot \sqrt{\frac{1}{6}} = \sqrt{\frac{5}{8} \cdot \frac{1}{6}} = \sqrt{\frac{5}{48}} = \frac{\sqrt{5}}{\sqrt{48}}$$

Now rationalize the denominator by multiplying by $\sqrt{3}$ (since $\sqrt{48} \cdot \sqrt{3} = \sqrt{48 \cdot 3} = \sqrt{144} = 12$).

$$\frac{\sqrt{5}}{\sqrt{48}} = \frac{\sqrt{5} \cdot \sqrt{3}}{\sqrt{48} \cdot \sqrt{3}} = \frac{\sqrt{15}}{\sqrt{144}} = \frac{\sqrt{15}}{12}$$

Work Quick Check 3 at the side.

Example 5 Rationalize the denominator of $\frac{\sqrt{4x}}{\sqrt{y}}$. Assume that x and y are positive.

Multiply numerator and denominator by $\sqrt{y}$.

$$\frac{\sqrt{4x}}{\sqrt{y}} = \frac{\sqrt{4x} \cdot \sqrt{y}}{\sqrt{y} \cdot \sqrt{y}} = \frac{\sqrt{4xy}}{y} = \frac{2\sqrt{xy}}{y}$$

Work Quick Check 4 at the side.

name date hour

9.4 EXERCISES

Perform the indicated operations. Write all answers in simplest form. Rationalize all denominators. See Examples 1–4.

1. $\frac{6}{\sqrt{5}}$ __________

2. $\frac{4}{\sqrt{2}}$ __________

3. $\frac{5}{\sqrt{5}}$ __________

4. $\frac{15}{\sqrt{15}}$ __________

5. $\frac{3}{\sqrt{7}}$ __________

6. $\frac{12}{\sqrt{10}}$ __________

7. $\frac{8\sqrt{3}}{\sqrt{5}}$ __________

8. $\frac{9\sqrt{6}}{\sqrt{5}}$ __________

9. $\frac{12\sqrt{10}}{8\sqrt{3}}$ __________

10. $\frac{9\sqrt{15}}{6\sqrt{2}}$ __________

11. $\frac{8}{\sqrt{27}}$ __________

12. $\frac{12}{\sqrt{18}}$ __________

13. $\frac{3}{\sqrt{50}}$ __________

14. $\frac{5}{\sqrt{75}}$ __________

15. $\frac{12}{\sqrt{72}}$ __________

16. $\frac{21}{\sqrt{45}}$ __________

17. $\frac{9}{\sqrt{32}}$ __________

18. $\frac{50}{\sqrt{125}}$ __________

19. $\frac{\sqrt{8}}{\sqrt{2}}$ __________

20. $\frac{\sqrt{27}}{\sqrt{3}}$ __________

21. $\frac{\sqrt{10}}{\sqrt{5}}$ __________

22. $\frac{\sqrt{6}}{\sqrt{3}}$ __________

23. $\frac{\sqrt{40}}{\sqrt{3}}$ __________

24. $\frac{\sqrt{5}}{\sqrt{8}}$ __________

25. $\sqrt{\frac{1}{2}}$ __________

26. $\sqrt{\frac{1}{8}}$ __________

27. $\sqrt{\frac{10}{7}}$ __________

28. $\sqrt{\frac{2}{3}}$ __________

29. $\sqrt{\frac{9}{5}}$ __________

30. $\sqrt{\frac{16}{7}}$ __________

31. $\sqrt{\frac{7}{5}} \cdot \sqrt{10}$ __________

32. $\sqrt{\frac{1}{3}} \cdot \sqrt{3}$ __________

33. $\sqrt{\frac{3}{4}} \cdot \sqrt{\frac{1}{5}}$ __________

34. $\sqrt{\frac{1}{10}} \cdot \sqrt{\frac{10}{3}}$ ____________ 35. $\sqrt{\frac{21}{7}} \cdot \sqrt{\frac{21}{8}}$ ____________ 36. $\sqrt{\frac{1}{11}} \cdot \sqrt{\frac{33}{16}}$ ____________

37. $\sqrt{\frac{2}{5}} \cdot \sqrt{\frac{3}{10}}$ ____________ 38. $\sqrt{\frac{9}{8}} \cdot \sqrt{\frac{7}{16}}$ ____________ 39. $\sqrt{\frac{16}{27}} \cdot \sqrt{\frac{1}{9}}$ ____________

Perform the indicated operations. Write all answers in simplest form. Rationalize all denominators. Assume that all variables represent positive real numbers. See Example 5.

40. $\sqrt{\frac{5}{x}}$ ____________ 41. $\sqrt{\frac{6}{p}}$ ____________ 42. $\sqrt{\frac{4r^3}{s}}$ ____________

43. $\sqrt{\frac{6p^3}{3m}}$ ____________ 44. $\sqrt{\frac{a^3 b}{6}}$ ____________ 45. $\sqrt{\frac{x^2}{4y}}$ ____________

46. $\sqrt{\frac{m^2 n}{2}}$ ____________ 47. $\sqrt{\frac{9a^2 r}{5}}$ ____________ 48. $\sqrt{\frac{2x^2 z^4}{3y}}$ ____________

OBJECTIVES

- Simplify radical expressions

What the "simplest" form of a radical is may not be always clear. In this book, a radical expression is simplified when the following five rules are satisfied.

1. If a radical represents a rational number, then that rational number should be used in place of the radical.

For example, $\sqrt{49}$ is simplified by writing 7; $\sqrt{64}$ as 8; $\sqrt{169/9}$ as 13/3.

2. If a radical expression contains products of radicals, the product rule for radicals, $\sqrt{x} \cdot \sqrt{y} = \sqrt{xy}$, should be used to get a single radical.

For example, $\sqrt{3} \cdot \sqrt{2}$ is simplified to $\sqrt{6}$; $\sqrt{5} \cdot \sqrt{x}$ to $\sqrt{5x}$.

3. If a radicand has a factor that is a perfect square, the radical should be expressed as the product of the positive square root of the perfect square and the remaining radical factor.

For example, $\sqrt{20}$ is simplified to $\sqrt{20} = \sqrt{4 \cdot 5} = \sqrt{4} \cdot \sqrt{5} = 2\sqrt{5}$; $\sqrt{75}$ as $5\sqrt{3}$.

4. If a radical expression contains sums or differences of radicals, the distributive property should be used to combine terms, if possible.

For example, $3\sqrt{2} + 4\sqrt{2}$ is combined as $7\sqrt{2}$; but $3\sqrt{2} + 4\sqrt{3}$ cannot be further combined.

5. Any radicals in the denominator should be changed to rational numbers.

For example, $5/\sqrt{3}$ is rationalized as

$$\frac{5}{\sqrt{3}} = \frac{5\sqrt{3}}{\sqrt{3} \cdot \sqrt{3}} = \frac{5\sqrt{3}}{3}.$$

Example 1 Simplify $\sqrt{16} + \sqrt{9}$.

We have $\sqrt{16} + \sqrt{9} = 4 + 3 = 7$.

Example 2 Simplify $5\sqrt{2} + 2\sqrt{18}$.

First simplify $\sqrt{18}$.

$$\begin{aligned} 5\sqrt{2} + 2\sqrt{18} &= 5\sqrt{2} + 2(\sqrt{9} \cdot \sqrt{2}) \\ &= 5\sqrt{2} + 2(3\sqrt{2}) \\ &= 5\sqrt{2} + 6\sqrt{2} \\ &= 11\sqrt{2} \end{aligned}$$

Work Quick Check 1 at the side.

QUICK CHECKS

1. Simplify each radical.

(a) $\sqrt{36} + \sqrt{25}$

(b) $3\sqrt{3} + 2\sqrt{27}$

(c) $4\sqrt{8} - 2\sqrt{32}$

(d) $2\sqrt{12} - 5\sqrt{48}$

Quick Check Answers

1. (a) 11 (b) $9\sqrt{3}$ (c) 0 (d) $-16\sqrt{3}$

2. Find each product. Simplify the answers.

(a) $\sqrt{7}(\sqrt{2}+\sqrt{5})$

(b) $\sqrt{2}(\sqrt{8}+\sqrt{20})$

(c) $(\sqrt{2}+5\sqrt{3})(\sqrt{3}-2\sqrt{2})$

(d) $(2\sqrt{7}+\sqrt{10})(3\sqrt{7}-2\sqrt{10})$

3. Find each product. Simplify the answers.

(a) $(3+\sqrt{5})(3-\sqrt{5})$

(b) $(\sqrt{3}-2)(\sqrt{3}+2)$

(c) $(\sqrt{5}+\sqrt{3})(\sqrt{5}-\sqrt{3})$

(d) $(\sqrt{11}+\sqrt{14})(\sqrt{11}-\sqrt{14})$

Quick Check Answers
2. (a) $\sqrt{14}+\sqrt{35}$ (b) $4+2\sqrt{10}$ (c) $11-9\sqrt{6}$ (d) $22-\sqrt{70}$
3. (a) 4 (b) -1 (c) 2 (d) -3

Example 3 Simplify $\sqrt{5}(\sqrt{8}-\sqrt{32})$.

Using the distributive property, we have

$$\begin{aligned}\sqrt{5}(\sqrt{8}-\sqrt{32}) &= \sqrt{5}\cdot\sqrt{8}-\sqrt{5}\cdot\sqrt{32}\\ &= \sqrt{40}-\sqrt{160}\\ &= \sqrt{4}\cdot\sqrt{10}-\sqrt{16}\cdot\sqrt{10}\\ &= 2\sqrt{10}-4\sqrt{10}\\ &= -2\sqrt{10}.\end{aligned}$$

Example 4 Simplify the product $(\sqrt{3}+2\sqrt{5})(\sqrt{3}-4\sqrt{5})$.

The products of these sums of radicals can be found in much the same way that we found the product of binomials in Chapter 4. The pattern of multiplication is the same.

$$\begin{aligned}&(\sqrt{3}+2\sqrt{5})(\sqrt{3}-4\sqrt{5})\\ &= \sqrt{3}\cdot\sqrt{3}+\sqrt{3}(-4\sqrt{5})+2\sqrt{5}\cdot\sqrt{3}+2\sqrt{5}(-4\sqrt{5})\\ &= 3-4\sqrt{15}+2\sqrt{15}-8\cdot 5\\ &= 3-2\sqrt{15}-40\\ &= -37-2\sqrt{15}\end{aligned}$$

Work Quick Check 2 at the side.

Just as we found certain special products of binomials in Chapter 5, there are special products of radicals. Examples 5 and 6 use the difference of two squares,

$$(a+b)(a-b)=a^2-b^2.$$

Example 5 Simplify the product $(4-\sqrt{3})(4+\sqrt{3})$.

Follow the pattern of binomial multiplication.

$$\begin{aligned}(4-\sqrt{3})(4+\sqrt{3}) &= 4\cdot 4+4\sqrt{3}-4\sqrt{3}\\ &\quad -\sqrt{3}\cdot\sqrt{3}\\ &= 16-3\\ &= 13\end{aligned}$$

Example 6 Simplify $(\sqrt{12}-\sqrt{6})(\sqrt{12}+\sqrt{6})$.

$$\begin{aligned}&(\sqrt{12}-\sqrt{6})(\sqrt{12}+\sqrt{6})\\ &= \sqrt{12}\cdot\sqrt{12}+\sqrt{12}\cdot\sqrt{6}-\sqrt{12}\cdot\sqrt{6}-\sqrt{6}\cdot\sqrt{6}\\ &= 12-6\\ &= 6\end{aligned}$$

Work Quick Check 3 at the side.

We can use products of radicals similar to those in Examples 5 and 6 to rationalize the denominators in more complicated expressions, such as

$$\frac{2}{4-\sqrt{3}}.$$

We saw in Example 5 that if we multiply this denominator, $4-\sqrt{3}$, by the radical $4+\sqrt{3}$, then the product $(4-\sqrt{3})(4+\sqrt{3})$ is the rational number 13. If we multiply numerator and denominator by $4+\sqrt{3}$, we get

$$\frac{2}{4-\sqrt{3}} = \frac{2(4+\sqrt{3})}{(4-\sqrt{3})(4+\sqrt{3})} = \frac{2(4+\sqrt{3})}{13}.$$

The denominator has now been rationalized—it contains no radical signs.

Example 7 Rationalize the denominator in the quotient

$$\frac{4}{3+\sqrt{5}}.$$

To eliminate the radical in the denominator, multiply numerator and denominator by $3-\sqrt{5}$.

$$\begin{aligned}\frac{4}{3+\sqrt{5}} &= \frac{4(3-\sqrt{5})}{(3+\sqrt{5})(3-\sqrt{5})}\\ &= \frac{4(3-\sqrt{5})}{9+3\sqrt{5}-3\sqrt{5}-5}\\ &= \frac{4(3-\sqrt{5})}{9-5}\\ &= \frac{4(3-\sqrt{5})}{4}\\ &= 3-\sqrt{5}\end{aligned}$$

The expressions $3+\sqrt{5}$ and $3-\sqrt{5}$ are called **conjugates** of each other.

Work Quick Check 4 at the side.

4. Rationalize each of the denominators.

(a) $\frac{5}{4+\sqrt{2}}$

(b) $\frac{3}{2-\sqrt{5}}$

(c) $\frac{1}{6+\sqrt{3}}$

(d) $\frac{\sqrt{3}}{3+\sqrt{3}}$

Quick Check Answers

4. (a) $\frac{5(4-\sqrt{2})}{14}$ (b) $-3(2+\sqrt{5})$ (c) $\frac{6-\sqrt{3}}{33}$ (d) $\frac{\sqrt{3}-1}{2}$

WORK SPACE

name date hour

9.5 EXERCISES

Simplify each expression. Use the five rules given in the text. See Examples 1–6.

1. $3\sqrt{5} + 8\sqrt{45}$ ________
2. $6\sqrt{2} + 4\sqrt{18}$ ________
3. $9\sqrt{50} - 4\sqrt{72}$ ________
4. $3\sqrt{80} - 5\sqrt{45}$ ________
5. $\sqrt{2}(\sqrt{8} - \sqrt{32})$ ________
6. $\sqrt{3}(\sqrt{27} - \sqrt{3})$ ________
7. $\sqrt{5}(\sqrt{3} + \sqrt{7})$ ________
8. $\sqrt{7}(\sqrt{10} - \sqrt{3})$ ________
9. $2\sqrt{5}(\sqrt{2} + \sqrt{5})$ ________
10. $3\sqrt{7}(2\sqrt{7} - 4\sqrt{5})$ ________
11. $-\sqrt{14} \cdot \sqrt{2} - \sqrt{28}$ ________
12. $\sqrt{6} \cdot \sqrt{3} - 2\sqrt{50}$ ________
13. $(2\sqrt{6} + 3)(3\sqrt{6} - 5)$ ________
14. $(4\sqrt{5} - 2)(2\sqrt{5} + 3)$ ________
15. $(5\sqrt{7} - 2\sqrt{3})(3\sqrt{7} + 3\sqrt{3})$ ________
16. $(2\sqrt{10} + 5\sqrt{2})(3\sqrt{10} - 4\sqrt{2})$ ________
17. $(3\sqrt{2} + 4)(3\sqrt{2} + 4)$ ________
18. $(4\sqrt{5} - 1)(4\sqrt{5} - 1)$ ________
19. $(2\sqrt{7} - 3)^2$ ________
20. $(3\sqrt{5} + 5)^2$ ________
21. $(3 - \sqrt{2})(3 + \sqrt{2})$ ________
22. $(7 - \sqrt{5})(7 + \sqrt{5})$ ________
23. $(2 + \sqrt{8})(2 - \sqrt{8})$ ________
24. $(3 + \sqrt{11})(3 - \sqrt{11})$ ________
25. $(\sqrt{6} - \sqrt{5})(\sqrt{6} + \sqrt{5})$ ________
26. $(\sqrt{11} + \sqrt{10})(\sqrt{11} - \sqrt{10})$ ________

Rationalize the denominators. See Example 7.

27. $\dfrac{1}{3 + \sqrt{2}}$ ________
28. $\dfrac{1}{4 - \sqrt{3}}$ ________

29. $\frac{5}{2+\sqrt{5}}$ ______

30. $\frac{6}{3+\sqrt{7}}$ ______

31. $\frac{7}{2-\sqrt{11}}$ ______

32. $\frac{38}{5-\sqrt{6}}$ ______

33. $\frac{\sqrt{2}}{1+\sqrt{2}}$ ______

34. $\frac{\sqrt{7}}{2-\sqrt{7}}$ ______

35. $\frac{\sqrt{5}}{1-\sqrt{5}}$ ______

36. $\frac{\sqrt{3}}{2+\sqrt{3}}$ ______

37. $\frac{\sqrt{12}}{\sqrt{3}+1}$ ______

38. $\frac{\sqrt{18}}{\sqrt{2}-1}$ ______

39. $\frac{2\sqrt{3}}{\sqrt{3}+5}$ ______

40. $\frac{\sqrt{12}}{2-\sqrt{2}}$ ______

OBJECTIVES

- Solve equations with radicals

How can we solve an equation involving radicals, such as

$$\sqrt{x+1} = 3?$$

The addition and multiplication properties of equality will not help us here; we need another property.

If $a = b$, then $a^2 = b^2$. For example, if $y = 4$, then we can square both sides of the equation to get

$$y^2 = 4^2 \quad \text{or} \quad y^2 = 16.$$

This last equation, $y^2 = 16$, has *two* solutions, $y = 4$ or $y = -4$, while the original equation, $y = 4$, has only *one* solution.

As shown by this example, squaring both sides of an equation can lead to a new equation with more solutions than the original equation. Because of this possibility, we need to check all proposed solutions in the *original* equation.

In summary, we have the **squaring property of equality.**

> If both sides of a given equation are squared, all solutions of the original equation are also solutions of the squared equation.

Example 1 Solve the equation $\sqrt{x+1} = 3$.

Use the squaring property of equality to square both sides of the equation.

$$(\sqrt{x+1})^2 = 3^2$$
$$x + 1 = 9$$
$$x = 8$$

Now check this answer in the original equation.

$$\sqrt{x+1} = 3$$
$$\sqrt{8+1} = 3$$
$$\sqrt{9} = 3$$
$$3 = 3 \qquad \text{True}$$

Since this statement is true, the solution of $\sqrt{x+1} = 3$ is the number 8.

Example 2 Solve $3\sqrt{x} = \sqrt{x+8}$.

Squaring both sides gives

$$(3\sqrt{x})^2 = (\sqrt{x+8})^2$$
$$3^2(\sqrt{x})^2 = (\sqrt{x+8})^2$$
$$9x = x + 8$$
$$8x = 8$$
$$x = 1.$$

QUICK CHECKS

1. Solve each equation.

(a) $\sqrt{x+3} = 1$

(b) $\sqrt{2x+1} = 5$

(c) $\sqrt{3x+9} = 2\sqrt{x}$

(d) $5\sqrt{x} = \sqrt{20x+5}$

2. Solve each equation that has a solution. (Hint: In (c) add -4 to both sides.)

(a) $\sqrt{x} = -5$

(b) $\sqrt{x} = 2$

(c) $\sqrt{x} + 4 = 0$

(d) $\sqrt{x} - 8 = 0$

Quick Check Answers
1. (a) -2 (b) 12 (c) 9 (d) 1
2. (a) no solution (b) 4 (c) no solution (d) 64

Check this proposed solution.

$$3\sqrt{x} = \sqrt{x+8}$$
$$3\sqrt{1} = \sqrt{1+8}$$
$$3(1) = \sqrt{9}$$
$$3 = 3 \qquad \text{True}$$

The solution of $3\sqrt{x} = \sqrt{x+8}$ is the number 1.

Work Quick Check 1 at the side.

Not all equations with radicals even have a solution as Example 3 shows.

Example 3 Solve the equation $\sqrt{x} = -3$.

Square both sides.

$$(\sqrt{x})^2 = (-3)^2$$
$$x = 9$$

Check this proposed answer in the original equation.

$$\sqrt{x} = -3$$
$$\sqrt{9} = -3$$
$$3 = -3 \qquad \text{False}$$

Since the statement $3 = -3$ is false, the number 9 is not a solution of the given equation. Thus, $\sqrt{x} = -3$ has no real number solutions at all.

Work Quick Check 2 at the side.

The next example uses the fact that

$$(a+b)^2 = a^2 + 2ab + b^2.$$

Example 4 Solve the equation $\sqrt{2y-3} = y - 3$.

To square both sides, we must square $y - 3$.

$$(y-3)^2 = (y-3)(y-3)$$
$$= y^2 - 6y + 9$$

Now we can square both sides.

$$(\sqrt{2y-3})^2 = (y-3)^2$$
$$2y - 3 = y^2 - 6y + 9$$

This equation is quadratic, since it has a y^2 term. To solve the equation, we must get it equal to 0. To do this, add $-2y$ and 3 to both sides.

$$0 = y^2 - 8y + 12$$

This equation can be solved by factoring.

$$0 = (y - 6)(y - 2)$$

Make each factor equal to 0.

$$y - 6 = 0 \quad \text{or} \quad y - 2 = 0$$
$$y = 6 \quad \text{or} \quad y = 2$$

Check both of these proposed solutions in the original equation.

If $y = 6$,	If $y = 2$,
$\sqrt{2y - 3} = y - 3$	$\sqrt{2y - 3} = y - 3$
$\sqrt{2(6) - 3} = 6 - 3$	$\sqrt{2(2) - 3} = 2 - 3$
$\sqrt{12 - 3} = 3$	$\sqrt{4 - 3} = -1$
$\sqrt{9} = 3$	$\sqrt{1} = -1$
$3 = 3$ True	$1 = -1$ False

Only $y = 6$ is a valid solution of the equation.

Work Quick Check 3 at the side.

3. Solve each equation.

(a) $\sqrt{6x + 6} = x + 1$

(b) $x + 3 = \sqrt{12x + 1}$

(c) $2x - 1 = \sqrt{10x + 9}$

Example 5 Solve the equation $\sqrt{x} + 1 = 2x$.

Square both sides.

$$(\sqrt{x} + 1)^2 = (2x)^2$$
$$x + 2\sqrt{x} + 1 = 4x^2$$

Squaring both sides of the given equation produced an equation still containing a radical. It would be better to rewrite the original equation so that the radical is alone on one side of the equals sign. To do this, add -1 to both sides.

$$\sqrt{x} = 2x - 1$$

Now square both sides.

$$(\sqrt{x})^2 = (2x - 1)^2$$
$$x = 4x^2 - 4x + 1$$

Add $-x$ to both sides.

$$0 = 4x^2 - 5x + 1$$

This equation is a quadratic equation, which can be solved by factoring.

$$0 = (4x - 1)(x - 1)$$
$$4x - 1 = 0 \quad \text{or} \quad x - 1 = 0$$
$$x = \frac{1}{4} \quad \text{or} \quad x = 1$$

Both of these proposed solutions must be checked in the original equation. For $x = \frac{1}{4}$, we obtain a *false* statement, while $x = 1$ leads to a *true* statement. Therefore, the only solution to the original equation is the number 1.

Work Quick Check 4 at the side.

4. Solve each equation.

(a) $\sqrt{x} - 3 = x - 15$

(b) $\sqrt{x + 5} + 2 = x + 5$

Quick Check Answers

3. (a) 5, −1 (b) 4, 2 (c) 4
4. (a) 16 (b) −1

WORK SPACE

name date hour

9.6 EXERCISES

Find all solutions for each equation. See Examples 1–3.

1. $\sqrt{x} = 2$ __________

2. $\sqrt{m} = 5$ __________

3. $\sqrt{y + 3} = 2$ __________

4. $\sqrt{z + 1} = 5$ __________

5. $\sqrt{t - 3} = 2$ __________

6. $\sqrt{r + 5} = 4$ __________

7. $\sqrt{n + 8} = 1$ __________

8. $\sqrt{k + 10} = 2$ __________

9. $\sqrt{m + 5} = 0$ __________

10. $\sqrt{y - 4} = 0$ __________

11. $\sqrt{z + 5} = -2$ __________

12. $\sqrt{t - 3} = -2$ __________

13. $\sqrt{k} - 2 = 5$ __________

14. $\sqrt{p} - 3 = 7$ __________

15. $\sqrt{y} + 4 = 2$ __________

16. $\sqrt{m} + 6 = 5$ __________

17. $\sqrt{5t - 9} = 2\sqrt{t}$ __________

18. $\sqrt{3n + 4} = 2\sqrt{n}$ __________

19. $3\sqrt{r} = \sqrt{8r + 16}$ __________

20. $2\sqrt{r} = \sqrt{3r + 9}$ __________

21. $\sqrt{5y - 5} = \sqrt{4y + 1}$ __________

22. $\sqrt{2x + 2} = \sqrt{3x - 5}$ __________

23. $\sqrt{x + 2} = \sqrt{2x - 5}$ __________

24. $\sqrt{3m + 3} = \sqrt{5m - 1}$ __________

25. $\sqrt{2t + 9} = \sqrt{t + 5}$ __________

26. $\sqrt{6z + 22} = \sqrt{2z + 10}$ __________

27. $\sqrt{2x + 6} = \sqrt{4x - 4}$ __________

28. $\sqrt{5x - 6} = \sqrt{4x - 3}$ __________

Find all solutions for each equation. See Examples 4 and 5. Remember that $(a + b)^2 = a^2 + 2ab + b^2$ *and* $(\sqrt{a})^2 = a$.

29. $\sqrt{2x + 1} = x - 7$ __________

30. $\sqrt{5x + 1} = x + 1$ __________

31. $\sqrt{3x + 10} = 2x - 5$ __________

32. $\sqrt{4x + 13} = 2x - 1$ __________

33. $\sqrt{x+1}-1=x$ __________

34. $\sqrt{3x+3}+5=x$ __________

35. $\sqrt{4x+5}-2=2x-7$ __________

36. $\sqrt{6x+7}-1=x+1$ __________

37. $3\sqrt{x+13}=x+9$ __________

38. $2\sqrt{x+7}=x-1$ __________

39. $\sqrt{4x}-x+3=0$ __________

40. $\sqrt{2x}-x+4=0$ __________

41. $\sqrt{3x}-4=x-10$ __________

42. $\sqrt{x}+9=x+3$ __________

Solve each problem.

43. The square root of the sum of a number and 4 is 5. Find the number. __________

44. A certain number is the same as the square root of the product of 8 and the number. Find the number. __________

45. Three times the square root of two equals the square root of the sum of some number and ten. Find the number. __________

46. The negative square root of a number equals that number decreased by two. Find the number. __________

47. To estimate the speed at which a car was traveling at the time of an accident, police sometimes use the following procedure: A police officer drives the car involved in the accident under conditions similar to those when the accident took place, and skids to a stop. If the car is driven at 30 miles per hour, then the speed at the time of the accident is given by

$$s=\sqrt{\frac{900a}{p}},$$

where a is the length of the skid marks left at the time of the accident, and p is the length of the skid marks in the police test. Find s if

(a) $a = 900$ feet and $p = 100$ feet. __________

(b) $a = 400$ feet and $p = 25$ feet. __________

(c) $a = 80$ feet and $p = 20$ feet. __________

(d) $a = 120$ feet and $p = 30$ feet. __________

name date hour

CHAPTER 9 TEST

Find each square root. Use the square root table if necessary.

1. $\sqrt{100}$ ______
2. $\sqrt{961}$ ______
3. $\sqrt{5776}$ ______
4. $\sqrt{77}$ ______
5. $\sqrt{190}$ ______
6. $\sqrt{480}$ ______

Simplify where possible.

7. $\sqrt{8}$ ______
8. $\sqrt{27}$ ______
9. $\sqrt{50}$ ______
10. $\sqrt{128}$ ______
11. $\sqrt{75}$ ______
12. $\sqrt{12}+\sqrt{48}$ ______
13. $\sqrt{20}-\sqrt{45}$ ______
14. $\sqrt{8}+2\sqrt{18}$ ______
15. $3\sqrt{6}+\sqrt{14}$ ______
16. $3\sqrt{21}+\sqrt{14}$ ______
17. $3\sqrt{27x}-4\sqrt{48x}$ ______
18. $\sqrt{32x^2y^3}$ ______
19. $(6-\sqrt{5})(6+\sqrt{5})$ ______
20. $(\sqrt{2}+\sqrt{3})(\sqrt{2}-\sqrt{3})$ ______
21. $(1-\sqrt{3})^2$ ______
22. $(\sqrt{5}+\sqrt{6})^2$ ______
23. $\dfrac{4}{\sqrt{3}}$ ______
24. $\dfrac{6\sqrt{30}}{2\sqrt{5}}$ ______
25. $\dfrac{3\sqrt{2}}{\sqrt{6}}$ ______
26. $\dfrac{2}{3-\sqrt{7}}$ ______
27. $\dfrac{-3}{4-\sqrt{3}}$ ______

Solve each equation.

28. $\sqrt{x} = 4$ ________________

29. $\sqrt{x} + 5 = 0$ ________________

30. $\sqrt{x + 2} = 5$ ________________

31. $\sqrt{2x + 8} = 2\sqrt{x}$ ________________

32. $\sqrt{2x + 11} = \sqrt{x + 6}$ ________________

33. $\sqrt{x + 2} - x = 2$ ________________

10 Quadratic Equations

10.1 SOLVING QUADRATIC EQUATIONS BY THE SQUARE ROOT METHOD

OBJECTIVES

- Solve equations involving only a squared term
- Solve equations of the form $(2x + 3)^2 = 25$

In Chapter 5, we solved quadratic equations (second-degree equations) by factoring. However, not all quadratic equations can be solved by factoring. To solve equations like

$$(x - 3)^2 = 16,$$

where the square of a binomial is equal to some number, we can use square roots.

If two positive numbers are equal, then they must have the same square roots or square roots which are additive inverses of each other. That is, if b is a positive number and

$$\text{if } a^2 = b, \text{ then } a = \sqrt{b} \text{ or } a = -\sqrt{b}.$$

Example 1 Solve the equations.

(a) $x^2 = 16$

By the statement above, if $x^2 = 16$, then

$$x = \sqrt{16} = 4 \quad \text{or} \quad x = -\sqrt{16} = -4.$$

(b) $p^2 = 9$

Taking square roots on both sides gives

$$p = 3 \quad \text{or} \quad p = -3.$$

(c) $z^2 = 5$

The solution is $z = \sqrt{5}$ or $z = -\sqrt{5}$.

(d) $m^2 = 8$

We have $m = \sqrt{8}$ or $m = -\sqrt{8}$. Since $\sqrt{8} = 2\sqrt{2}$, $m = 2\sqrt{2}$ or $m = -2\sqrt{2}$.

(e) $y^2 = -4$

Since -4 is a negative number, and since the square of a real number cannot be negative, there is no solution for this equation.

QUICK CHECKS

1. Solve each equation.

(a) $k^2 = 49$

(b) $b^2 = 11$

(c) $c^2 = 12$

(d) $x^2 = -9$

2. Solve each equation.

(a) $(m + 2)^2 = 36$

(b) $(p - 4)^2 = 3$

3. Solve.

$(2x - 5)^2 = 18$

4. Solve.

$(x + 8)^2 = -25$

Quick Check Answers

1. (a) $7, -7$ (b) $\sqrt{11}, -\sqrt{11}$ (c) $2\sqrt{3}, -2\sqrt{3}$ (d) no real number solutions
2. (a) $4, -8$ (b) $4 + \sqrt{3}$, $4 - \sqrt{3}$
3. $\frac{5 + 3\sqrt{2}}{2}, \frac{5 - 3\sqrt{2}}{2}$
4. no real number solutions

Work Quick Check 1 at the side.

We can solve the equation $(x - 3)^2 = 16$ in the same way we solved the equations in Example 1.

$$\text{If } (x - 3)^2 = 16, \text{ then } x - 3 = 4 \text{ or } x - 3 = -4.$$

From the last two equations we get

$$x = 7 \quad \text{or} \quad x = -1.$$

Check both answers in the original equation.

$$(7 - 3)^2 = 4^2 = 16 \text{ and } (-1 - 3)^2 = (-4)^2 = 16.$$

Both 7 and -1 are solutions.

Example 2 Solve $(x - 1)^2 = 6$.

Take the square root on both sides.

$$x - 1 = \sqrt{6} \quad \text{or} \quad x - 1 = -\sqrt{6}$$
$$x = 1 + \sqrt{6} \quad \text{or} \quad x = 1 - \sqrt{6}$$

Check: $(1 + \sqrt{6} - 1)^2 = (\sqrt{6})^2 = 6;$
$(1 - \sqrt{6} - 1)^2 = (-\sqrt{6})^2 = 6.$

The solutions are $1 + \sqrt{6}$ and $1 - \sqrt{6}$.

Work Quick Check 2 at the side.

Example 3 Solve the equation $(3r - 2)^2 = 27$.

Taking square roots gives

$$3r - 2 = \sqrt{27} \quad \text{or} \quad 3r - 2 = -\sqrt{27}.$$

Now simplify the radical: $\sqrt{27} = \sqrt{9 \cdot 3} = \sqrt{9} \cdot \sqrt{3} = 3\sqrt{3}$, so

$$3r - 2 = 3\sqrt{3} \quad \text{or} \quad 3r - 2 = -3\sqrt{3}$$
$$3r = 2 + 3\sqrt{3} \quad \text{or} \quad 3r = 2 - 3\sqrt{3}$$
$$r = \frac{2 + 3\sqrt{3}}{3} \qquad r = \frac{2 - 3\sqrt{3}}{3}.$$

The solutions are

$$\frac{2 + 3\sqrt{3}}{3} \quad \text{and} \quad \frac{2 - 3\sqrt{3}}{3}.$$

Work Quick Check 3 at the side.

Example 4 Solve $(x + 3)^2 = -9$.

The square root of -9 is not a real number. Hence there is no real number solution.

Work Quick Check 4 at the side.

name date hour

10.1 EXERCISES

Solve each equation by taking the square root of both sides. Express all radicals in simplest form. See Example 1.

1. $x^2 = 25$ ______
2. $x^2 = 100$ ______
3. $x^2 = 64$ ______
4. $x^2 = 81$ ______
5. $x^2 = 13$ ______
6. $x^2 = 7$ ______
7. $x^2 = 2$ ______
8. $x^2 = 6$ ______
9. $x^2 = 24$ ______
10. $x^2 = 27$ ______

Solve each equation by taking the square root of both sides. Express all square roots in simplest form. See Examples 2–4.

11. $(x-2)^2 = 16$ ______
12. $(r+4)^2 = 25$ ______
13. $(a+4)^2 = 10$ ______
14. $(r-3)^2 = 15$ ______
15. $(x-1)^2 = 32$ ______
16. $(y+5)^2 = 28$ ______
17. $(2m-1)^2 = 9$ ______
18. $(3y-7)^2 = 4$ ______
19. $(3z+5)^2 = 9$ ______
20. $(2y-7)^2 = 49$ ______
21. $(6m-2)^2 = 121$ ______
22. $(7m-10)^2 = 144$ ______
23. $(2a-5)^2 = 30$ ______
24. $(2y+3)^2 = 45$ ______
25. $(3p-1)^2 = 18$ ______
26. $(5r-6)^2 = 75$ ______
27. $(2k-5)^2 = 98$ ______
28. $(4x-1)^2 = 48$ ______
29. $(3m+4)^2 = 8$ ______
30. $(5y-3)^2 = 50$ ______

31. One expert at marksmanship can hold a silver dollar at forehead level, drop it, draw his gun, and shoot the coin as it passes waist level. The distance traveled by a falling object is given by $d = 16t^2$, where d is the distance the object falls in t seconds. If the coin falls about 4 feet, estimate the time that elapses between the dropping of the coin and the shot. ______________

10.2 SOLVING QUADRATIC EQUATIONS BY COMPLETING THE SQUARE

OBJECTIVES

- Solve quadratic equations by completing the square

Consider the equation

$$x^2 + 6x + 7 = 0.$$

In the preceding section, we learned to solve equations of the type

$$(x + 3)^2 = 2.$$

If we can rewrite the equation

$$x^2 + 6x + 7 = 0$$

in a form like

$$(x + 3)^2 = 2,$$

we can solve it by taking square roots of both sides. We show how to do this in the following example.

Example 1 Solve $x^2 + 6x + 7 = 0$.

To start, add -7 to both sides of the equation to get

$$x^2 + 6x = -7.$$

We want the quantity on the left-hand side of $x^2 + 6x = -7$ to be a perfect square trinomial. Note that $x^2 + 6x + 9$ is a perfect square, since

$$x^2 + 6x + 9 = (x + 3)^2.$$

Hence, if we add 9 to both sides, we will have an equation with a perfect square trinomial on the left-hand side, as desired.

$$x^2 + 6x + 9 = -7 + 9$$
$$(x + 3)^2 = 2$$

Now take the square root of both sides of the equation to complete the solution.

$$x + 3 = \sqrt{2} \quad \text{or} \quad x + 3 = -\sqrt{2}$$
$$x = \sqrt{2} - 3 \quad \text{or} \quad x = -\sqrt{2} - 3.$$

The solutions of the original equation are $\sqrt{2} - 3$ and $-\sqrt{2} - 3$. Verify this by substituting $\sqrt{2} - 3$ and $-\sqrt{2} - 3$ for x in the equation.

The process of changing the form of the equation in Example 1 from

$$x^2 + 6x + 7 = 0 \quad \text{to} \quad (x + 3)^2 = 2$$

is called **completing the square**.

Work Quick Check 1 at the side.

QUICK CHECKS

1. Find the number that should be added to

$$y^2 + 14y$$

to make it a perfect square trinomial.

Example 2 Find the solutions of the quadratic equation

$$m^2 - 5m = 2.$$

Quick Check Answers
1. 49

2. Solve by completing the square.

(a) $x^2 - 14x = -40$

(b) $a^2 + 3a = 1$

(c) $x^2 + 5x + 3 = 0$

We need to add a suitable number to both sides of the equation to make the left side a perfect square. To decide which number to use, first make sure that the coefficient of the m^2 term is 1, and if it is not, multiply both sides by an appropriate number to make it 1. Next, take half of the coefficient of m and square it. Then add this squared number to both sides of the equation. (We do this because the coefficient of the middle term of the perfect square trinomial $a^2 + 2ab + b^2$ is the product of ab times 2.)

In the equation above, the coefficient of m is -5, and half of -5 is $-5/2$. If we square $-5/2$, we get $25/4$, which is the number to be added to both sides.

$$m^2 - 5m + \frac{25}{4} = 2 + \frac{25}{4}$$

The trinomial $m^2 - 5m + 25/4$ is a perfect square trinomial.

$$m^2 - 5m + \frac{25}{4} = \left(m - \frac{5}{2}\right)^2.$$

Thus,

$$\left(m - \frac{5}{2}\right)^2 = 2 + \frac{25}{4}$$
$$= \frac{33}{4}.$$

Now take square roots.

$$m - \frac{5}{2} = \sqrt{\frac{33}{4}} \quad \text{or} \quad m - \frac{5}{2} = -\sqrt{\frac{33}{4}}$$
$$m = \frac{5}{2} + \sqrt{\frac{33}{4}} \qquad\qquad m = \frac{5}{2} - \sqrt{\frac{33}{4}}$$

Simplify the radical: $\sqrt{\frac{33}{4}} = \frac{\sqrt{33}}{\sqrt{4}} = \frac{\sqrt{33}}{2}$.

The solutions are

$$\frac{5 + \sqrt{33}}{2} \quad \text{and} \quad \frac{5 - \sqrt{33}}{2}.$$

Work Quick Check 2 at the side.

Example 3 Solve the equation $2x^2 - 7x = 9$.

To begin, we need the coefficient 1 for the x^2 term. Multiply both sides of the equation by $1/2$. The result is

$$x^2 - \frac{7}{2}x = \frac{9}{2}.$$

Now take half the coefficient of x and square it. Half of $-7/2$ is $-7/4$, and $-7/4$ squared is $49/16$. Add $49/16$ to both sides of the equation, and write the left side as a perfect square.

Quick Check Answers

2. (a) 4, 10 (b) $\frac{-3 + \sqrt{13}}{2}$, $\frac{-3 - \sqrt{13}}{2}$ (c) $\frac{-5 + \sqrt{13}}{2}$, $\frac{-5 - \sqrt{13}}{2}$

$$x^2 - \frac{7}{2}x + \frac{49}{16} = \frac{9}{2} + \frac{49}{16}$$

$$\left(x - \frac{7}{4}\right)^2 = \frac{121}{16}$$

Take the square root of both sides.

$$x - \frac{7}{4} = \sqrt{\frac{121}{16}} \quad \text{or} \quad x - \frac{7}{4} = -\sqrt{\frac{121}{16}}$$

Since $\sqrt{\frac{121}{16}} = \frac{11}{4}$, we can write

$$x - \frac{7}{4} = \frac{11}{4} \quad \text{or} \quad x - \frac{7}{4} = \frac{-11}{4}$$

$$x = \frac{18}{4} \qquad\qquad x = \frac{-4}{4}$$

$$= \frac{9}{2} \qquad\qquad = -1.$$

The solutions are 9/2 and −1. Since the solutions are rational numbers, the original equation could have been factored as follows.

$$2x^2 - 7x - 9 = (2x - 9)(x + 1).$$

Work Quick Check 3 at the side.

3. Solve by completing the square.

$3x^2 + 5x - 2 = 0$

Example 4 Use the method of completing the square to solve the equation

$$4p^2 + 8p + 5 = 0.$$

First multiply both sides by 1/4 to get the coefficient 1 for the p^2 term. The result is

$$p^2 + 2p + \frac{5}{4} = 0.$$

Add −5/4 to both sides, which gives

$$p^2 + 2p = \frac{-5}{4}.$$

The coefficient of p is 2. Take half of 2, square the result, and add it to both sides. The left-hand side can then be written as a perfect square.

$$p^2 + 2p + 1 = \frac{-5}{4} + 1$$

$$(p + 1)^2 = \frac{-1}{4}$$

At this point, we should take the square root of both sides. However, in the real number system, the square root of −1/4 has no meaning. This equation has no solution in the real number system.

Work Quick Check 4 at the side.

4. Solve by completing the square.

$a^2 + 3a + 10 = 0$

Quick Check Answers

3. $-2, \frac{1}{3}$

4. no real number solutions

Summary: Completing the Square in a Quadratic Equation

1. If the coefficient of the squared term is 1, proceed to step 2. If the coefficient of the squared term is not 1, but some other number a, multiply both sides of the equation by the reciprocal of a, $1/a$. This gives an equation which has 1 as coefficient of the squared term.
2. Make sure all terms with variables are on one side of the equals sign, and all numbers are on the other side.
3. Take half the coefficient of x and square it. Add the square to both sides of the equation. The side containing the variables can now be written as a perfect square.
4. Take the square root of both sides. The solutions are determined by solving the two resulting equations.

name date hour

10.2 EXERCISES

Find the number that should be added to each of the following to make it a perfect square. See Example 1.

1. $x^2 + 2x$ ________

2. $y^2 - 4y$ ________

3. $x^2 + 18x$ ________

4. $m^2 - 3m$ ________

5. $z^2 + 9z$ ________

6. $p^2 + 22p$ ________

7. $x^2 + 14x$ ________

8. $r^2 + 7r$ ________

9. $y^2 + 5y$ ________

10. $q^2 - 8q$ ________

Solve each of the following equations by completing the square. See Examples 2–4. You may have to simplify first.

11. $x^2 + 4x = -3$ ________

12. $y^2 - 4y = 0$ ________

13. $a^2 + 2a = 5$ ________

14. $m^2 + 4m = 12$ ________

15. $z^2 + 6z = -8$ ________

16. $q^2 - 8q = -16$ ________

17. $x^2 - 6x + 1 = 0$ ________

18. $b^2 - 2b - 2 = 0$ ________

19. $c^2 + 3c = 2$ ________

20. $k^2 + 5k - 3 = 0$ ________

21. $2m^2 + 4m = -7$ ________

22. $3y^2 - 9y + 5 = 0$ ________

23. $6q^2 - 8q + 3 = 0$ ________

24. $4y^2 + 4y - 3 = 0$ ________

25. $-x^2 + 6x = 4$ ________

26. $3y^2 - 6y - 2 = 0$ ________

27. $2m^2 - 4m - 5 = 0$ ________

28. $-x^2 + 4 = 2x$ ________

29. $3x^2 - 2x = 1$ ________

30. $-x^2 - 4 = 2x$ ________

31. $m^2 - 4m + 8 = 6m$ ____________

32. $2z^2 = 8z + 5 - 4z^2$ ____________

33. $3r^2 - 2 = 6r + 3$ ____________

34. $4p - 3 = p^2 + 2p$ ____________

35. $(x + 1)(x + 3) = 2$ ____________

36. $(x - 3)(x + 1) = 1$ ____________

10.3 SOLVING QUADRATIC EQUATIONS BY THE QUADRATIC FORMULA

OBJECTIVES

- Identify the letters a, b, and c in a quadratic equation
- Use the quadratic formula to solve quadratic equations
- Solve quadratic equations involving fractions

Completing the square can be used to solve any quadratic equation, but the method is not very handy. In this section, we will work out a general formula, the quadratic formula, which gives the solution for any quadratic equation.

To get the quadratic formula, we start with the general form of a quadratic equation,

$$ax^2 + bx + c = 0, \qquad a \neq 0.$$

The restriction $a \neq 0$ is important to make sure that the equation is in fact quadratic. If $a = 0$, then the equation becomes $0x^2 + bx + c = 0$, or $bx + c = 0$. This is a linear, not a quadratic, equation.

Example 1 Match the coefficients of each of the following quadratic equations with the letters a, b, and c of the general quadratic equation

$$ax^2 + bx + c = 0.$$

(a) $2x^2 + 3x - 5 = 0$

In this example $a = 2$, $b = 3$, and $c = -5$.

(b) $-x^2 + 2 = 6x$

First rewrite the equation with 0 on one side to match the general form $ax^2 + bx + c = 0$.

$$-x^2 + 2 = 6x$$
$$-x^2 - 6x + 2 = 0$$

Now we can identify $a = -1$, $b = -6$, and $c = 2$.

(c) $2(x + 3)(x - 1) = 0$

Here we must multiply first.

$$2(x + 3)(x - 1) = 2(x^2 + 2x - 3)$$
$$= 2x^2 + 4x - 6.$$

Thus, the equation becomes

$$2x^2 + 4x - 6 = 0$$

and $a = 2$, $b = 4$, $c = -6$.

Work Quick Check 1 at the side.

QUICK CHECKS

1. Match the coefficients of each of the following quadratic equations with the letters a, b, and c of the general quadratic equation $ax^2 + bx + c = 0$.

(a) $5x^2 + 2x - 1 = 0$

(b) $3x^2 = x - 2$

(c) $x(x + 5) = 4$

We solve the equation $ax^2 + bx + c = 0$ by completing the square. First, we need the coefficient 1 for the x^2 term. To get this, multiply both sides by $1/a$, which gives

$$x^2 + \frac{b}{a}x + \frac{c}{a} = 0.$$

Quick Check Answers

1. (a) $a = 5, b = 2, c = -1$
(b) $a = 3, b = -1, c = 2$
(c) $a = 1, b = 5, c = -4$

Next add $-c/a$ to both sides, to get

$$x^2 + \frac{b}{a}x = -\frac{c}{a}$$

Now complete the square on the left. To do this, take half the coefficient of x, that is, $b/2a$. Square $b/2a$ to get $b^2/4a^2$. Next add $b^2/4a^2$ to both sides of the equation.

$$x^2 + \frac{b}{a}x + \frac{b^2}{4a^2} = \frac{-c}{a} + \frac{b^2}{4a^2}$$

Rewrite the left-hand side as a perfect square.

$$\left(x + \frac{b}{2a}\right)^2 = \frac{-c}{a} + \frac{b^2}{4a^2}$$

Now simplify the right-hand side of the equation.

$$\begin{aligned}\left(x + \frac{b}{2a}\right)^2 &= \frac{b^2}{4a^2} + \frac{-c}{a} \\ &= \frac{b^2}{4a^2} + \frac{-4ac}{4a^2} \\ &= \frac{b^2 - 4ac}{4a^2}\end{aligned}$$

Take the square root of both sides.

$$x + \frac{b}{2a} = \sqrt{\frac{b^2 - 4ac}{4a^2}} \quad \text{or} \quad x + \frac{b}{2a} = -\sqrt{\frac{b^2 - 4ac}{4a^2}}$$

Simplify the radical.

$$\sqrt{\frac{b^2 - 4ac}{4a^2}} = \frac{\sqrt{b^2 - 4ac}}{\sqrt{4a^2}} = \frac{\sqrt{b^2 - 4ac}}{2a}$$

Now we can write the solutions as follows:

$$\begin{aligned} x + \frac{b}{2a} &= \frac{\sqrt{b^2 - 4ac}}{2a} & \text{or} \quad x + \frac{b}{2a} &= \frac{-\sqrt{b^2 - 4ac}}{2a} \\ x &= \frac{-b}{2a} + \frac{\sqrt{b^2 - 4ac}}{2a} & x &= \frac{-b}{2a} - \frac{\sqrt{b^2 - 4ac}}{2a} \\ x &= \frac{-b + \sqrt{b^2 - 4ac}}{2a} & x &= \frac{-b - \sqrt{b^2 - 4ac}}{2a}. \end{aligned}$$

Hence, the solutions of the general quadratic equation $ax^2 + bx + c = 0$ $(a \neq 0)$ are

$$\frac{-b + \sqrt{b^2 - 4ac}}{2a} \quad \text{and} \quad \frac{-b - \sqrt{b^2 - 4ac}}{2a}.$$

For convenience, the solutions are often expressed in compact form by using the symbol ± (read "plus or minus"). The result is called the **quadratic formula.**

$$x = \frac{-b \pm \sqrt{b^2 - 4ac}}{2a}$$

Example 2 Use the quadratic formula to solve $2x^2 - 7x - 9 = 0$.

To begin, match the coefficients of the variables with the letter symbols of the general quadratic equation

$$ax^2 + bx + c = 0.$$

Here, $a = 2$, $b = -7$, and $c = -9$. Substitute these numbers into the quadratic formula and simplify the result.

$$x = \frac{-b \pm \sqrt{b^2 - 4ac}}{2a}$$

$$= \frac{-(-7) \pm \sqrt{(-7)^2 - 4(2)(-9)}}{2(2)}$$

$$= \frac{7 \pm \sqrt{49 + 72}}{4}$$

$$= \frac{7 \pm \sqrt{121}}{4}$$

Since $\sqrt{121} = 11$,

$$x = \frac{7 \pm 11}{4}.$$

To write the two solutions separately, first take the plus sign:

$$x = \frac{7 + 11}{4} = \frac{18}{4} = \frac{9}{2}.$$

Then take the minus sign:

$$x = \frac{7 - 11}{4} = \frac{-4}{4} = -1.$$

The solutions of $2x^2 - 7x - 9 = 0$ are $9/2$ and -1.

Work Quick Check 2 at the side.

2. Solve using the quadratic formula.

(a) $2x^2 + 3x - 5 = 0$

(b) $6p^2 + p = 1$

(c) $81k^2 - 36k + 4 = 0$

Example 3 Solve $x^2 = 2x + 1$.

To find a, b, and c, add $-2x - 1$ to both sides of the equation to get

$$x^2 - 2x - 1 = 0.$$

Then $a = 1$, $b = -2$, and $c = -1$. The solution is found by substituting these values into the quadratic formula.

$$x = \frac{-b \pm \sqrt{b^2 - 4ac}}{2a}$$

$$= \frac{-(-2) \pm \sqrt{(-2)^2 - 4(1)(-1)}}{2(1)}$$

$$= \frac{2 \pm \sqrt{4 + 4}}{2}$$

$$= \frac{2 \pm \sqrt{8}}{2}$$

Quick Check Answers

2. (a) $1, -\frac{5}{2}$ (b) $-\frac{1}{2}, \frac{1}{3}$ (c) $\frac{2}{9}$

3. Solve using the quadratic formula.

(a) $-y^2 = 8y + 1$

(b) $x^2 - 2x = 4$

(c) $4m^2 - 12m + 2 = 0$

Since $\sqrt{8} = \sqrt{4 \cdot 2} = \sqrt{4} \cdot \sqrt{2} = 2\sqrt{2}$,

$$x = \frac{2 \pm 2\sqrt{2}}{2}.$$

And since $2 \pm 2\sqrt{2}$ factors as $2(1 \pm \sqrt{2})$,

$$x = \frac{2(1 \pm \sqrt{2})}{2} = 1 \pm \sqrt{2}.$$

Therefore, the two solutions of this equation are

$$1 + \sqrt{2} \quad \text{and} \quad 1 - \sqrt{2}.$$

Work Quick Check 3 at the side.

Example 4 Solve the equation $x^2 + 5x + 8 = 0$.

Here $a = 1$, $b = 5$, and $c = 8$. Substitute into the quadratic formula and simplify the result.

$$x = \frac{-5 \pm \sqrt{5^2 - 4(1)(8)}}{2(1)}$$

$$= \frac{-5 \pm \sqrt{25 - 32}}{2}$$

$$= \frac{-5 \pm \sqrt{-7}}{2}$$

The radical $\sqrt{-7}$ is not a real number, so the equation has no real number solutions.

Work Quick Check 4 at the side.

4. Solve using the quadratic formula.

(a) $3x^2 - 4x + 2 = 0$

(b) $4k^2 - 20k + 26 = 0$

Example 5 Solve the equation

$$\frac{1}{10}t^2 = \frac{2}{5} - \frac{1}{2}t.$$

To eliminate the denominators, multiply both sides of the equation by the common denominator 10.

$$10\left(\frac{1}{10}t^2\right) = 10\left(\frac{2}{5} - \frac{1}{2}t\right)$$

$$t^2 = 4 - 5t$$

Add $-4 + 5t$ to both sides of the equation to get

$$t^2 + 5t - 4 = 0.$$

In this form, we can identify $a = 1$, $b = 5$, and $c = -4$. Use the quadratic formula to complete the solution.

Quick Check Answers

3. (a) $-4 + \sqrt{15}$, $-4 - \sqrt{15}$ (b) $1 + \sqrt{5}$, $1 - \sqrt{5}$ (c) $\frac{3 + \sqrt{7}}{2}$, $\frac{3 - \sqrt{7}}{2}$

4. (a) no real number solutions (b) no real number solutions

$$t = \frac{-5 \pm \sqrt{25 - 4(1)(-4)}}{2(1)}$$

$$= \frac{-5 \pm \sqrt{25 + 16}}{2}$$

$$= \frac{-5 \pm \sqrt{41}}{2}$$

The solutions are

$$\frac{-5 + \sqrt{41}}{2} \quad \text{and} \quad \frac{-5 - \sqrt{41}}{2}.$$

Work Quick Check 5 at the side.

5. Solve using the quadratic formula.

(a) $\frac{1}{6}a^2 = \frac{1}{3} + \frac{5}{6}a$

(b) $m^2 = \frac{2}{3}m + \frac{4}{9}$

(c) $\frac{2}{7}r^2 - r = \frac{8}{7}$

Quick Check Answers

5. (a) $\frac{5 + \sqrt{33}}{2}, \frac{5 - \sqrt{33}}{2}$

(b) $\frac{1 + \sqrt{5}}{3}, \frac{1 - \sqrt{5}}{3}$

(c) $\frac{7 + \sqrt{113}}{4}, \frac{7 - \sqrt{113}}{4}$

WORK SPACE

name date hour

10.3 EXERCISES

For each equation, identify the letters a, b, and c of the general quadratic equation, $ax^2 + bx + c = 0$. Do not try to solve. See Example 1.

1. $3x^2 + 4x - 8 = 0$ ______
2. $9x^2 + 2x - 3 = 0$ ______
3. $-8x^2 - 2x - 3 = 0$ ______
4. $-2x^2 + 3x - 8 = 0$ ______
5. $2x^2 = 3x - 2$ ______
6. $9x^2 - 2 = 4x$ ______
7. $x^2 = 2$ ______
8. $x^2 - 3 = 0$ ______
9. $3x^2 - 8x = 0$ ______
10. $5x^2 = 2x$ ______
11. $(x - 3)(x + 4) = 0$ ______
12. $(x + 6)^2 = 3$ ______
13. $9(x - 1)(x + 2) = 8$ ______
14. $(3x - 1)(2x + 5) = x(x - 1)$ ______

Use the quadratic formula to solve each equation. Write all radicals in simplified form. Reduce answers to lowest terms. See Examples 2 and 3.

15. $x^2 + 2x - 2 = 0$ ______
16. $6x^2 + 6x + 1 = 0$ ______
17. $x^2 + 4x + 4 = 0$ ______
18. $3x^2 - 5x + 1 = 0$ ______
19. $x^2 = 13 - 12x$ ______
20. $x^2 = 8x + 9$ ______
21. $2x^2 + 12x + 5 = 0$ ______
22. $x^2 = -19 + 20x$ ______
23. $5x^2 + 4x - 1 = 0$ ______
24. $5x^2 + x - 1 = 0$ ______
25. $2x^2 = 3x + 5$ ______
26. $-2x^2 + 7x = -30$ ______
27. $x^2 + 6x + 9 = 0$ ______
28. $x^2 - 2x + 1 = 0$ ______

29. $5x^2 + 5x = 0$ __________

30. $x^2 = 20$ __________

31. $2x^2 + 2x + 4 = 4 - 2x$ __________

32. $3x^2 - 4x + 3 = 8x - 1$ __________

Use the quadratic formula to solve each equation. See Examples 4 and 5.

33. $2x^2 + x + 7 = 0$ __________

34. $x^2 + x + 1 = 0$ __________

35. $2x^2 = 3x - 2$ __________

36. $x^2 = 5x - 20$ __________

37. $\frac{1}{2}x^2 = -\frac{1}{6}x + 1$ __________

38. $\frac{3}{2}x^2 - x = \frac{4}{3}$ __________

39. $\frac{2}{3}x^2 - \frac{4}{9}x - \frac{1}{3} = 0$ __________

40. $-\frac{2}{5}x^2 + \frac{3}{5}x = -1$ __________

OBJECTIVES

- Identify parabolas
- Graph quadratic equations

In Chapter 7, we graphed straight lines to represent the solutions of linear equations. Now we investigate the graphs of quadratic equations in two variables which have the form $y = ax^2 + bx + c$. Perhaps the simplest such quadratic equation is

$$y = x^2$$

(which is the same as $y = 1x^2 + 0x + 0$). The graph of this equation cannot be a straight line since only linear equations of the form

$$ax + by + c = 0$$

have graphs which are straight lines. However, we can graph $y = x^2$ in much the same way as we graphed straight lines, by selecting values for x and then finding the corresponding y-values.

Example 1 Graph $y = x^2$.

If $x = 2$ in the equation $y = x^2$, then

$$y = 2^2 = 4.$$

Thus the point (2, 4) belongs to the graph of $y = x^2$. (Recall that in an ordered pair such as (2, 4), the x-value comes first and the y-value second.) We can complete a chart showing values of y for some values of x (which we choose arbitrarily).

Equation	Ordered pairs	
$y = x^2$	$(-3, 9)$	$(1, 1)$
	$(-2, 4)$	$(2, 4)$
	$(-1, 1)$	$(3, 9)$
	$(0, 0)$	

If we plot these points on a coordinate system and draw a smooth curve through them, we get the graph shown in Figure 1 called a **parabola**.

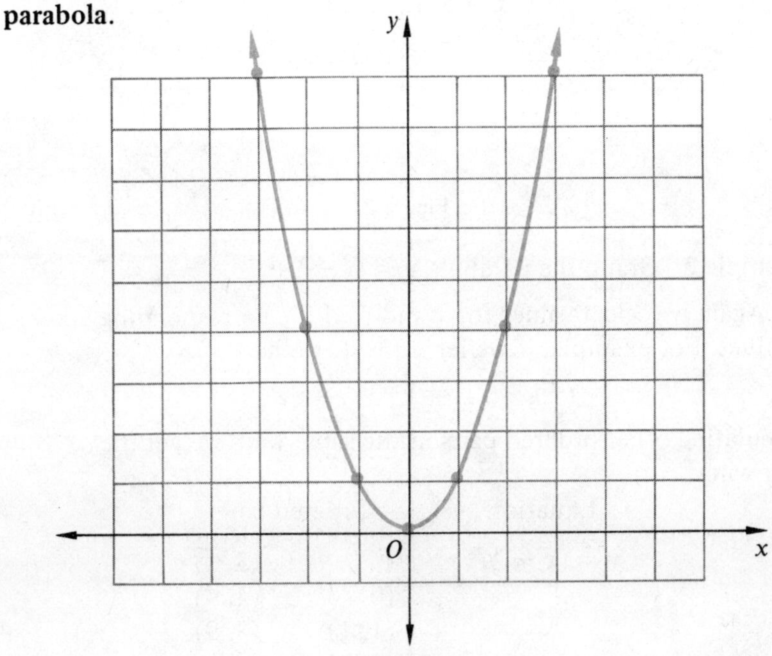

Figure 1

Every quadratic equation of the form

$$y = ax^2 + bx + c$$

has a graph which is a parabola. Because of its many useful properties, the parabola occurs frequently in real-life applications. For example, if an object is thrown into the air, the path that the object follows is a parabola (discounting wind resistance). The cross-sections of radar, spotlight, and telescope reflectors also form parabolas.

Example 2 Graph the parabola $y = -x^2$.

We could select values for x and then find the corresponding y-values. But note that for a given x-value, the y-value will be the negative of the corresponding y-value of the parabola $y = x^2$ discussed above. Hence this new parabola has the same shape as the one in the preceding figure, but is turned in the opposite direction. We say it opens downward (the graph in Figure 1 opens upward), as shown in Figure 2.

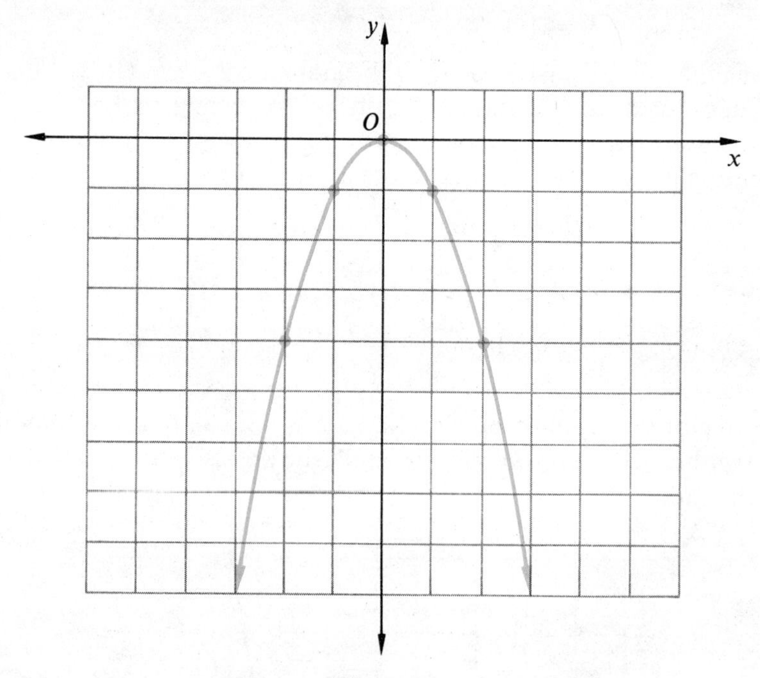

Figure 2

Example 3 Graph the parabola $y = (x - 2)^2$.

Again we select values for x and find the corresponding y-values. For example, if we let $x = -1$, we have

$$y = (-1 - 2)^2 = (-3)^2 = 9.$$

Calculating other ordered pairs in the same way, we get the following:

Equation	Ordered pairs	
$y = (x - 2)^2$	$(-1, 9)$	$(3, 1)$
	$(0, 4)$	$(4, 4)$
	$(1, 1)$	$(5, 9)$
	$(2, 0)$	

Plotting these points and joining them gives the graph shown in Figure 3. Note that the parabola of the figure has the same shape as our original parabola ($y = x^2$), but is shifted two units to the right.

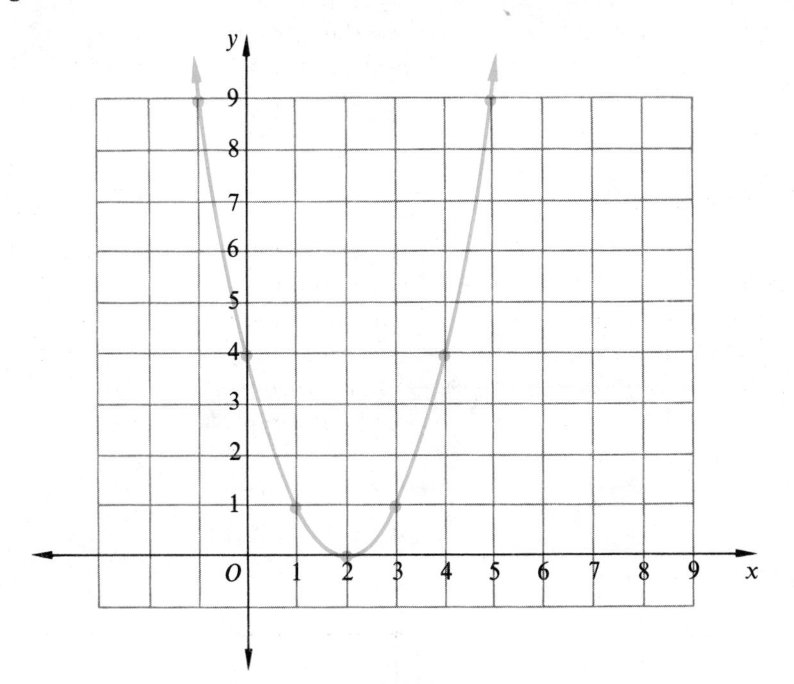

Figure 3

Work Quick Check 1 at the side.

Example 4 Graph the parabola $y = -3(x + 4)^2$.

If we make a table of values and plot the points, we get the graph shown in Figure 4. The graph opens downward, is shifted four units to the left, and is narrower than the graph of $y = x^2$.

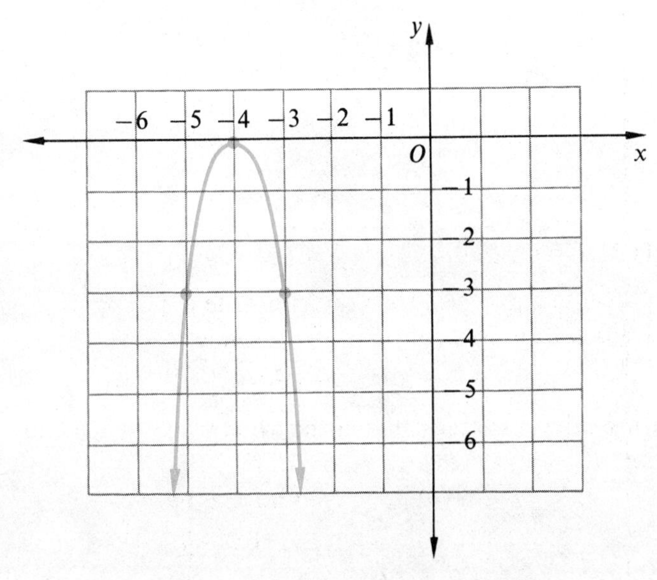

Figure 4

Work Quick Check 2 at the side.

QUICK CHECKS

1. Graph the parabolas.

(a) $y = (x - 3)^2$

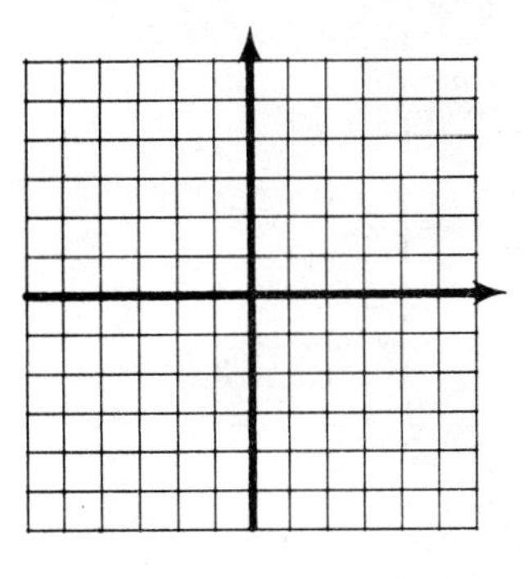

(b) $y = (x + 3)^2$

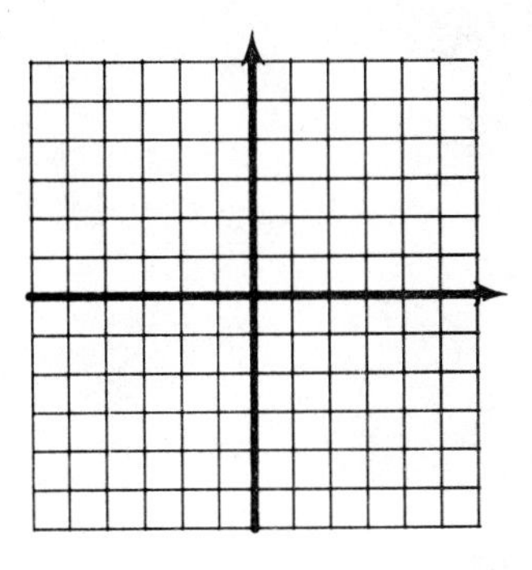

2. Graph the parabola $y = -2(x + 1)^2$.

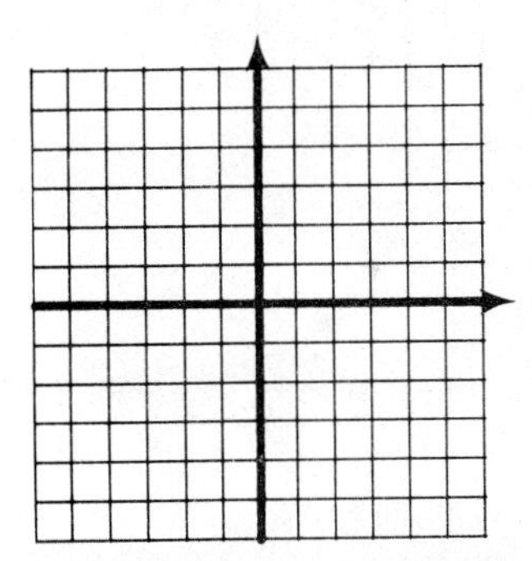

Quick Check Answers

1. (a) (b)

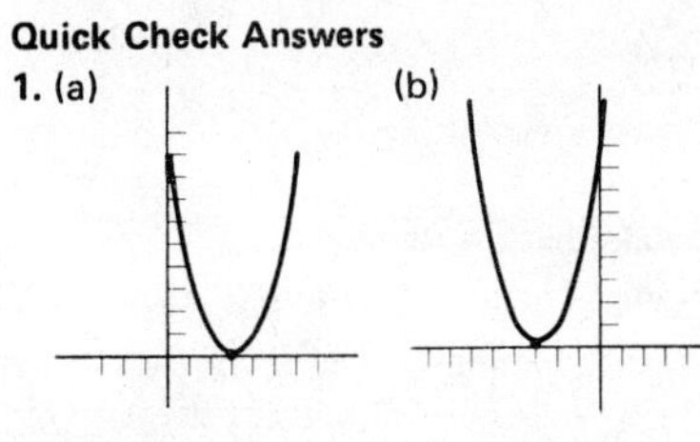

2.

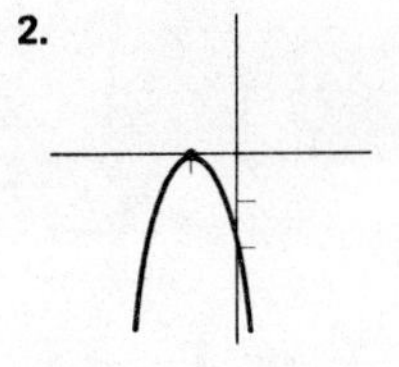

3. Graph the parabolas.

(a) $y = x^2 + 2$

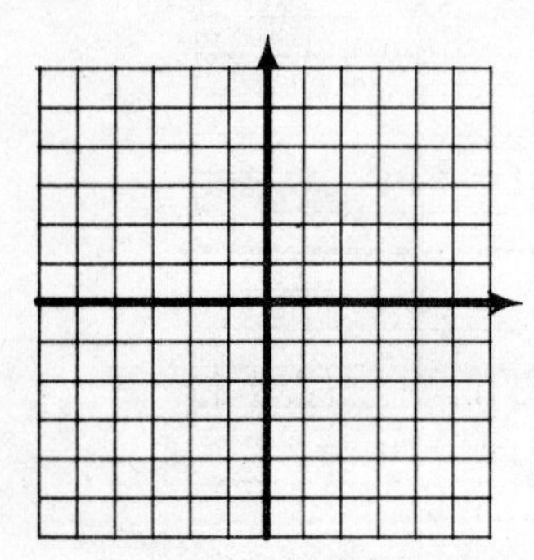

(b) $y = -x^2 + 2$

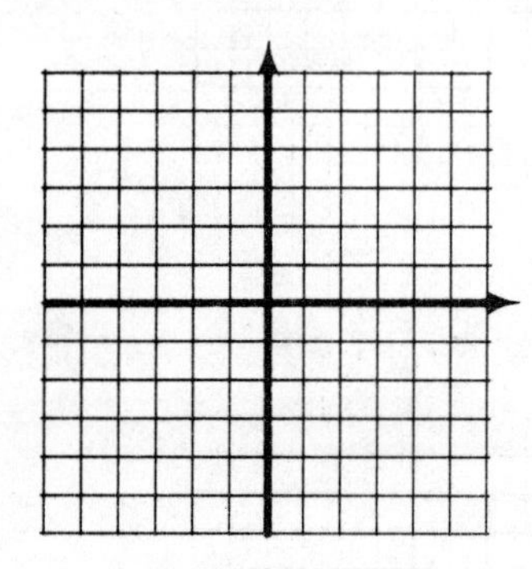

Example 5 Graph the parabola $y = x^2 - 3$.

The graph is shown in Figure 5. Note that this time the graph is shifted three units downward, as compared to the graph of $y = x^2$.

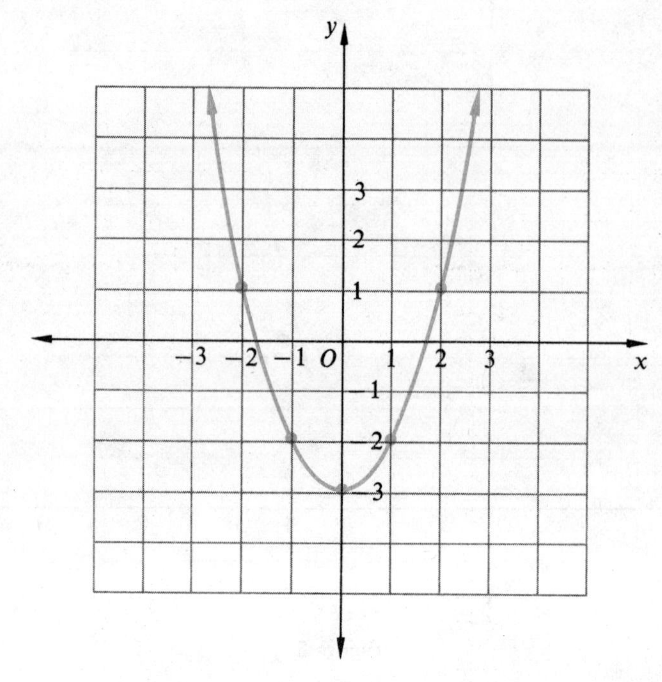

Figure 5

Work Quick Check 3 at the side.

Quick Check Answers

3. (a)

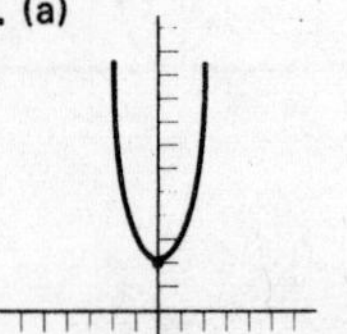

(b)

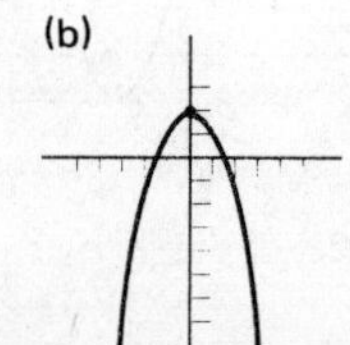

name date hour

10.4 EXERCISES

Sketch the graph of each equation. See Examples 1–5.

1. $y = 2x^2$

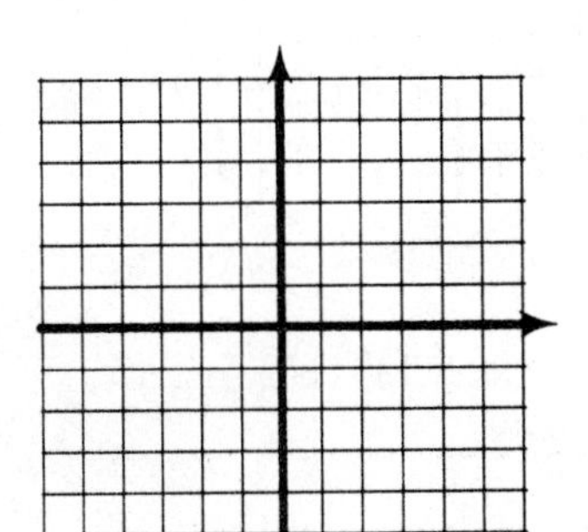

2. $y = -2x^2$

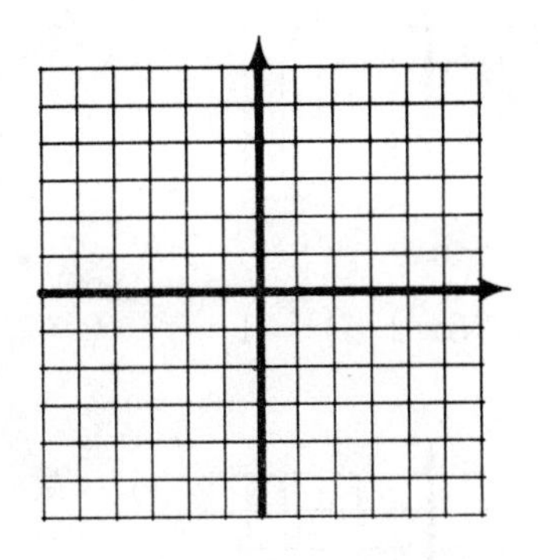

3. $y = (x + 1)^2$

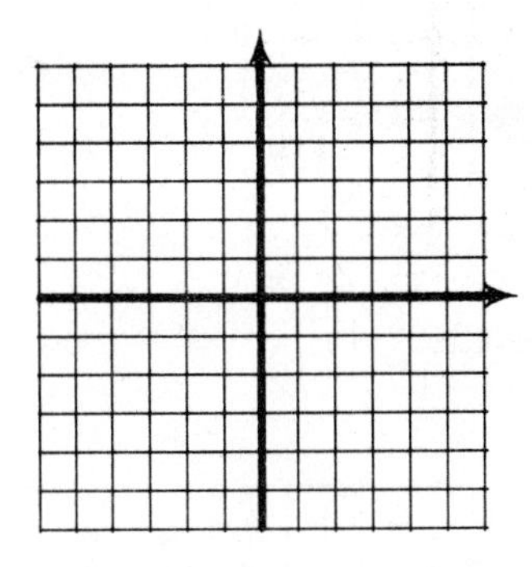

4. $y = (x - 2)^2$

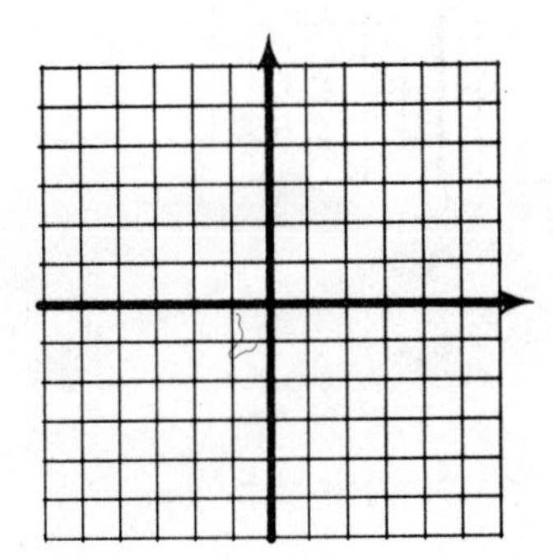

5. $y = -(x + 1)^2$

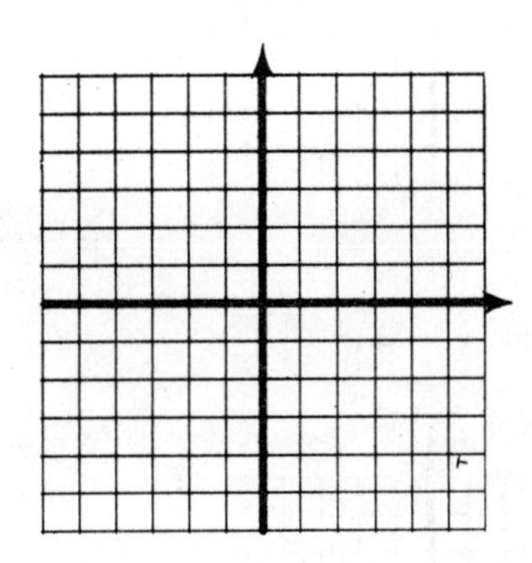

6. $y = -(x - 2)^2$

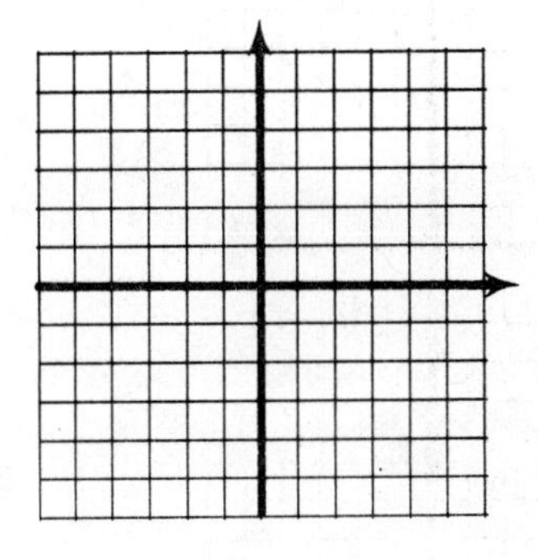

7. $y = x^2 + 1$

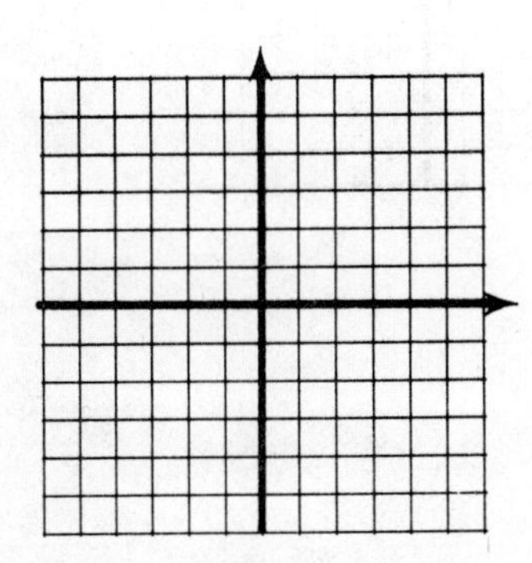

8. $y = -x^2 - 2$

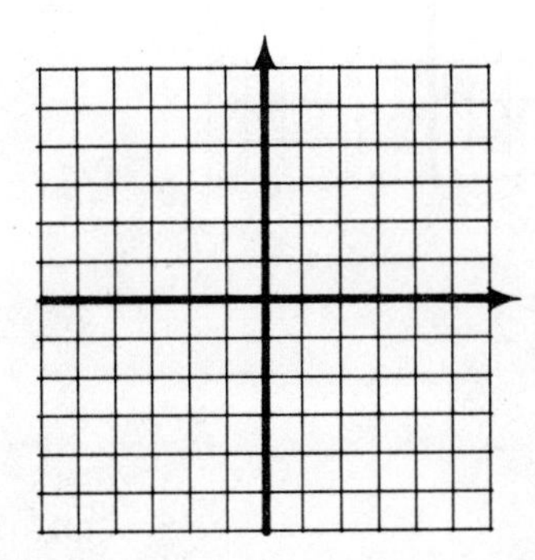

9. $3y = -2x^2$

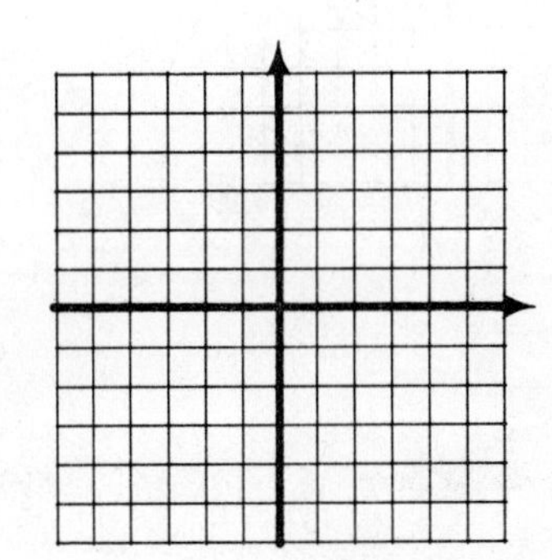

10. $2y = 3x^2$

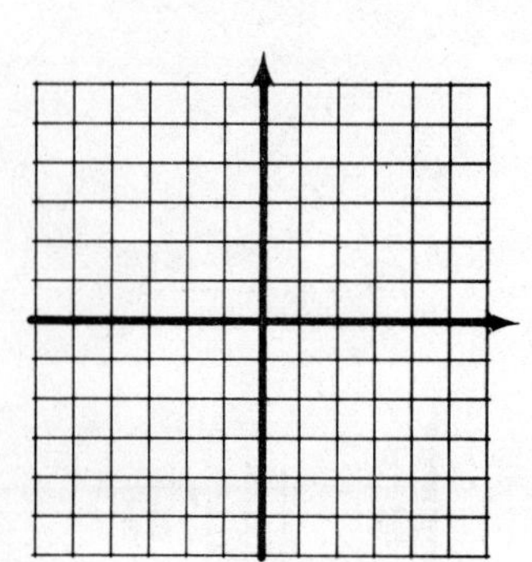

11. $2y = -x^2 + 1$

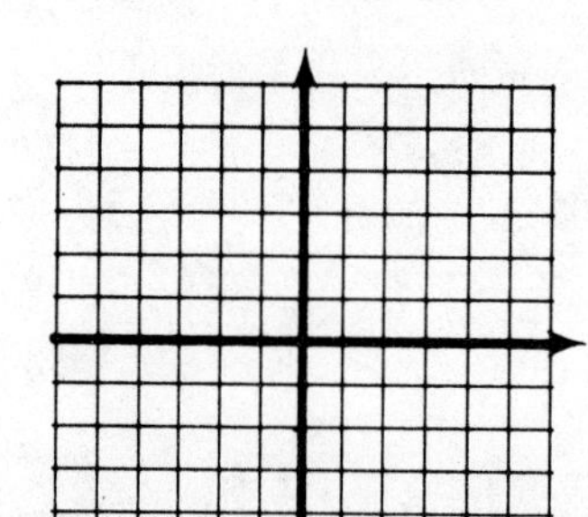

12. $3y = -x^2 + 3$

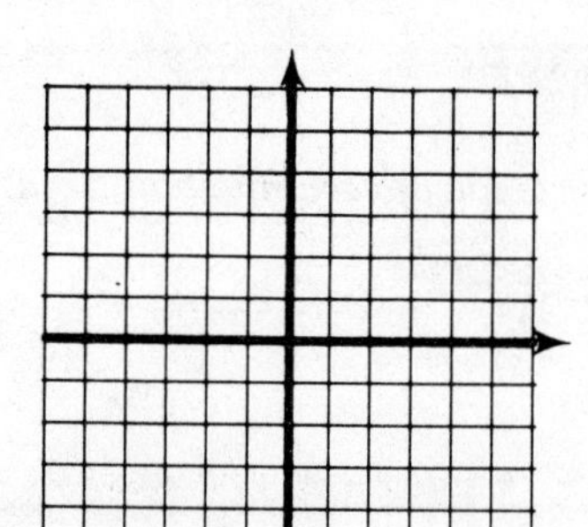

13. $y = (2x - 1)^2$

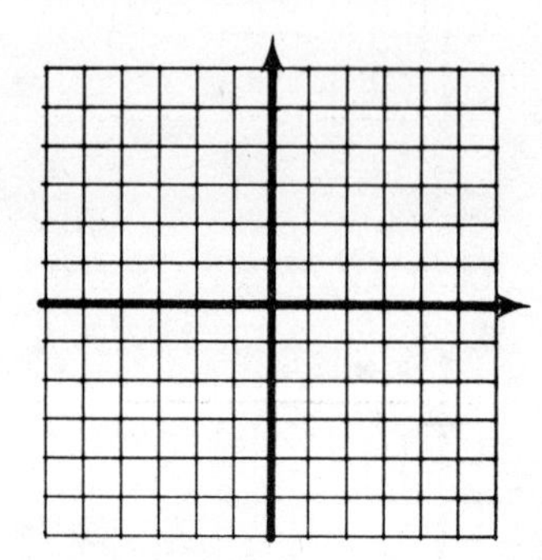

14. $y = (8x + 5)^2$

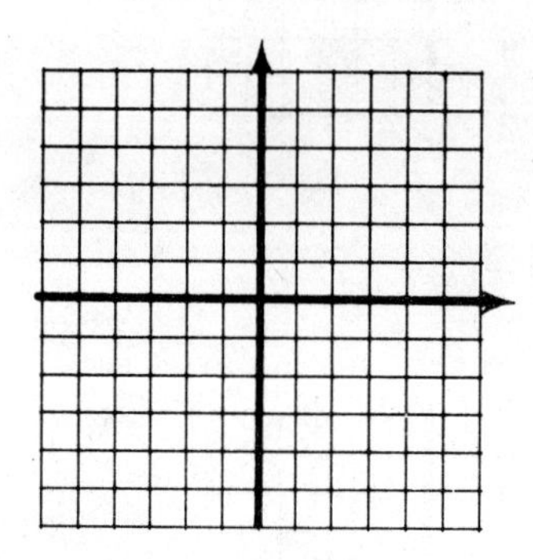

15. $y = (x + 1)^2 + 2$

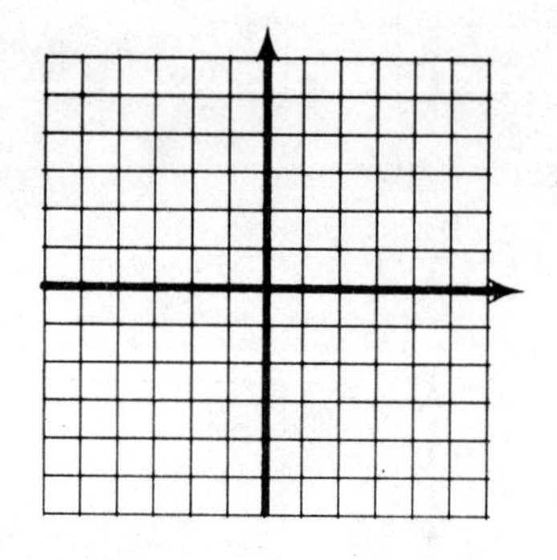

16. $y = (x - 2)^2 - 1$

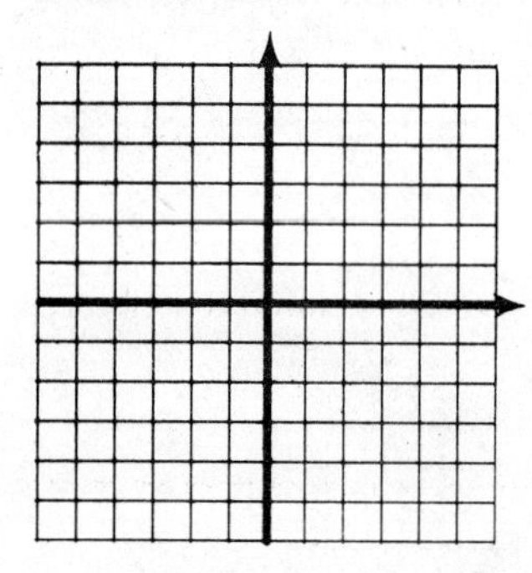

17. $y = (x - 4)^2 - 1$

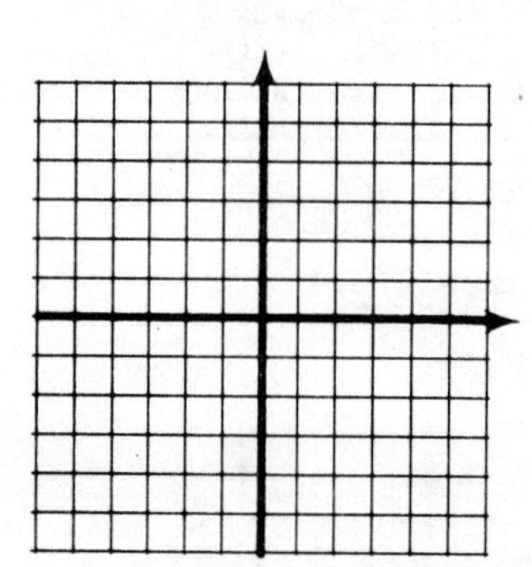

18. $y = (x + 3)^2 + 3$

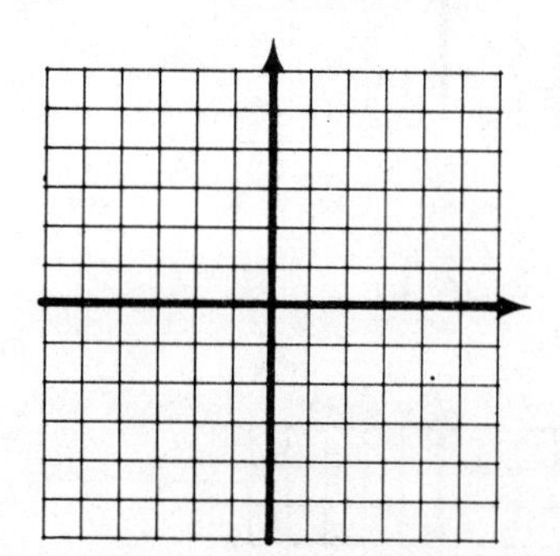

name date hour

CHAPTER 10 TEST

Solve by completing the square.

1. $2x^2 - 5x = 0$ ________

2. $5x^2 = 2 - 9x$ ________

Solve using the square root method.

3. $x^2 = 5$ ________

4. $(x - 3)^2 = 49$ ________

5. $(3x - 2)^2 = 35$ ________

Solve using the quadratic formula.

6. $m^2 = 3m + 10$ ________

7. $2k^2 + 5k = 3$ ________

8. $y^2 - 2y = 1$ ________

9. $x^2 = 6x - 2$ ________

10. $p^2 + 16 = 8p$ ________

11. $3z^2 + 2 = 7z$ ________

12. $4n^2 + 8n + 5 = 0$ ________

13. $y^2 - \frac{5}{3}y + \frac{1}{3} = 0$ ________

Draw the graph of each equation.

14. $y = -3x^2$

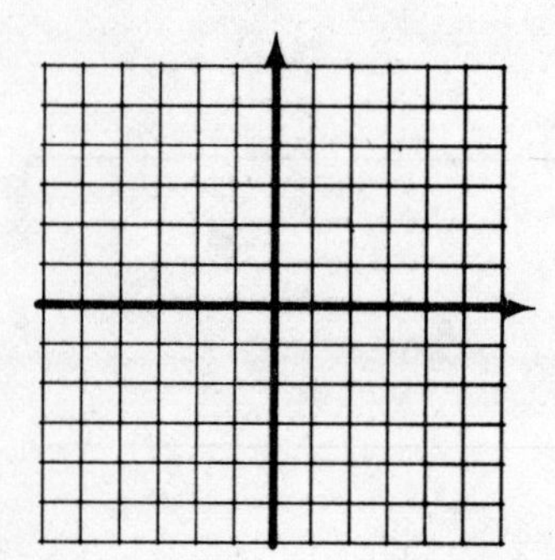

15. $y = 3x^2 - 2$

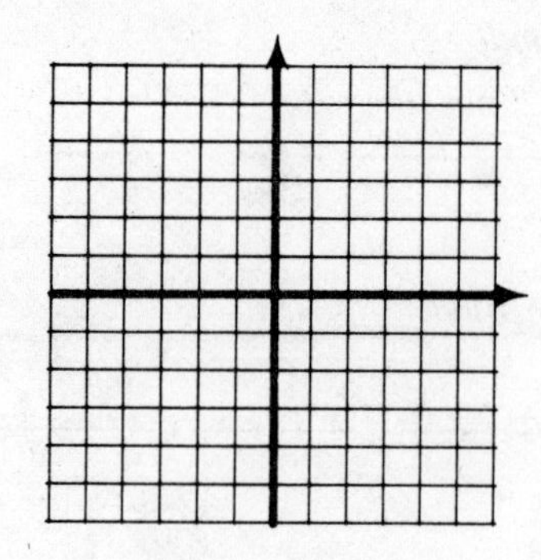

name date hour

FINAL EXAMINATION

The questions on this examination are made up of worked out examples from the textbook. If you have difficulty with any of these questions, look on the indicated page. There you will see the problem worked out in detail.

1. Find $\frac{3}{4} \div \frac{8}{5}$. (page 3)
2. Multiply 29.3 and 4.52. (page 12)
3. Find the value of $5 + 2 \cdot 3$. (page 21)
4. Find the value of $2x + 5y$ when $x = 5$ and $y = 3$. (page 25)
5. Is it true that $-3 < -1$? (page 37)
6. Simplify $-|-14|$. (page 38)
7. Add $-1/2$ and $1/8$. (page 43)
8. Subtract $-3 - (-5)$. (page 49)
9. Find $(-9)(2) - (-3)(2)$. (page 56)
10. Name the property that makes $[2 + (-7)] + 6 = 2 + (-7 + 6)$ true. (page 70)
11. Use the distributive property to simplify $3(k - 9)$. (page 72)
12. Solve the equation $3k + 17 = 4k$. (page 86)
13. Solve $a/4 = 3$. (page 90)
14. Solve $3r + 4 - 2r - 7 = 4r + 3$. (page 95)
15. Solve $6a - (3 + 2a) = 3a + 1$. (page 97)
16. If three times the sum of a number and 4 is decreased by twice the number, the result is -6. Find the number. (page 105)
17. Solve $-x \leq -11$. (page 128)
18. Simplify $6^3 \cdot 6^5$. (page 142)
19. Simplify $(2^5)^3$. (page 143)
20. Find the product of $3p - 5$ and $2p + 6$. (page 162)
21. Divide $8x^3 - 4x^2 - 14x + 15$ by $2x + 3$. (page 177)
22. Write .0000762 in scientific notation. (page 187)
23. Factor out the greatest common factor: $20m^5 + 10m^4 + 15m^3$. (page 195)
24. Factor $p^2 - 2p - 15$. (page 200)
25. Factor $8x^2 + 6x - 9$. (page 206)
26. Factor $9m^2 - 24m + 16$. (page 212)
27. Solve the equation $2p^2 - 13p + 20 = 0$. (page 216)
28. The width of a rectangle is 4 centimeters less than the length. The area is 96 square centimeters. Find the length and width. (page 221)

29. Reduce $\frac{3x-12}{5x-20}$ to lowest terms. (page 239)

30. Find the quotient: $\frac{(3m)^2}{(2n)^2} \div \frac{6m^3}{16n^2}$. (page 244)

31. Add $\frac{x}{x^2-1} + \frac{x}{x+1}$. (page 250)

32. Solve $\frac{2m}{m^2-4} + \frac{1}{m-2} = \frac{2}{m+2}$. (page 256)

33. The Big Muddy River has a current of 3 miles per hour. A motorboat takes as long to go 12 miles downstream as to go 8 miles upstream. What is the speed of the boat in still water? (page 261)

34. Complete the following ordered pairs for the equation $5x - y = 24$: (5,), (−3,), (0,). (page 276)

35. Graph (−2, 3), (−1, −4), and (3/2, 2). (page 281)

36. Graph $4x - 5y = 20$. (page 288)

37. Graph $y = -4$. (page 290)

38. Graph $2x + 3y \leq 6$. (page 297)

39. Use the graphical method to solve the system $\begin{aligned} 2x + 3y &= 4. \\ 3x - y &= -5 \end{aligned}$ (page 308)

40. Solve the system $\begin{aligned} -2x + y &= -11. \\ 5x - y &= 26 \end{aligned}$ (page 316)

41. Solve the system $\begin{aligned} 3x - y &= 4. \\ -9x + 3y &= -12 \end{aligned}$ (page 321)

42. Use substitution to solve the system $\begin{aligned} 2x + 5y &= 7. \\ x &= -1 - y \end{aligned}$ (page 325)

43. Admission prices at a football game were $1.25 for adults and $.50 for children. The total receipts from the game were $530.75. Tickets were sold to 454 people. How many adults and how many children attended the game? (page 331)

44. Is $\sqrt{17}$ rational or irrational? (page 348)

45. Simplify $\sqrt{300}$. (page 353)

46. Simplify $2\sqrt{12} + 3\sqrt{75}$. (page 361)

47. Rationalize the denominator of $12/\sqrt{8}$. (page 365)

48. Solve the equation $3\sqrt{x} = \sqrt{x+8}$. (page 375)

49. Use the quadratic formula to solve $\frac{1}{10}t^2 = \frac{2}{5} - \frac{1}{2}t$. (page 396)

50. Graph $y = x^2 - 3$. (page 404)

Appendices

APPENDIX A SETS

OBJECTIVES

- List the elements of a set
- Decide whether a given object is an element of a given set
- Decide whether a set is finite or infinite
- Decide whether a given set is a subset of another set
- Find the complement of a set
- Find the union and the intersection of two sets

A **set** is a collection of things. The objects in a set are called the **elements** of the set. A set can be represented by listing its elements between **set braces**, { }. The order in which the elements of a set are listed is unimportant.

Example 1 Represent the following sets by listing the elements.

(a) The set of states in the United States which border on the Pacific Ocean = {California, Oregon, Washington, Hawaii, Alaska}.

(b) The set of all counting numbers less than six = $\{1, 2, 3, 4, 5\}$.

(c) The set of all coins currently issued in the United States = {penny, nickel, dime, quarter, half-dollar, dollar}.

Work Quick Check 1 at the side.

Capital letters are used to name sets. To state that 5 is an element of

$$S = \{1, 2, 3, 4, 5\},$$

we write $5 \in S$. The statement $6 \notin S$ means that 6 is not an element of S.

Work Quick Check 2 at the side.

A set with no elements is called the **empty** set or the **null** set. The symbols $\emptyset$ or { } are used for the empty set. If we let A be the set of all cats that fly, then A is the empty set and we write

$$A = \emptyset \quad \text{or} \quad A = \{\ \}.$$

In any discussion of sets, there is some set which includes all the elements under consideration. This set is called the **universal** set for that situation. For example, if the discussion is about presidents of the United States, then the set of all presidents of the United States is the universal set. The universal set is denoted U.

In Example 1, there are five elements in the set in part (a), five in part (b), and six in part (c). Sets in which the number of elements can be determined are called **finite** sets. On the other hand, the set of natural numbers, for example, is an **infinite** set,

QUICK CHECKS

1. Represent the following sets by listing the elements.

(a) The set of whole numbers between 2.5 and 4.8

(b) The set of all the days of the week

2. Decide whether each statement is true or false for the set $T = \{m, n, p, q\}$.

(a) $m \in T$

(b) $n \in T$

(c) $k \notin T$

(d) $h \notin T$

Quick Check Answers

1. (a) $\{3, 4\}$ (b) {Sunday, Monday, Tuesday, Wednesday, Thursday, Friday, Saturday}
2. (a) True (b) True (c) True (d) True

3. List the elements of each set, if possible. Decide whether each set is finite or infinite.

(a) The set of integers between −2 and 2

(b) The set of all even numbers

(c) The set of all numbers between 0 and 1

4. Let $P = \{5, 10, 15\}$, $Q = \{5\}$, $R = \{10, 15\}$, $S = \{15, 10, 5\}$. Use $=$, $\subset$, or $\not\subset$ to make each statement true.

(a) $Q \quad P$

(b) $R \quad P$

(c) $S \quad P$

(d) $P \quad Q$

5. Find all subsets of $\{2, 4, 10\}$.

Quick Check Answers

3. (a) $\{-1, 0, 1\}$; finite
(b) $\{\ldots, -2, 0, 2, 4, \ldots\}$; infinite
(c) Cannot be listed; infinite

4. (a) $\subset$ (b) $\subset$ (c) $=$
(d) $\not\subset$

5. $\{2, 4, 10\}$, $\{2, 4\}$, $\{2, 10\}$, $\{4, 10\}$, $\{2\}$, $\{4\}$, $\{10\}$, $\emptyset$

because there is no final number. We can list the elements of the set of natural numbers as

$$N = \{1, 2, 3, 4, \ldots\},$$

where the three dots indicate that the set continues indefinitely. Not all infinite sets can be listed in this way. For example, there is no way to list the elements in the set of all numbers between one and two.

Example 2 List the elements of each set, if possible. Decide whether each set is finite or infinite.

(a) The set of all integers

One way to list the elements is $\{\ldots, -2, -1, 0, 1, 2, \ldots\}$. The set is infinite.

(b) The set of all natural numbers between 0 and 5

$\{1, 2, 3, 4\}$ The set is finite.

(c) The set of all rational numbers

This is an infinite set whose elements cannot be listed.

Work Quick Check 3 at the side.

Two sets are **equal** if they have exactly the same elements. Thus, the set of natural numbers and the set of positive integers are equal sets. Also, the sets

$$\{1, 2, 4, 7\} \quad \text{and} \quad \{4, 2, 7, 1\}$$

are equal. The order of the elements does not make a difference.

If all the elements of a set A are also elements of some set B, then we say A is a **subset** of B, written $A \subset B$. We use the symbol $A \not\subset B$ to mean A is not a subset of B.

Example 3 Let $A = \{1, 2, 3, 4\}$, $B = \{1, 4\}$, and $C = \{1\}$. Then $B \subset A$, $C \subset A$, $C \subset B$, but $A \not\subset B$, $A \not\subset C$, and $B \not\subset C$.

Work Quick Check 4 at the side.

The set $M = \{a, b\}$ has four subsets: $\{a, b\}$, $\{a\}$, $\{b\}$, and $\emptyset$. The empty set is a subset of any set. How many subsets does $N = \{a, b, c\}$ have? There is one subset with 3 elements: $\{a, b, c\}$. There are three subsets with 2 elements:

$$\{a, b\}, \quad \{a, c\}, \quad \text{and} \quad \{b, c\}.$$

There are three subsets with 1 element:

$$\{a\}, \quad \{b\}, \quad \text{and} \quad \{c\}.$$

There is one subset with 0 elements: $\emptyset$. Thus, set N has eight subsets.

In general, a set with n elements has 2^n subsets.

Work Quick Check 5 at the side.

To illustrate the relationships between sets, **Venn diagrams** are often used. A rectangle represents the universal set, U. The sets under discussion are represented by regions within the rectangle. The Venn diagram in Figure 1 shows that $B \subset A$.

6. Given $U = \{1, 2, 3, 4, 5, 6\}$, $A = \{2, 4, 6\}$, $B = \{1, 2, 3, 4\}$, and $C = \{1, 2, 5, 6\}$, find A', B', and C'.

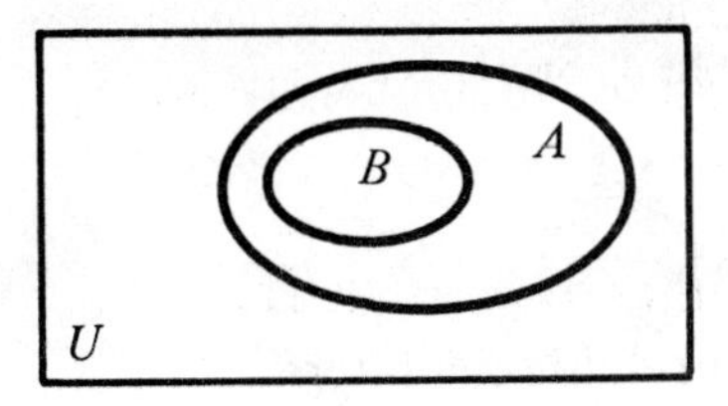

Figure 1

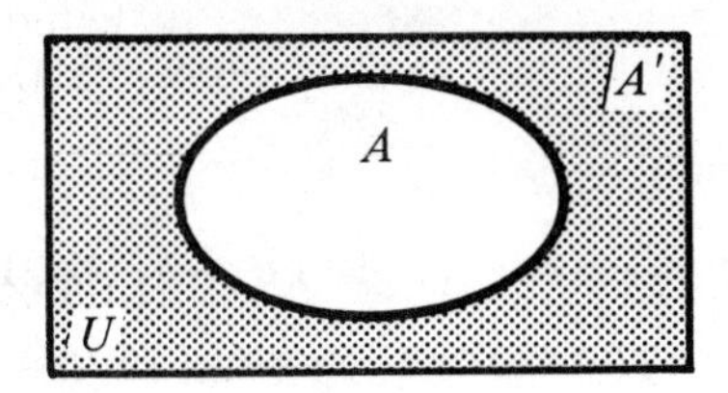

Figure 2

For every set A, there is a set A', the **complement** of A, which contains all the elements of U that are not in A. The shaded region in the Venn diagram of Figure 2 represents A'.

Example 4 Given $U = \{a, b, c, d, e, f, g\}$, $A = \{a, b, c\}$, $B = \{a, d, f, g\}$, and $C = \{d, e\}$.

(a) $A' = \{d, e, f, g\}$

(b) $B' = \{b, c, e\}$

(c) $C' = \{a, b, c, f, g\}$

Work Quick Check 6 at the side.

7. Given $A = \{2, 4, 6\}$ and $B = \{1, 2, 3, 4\}$, find $A \cup B$.

The **union** of two sets A and B, written $A \cup B$, is the set of all elements of A plus all elements of B. Thus, for the sets in Example 4,

$$A \cup B = \{a, b, c, d, f, g\}$$

and

$$A \cup C = \{a, b, c, d, e\}.$$

In Figure 3 the shaded region is the union of sets A and B.

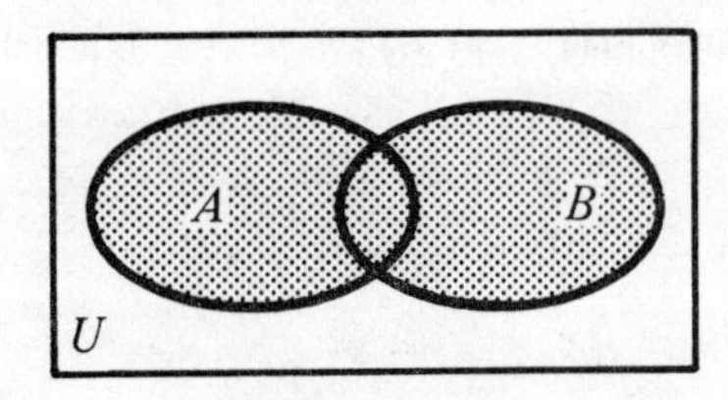

Figure 3

Example 5 If $M = \{2, 5, 7\}$ and $N = \{1, 2, 3, 4, 5\}$ then

$$M \cup N = \{1, 2, 3, 4, 5, 7\}.$$

Work Quick Check 7 at the side.

Quick Check Answers

6. $A' = \{1, 3, 5\}$, $B' = \{5, 6\}$, $C' = \{3, 4\}$
7. $\{1, 2, 3, 4, 6\}$

8. Let $S = \{a, b, c, d, e, f\}$, $T = \{a, c, k\}$, $W = \{d, g, h\}$. Find

(a) $S \cap T$

(b) $S \cap W$

(c) $T \cap W$

9. Use the sets of Example 7 to find

(a) $A \cap C$

(b) $A \cup C$

(c) C'.

Quick Check Answers

8. (a) $\{a, c\}$ (b) $\{d\}$ (c) $\emptyset$
9. (a) $\{2\}$ (b) $\{2, 5, 7, 10, 14, 20\}$ (c) $\{10, 14, 20\}$

The **intersection** of two sets A and B, written $A \cap B$, is the set of all elements which belong to both A and B. For example, if

$$A = \{\text{Jose, Ellen, Marge, Kevin}\}$$

and

$$B = \{\text{Jose, Patrick, Ellen, Sue}\},$$

then

$$A \cap B = \{\text{Jose, Ellen}\}.$$

The shaded region in Figure 4 represents the intersection of the two sets A and B.

Example 6 Suppose $P = \{3, 9, 27\}$, $Q = \{2, 3, 10, 18, 27, 28\}$, and $R = \{2, 10, 28\}$.

(a) $P \cap Q = \{3, 27\}$

(b) $Q \cap R = \{2, 10, 28\} = R$

(c) $P \cap R = \emptyset$

A pair of sets like P and R in Example 6 which have no elements in common are called **disjoint** sets. The Venn diagram in Figure 5 shows a pair of disjoint sets.

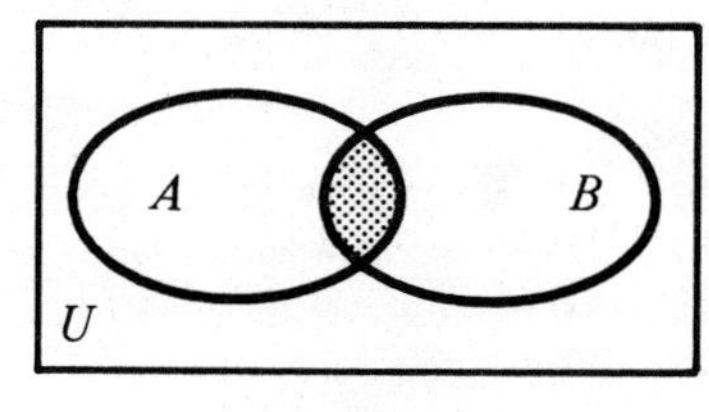

Figure 4

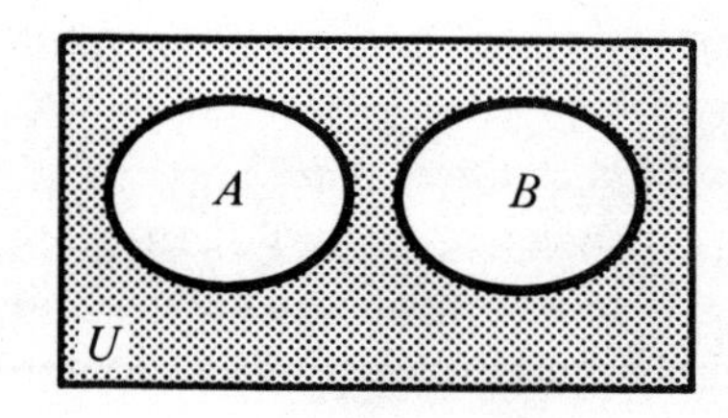

Figure 5

Work Quick Check 8 at the side.

Example 7 Let $U = \{2, 5, 7, 10, 14, 20\}$, $A = \{2, 10, 14, 20\}$, $B = \{5, 7\}$, and $C = \{2, 5, 7\}$.

(a) $A \cup B = \{2, 5, 7, 10, 14, 20\}$

(b) $A \cap B = \emptyset$

(c) $B \cup C = \{2, 5, 7\} = C$

(d) $B \cap C = \{5, 7\} = B$

(e) $A' = \{5, 7\}$

Work Quick Check 9 at the side.

name date hour

APPENDIX A EXERCISES

List the elements of each of the following sets. See Examples 1 and 2.

1. The set of all natural numbers less than eight. ______

2. The set of all integers between four and ten. ______

3. The set of seasons. ______

4. The set of months of the year. ______

5. The set of women presidents of the United States. ______

6. The set of all living humans more than 200 years old. ______

7. The set of letters of the alphabet between K and M. ______

8. The set of letters of the alphabet between D and H. ______

9. The set of positive even numbers. ______

10. The set of multiples of five. ______

11. Which of the sets described in Exercises 1–10 are infinite sets? ______

12. Which of the sets described in Exercises 1–10 are finite sets? ______

Write true or false for each of the following.

13. $5 \in \{1, 2, 5, 8\}$ ______

14. $6 \in \{1, 2, 3, 4, 5\}$ ______

15. $2 \in \{1, 3, 5, 7, 9\}$ ______

16. $1 \in \{6, 2, 5, 1\}$ ______

17. $7 \notin \{2, 4, 6, 8\}$ ______

18. $7 \notin \{1, 3, 5, 7\}$ ______

19. $\{2, 4, 9, 12, 13\} = \{13, 12, 9, 4, 2\}$ ______

20. $\{7, 11, 4\} = \{7, 11, 4, 0\}$ ______

Let $A = \{1, 3, 4, 5, 7, 8\}$
$B = \{2, 4, 6, 8\}$
$C = \{1, 3, 5, 7\}$
$D = \{1, 2, 3\}$
$E = \{3, 7\}$
$U = \{1, 2, 3, 4, 5, 6, 7, 8, 9, 10\}$

Write true or false for each of the following. See Examples 3, 5, 6, and 7.

21. $A \subset U$ ______

22. $D \subset A$ ______

23. $\emptyset \subset A$ ______

24. $\{1, 2\} \subset D$ ______

25. $C \subset A$ ______

26. $A \subset C$ ______

27. $D \subset B$ ______

28. $E \subset C$ ______

29. $D \not\subset E$ ______

30. $E \not\subset A$ ______

31. There are exactly 4 subsets of E. ______

32. There are exactly 8 subsets of D. ______

33. There are exactly 12 subsets of C. ______

34. There are exactly 16 subsets of B. ______

35. $\{4, 6, 8, 12\} \cap \{6, 8, 14, 17\} = \{6, 8\}$ ______

36. $\{2, 5, 9\} \cap \{1, 2, 3, 4, 5\} = \{2, 5\}$ ______

37. $\{3, 1, 0\} \cap \{0, 2, 4\} = \{0\}$ ______

38. $\{4, 2, 1\} \cap \{1, 2, 3, 4\} = \{1, 2, 3\}$ ______

39. $\{3, 9, 12\} \cap \emptyset = \{3, 9, 12\}$ ______

name date hour

40. $\{3, 9, 12\} \cup \emptyset = \emptyset$ ______

41. $\{4, 9, 11, 7, 3\} \cup \{1, 2, 3, 4, 5\} = \{1, 2, 3, 4, 5, 7, 9, 11\}$ ______

42. $\{1, 2, 3\} \cup \{1, 2, 3\} = \{1, 2, 3\}$ ______

43. $\{3, 5, 7, 9\} \cup \{4, 6, 8\} = \emptyset$ ______

44. $\{5, 10, 15, 20\} \cup \{5, 15, 30\} = \{5, 13\}$ ______

Let $U = \{a, b, c, d, e, f, g, h\}$
$A = \{a, b, c, d, e, f\}$
$B = \{a, c, e\}$
$C = \{a, f\}$
$D = \{d\}$
List the elements in the following sets.

45. A' ______

46. B' ______

47. C' ______

48. D' ______

49. $A \cap B$ ______

50. $B \cap A$ ______

51. $A \cap D$ ______

52. $B \cap D$ ______

53. $B \cap C$ ______

54. $A \cup B$ ______

55. $B \cup D$ ______

56. $B \cup C$ ______

57. $C \cup B$ ______

58. $C \cup D$ ______

59. $A \cap \emptyset$ ______

60. $B \cup \emptyset$ ______

61. Name any pair of disjoint sets among sets *A–D* above.

WORK SPACE

In the United States today, length is measured in inches, feet, or miles. Weight is measured in ounces (there are two kinds of ounces), pounds, or tons (there are long tons and short tons). Volume is measured in pints, quarts, or gallons. These weights and measures make up the **English system** of measures. The United States will gradually switch to the **metric system**, which is used in almost every country in the world. Many industries are now working on the switch to the metric system, which will be in effect in the U.S. by the mid-1980's. Many automobiles are now made to metric specifications.

The metric system was developed in France in 1789. The name comes from the basic unit of length, the **meter** (abbreviated m), which is a little longer than a yard.

The meter can be subdivided into smaller parts for measuring shorter distances. The most common subdivisions are the **millimeter**, or 1/1000 meter, and the **centimeter**, 1/100 meter. Millimeters are abbreviated mm and centimeters are abbreviated cm.

A dime is about 2 mm thick. Film for instamatic cameras is 35 mm wide. Many common parts of the body can be measured with centimeters and millimeters. A man is perhaps 150 cm tall, and a woman's waist might be 66 cm around. Some manufacturers of clothing use metric sizes–a size 40 suit becomes a size 102 suit in metric.

A major advantage of the metric system of measurement is the ease in converting from one unit of measure to another. This is illustrated in Examples 1–4.

Example 1 Convert as indicated.

(a) 28 cm to mm

One centimeter is made up of 10 mm. Thus, 28 cm is

$$28 \times 10 = 280 \text{ mm}$$

(b) 6 m to cm

$$6 \text{ m} = 6 \cdot 100 \text{ cm} = 600 \text{ cm}$$

(c) 3.823 m to mm

$$3.823 \text{ m} = 3.823(1000 \text{ mm}) = 3823 \text{ mm}$$

Work Quick Check 1 at the side.

Example 2 Convert as indicated.

(a) 250 mm to m

One meter is 1000 mm.

$$250 \text{ mm} = \frac{250}{1000} \text{ m} = .250 \text{ m}$$

OBJECTIVES

- Convert from one metric unit of measure to another
- Convert measurements from English to metric and from metric to English

QUICK CHECKS

1. Convert as indicated.

(a) 47 cm to mm

(b) 174 cm to mm

(c) 9 m to cm

(d) 8.2 m to cm

(e) 3.27 m to mm

Quick Check Answers

1. (a) 470 (b) 1740 (c) 900 (d) 820 (e) 3270

(b) 375 mm to cm

$$375 \text{ mm} = \frac{375}{10} \text{ cm} = 37.5 \text{ cm}$$

(c) 8.42 cm to m

$$8.42 \text{ cm} = \frac{8.42}{100} \text{ m} = .0842 \text{ m}$$

Work Quick Check 2 at the side.

Longer distances are measured in **kilometers** (km), or one thousand meters. A kilometer is about 5/8 of a mile. The table here shows distances in kilometers between various cities.

Some Highway Distances in Kilometers

	Seattle	Los Angeles	Denver	Houston	St. Louis	Chicago	Cleveland	Atlanta	Miami	Washington	New York	Boston
Seattle		1842	2167	3704	3393	3319	3868	4434	5504	4422	4673	4885
Los Angeles	1842		1825	2499	2973	3371	3894	3535	4364	4254	4690	4911
Denver	2167	1825		1651	1377	1635	2183	2254	3290	2729	4580	3200
Houston	3704	2499	1651		1278	1746	2066	1310	1957	2269	2632	3083
St. Louis	3393	2973	1377	1278		463	879	898	1968	1289	1554	1895
Chicago	3319	3371	1635	1746	463		552	1175	2188	1146	1352	1569
Cleveland	3868	3894	2183	2066	879	552		1104	2097	565	816	1017
Atlanta	4434	3535	2254	1310	898	1175	1104		1070	1014	1376	1718
Miami	5504	4364	3290	1957	1968	2188	2097	1070		1778	2140	2481
Washington	4422	4254	2729	2269	1289	1146	565	1014	1778		368	703
New York	4673	4690	4580	2632	1554	1352	816	1376	2140	368		348
Boston	4885	4911	3200	3083	1895	1569	1017	1718	2481	703	348	

Example 3 Convert as indicated.

(a) 9.7 km to m

Since 1000 m = 1km, we have

$$9.7 \text{ km} = 9.7(1000) = 9700 \text{ m}$$

(b) 8680 m to km

$$8680 \text{ m} = \frac{8680}{1000} \text{ km} = 8.680 \text{ km}$$

(c) 25,600 m to km

$$25{,}600 \text{ m} = 25.6 \text{ km}$$

Work Quick Check 3 at the side.

Weights in the metric system are based on the **gram** (g). A nickel weighs about 5 g, for example. Since a gram is such a small

2. Convert as indicated.

(a) 590 mm to cm

(b) 8600 mm to m

(c) 920 mm to cm

(d) 3720 cm to m

3. Convert as indicated.

(a) 5.63 km to m

(b) 24,000 m to km

(c) 99,460 m to km

Quick Check Answers
2. (a) 59 (b) 8.6 (c) 92 (d) 37.2
3. (a) 5630 (b) 24 (c) 99.46

weight, milligrams (1/1000 g) and centigrams (1/100 g) are mainly used to measure very small weights in science. A **kilogram**, or 1000 grams, is about 2.2 pounds. Kilograms (abbreviated kg) are sometimes called *kilos.*

Example 4 Convert as indicated.

(a) 2500 g to kg

1000 g = 1 kg. Thus,

$$2500 \text{ g} = \frac{2500}{1000} \text{ kg} = 2.5 \text{ kg}$$

(b) .38 kg to g

$$.38 \text{ kg} = .38 \cdot 100 \text{ g} = 380 \text{ g}$$

(c) 275 mg to g

$$275 \text{ mg} = \frac{275}{1000} \text{ g} = .275 \text{ g}$$

(d) 896 cg to g

$$896 \text{ cg} = \frac{896}{100} \text{ g} = 8.96 \text{ g}$$

Work Quick Check 4 at the side.

Volume is measured in **liters** (ℓ). A liter is about a quart. Milliliters (1/1000 liter), centiliters (1/100 liter), and kiloliters (1000 liters) are used mainly in science.

In summary, the metric system uses the following four common prefixes:

Prefix	Definition	Example
milli-	1/1000	1 millimeter = 1/1000 meter
centi-	1/100	1 centiliter = 1/100 liter
deci-	1/10	1 decigram = 1/10 gram
kilo-	1000	1 kilogram = 1000 grams

Eventually, people will think in the metric system as easily as they now think in the English system. To help you "think metric," you should get in the habit of estimating in the metric system. As an aid, use the *approximate* conversion table shown on page 422.

Example 5 Convert as indicated.

(a) 2 yards to meters

Look in the table, for "From yards to meters." You should find the number .9144. Multiply

$$2 \text{ yards} = 2(.9144) \text{ meters} = 1.8288 \text{ m.}$$

4. Convert as indicated.

(a) 3700 g to kg

(b) 924 g to kg

(c) 1.03 kg to g

(d) 1382 mg to g

(e) 769.8 cg to g

Quick Check Answers

4. (a) 3.7 (b) .924 (c) 1030 (d) 1.382 (e) 7.698

5. Convert as indicated.

(a) 350 yards to m

(b) 2500 feet to m

(c) 74 inches to m

(d) 210 miles to km

(e) 80 pounds to g

(f) 20 quarts to ℓ

6. Convert as indicated.

(a) 390 m to yards

(b) 700 m to feet

(c) 986 km to miles

(d) 4750 g to pounds

(e) 7.24 kg to pounds

(f) 370 ℓ to quarts

(g) 1000 ℓ to gallons

Quick Check Answers

5. (a) 320.04 (b) 762
(c) 1.8796 (d) 337.953
(e) 36,320 (f) 18.92

6. (a) 426.66 (b) 2296.7
(c) 612.7 (d) 10.45 (e) 15.928
(f) 391.09 (g) 264

METRIC TO ENGLISH			ENGLISH TO METRIC		
From	to	Multiply by	From	to	Multiply by
meters	yards	1.094	yards	meters	.9144
meters	feet	3.281	feet	meters	.3048
meters	inches	39.37	inches	meters	.0254
kilometers	miles	.6214	miles	kilometers	1.6093
grams	pounds	.00220	pounds	grams	454
kilograms	pounds	2.20	pounds	kilograms	.454
liters	quarts	1.057	quarts	liters	.946
liters	gallons	.264	gallons	liters	3.785

(b) 120 miles to km

Look at "From miles to kilometers," finding 1.6093.

$$120 \text{ miles} = 120 \cdot 1.6093 \text{ km} = 193.116 \text{ km}$$

(c) 25 gallons to liters

$$25 \text{ gallons} = 25 \cdot 3.785 \text{ liters} = 94.625 \text{ liters}$$

(d) 58 pounds to kilograms

$$58 \text{ pounds} = (58)(.454) \text{ kg} = 26.332 \text{ kg}$$

Work Quick Check 5 at the side.

Example 6 Convert as indicated.

(a) 72 meters to yards

Look for "From meters to yards" in the table above.

$$72 \text{ m} = (72)1.094 \text{ yards} = 78.768 \text{ yards}$$

(b) 400 km to miles

$$400 \text{ km} = 400 \cdot .6214 \text{ miles} = 248.56 \text{ miles}$$

(c) 2000 grams to pounds

$$2000 \text{ g} = (2000)(.00220) \text{ pounds} = 4.4 \text{ pounds}$$

(d) 850 liters to quarts

$$850 \text{ ℓ} = (850)1.057 \text{ quarts} = 898.45 \text{ quarts}$$

Work Quick Check 6 at the side.

name date hour

APPENDIX B EXERCISES

Make the indicated conversions within the metric system. See Examples 1–4.

1. 20 m to mm ______
2. 7.6 m to cm ______
3. 7 cm to mm ______
4. 9.63 cm to mm ______
5. 80 mm to cm ______
6. 500 mm to cm ______
7. 320 mm to m ______
8. 9760 cm to m ______
9. 5200 m to km ______
10. 15,000 m to km ______
11. 7.8 km to m ______
12. 49.8 km to m ______
13. 6 kg to g ______
14. 15.9 kg to g ______
15. 1.92 kg to g ______
16. 3.24 kg to g ______
17. 8200 g to kg ______
18. 16,200 g to kg ______
19. 69.4 mg to cg ______
20. 1749 cg to g ______
21. 8.1 g to cg ______
22. .042 g to mg ______
23. 9 ℓ to ml ______
24. 2.98 ℓ to ml ______
25. 57,000 ml to ℓ ______
26. 800 ml to ℓ ______
27. 29.6 ml to cl ______
28. 34.1 ml to cl ______

Use the table in the text to make each conversion. Round to the nearest tenth. See Examples 5 and 6.

29. 12 yards to m ______
30. 32.1 yards to m ______

31. 6.7 feet to m ________________

32. 46 feet to m ________________

33. 25 inches to m ________________

34. 77 inches to m ________________

35. 122 miles to km ________________

36. 400 miles to km ________________

37. 8.4 pounds to g ________________

38. 1.3 pounds to g ________________

39. 110 pounds to kg ________________

40. 680 pounds to kg ________________

41. 8 quarts to ℓ ________________

42. 13 quarts to ℓ ________________

43. 76 gallons to ℓ ________________

44. 12 gallons to ℓ ________________

45. 36 m to yards ________________

46. 80 m to yards ________________

47. 40 m to feet ________________

48. 11 m to feet ________________

49. 600 km to miles ________________

50. 850 km to miles ________________

51. 680 g to pounds ________________

52. 12,700 g to pounds ________________

53. 4.9 kg to pounds ________________

54. 10.1 kg to pounds ________________

55. 8 ℓ to quarts ________________

56. 14.1 ℓ to quarts ________________

57. 76.8 ℓ to gallons ________________

58. 130 ℓ to gallons ________________

Estimate each of the following.

59. Find your height in cm. ________________

60. Find your height in mm. ________________

61. Find the length of your longest finger in cm. ________________

Graphs and charts occur often in newspapers and magazines. There is hardly an issue of *Time* or *Newsweek* that does not contain a graph. The first step in drawing a graph is to make a **frequency distribution**, as shown in the first example.

OBJECTIVES

- Make frequency distributions
- Make bar graphs
- Make circle graphs
- Read graphs

Example 1 Mickey, Pat, and Terry are three roommates who decided to keep track of their food expenses for a year. They bought food once a week. After fifty weeks, they listed each week's expenses (to the nearest dollar) in columns as follows.

22	16	14	16	17	21	15	13	22	25
19	17	16	14	11	22	14	17	23	28
20	17	20	18	25	17	13	19	24	27
14	15	18	17	18	14	12	18	22	26
17	19	19	16	16	12	15	23	27	27

To describe their year's food expenses, they made the **frequency distribution table** shown below. By looking at the list of weekly expenses, they found that $11 was the smallest expense and $28 was the largest expense. They set up the table below, in which all expenses were tallied. By counting the tallies, they found the frequency of each weekly expense.

Weekly Expense	*Tally*	*Frequency*	*Weekly Expense*	*Tally*	*Frequency*
$11	|	1	$20	||	2
$12	||	2	$21	|	1
$13	||	2	$22	||||	4
$14	~~||||~~	5	$23	||	2
$15	|||	3	$24	|	1
$16	~~||||~~	5	$25	||	2
$17	~~||||~~ ||	7	$26	|	1
$18	||||	4	$27	|||	3
$19	||||	4	$28	|	1

The weekly expense occurring most often is $17. We show the frequency of each expense in the **bar graph** shown in Figure 1. The height of each bar represents the number of weeks that a particular expense occurred. (See page 426.)

The frequency distribution table above contains a great deal of information. Since it may be a little hard to "digest" all the numbers in the table, we can combine the expenses into groups to form **grouped data**.

Work Quick Check 1 at the side.

We also can make a bar graph of this data as shown in Figure 2. We see from this bar graph that the most frequent weekly expense fell in the $17 to $19 range, even though we no longer can identify $17 as the most frequent expense.

QUICK CHECKS

1. Use the information in the table to complete the chart.

Weekly Expense	*Tally*	*Frequency*
$11–13		
$14–16		
$17–19		
$20–22		
$23–25		
$26–28		

Quick Check Answers

1. 5, 13, 15, 7, 5, 5

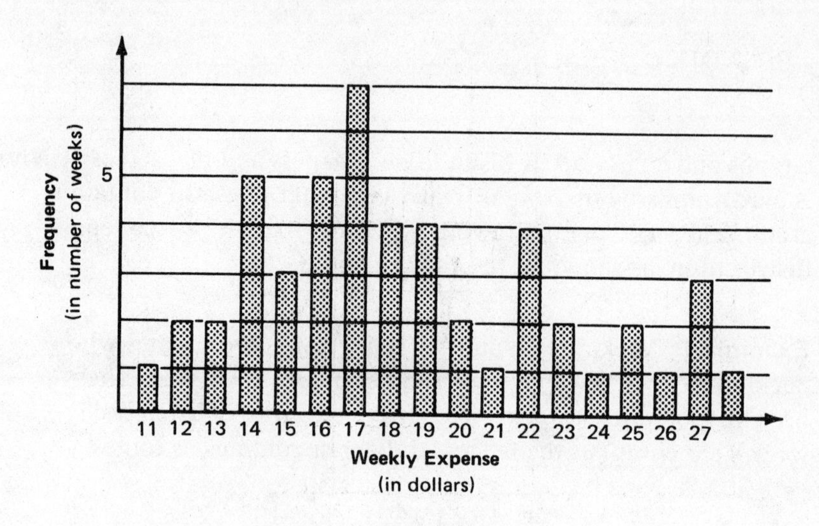

Figure 1

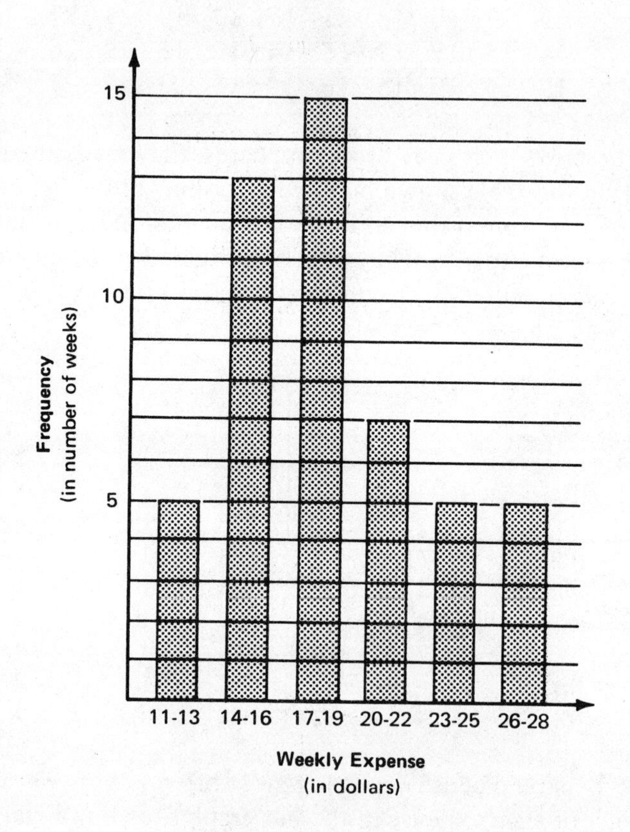

Figure 2

Notice that frequency distributions show which numbers occurred, and how many times, but do not show the order in which they occurred. To discover any trends that may have occurred during the year, we draw a chronological **line graph** using the original data. (See Figure 3.)

Item	*Percent of Total*
Food	30%
Rent	25%
Entertainment	15%
Clothing	10%
Books	10%
Other	10%

Example 2 Marna Nitekman, a student, kept track of her expenses for a period of time. After these expenses were converted to percents of the total expense, the results were as shown in the table at the side.

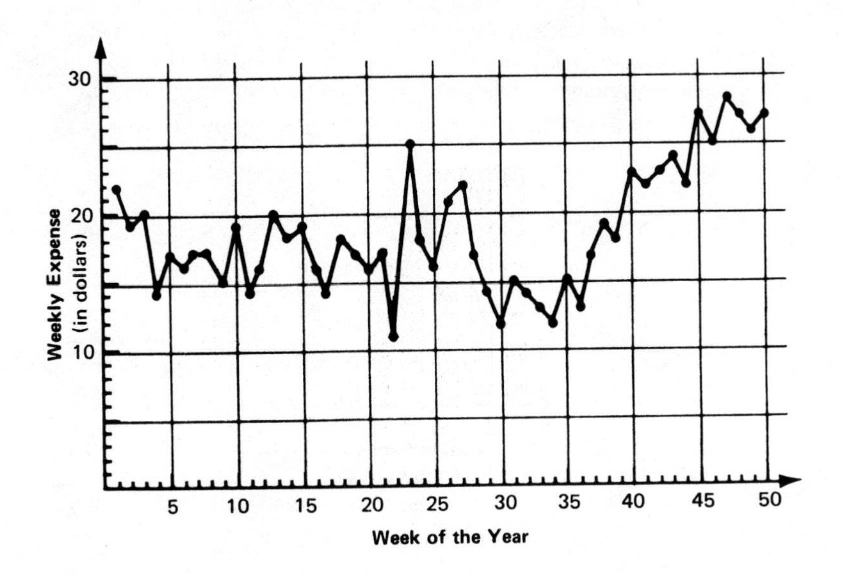

Figure 3

2. Find the number of degrees for each of the following.

(a) food

(b) entertainment

(c) books

(d) other

She decided to show these percents by using a **circle graph**. A circle has 360 degrees (written 360°). The 360° represents the total expenses. Since clothing is 10% of the total expense, she used

$$360° \times 10\% = 360° \times .10 = 36°$$

to represent her clothing expense. Since rent is 25% of the total expenses, she used

$$360° \times 25\% = 90°$$

to represent rent.

Work Quick Check 2 at the side.

After she found the degrees that represent each of her expenses, she drew the circle graph shown in Figure 4.

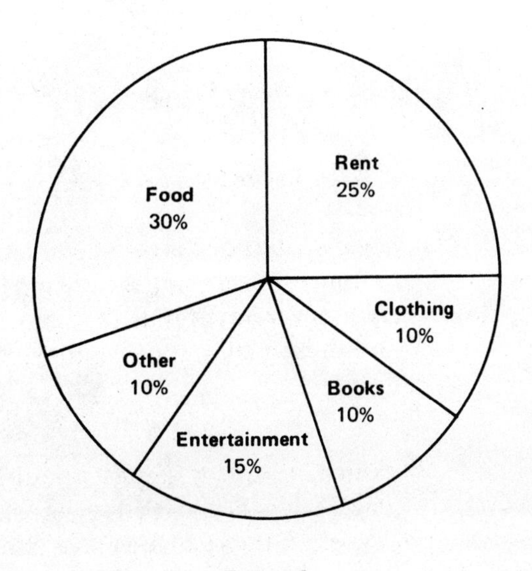

Figure 4

Quick Check Answers
2. (a) 108° (b) 54° (c) 36° (d) 36°

3. Find the following.

(a) the marriage rate in 1975

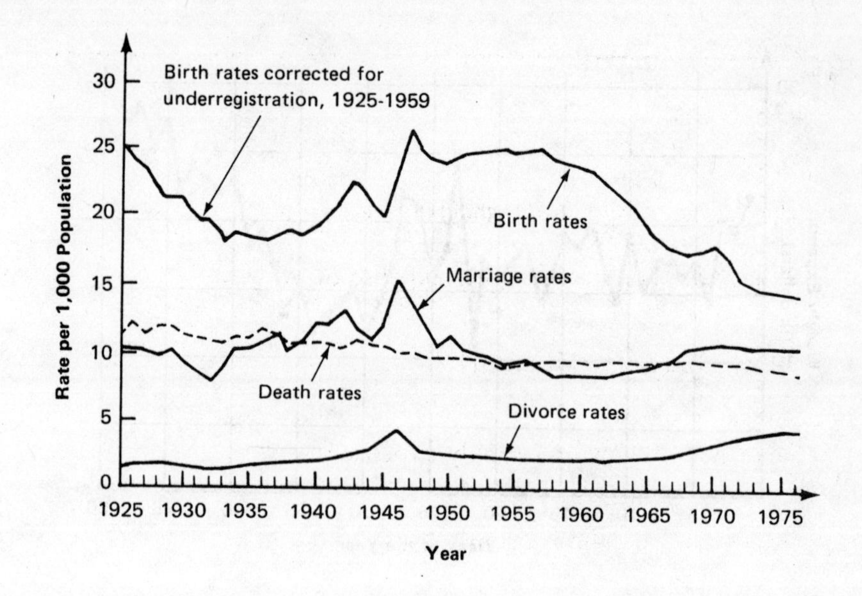

Figure 5

We sometimes use a **comparison graph** to compare different items. A comparison graph shows several items on one graph. We then can make comparisons among the various items shown on the graph.

(b) the death rate in 1970

Example 3 The line graph in Figure 5 shows four different vital statistics rates in one graph.

(a) To find the divorce rate in 1970, locate 1970 on the horizontal line. Then draw a line straight up until it intersects the line for divorces. Read across to the line marked "Rate per 1000 Population." You will have to estimate the answer; we get about 3.

(b) To find the birth rate in 1965, locate 1965 on the horizontal line. Read up to the birth rate line. Then read across to the vertical scale. The rate is about 20 births per 1000 people per year.

Work Quick Check 3 at the side.

Quick Check Answers

3. (a) about 10 (b) about 9

name date hour

APPENDIX C EXERCISES

In each problem, refer to the line graph in Figure 3.

1. What was the lowest weekly expense of the year? __________

2. In which week did it occur? __________

3. What was the highest weekly expense of the year? __________

4. In which week did it occur? __________

5. The three students whose weekly food expenses are shown in the graph had finals in the fourth and twenty-second weeks of the year. What happened to their food expenses in these weeks? __________

6. Is there an overall trend in food prices that you can see from the graph? __________

The following list shows the number of college units completed by 30 of the employees of the EZ Life Insurance Company.

74	133	4	127	20	30
103	27	139	118	138	121
149	132	64	141	130	76
42	50	95	56	65	104
4	140	12	88	119	64

7. Use these numbers to complete the following table.

Number of Units	*Frequency*
0–24	______
25–49	______
50–74	______
75–99	______
100–124	______
125–149	______

8. Complete this bar graph using the frequencies that you found.

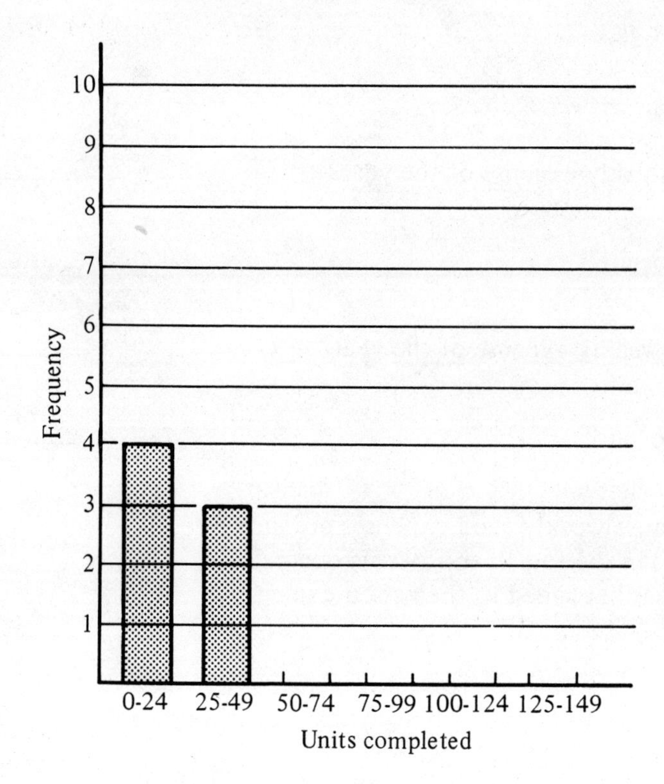

The daily high temperatures in Phoenix for the month of July one year were as follows. (The numbers are in chronological order as you read across. For example, the temperature on July 5 was 102°, and on July 11 it was 104°.)

79°	84°	88°	96°	102°	104°	110°	108°	106°	106°
104°	99°	97°	92°	94°	90°	82°	74°	72°	83°
85°	92°	100°	99°	101°	107°	111°	102°	97°	94°
92°									

9. Use these numbers to complete the following table.

Temperature	*Frequency*	*Temperature*	*Frequency*
70°–74°	________	95°–99°	________
75°–79°	________	100°–104°	________
80°–84°	________	105°–109°	________
85°–89°	________	110°–114°	________
90°–94°	________		

name date hour

10. Make a bar graph showing the results given in the table in Exercise 9.

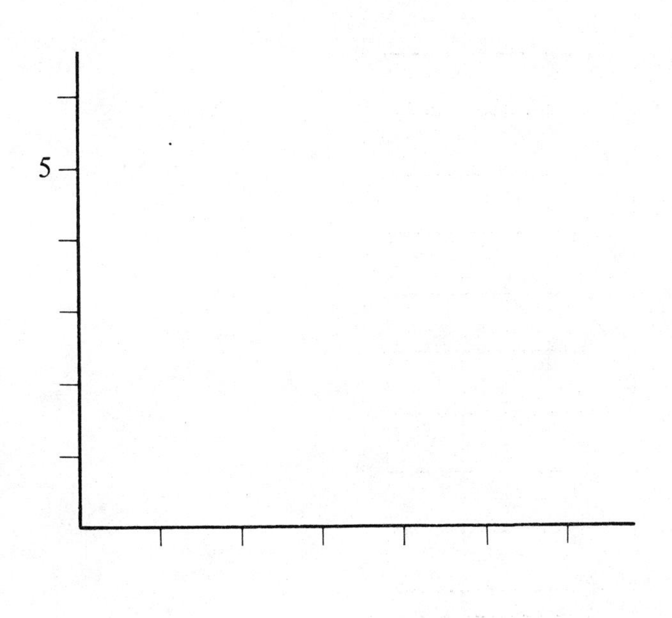

11. Make a line graph using the original numbers.

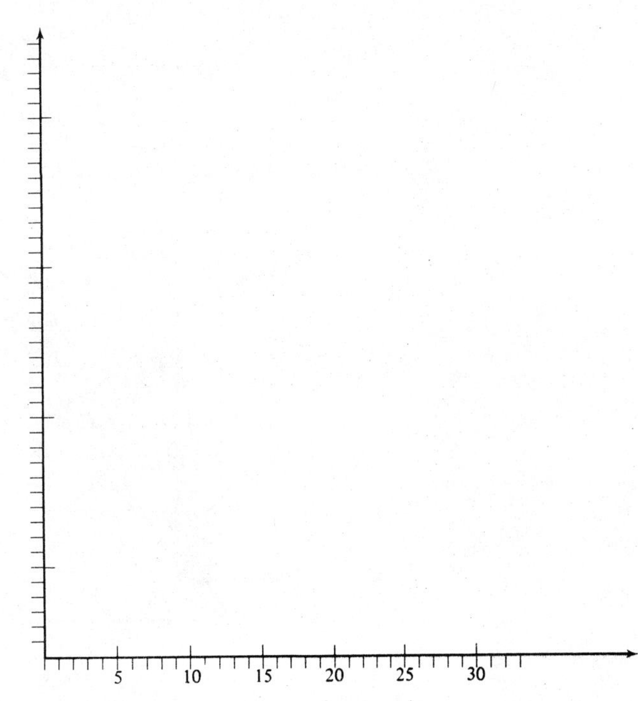

Use the line graph in Figure 5 (page 428) to answer each question.

12. Find the birth rate in 1975. ____________

13. Find the marriage rate in 1946. ____________

14. Find the divorce rate in 1930. ____________

15. Find the death rate in 1964. ____________

16. Which of the four rates seems most stable? ____________

17. Which rate is decreasing most rapidly? ____________

The graph below is used to estimate the hourly acreage covered by a farm implement when its width and speed of travel are known. For example, a 7 1/2 foot (90-inch) mower blade moving 4 miles per hour would cover about 3 5/8 acres per hour. This is found by going across the graph from the working width (90 inches) to the diagonal line for speed (4 mph), then down to the bottom to find hourly acreage.

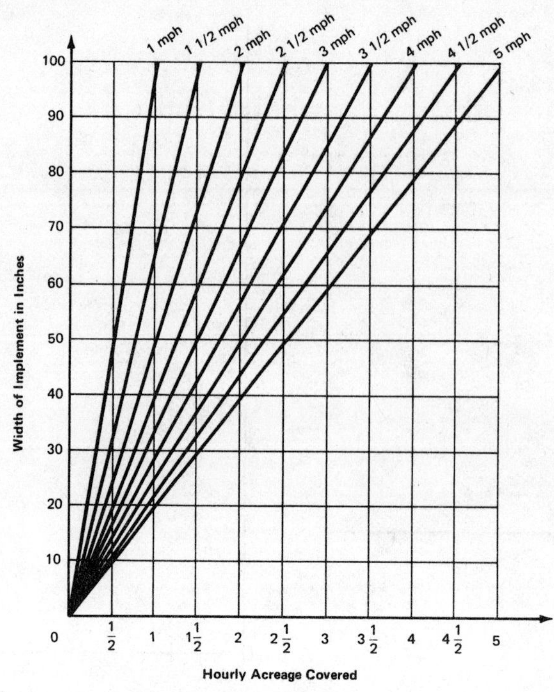

Courtesy of John Deere and Company.

18. What is the hourly acreage for a 36-inch implement moving 2 1/2 miles per hour? ________________

19. What is the hourly acreage for an 8-foot-wide combine moving 4 miles per hour? ________________

20. How fast must a tractor pull a 48-inch plow in order to plow one acre per hour? ________________

21. How wide a spray pattern is needed in order to spray 4 1/2 acres per hour at a speed of 4 1/2 miles per hour? ________________

Symbols

Symbol	Meaning
$+$	Plus sign, addition
$-$	Minus sign, subtraction
$\times$	Multiplication
$a(b)$, $(a)b$, or $(a)(b)$	Multiplication
$a \cdot b$	Multiplication
ab	Multiplication
$\div$	Division
a/b	Division
$=$	Equals
$\neq$	Is not equal to
$\approx$	Approximately equal to
$<$	Less than
$\nless$	Not less than
$\leq$	Less than or equal to
$>$	Greater than
$\ngtr$	Not greater than
$\geq$	Greater than or equal to
$\lvert x \rvert$	Absolute value of x
$\{a, b, c\}$	The set containing the elements a, b, and c
$\{x \mid P\}$	The set of all x satisfying property P
$\varnothing$	Empty set, or null set
(x, y)	Ordered pair
x^2	x squared; $x \cdot x$
x^3	x cubed; $x \cdot x \cdot x$
x^n	x to the power n; x to the nth
x^{-n}	x to the negative n; $x^{-n} = 1/x^n$
x^0	x to the power 0; $x^0 = 1$ if $x \neq 0$
$P(x)$	Polynomial having the variable x
$\sqrt{a}$	Positive square root of a; $\sqrt{a} \cdot \sqrt{a} = a$
$\sqrt[n]{a}$	nth root of a
i	Imaginary number; $i^2 = -1$
$a + bi$	Complex number

Formulas

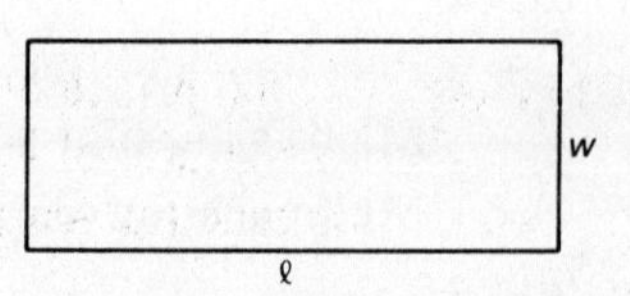

Rectangle The length is l, the width is w.

Perimeter $P = 2l + 2w$

Area $A = lw$

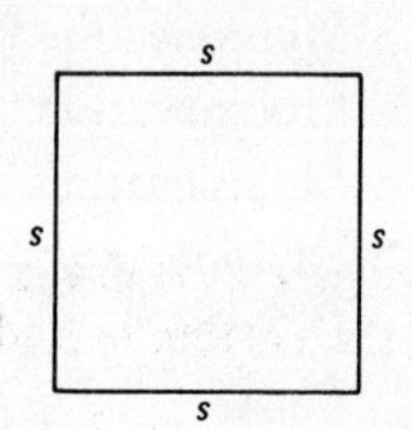

Square Side is s.

Perimeter $P = 4s$

Area $A = s^2$

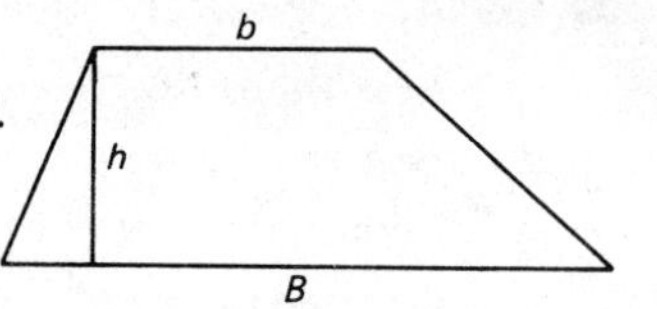

Trapezoid The two parallel sides are b and B. Altitude (height) is h.

Area $A = \frac{1}{2}(b + B)h$

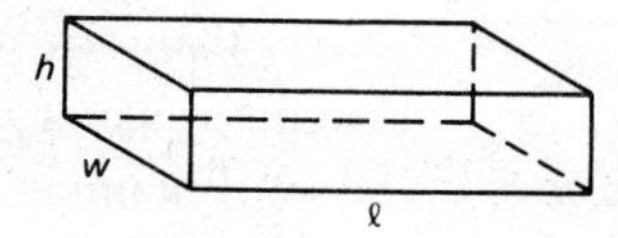

Rectangular solid The height is h.

Volume $V = lwh$

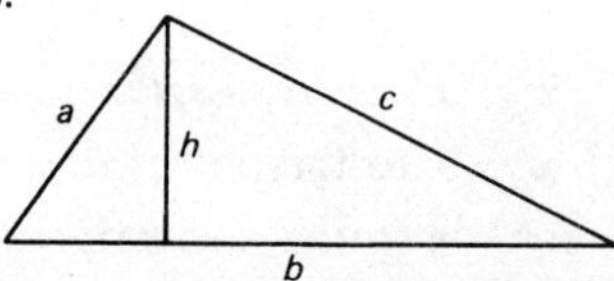

Triangle Sides are a, b, c, where b is the base. Altitude is h.

Perimeter $P = a + b + c$

Area $A = \frac{1}{2}bh$

Distance r is rate (or speed), t is time.

$d = rt$

Simple interest p is the principal (amount of money),
r is the rate of interest (expressed as percent),
t is time (in years).

$I = prt$

Circle The diameter is d, the radius is r.
Use the value $\pi = 3.14$.

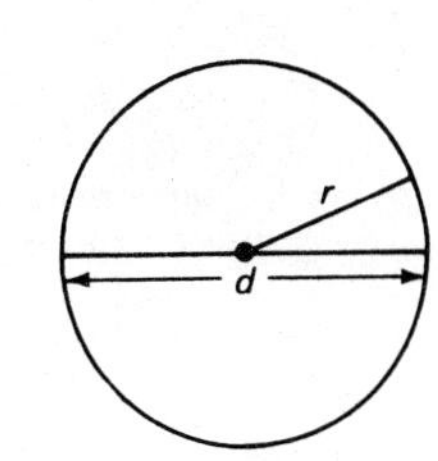

Diameter	$d = 2r$
Circumference	$C = 2\pi r$ or $C = \pi d$
Area	$A = \pi r^2$

Right circular cylinder The radius, r, is that of the top or bottom circle. The height is h.

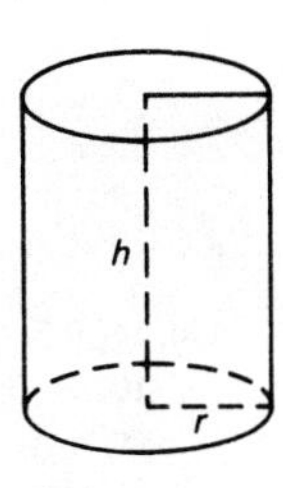

Surface area	$S = 2\pi r^2 + 2\pi rh$
Volume	$V = \pi r^2 h$

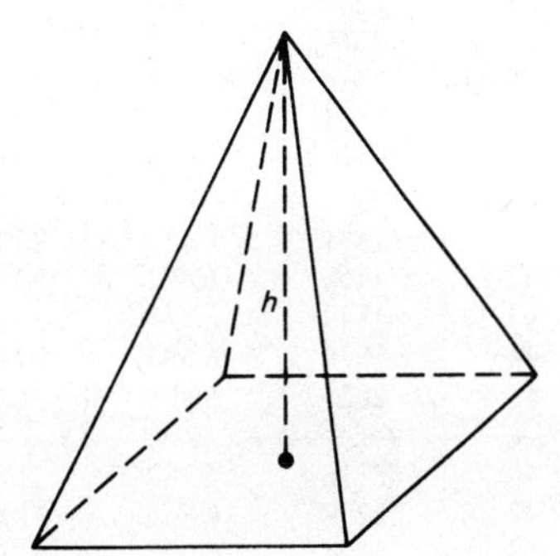

Right pyramid The area of the base is B.
The height is h.

Volume	$V = \frac{1}{3}Bh$

Squares and Square Roots

n	n^2	$\sqrt{n}$	$\sqrt{10n}$	n	n^2	$\sqrt{n}$	$\sqrt{10n}$
1	1	1.000	3.162	51	2601	7.141	22.583
2	4	1.414	4.472	52	2704	7.211	22.804
3	9	1.732	5.477	53	2809	7.280	23.022
4	16	2.000	6.325	54	2916	7.348	23.238
5	25	2.236	7.071	55	3025	7.416	23.452
6	36	2.449	7.746	56	3136	7.483	23.664
7	49	2.646	8.367	57	3249	7.550	23.875
8	64	2.828	8.944	58	3364	7.616	24.083
9	81	3.000	9.487	59	3481	7.681	24.290
10	100	3.162	10.000	60	3600	7.746	24.495
11	121	3.317	10.488	61	3721	7.810	24.698
12	144	3.464	10.954	62	3844	7.874	24.900
13	169	3.606	11.402	63	3969	7.937	25.100
14	196	3.742	11.832	64	4096	8.000	25.298
15	225	3.873	12.247	65	4225	8.062	25.495
16	256	4.000	12.649	66	4356	8.124	25.690
17	289	4.123	13.038	67	4489	8.185	25.884
18	324	4.243	13.416	68	4624	8.246	26.077
19	361	4.359	13.784	69	4761	8.307	26.268
20	400	4.472	14.142	70	4900	8.367	26.458
21	441	4.583	14.491	71	5041	8.426	26.646
22	484	4.690	14.832	72	5184	8.485	26.833
23	529	4.796	15.166	73	5329	8.544	27.019
24	576	4.899	15.492	74	5476	8.602	27.203
25	625	5.000	15.811	75	5625	8.660	27.386
26	676	5.099	16.125	76	5776	8.718	27.568
27	729	5.196	16.432	77	5929	8.775	27.749
28	784	5.292	16.733	78	6084	8.832	27.928
29	841	5.385	17.029	79	6241	8.888	28.107
30	900	5.477	17.321	80	6400	8.944	28.284
31	961	5.568	17.607	81	6561	9.000	28.460
32	1024	5.657	17.889	82	6724	9.055	28.636
33	1089	5.745	18.166	83	6889	9.110	28.810
34	1156	5.831	18.439	84	7056	9.165	28.983
35	1225	5.916	18.708	85	7225	9.220	29.155
36	1296	6.000	18.974	86	7396	9.274	29.326
37	1369	6.083	19.235	87	7569	9.327	29.496
38	1444	6.164	19.494	88	7744	9.381	29.665
39	1521	6.245	19.748	89	7921	9.434	29.833
40	1600	6.325	20.000	90	8100	9.487	30.000
41	1681	6.403	20.248	91	8281	9.539	30.166
42	1764	6.481	20.494	92	8464	9.592	30.332
43	1849	6.557	20.736	93	8649	9.644	30.496
44	1936	6.633	20.976	94	8836	9.695	30.659
45	2025	6.708	21.213	95	9025	9.747	30.822
46	2116	6.782	21.448	96	9216	9.798	30.984
47	2209	6.856	21.679	97	9409	9.849	31.145
48	2304	6.928	21.909	98	9604	9.899	31.305
49	2401	7.000	22.136	99	9801	9.950	31.464
50	2500	7.071	22.361	100	10000	10.000	31.623

Perfect squares are shown in color.

Selected Answers and Solutions

CHAPTER 1

Section 1.1 (page 7)

1. $\frac{7}{14} = \frac{7 \div 7}{14 \div 7} = \frac{1}{2}$ **2.** $\frac{1}{3}$ **3.** $\frac{5}{6}$ **5.** $\frac{8}{9}$ **6.** $\frac{7}{10}$
7. $\frac{2}{3}$ **9.** $\frac{2}{3}$ **10.** $\frac{4}{5}$ **11.** $\frac{3}{4} \cdot \frac{9}{5} = \frac{3 \cdot 9}{4 \cdot 5} = \frac{27}{20}$ or $1\frac{7}{20}$ **13.** $\frac{3}{25}$ **14.** $\frac{2}{3}$ **15.** $\frac{6}{5}$ or $1\frac{1}{5}$ **17.** $\frac{3}{10}$
18. $\frac{3}{2}$ or $1\frac{1}{2}$ **19.** $\frac{1}{9}$ **21.** $\frac{3}{20}$ **22.** 16 **23.** $\frac{5}{12}$
25. $\frac{13}{3}$ or $4\frac{1}{3}$ **26.** $\frac{18}{7}$ or $2\frac{4}{7}$ **27.** 28 **29.** $\frac{1}{12} + \frac{3}{12} = \frac{1+3}{12} = \frac{4}{12} = \frac{4 \div 4}{12 \div 4} = \frac{1}{3}$ **30.** $\frac{4}{3}$ or $1\frac{1}{3}$
31. $\frac{7}{10}$ **33.** $\frac{10}{9}$ or $1\frac{1}{9}$ **34.** $\frac{11}{15}$ **35.** $\frac{19}{22}$
37. $\frac{1}{15}$ **38.** $\frac{1}{9}$ **39.** $\frac{5}{6} - \frac{3}{10} = \frac{5 \cdot 5}{6 \cdot 5} - \frac{3 \cdot 3}{10 \cdot 3} = \frac{25}{30} - \frac{9}{30} = \frac{25-9}{30} = \frac{16}{30} = \frac{8}{15}$ **41.** $9\frac{3}{8}$ **42.** $5\frac{11}{12}$
43. $5\frac{1}{6}$ **45.** $1\frac{5}{12}$ **46.** $1\frac{4}{45}$ **47.** $\frac{49}{30}$ or $1\frac{19}{30}$
49. $\frac{13}{28}$ **50.** $\frac{1}{3}$ **51.** $\frac{1}{8} + \frac{1}{3} + \frac{1}{4} = \frac{3}{24} + \frac{8}{24} + \frac{6}{24} = \frac{17}{24}$ **53.** $14\frac{7}{16}$ **54.** $618\frac{3}{4}$ **55.** $8\frac{23}{24}$
57. $16 \times 2\frac{1}{4} = 16 \times \frac{9}{4} = 36$ **58.** 56

Section 1.2 (page 15)

1. 80 + 6 **2.** 10 + 5 **3.** 600 + 90 + 4
5. 5000 + 200 + 30 + 7 **6.** 4000 + 700 + 60 + 1
7. 30 + 6 + .8 + .01 **9.** .5 + .06 + .007
10. .1 + .04 + .006 **11.** $.2 = \frac{2}{10} = \frac{1}{5}$ **13.** $\frac{9}{25}$
14. $\frac{18}{25}$ **15.** $\frac{42}{125}$ **17.** $\frac{161}{200}$ **18.** $\frac{5}{8}$

19.
```
    .625
8/5.0000
  4 8
    20
    16
     40
     40
```
21. .75 **22.** .4375 or .438 (rounded)

23. .5625 or .563 (rounded) **25.** .667 **26.** .833
27. .778

29.
```
   14.230
    9.810
   74.630
 + 18.715
 --------
  117.385
```
30. 887.859

31. 13.21 **33.** 96.101 **34.** 31.48 **35.** 4.849

37.
```
 39.6 ← one decimal place
  4.2 ← one decimal place
  79 2
 1584
166.32 ← two decimal places
```
38. 43.01

39. 164.19 **41.** .1344 **42.** 1.6559 **43.** 4.14
45. 2.23 **46.** 6.14 **47.** 4.8 **49.** .80 = 80(.01) = 80% **50.** 75% **51.** .7% **53.** 67% **54.** .3%
55. 12.5% **57.** 53% = 53 · 1% = 53(.01) = .53
58. .38 **59.** 1.29 **61.** .96 **62.** .11 **63.** .009
65. .14(780) = 109.2 **66.** 42 **67.** 238.92
69. 4 **70.** 2500 **71.** 10,185 **73.** 121
74. 12,800 **75.** 148.44 **77.** 5337 **78.** 6.94
79. 10.5 **81.** 12% · \$23,000 = .12(\$23,000) = \$2760 **82.** \$150(12) = \$1800 **83.** \$5760
85. \$165 **86.** \$37.50 **87.** 249 **89.** 63
90. \$5040

Section 1.3 (page 23)

1. True **2.** True **3.** False **5.** True **6.** True
7. True **9.** False **10.** False **11.** False
13. True **14.** False **15.** False **17.** True
18. False **19.** 3 · 4 + 7 = 12 + 7 = 19, which is not less than 10. False **21.** True **22.** True
23. True **25.** False **26.** False **27.** False
29. True **30.** True **31.** 6 · 5 + 3 · 10 = 30 + 30 = 60, which is not less than 0. False **33.** True
34. False **35.** True **37.** False **38.** False
39. False **41.** False **42.** False
43. $\frac{2(5+1) - 3(1+1)}{5(8-6) - 4 \cdot 2} = \frac{2(6) - 3(2)}{5(2) - 8} = \frac{12-6}{10-8} = \frac{6}{2} = 3$, which is less than or equal to 3. True
45. 10 − (7 − 3) = 6 **46.** 3 · (5 + 7) = 36
47. (3 · 5) + 7 = 22, or use no parentheses at all
49. (3 · 5) − 4 = 11, or use no parentheses at all
50. (3 · 5) + (2 · 4) = 23, or use no parentheses at all
51. 3 · (5 + 2) · 4 = 84 **53.** 3 · (5 − 2) · 4 = 36
54. (3 · 5) − (2 · 4) = 7, or use no parentheses at all
55. (100 ÷ 20) ÷ 5 = 1, or use no parentheses at all
56. (360 ÷ 18) ÷ 4 = 5, or use no parentheses at all

Section 1.4 (page 29)

1. $x + 9 = 3 + 9 = 12$ **2.** 1 **3.** 15 **5.** 14
6. 22 **7.** $x + y = 4 + 2 = 6$ **9.** 43 **10.** 22
11. 24 **13.** 10 **14.** 16 **15.** 2 **17.** $\frac{10}{12}$ or $\frac{5}{6}$
18. $\frac{21}{4}$ or $5\frac{1}{4}$ **19.** If $x = 9$, then $x + 6 = 15$ becomes $9 + 6 = 15$, which is true, so 9 is a solution. **21.** No
22. No **23.** Yes **25.** Yes **26.** Yes **27.** No
29. Try $x = 2$, $x = 4$, and $x = 6$. Only $x = 2$ produces a true statement, so that the solution is 2. 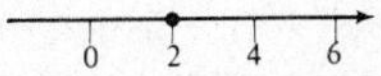 **30.** 3 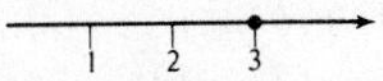
31. 4 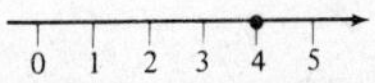 **33.** 6 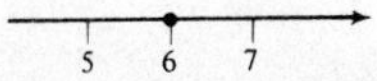
34. 5  **35.** 4 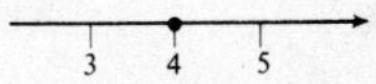
37. No solutions **38.** 2

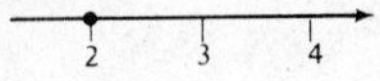

39. 3, 4, 5 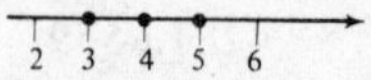

41. 1, 2, 3 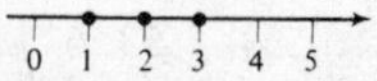

42. 1, 2 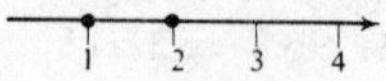

43. 2, 3, 4 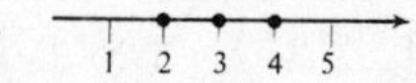

45. No solutions **46.** No solutions

47. 2, 3, 5, 6 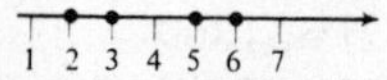 **49.** $2x$ **50.** $5x$

51. $6 + x$ or $x + 6$ **53.** $8 - x$ **54.** $x - 9$ **55.** $8 + 3x$ or $3x + 8$ **57.** $15 - 2x$ **58.** $52 + 8x$, or $8x + 52$ **59.** $x + 8 = 12$; 4 **61.** $x + 2 < 11$; 0, 2, 4, 6, 8 **62.** $2x + 6 < 10$ or $6 + 2x < 10$; 0 **63.** $2x + 5 < 10$ or $5 + 2x < 10$; 0, 2 **65.** $\frac{x+2}{4} = 3$; 10 **66.** $3x = 2 + 2x$; 2 **67.** $\frac{12}{x} = 3x$; 2

Chapter 1 Test (page 33)

1. $\frac{3}{8}$ **2.** $\frac{7}{11}$ **3.** $\frac{2}{3}$ **4.** $\frac{61}{40}$ or $1\frac{21}{40}$ **5.** $\frac{203}{120}$ or $1\frac{83}{120}$ **6.** $14\frac{3}{8}$ **7.** $\frac{2}{3}$ **8.** $\frac{18}{19}$ **9.** $58\frac{4}{9}$ **10.** $\frac{5}{8}$
11. .5625 **12.** 21.77 **13.** 24.7 **14.** 15.8256 **15.** 11.56 **16.** 19% **17.** .762 **18.** 13.6 **19.** 653.2 **20.** .85(240) = 204 **21.** .20(48) = \$9.60, the discount **22.** True **23.** $4[5(1) - 3] = 4[5 - 3] = 4[2] = 8$, which is not less than 8. False

24. True **25.** False **26.** 6, 8, 10 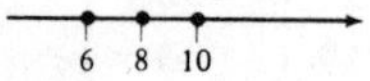

27. 0, 1, 2, 3 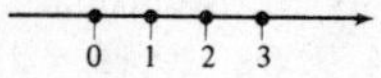

28. 2, 3, 4 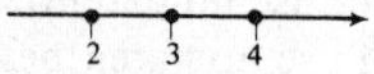

CHAPTER 2

Section 2.1 (page 39)

1. −8 **2.** −12 **3.** 9 **5.** 2 **6.** 3 **7.** −15 **9.** The value of $|8|$ is 8; the additive inverse of this number is −8. **10.** −5 **11.** −3 **13.** −14 **14.** −8 **15.** −16 **17.** $|-2|$ **18.** $|-3|$ **19.** $|-8|$ **21.** $-|-3|$ **22.** True **23.** True **25.** True **26.** False **27.** False **29.** False **30.** True **31.** True **33.** False **34.** False

35. True **37.** 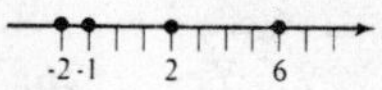

38. 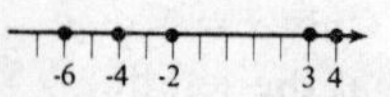 **39.** 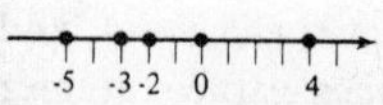

41. Try all the numbers in the list to see which ones work. The only numbers that make $|x| = 3$ true are 3 and −3. **42.** 4, −4 **43.** −1, 0, 1 **45.** −1, 0, 1 **46.** −2, −1, 0, 1, 2 **47.** −4, −3, −2, 2, 3, 4 **49.** 0 **50.** −4, 4 **51.** All the numbers

Section 2.2 (page 45)

1. 2 **2.** 3 **3.** −2 **5.** −8 **6.** −11 **7.** −11 **9.** −12 **10.** −16 **11.** 4 **13.** $4 + [13 + (-5)] = 4 + (8) = 12$ **14.** −5 **15.** 5 **17.** 2 **18.** −1 **19.** −9 **21.** 13 **22.** 4 **23.** −11 **25.** $\frac{1}{2}$ **26.** $\frac{3}{10}$ **27.** $-\frac{19}{24}$ **29.** $-\frac{3}{4}$ **30.** $2\frac{1}{8}$ **31.** −.5

33. −7.7 **34.** −16.6 **35.** −8 **37.** 0 **38.** −19 **39.** $-10 + (-11) + 1 = -20$ **41.** True **42.** True **43.** False **45.** False **46.** False **47.** True **49.** True **50.** False **51.** False **53.** True **54.** True **55.** Try all numbers in the domain. The only number that works, which is the solution, is −2. **57.** −3 **58.** −3 **59.** −2 **61.** −2 **62.** −3 **63.** 2 **65.** \$15 − \$6 = \$9 **66.** 6000 − 4000 = 2000 **67.** −15 − 120 = −135 feet, or 135 feet below sea level **69.** 90 **70.** −26°

Section 2.3 (page 51)

1. $3 - 6 = 3 + (-6) = -3$ **2.** −5 **3.** −4 **5.** −8 **6.** −15 **7.** −14 **9.** 9 **10.** 13 **11.** $5 - (-12) = 5 + (12) = 17$ **13.** −4 **14.** −2

15. $2 - (-2) = 2 + (2) = 4$ **17.** 1 **18.** 4 **19.** $\frac{3}{4}$ **21.** $-\frac{11}{8}$ or $-1\frac{3}{8}$ **22.** $-\frac{4}{3}$ or $-1\frac{1}{3}$ **23.** $\frac{15}{8}$ or $1\frac{7}{8}$

25. 11.6 **26.** 17.3 **27.** −9.9 **29.** $(4 - 6) + 12 = (-2) + 12 = 10$ **30.** 0 **31.** −5 **33.** 11 **34.** 12 **35.** −10 **37.** $(-5 - 6) - (9 - 2) = (-11) - 7 = -18$ **38.** −17 **39.** 2 **41.** −16 **42.** 4 **43.** −12 **45.** $12 - (-6)$; 18 **46.** $15 - (-8)$; 23 **47.** $-25 - (-4)$; −21 **49.** $-24 - (-27)$; 3 **50.** $8 - (-5)$; 13 **51.** $-5 - 10 = -15$ **53.** $14{,}494 - (-282) = 14{,}494 + 282 = 14{,}776$ **54.** \$15 − (−\$12) = \$27

Section 2.4 (page 59)

1. 12 **2.** −12 **3.** −12 **5.** 5 **6.** 45 **7.** 44 **9.** 120 **10.** −45 **11.** −48 **13.** −30 **14.** 0 **15.** −65 **17.** 0 **18.** −45 **19.** −165 **21.** $\frac{5}{12}$ **22.** $-\frac{1}{2}$ **23.** −.102 **25.** $9(6 - 10) = 9(-4) = -36$

26. −15 **27.** 12 **29.** 5 **30.** 15 **31.** $(2 - 5)(3 - 7) = (-3)(-4) = 12$ **33.** 18 **34.** 27 **35.** −14 **37.** $-12 - (-2) = -10$ **38.** −1 **39.** 45 **41.** 12 **42.** 42 **43.** $(2)(-4) + (6)(4) = -8 + 24 = 16$ **45.** $2x + 7y = 2(-2) + 7(3) = -4 + 21 = 17$ **46.** 9 **47.** −28 **49.** $(2x + y)(3a) = (-4 + 3)(3 \cdot -4) = (-1)(-12) = 12$ **50.** −128 **51.** −360 **53.** −2 **54.** −2 **55.** 0 **57.** −2 **58.** −3

Section 2.5 (page 65)

1. $\frac{1}{9}$ **2.** $\frac{1}{8}$ **3.** $\frac{1}{-4}$ or $-\frac{1}{4}$ **5.** $\frac{3}{2}$ **6.** $\frac{4}{3}$ **7.** $-\frac{10}{9}$ **9.** No reciprocal **10.** −2 **11.** −4 **13.** −10 **14.** −6 **15.** −4 **17.** −10 **18.** 2 **19.** 5

21. 14 22. 36 23. 50 25. 0 26. $\frac{2}{3}$
27. $\frac{10}{3}$ or $3\frac{1}{3}$ 29. 1.4 30. −5 31. −20
33. −3 34. −10 35. −10 37. 10 38. −4
39. −9 41. $\frac{-200}{-6-(-4)} = \frac{-200}{-2} = 100$ 42. −6
43. −3 45. −1 46. 4 47. $\frac{-12(-3)}{-15-(-3)} = \frac{36}{-12} =$
-3 49. 20 50. −3 51. 4 53. 2 54. −2
55. 1 57. 2 58. No such number (denominator is 0) 59. No such number (denominator is 0)
61. −32, −16, −8, −4, −2, −1, 1, 2, 4, 8, 16, 32
62. −25, −5, −1, 1, 5, 25 63. −14, −7, −2, −1, 1, 2, 7, 14 65. −50, −25, −10, −5, −2, −1, 1, 2, 5, 10, 25, 50 66. −17, −1, 1, 17 67. −13, −1, 1, 13
69. −37, −1, 1, 37 70. −8 71. −2 73. 4
74. 0 75. 0 77. −2 78. −4 79. −6
81. $4x = -32; -8$ 82. $\frac{x}{5} = 15; 75$ 83. $\frac{x}{6} = -3;$ -18 85. $\frac{x}{-3} = -4; 12$ 86. $\frac{x}{2} = -6; -12$
87. $\frac{x}{-9} = 2; -18$

Section 2.6 (page 73)

1. Commutative 2. Associative 3. Associative
5. Commutative 6. Associative 7. Commutative
9. Closure 10. Closure 11. Inverse
13. Identity 14. Identity 15. Inverse
17. Identity 18. Identity 19. Distributive
21. $k + 9$ 22. $5 + z$ 23. m 25. $3r + 3m$
26. $11k + 11z$ 27. 1 29. 0 30. 0 31. $-5 + 5$, or 0 33. $-3r - 6$ 34. $4k - 20$ 35. 9
37. $k + [5 + (-6)] = k - 1$ 38. $m + [4 + (-2)] = m + 2$ 39. $4z + (2r + 3k)$ 41. $5m + 10$
42. $6k + 30$ 43. $-4r - 8$ 45. $-8k + 16$
46. $-4z + 20$ 47. $-9a - 27$ 49. $4r + 32$
50. $6m + 72$ 51. $-16 + 2k$ 53. $10r + 12m$
54. $10a + 20b$ 55. $-12x + 16y$ 57. $5(8 + 9) = 5(17) = 85$ 58. $4(3 + 9) = 4(12) = 48$
59. $7(2 + 8) = 7(10) = 70$ 61. $9(p + q)$
62. $8(2x + 3y)$ 63. $5(7z + 8w)$ 65. (a) yes (b) no (c) yes 66. (a) yes (b) no (c) no
67. (a) yes (b) yes (c) yes 69. (a) yes (b) no (c) yes 70. (a) yes (b) no (c) yes

Chapter 2 Test (page 77)

1. 4 2. −9 3. −1 4. 32 5. −8 6. −4
7. 10 8. 6 9. −72 10. −3 11. −9 12. −1
13. 2 14. No such number 15. 0 16. 0
17. −5 18. −33 19. −15 20. 2 21. 18
22. 25 23. −19 24. 1 25. $\frac{-7-(-5+1)}{-4-(-3)} = \frac{-7-(-4)}{-4-(-3)} = \frac{-7+(+4)}{-4+(+3)} = \frac{-3}{-1} = 3$ 26. 4
27. Denominator is 0; no such number 28. 3
29. −1 30. 9 31. −5 32. −4 33. −2
34. −5 35. $-|-8|$ 36. 0 37. 3 38. −6
39. H, J 40. B 41. F, G 42. C, D
43. A, I 44. E 45. Yes

CHAPTER 3

Section 3.1 (page 83)

1. Let $x = 5$; $2 \cdot 5 + 3 = 10 + 3 = 13$; 5 is a solution
2. Yes 3. No 5. Yes 6. Yes 7. Yes
9. No 10. Yes 11. $2k + 5k + 9 + 6 = 7k + 15$
13. $m - 1$ 14. $-15 - 12x$ 15. $-4y$ 17. $2x + 6$
18. $-5r - 10$ 19. $20 - 7m$ 21. $7x = 14; 2$
22. $x + 11 = 20; 9$ 23. $k + 5 = 13; 8$ 25. $z + 7 = 13; 6$ 26. $17 = -3 + k; 20$ 27. $x - 17 = 2; 19$
29. $5x + 3x + 1 = 49; 6$ 30. $4x - 3x + 8 = 11; 3$
31. $x + 3x = 24$; x is 6, so Joann is $3 \cdot 6 = 18$ years old

Section 3.2 (page 87)

1. $x - 3 = 7; x - 3 + 3 = 7 + 3; x = 10$; solution is 10
2. 8 3. −2 5. 10 6. 3 7. −8 9. −5
10. 7 11. −2 13. 4 14. −3 15. −5
17. −11 18. −26 19. −6 21. $4x + 2x - 5x + 3 = 10; x + 3 = 10; x = 7$; solution is 7 22. 19
23. 8 25. −10 26. −8 27. −9 29. 0
30. 0 31. $3x = 17 + 2x; 17$ 33. $5x + 3x = 7x + 9; 9$ 34. $x + (x + 1) = 3x - 13$; the integers are 14 and 15

Section 3.3 (page 93)

1. $5x = 25; \frac{1}{5}(5x) = \frac{1}{5}(25); x = 5$; solution is 5 2. 4
3. 25 5. −8 6. −12 7. −7 9. −4 10. −4
11. −9 13. 0 14. 0 15. −6 17. 4 18. 8
19. $5x = 20$; solution is 4 21. 8 22. 7 23. 20
25. −3 26. −3 27. 4 29. −3 30. −5
31. 7 33. $\frac{m}{2} = 16; 2\left(\frac{m}{2}\right) = 2(16); m = 32$; solution is 32 34. 15 35. 49 37. 9 38. 24 39. 8
41. −80 42. −64 43. $\frac{15}{2}$ 45. $\frac{49}{2}$ 46. $\frac{22}{3}$
47. 3 49. −6 50. 4 51. $\frac{x}{4} = 6; 24$
53. $\frac{x}{4} = 62$; \$248 54. $\frac{2x}{5} = 4; 10$

Section 3.4 (page 99)

1. $3(k - 6) = 3 \cdot k - 3 \cdot 6 = 3k - 18$ 2. $5m + 20$
3. $30t + 66$ 5. $-3n - 15$ 6. $-4v + 32$
7. $-6x + 8$ 9. $4r + 14$ 10. $8m - 18$
11. $-5(2r - 3) + 2(5r + 3) = -5(2r) + (-5)(-3) + 2(5r) + 2(3) = -10r + 15 + 10r + 6 = 15 + 6 = 21$
13. $12k - 13$ 14. $13p - 13$ 15. $13k - 21$
17. $4h + 8 = 16; 4h + 8 + (-8) = 16 + (-8); 4h = 8; \frac{1}{4}(4h) = \frac{1}{4}(8); h = 2$; solution is 2 18. 8 19. −4
21. 9 22. 5 23. −2 25. −1 26. 5 27. −4
29. $-4 - 3(2x + 1) = 11; -4 - 6x - 3 = 11; -7 - 6x = 11; -6x = 18; x = -3$; solution is −3 30. 2 31. 18
33. 8 34. 2 35. −2 37. 6 38. 3 39. −5
41. 0 42. 3 43. 5 45. 0 46. 1 47. $-\frac{1}{5}$

$49.\ -\frac{5}{7}$ $50.\ -\frac{29}{2}$ **51.** $3(x-17)=102; 51$

53. $8-3(x+4)=2; -2$ **54.** $5(x+3)+5=-5; -5$

Section 3.5 (page 107)

1. $8+x$ or $x+8$ **2.** $-6+x$ or $x+(-6)$ **3.** $-1+x$ or $x+(-1)$ **5.** $x+(-18)$ **6.** $x+12$ or $12+x$ **7.** $x-5$ **9.** $x-9$ **10.** $x-16$ **11.** $9x$ **13.** $3x$ **14.** $\frac{3}{5}x$ **15.** $\frac{x}{6}$ **17.** $\frac{x}{-4}$ **18.** $\frac{7}{x}$ **19.** $8(x+3)$

21. $\frac{1}{x}-x$ **22.** $3\left(\frac{x}{2}\right)$ **23.** $8(x-8)$ **25.** Let x be the number; $3x-2=22; 8$ **26.** Let x be the number; $4x+6=42; 9$ **27.** Let x be the number; $4(x+3)=36; 6$ **29.** Let x be the number; $2x+x=90; 30$ **30.** Let x be the number; $-2(x+8)=-8; -4$ **31.** Let x be the number; $x-6=7x; -1$

33. Let x be the number; $5x+2x=10; \frac{10}{7}$ **34.** Let x be the number; $11x-7x=9; \frac{9}{4}$ **35.** Let x be the length of the shorter piece; $x+(x+6)=44$; 19 inches **37.** Let x be the lowest grade; $x+(x+42)=138; 48$ **38.** Let x be the number that Chuck did; $x+(x+25)=173; 74$ **39.** Let x be the number of tranquilizers; $x+(x+12)=84; 36$ **41.** Let x be the number of gallons that Bossie gave; $x+(x+238)=1464; 613$ **42.** Let x be Bob's age; then $3x$ is Kevin's age; three years ago Bob was $x-3$ years old, and Kevin was $3x-3$ years old; the sum of their ages then was 22, so that $(x-3)+(3x-3)=22$; x is 7, so that Bob is 7 and Kevin is $3\cdot 7=21$. **43.** 36 quart cartons

Section 3.6 (page 117)

1. $P=4s; s=32$, so $P=4(32)=128$ **2.** 16 **3.** 36 **5.** 8 **6.** 10 **7.** 14 **9.** 21 **10.** 33 **11.** $d=rt; d=8, r=2$ so $8=2t; t=4$ **13.** 1.5 **14.** 4

15. $A=lw$; $(1/w)(A)=(1/w)(lw)$; $A/w=l$ **17.** $t=\frac{d}{r}$

18. $w=\frac{V}{lh}$ **19.** $h=\frac{V}{lw}$ **21.** $t=\frac{I}{pr}$ **22.** $r=\frac{C}{2\pi}$

23. $b=\frac{2A}{h}$ **25.** $w=(P-2l)/2$ **26.** $b=p-a-c$ or $b=p-(a+c)$ **27.** $A=\frac{1}{2}(b+B)h$; $2A=2\cdot\frac{1}{2}(b+B)h$; $2A=(b+B)h$; $\frac{2A}{h}=b+B$ (after multiplying both sides by $\frac{1}{h}$); $\frac{2A}{h}+(-B)=b+B+(-B)$; $\frac{2A}{h}-B=b$ **29.** $A=lw$; $60=6l$; l is 10 meters

30. 20 centimeters **31.** $C=2\pi r$; $C=2(3.14)(6)$; $C=37.68$ feet **33.** 24 kilometers **34.** 40 miles per hour **35.** 385 miles **37.** 4 **38.** 2 meters **39.** Let w be the width; then the perimeter is $16w$; the length is $w+12$; the formula is $P=2l+2w$, so $16w=2(w+12)+2w$; $w=2$, so that the rectangle is 2 centimeters by $2+12=14$ centimeters. **41.** 3

42. $2\frac{1}{2}$ hours **43.** 2 hours **45.** Let x be the number of days; then her total costs are $28+45x$, since the \$28 is paid only once, and the \$45 is paid each day. Then $28+45x=163$, and $x=3$, so that she can stay 3 days. **46.** 8 hamburgers, 4 fries

Section 3.7 (page 125)

1.  4 **2.**  -3

3.  -5 **5.** 3

6.  4 **7.**  -2 5

9.  3 5 **10.** 0 10

11. $a<2$ **13.** $z\geq 1$ **14.** $p\geq -8$ **15.** $p\leq 12$ **17.** $k\geq 5$ **18.** $y<-2$ **19.** $x\leq 0$ **21.** $x<6$

22. $x\leq -8$ **23.** $n\leq -11$;  -11

25. $z>5$ 5

26. $x\leq -11$  -11

27. $y<5$ 5

29. $k\geq 44$ 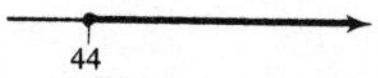 44

30. $m<-31$ -31

31. $4x+8<3x+5; x<-3$ **33.** $2x-x\leq 7$; 7 meters **34.** $18+13+x\leq 55$; 24 centimeters

Section 3.8 (page 131)

1. $x<9$ 9

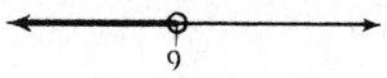

2. $h\geq 4$ 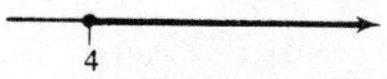 4

3. $r\geq -3$ -3

5. $k\geq -6$  -6

6. $v<-2$ -2

7. $y<-9$ -9

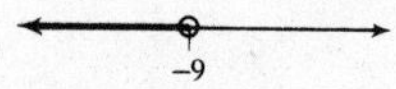

9. $m<7$ 7

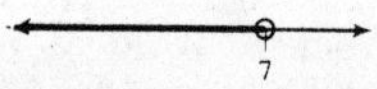

10. $x\geq 2$  2

11. $r>\frac{8}{3}$ 8/3

13. $4k+1\geq 2k-9$; $4k+1+(-1)\geq 2k-9+(-1)$; $4k\geq 2k-10$; $4k+(-2k)\geq 2k-10+(-2k)$; $2k\geq -10$; $\frac{1}{2}(2k)\geq\frac{1}{2}(-10)$; $k\geq -5$ -5

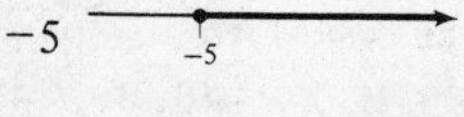

14. $y<3$ 3

15. $r < 10$

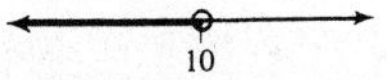

17. $q > -2$

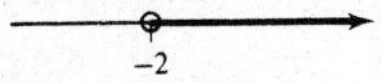

18. $x \leqslant 2$

19. $p > -5$

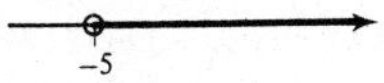

21. $k \leqslant 0$

22. $y > 0$

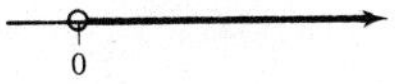

23. $x < -11$

25. $r \geqslant -1$

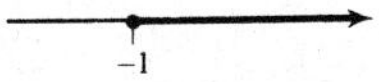

26. $y < \frac{1}{2}$

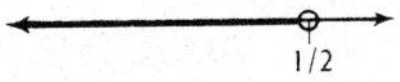

27. $p \leqslant \frac{8}{3}$

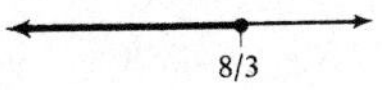

29. Let x be the minimum grade. Then $\frac{75 + 82 + x}{3} \geqslant 80$; solve to get $x \geqslant 83$, so that the student must score 83 or more. **30.** 85.7; 52.3 **31.** $2x + 3(x + 2) > 17$; any number greater than $\frac{11}{5}$

22. $m > 7$

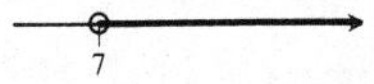

23. $k > 2$

24. $k \leqslant 1$

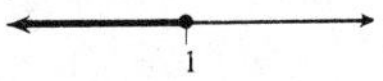

25. $r \leqslant 4$

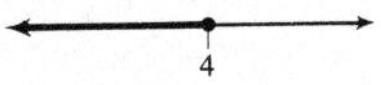

26. Let l be the length. Then the perimeter is $3l - 2$. Since the perimeter is also given as 190, we have $3l - 2 = 190$, or $l = 64$. To find the width, use the formula $P = 2l + 2w$, with $P = 190$ and $l = 64$. This gives $190 = 2(64) + 2w$, or $190 = 128 + 2w$, or $62 = 2w$, or $w = 31$. The rectangle is 31 inches by 64 inches. **27.** 3 **28.** Let x be his speed for the first 3 hours, so that $x - 2$ is his speed for the last 2 hours. Since $d = rt$, his distance at the faster speed is $3x$ (3 hours at x miles per hour) and his distance at the slower speed is $2(x - 2)$. The total distance is 21 miles, so that $3x + 2(x - 2) = 21$. Solve this to find that x is 5 miles per hour. **29.** Let x be Hank's age now. Then Don's age now is $x + 20$. In 5 years, Hank will be $x + 5$ years old, and Don will be $x + 20 + 5$, or $x + 25$ years old. In 5 years, Don will be twice as old as Hank, or $x + 25 = 2(x + 5)$, or $x = 15$. So Hank is now 15 years old, and Don is $x + 20 = 15 + 20 = 35$. **30.** 60% **31.** $90/75 = 1.2 = 120\%$

Section 3.9 (page 137)

1. .27 **2.** .39 **3.** .96 **5.** 1.37 **6.** 2.41 **7.** .002 **9.** 70% **10.** 30% **11.** 42% **13.** 192% **14.** 381% **15.** $\frac{625}{125\%} = \frac{625}{1.25} = 500$ **17.** 83 **18.** 40,000 **19.** 562.5 **21.** 6400 **22.** 600 **23.** $\frac{1318}{2636} = .50 = 50\%$ **25.** 48% **26.** 20% **27.** 7960 **29.** $\frac{\$17.80}{12\%} = \frac{\$17.80}{.12} = \$148.33$ **30.** 2325 **31.** 2% **33.** $\frac{\$2013}{110\%} = \frac{\$2013}{1.10} = \$1830$ **34.** 358,000 **35.** $\frac{\$1648.40}{104\%} = \frac{\$1648.40}{1.04} = \$63.40$

Chapter 3 Test (page 139)

1. $4x + 2$ **2.** k **3.** $-r$ **4.** $z - 10$ **5.** $7m - 12$ **6.** $12z - 22$ **7.** 3 **8.** 4 **9.** 5 **10.** -5 **11.** $\frac{11}{3}$ **12.** $4k - 6k + 8k - 24 = -2k - 24$; $6k - 24 = -2k - 24$; $8k = 0$; $k = 0$ **13.** -5 **14.** 10 **15.** $\left(\frac{3}{2}\right)\left(\frac{2}{3}z\right) = \left(\frac{3}{2}\right)(18)$; $z = 27$ **16.** $-\frac{11}{2}$ **17.** -3 **18.** $p = \frac{I}{rt}$ **19.** $h = 2A/(b + B)$

20. $x \leqslant 4$

21. $z \geqslant -2$

CHAPTER 4

Section 4.1 (page 147)

1. 5^{12} ← exponent; base **2.** Base is a, exponent is 6 **3.** Base is $3m$, exponent is 4 **5.** -125^3 ← exponent; base is 125 (not -125) **6.** Base is -1 (because of the parentheses), exponent is 8 **7.** Base is -24, exponent is 2 **9.** Base is m (exponent refers only to the m, *not* to $3m$), exponent is 2 **10.** Base is y, exponent is 3 **11.** $3 \cdot 3 \cdot 3 \cdot 3 \cdot 3 = 3^5$, since 3 is a factor 5 times **13.** 5^4 **14.** 3^9 **15.** $(-2)^5$, since -2 appears as a factor 5 times **17.** $1/4^5$, since 4 is a factor in the denominator 5 times **18.** $1/(-2)^3$ **19.** $1/3^4$ **21.** $3^2 + 3^4 = 9 + 81 = 90$ **22.** 192 **23.** 80 **25.** 36 **26.** 20 **27.** $4^0 + 5^0 = 1 + 1 = 2$ **29.** 2 **30.** 2 **31.** $4^2 \cdot 4^3 = 4^{2+3} = 4^5$ **33.** 9^8 **34.** 8^{10} **35.** 3^{11} **37.** $4^3 \cdot 4^5 \cdot 4 = 4^{3+5+1} = 4^9$ **38.** 2^{12} **39.** $(-3)^5$ **41.** $(-2)^9$ **42.** $(-3)^{10}$ **43.** $4^{3-2} = 4^1 = 4$ **45.** $4^2/4^4 = 1/4^{4-2} = 1/4^2$ **46.** $1/14^4$ **47.** 8^6 **49.** $6^1 = 6$ **50.** 7^7 **51.** $(-14)^1 = -14$ **53.** $-19^0/(-18)^0 = -1/1 = -1$ **54.** -1 **55.** $x^4 \cdot x^5 = x^{4+5} = x^9$ **57.** r^{11} **58.** p^{14} **59.** $(y^3)^3/(y^2)^2 = y^{3 \cdot 3}/y^{2 \cdot 2} = y^9/y^4 = y^5$ **61.** a^3 **62.** s^4 **63.** k^6 **65.** $(5m)^3 = 5^3m^3 = 125m^3$ **66.** $2^4x^4y^4$ or $16x^4y^4$ **67.** $3^4m^4n^4 = 81m^4n^4$ **69.** $(-3x^5)^2 = (-3)^2(x^5)^2 = 9x^{5 \cdot 2} = 9x^{10}$ **70.** $4^4m^{12}n^8$ or $256m^{12}n^8$ **71.** $5^3p^6q^3 = 125p^6q^3$ **73.** $a^3/5^3 = a^3/125$ **74.** $81/x^2$ **75.** $243m^5n^5/32$

77. $\dfrac{x^7x^8(x^3)^2}{x^9x^7} = \dfrac{x^{7+8} \cdot x^{3\cdot 2}}{x^{9+7}} = \dfrac{x^{15}x^6}{x^{16}} = \dfrac{x^{15+6}}{x^{16}} = \dfrac{x^{21}}{x^{16}} = x^{21-16} = x^5$ 78. $\dfrac{1}{m^5}$ 79. $\dfrac{b^{11}(b^2)^4}{(b^3)^3(b^2)^6} = \dfrac{b^{11}b^{2\cdot 4}}{b^{3\cdot 3}b^{2\cdot 6}} = \dfrac{b^{11}b^8}{b^9b^{12}} = \dfrac{b^{11+8}}{b^{9+12}} = \dfrac{b^{19}}{b^{21}} = \dfrac{1}{b^{21-19}} = \dfrac{1}{b^2}$

Section 4.2 (page 155)

1. For $x = 2$: $2(2)^2 - 4(2) = 2 \cdot 4 - 8 = 8 - 8 = 0$; for $x = -1$: $2(-1)^2 - 4(-1) = 2(1) + 4 = 2 + 4 = 6$ 2. $38; -1$ 3. $36; -12$ 5. $19; -2$ 6. $14; -19$ 7. $-5; 1$ 9. $-12; 0$ 10. $12; 9$ 11. $P(-1) = (-1)^3 - 3(-1)^2 + 2(-1) - 3 = -1 - 3 - 2 - 3 = -9$ 13. -3 14. 15 15. -27 17. $Q(-2) = (-2)^4 - 1 = 16 - 1 = 15$ 18. First find $P(-2)$, which is -27. Then find $Q(-2)$, which is 15. So $P(-2) + Q(-2) = -27 + 15 = -12$.   19. First find $P(-1)$, which is -9. Then find $Q(-2)$, which is 15. So $P(-1) \cdot Q(-2) = -9 \cdot 15 = -135$.   21. $3m^5 + 5m^5 = (3 + 5)m^5 = 8m^5$ 22. $-1y^3 = -y^3$ 23. $-r^5$ 25. Cannot be simplified, since the terms are not like terms (exponents are different)   26. Cannot be simplified   27. $3x^5 + 2x^5 - 4x^5 = (3 + 2 - 4)x^5 = x^5$ 29. $-p^7$ 30. 0   31. $6y^2$ 33. $5x^4 - 8x$ cannot be simplified further; it is of degree 4 since the highest exponent is 4, and it is a binomial since it contains two terms.   34. Simplifies to $-4y$, degree 1, monomial   35. Simplified already, degree 9, trinomial   37. Degree 8, trinomial   38. Simplifies to $1x^5$ or just x^5, degree 5, monomial   39. Degree 2, monomial   41. Always true, since a binomial is defined to be a *polynomial* of two terms   42. Sometimes   43. Never   45. Never   46. Sometimes

Section 4.3 (page 159)

1. $5m^2 + 3m$ 2. $10a^3 + a^2$ 3. Change the problem to
$$\begin{array}{r} 12x^4 - x^2 \\ \underline{-8x^4 - 3x^2} \end{array}.$$
Then add to get $4x^4 - 4x^2$.

5. Change all the signs on the second row, getting $-3n^5 - 7n^3 - 8$. Then add, getting $-n^5 - 12n^3 - 2$.   6. $-4r^2 - 6r + 5$ 7. $12m^3 + m^2 + 12m - 14$ 9. $15m^2 - 3m + 4$ 10. $5a^4 - 4a^3 + 3a^2 - a + 1$ 11. $8b^2 + 2b + 7$ 13. $(2r^2 + 3r) - (3r^2 + 5r) = 2r^2 + 3r - 3r^2 - 5r = -r^2 - 2r$ 14. $-2r^2 + 7r - 6$ 15. $5m^2 - 14m$ 17. $-6s^2 + 5s + 1$ 18. $3 + s$ 19. $4s + 2s^2$ 21. $4x^3 + 2x^2 + 5x$ 22. $-3b^6 + b^4 - 3b^2$ 23. $-11y^4 + 8y^2 + 3y$ 25. $a^4 - a^2 + 1$ 26. $-7m^2 + 6m + 1$ 27. $[(8m^2 + 4m - 7) - (2m^2 - 5m + 2)] - (m^2 + m + 1)$
$= [8m^2 + 4m - 7 - 2m^2 + 5m - 2] - (m^2 + m + 1)$
$= [8m^2 - 2m^2 + 4m + 5m - 7 - 2] - (m^2 + m + 1)$
$= (6m^2 + 9m - 9) - (m^2 + m + 1)$
$= 6m^2 + 9m - 9 - m^2 - m - 1$
$= 5m^2 + 8m - 10$
29. $4 + x^2 > 8$ 30. $(5 + 2x) - (6 + 3x) > 8x + x^2$ 31. $(5 + x^2) + (3 - 2x) \neq 5$

Section 4.4 (page 165)

1. $(-4x^5)(8x^2) = (-4 \cdot 8)(x^5 \cdot x^2) = -32x^{5+2} = -32x^7$ 2. $-6x^{12}$ 3. $15y^{11}$ 5. $30a^9$ 6. $15m^{10}$ 7. $2m(3m + 2) = (2m)(3m) + (2m)(2) = 6m^2 + 4m$ 9. $-6p^4 + 12p^3$ 10. $12x + 8x^2 + 20x^4$ 11. $-8z(2z + 3z^2 + 3z^3) = (-8z)(2z) + (-8z)(3z^2) + (-8z)(3z^3) = -16z^2 - 24z^3 - 24z^4$ 13. $6y + 4y^2 + 10y^5$ 14. $-6m^6 - 10m^5 - 12m^4$ 15. $(m + 7)(m + 5) = (m + 7)m + (m + 7)5 = m^2 + 7m + 5m + 35 = m^2 + 12m + 35$ 17. $x^2 - 25$ 18. $y^2 - 64$ 19. $t^2 - 16$ 21. $6p^2 - p - 5$ 22. $12x^2 + 10x - 12$ 23. $16m^2 - 9$ 25. $6b^2 + 46b - 16$ 26. $10a^2 + 37a + 7$

27.
$$\begin{array}{rl} 2x^2 + 4x + 1 & \\ \underline{6x + 1} & \\ 2x^2 + 4x + 1 & \text{(multiply by 1)} \\ \underline{12x^3 + 24x^2 + 6x \phantom{{}+1}} & \text{(multiply by } 6x\text{)} \\ 12x^3 + 26x^2 + 10x + 1 & \text{(add)} \end{array}$$

29. $81a^3 + 27a^2 + 11a + 2$ 30. $6r^3 + 5r^2 - 12r + 4$ 31. $20m^4 - m^3 - 8m^2 - 17m - 15$

33.
$$\begin{array}{rl} 3x^5 - 2x^3 + x^2 - 2x + 3 & \\ \underline{2x - 1} & \\ -3x^5 \phantom{{}-4x^4} + 2x^3 - x^2 + 2x - 3 & \text{(multiply by } -1\text{, leaving a space for the missing } x^4 \text{ term)} \\ \underline{6x^6 \phantom{{}-3x^5} - 4x^4 + 2x^3 - 4x^2 + 6x \phantom{{}-3}} & \text{(multiply by } 2x\text{)} \\ 6x^6 - 3x^5 - 4x^4 + 4x^3 - 5x^2 + 8x - 3 & \text{(add)} \end{array}$$

34. $2a^5 + a^4 - a^3 + a^2 - a + 3$ 35. $(x + 7)^2 = (x + 7)(x + 7) = x^2 + 14x + 49$ 37. $a^2 - 8a + 16$ 38. $b^2 - 20b + 100$ 39. $(m - 5)^3 = (m - 5)^2 \cdot (m - 5) = (m^2 - 10m + 25)(m - 5) = m^3 - 15m^2 + 75m - 125$ 41. $k^4 + 4k^3 + 6k^2 + 4k + 1$ 42. $r^4 - 4r^3 + 6r^2 - 4r + 1$

Section 4.5 (page 171)

1. $(r - 1)(r + 3) = r^2 - r + 3r - 3 = r^2 + 2r - 3$ 2. $x^2 - 3x - 10$ 3. $x^2 - 10x + 21$ 5. $(2x - 1)(3x + 2) = 6x^2 - 3x + 4x - 2 = 6x^2 + x - 2$ 6. $8y^2 - 6y - 5$ 7. $6z^2 - 13z - 15$ 9. $2a^2 + 9a + 4$ 10. $6x^2 + 7x - 3$ 11. $8r^2 + 2r - 3$ 13. $6a^2 + 8a - 8$ 14. $110m^2 + 21m - 110$ 15. $(4 + 5x)(5 - 4x) = 20 + 25x - 16x - 20x^2 = 20 + 9x - 20x^2$ 17. $-12 + 5r + 2r^2$ 18. $-10 + 17z - 6z^2$ 19. $15 + a - 2a^2$ 21. $(m + 2)^2 = m^2 + 2(m)(2) + 2^2 = m^2 + 4m + 4$ 22. $x^2 + 16x + 64$ 23. $25 + 10x + x^2$ 25. $x^2 + 4xy + 4y^2$ 26. $9m^2 - 6mn + n^2$ 27. $4z^2 - 20zx + 25x^2$ 29. $25p^2 + 20pq + 4q^2$ 30. $64a^2 - 48ab + 9b^2$ 31. $16a^2 + 40ab + 25b^2$ 33. $(m - n)(m + n) = m^2 - n^2$ 34. $p^2 - q^2$ 35. $r^2 - z^2$ 37. $36a^2 - p^2$ 38. $25y^2 - 9x^2$ 39. $4m^2 - 25$ 41. $49y^2 - 100$ 42. $36x^2 - 9$ 43. $(3 + x)^2 = 5$ 45. $(3 + x)(x - 4) > 7$ 46. $(6x)(2x + 4) - 5 = 8$

Section 4.6 (page 175)

1. $\dfrac{4x^2}{2x} = \dfrac{4}{2} \cdot \dfrac{x^2}{x} = 2x$ 2. $4m^4$ 3. $2a^2$ 5. $\dfrac{9k^3}{m}$

6. $6x^3y^4$ 7. $\frac{60m^4 - 20m^2}{2m} = \frac{60m^4}{2m} - \frac{20m^2}{2m} =$ $30m^3 - 10m$ 9. $60m^5 - 30m^2 + 40m$
10. $5m^4 - 8m + 4m^2$ 11. $4m^2 - 2m + 3$
13. $\frac{m^2}{2m} + \frac{m}{2m} + \frac{1}{2m} = \frac{m}{2} + \frac{1}{2} + \frac{1}{2m}$ 14. $m - 1 + \frac{5}{2m}$
15. $\frac{3x^4 + 9x^3 + 3x^2 + 6x}{3x} = \frac{3x^4}{3x} + \frac{9x^3}{3x} + \frac{3x^2}{3x} + \frac{6x}{3x} =$ $x^3 + 3x^2 + x + 2$ 17. $4x^3 - x^2 + 1$
18. $15x^2 + 5x - 3$ 19. $9x^2 - 3x^3 + 6x^4$
21. $\frac{x^2}{3} + 2x - \frac{1}{3}$ 22. $\frac{4x^3}{3} - x^2 + \frac{2}{3}$
23. $\frac{8k^4 - 12k^3 - 2k^2 + 7k - 3}{2k} = \frac{8k^4}{2k} - \frac{12k^3}{2k} - \frac{2k^2}{2k} + \frac{7k}{2k} - \frac{3}{2k} = 4k^3 - 6k^2 - k + \frac{7}{2} - \frac{3}{2k}$ 25. $10p^3 - 5p^2 + 3p - \frac{3}{p}$ 26. $\frac{25}{p} + 10 + p$ 27. $\frac{2}{x^3} + \frac{4}{x^2} + \frac{5}{2x}$
29. $4y^3 - 2 + \frac{3}{y}$ 30. $\frac{4}{3} - a + \frac{5}{3a}$ 31. $\frac{12}{x} - \frac{6}{x^2} + \frac{14}{x^3} - \frac{10}{x^4}$ 33. $-24m^5 + 16m^4$
34. $-63y^4 - 21y^3 - 35y^2 + 14y$

Section 4.7 (page 181)

1. $$\begin{array}{r} x+2 \\ x-3\overline{)x^2 - x - 6} \\ \underline{x^2 - 3x} \\ 2x - 6 \\ \underline{2x - 6} \end{array}$$
2. $m - 6$ 3. $2y - 5$

5. $p - 4$ 6. $x + 3$ 7. $r - 5$ 9. $6m - 1$
10. $y - 3$ 11. $a - 7$

13. $$\begin{array}{r} x+2 \\ 2x+1\overline{)2x^2 + 5x + 3} \\ \underline{2x^2 + x} \\ 4x + 3 \\ \underline{4x + 2} \\ 1 \end{array}$$
Quotient: $x + 2 + \frac{1}{2x + 1}$

14. $2m - 1 + \frac{4}{2m - 1}$ 15. $a - 2 + \frac{6}{2a + 1}$

17. $d - 3 + \frac{17}{2d + 4}$ 18. $4m - 1 + \frac{-5}{m + 3}$

19. $$\begin{array}{r} x^2 - x + 2 \\ 2x+1\overline{)2x^3 - x^2 + 3x + 2} \\ \underline{2x^3 + x^2} \\ -2x^2 + 3x \\ \underline{-2x^2 - x} \\ 4x + 2 \\ \underline{4x + 2} \end{array}$$
21. $4k^3 - k + 2$

22. $9r^3 - 2r + 6$ 23. $$\begin{array}{r} 3y + 1 \\ y^2+1\overline{)3y^3 + y^2 + 3y + 1} \\ \underline{3y^3 + 3y} \\ y^2 + 1 \\ \underline{y^2 + 1} \end{array}$$

25. $x^2 + 1 + \frac{-6x + 2}{x^2 - 2}$ 26. $x^2 - 1 + 4/(x^2 - 1)$

27. $x^2 + x + 1$ 29. $x^2 - 1$ 30. $x^3 + x + \frac{x - 1}{x^3 - 1}$

Section 4.8 (page 185)

1. $3^{-3} = \frac{1}{3^3} = \frac{1}{27}$ 2. $\frac{1}{16}$ 3. $\frac{1}{25}$ 5. $\frac{1}{9}$
6. $(-12)^{-1} = \frac{1}{(-12)^1} = \frac{1}{-12} = -\frac{1}{12}$ 7. $\frac{1}{36}$ 9. $\frac{1}{7}$
10. $\frac{1}{144}$ 11. $\left(\frac{1}{2}\right)^{-5} = \frac{1}{\left(\frac{1}{2}\right)^5} = \frac{1}{\frac{1}{32}} = 1 \div \frac{1}{32} =$ $1 \times \frac{32}{1} = 32$ 13. 2 14. $\frac{4}{3}$ 15. $\frac{27}{8}$
17. $2^{-1} + 3^{-1} = \frac{1}{2} + \frac{1}{3} = \frac{5}{6}$ 18. $\frac{1}{3} - \frac{1}{4} = \frac{1}{12}$ 19. $\frac{9}{20}$
21. $3^4 \cdot 3^{-5} = 3^{4+(-5)} = 3^{-1} = \frac{1}{3}$ 22. 5 23. 9
25. $\frac{1}{9}$ 26. $\frac{1}{6^3}$ 27. 8^3 29. $\frac{4^3 \cdot 4^{-5}}{4^7} = \frac{4^{3+(-5)}}{4^7} =$ $\frac{4^{-2}}{4^7} = 4^{-2-7} = 4^{-9} = \frac{1}{4^9}$ 30. 2^2 or 4 31. $\frac{1}{5^9}$
33. m^5 34. $\frac{1}{p^7}$ 35. $\frac{m^{11} \cdot m^{-7}}{m^5} = \frac{m^4}{m^5} = m^{4-5} =$ $m^{-1} = \frac{1}{m}$ 37. $\frac{1}{r}$ 38. $\frac{1}{x^4}$ 39. $\frac{a^6 \cdot a^{-3}}{a^{-5} \cdot a} = \frac{a^3}{a^{-4}} =$ $a^{3-(-4)} = a^7$ 41. $\frac{9}{x^{10}}$ 42. $\frac{p^8}{25}$ 43. $(9^{-1}y^5)^{-2} =$ $9^{(-1)(-2)}y^{5(-2)} = 9^2y^{-10} = 81y^{-10} = \frac{81}{y^{10}}$ 45. $\frac{b}{a}$
46. $\frac{9}{4a^2}$ 47. $\frac{5^2m^{-4}}{m^{-2}} = 5^2m^{-4-(-2)} = 5^2m^{-2} =$ $25m^{-2} = \frac{25}{m^2}$ 49. $\frac{3^{-2}x^{-4}5^3x^{-3}}{3x^{-5}} = \frac{3^{-2}x^{-7}5^3}{3^1x^{-5}} =$ $3^{-2-1}x^{-7-(-5)}5^3 = 3^{-3}x^{-2}5^3 = \frac{5^3}{3^3x^2} = \frac{125}{27x^2}$
50. $108/y^5z^3$ 51. $a^{11}/2b^5$

Section 4.9 (page 189)

1. 6.835×10^9 2. 3.21×10^{14} 3. 8.36×10^{12}
5. 2.15×10^2 6. 6.83×10^2 7. 2.5×10^4
9. 3.5×10^{-2} 10. 5×10^{-3} 11. 1.01×10^{-2}
13. 1.2×10^{-5} 14. 9.82×10^{-7}
15. 8,100,000,000 17. 9,132,000 18. 2.14
19. 324,000,000 21. .00032 22. .0000576
23. .041 25. $(2 \times 10^8) \times (4 \times 10^{-3}) = (2 \times 4) \times (10^8 \times 10^{-3}) = 8 \times 10^5 = 800{,}000$ 26. 1500
27. .000004 29. $\frac{9}{3} \times 10^{5-(-1)} = 3 \times 10^6 =$ 3,000,000 30. .00000003 31. .2 33. 4×10^{-4}; 8×10^{-4} 34. 4.037×10^9; 3×10^{-4}
35. 3.68×10^{15} 37. 1000; .06102
38. 400,000,000,000,000

Chapter 4 Test (page 191)

1. $P(2) = 2^4 + 2 \cdot 2^2 - 7 \cdot 2 + 2 = 12$ 2. 12
3. $4^5 = 1024$ 4. $\frac{1}{6^2} = \frac{1}{36}$ 5. $(-2)^5 = -32$

6. $2^4 = 16$ 7. $2^6 = 64$ 8. $\frac{2^2}{3^2} = \frac{4}{9}$ 9. $\frac{1}{8^2} = \frac{1}{64}$ 10. $8^1 = 8$ 11. $(2x^2y^3)^{-3} = 2^{-3}x^{-6}y^{-9} = \frac{1}{8x^6y^9}$ 12. Degree 2; binomial (simplified form is $-x^2 + 6x$) 13. Degree 4; trinomial 14. Degree 3; none of these 15. Degree 0; monomial 16. $10x^3 - 2x^2 - 8x$ 17. $2x^5 + 3x^3 - 4x + 7 - x^5 + 3x^3 - x^2 + 2x + 5 = x^5 + 6x^3 - x^2 - 2x + 12$ 18. $3y^2 - 2y - 2$ 19. $(10)(-4)x^3x^2 = -40x^{3+2} = -40x^5$ 20. $r^2 - 3r - 10$ 21. $6t^2 - t - 12$ 22. $4p^2 + 20p + 25$ 23. $x^2 - 64$ 24. $\frac{15r^4}{5r} - \frac{10r^3}{5r} + \frac{25r^2}{5r} - \frac{15r}{5r} = 3r^3 - 2r^2 + 5r - 3$ 25. $3x^4 + 18x^3 + 27x^2 + 60x + 60$ 26. 6×10^6 27. 2.45×10^8 28. 48,000 29. 291,000,000 30. .0645 31. .0000103 32. 30,000,000 33. .00002

CHAPTER 5

Section 5.1 (page 197)

1. Factors of 14 are the integers that divide into 14. They are −14, −7, −2, −1, 1, 2, 7, 14. 2. −18, −9, −6, −3, −2, −1, 1, 2, 3, 6, 9, 18 3. −27, −9, −3, −1, 1, 3, 9, 27 5. −45, −15, −9, −5, −3, −1, 1, 3, 5, 9, 15, 45 6. −50, −25, −10, −5, −2, −1, 1, 2, 5, 10, 25, 50 7. −60, −30, −20, −15, −12, −10, −6, −5, −4, −3, −2, −1, 1, 2, 3, 4, 5, 6, 10, 12, 15, 20, 30, 60 9. −100, −50, −25, −20, −10, −5, −4, −2, −1, 1, 2, 4, 5, 10, 20, 25, 50, 100 10. −130, −65, −26, −13, −10, −5, −2, −1, 1, 2, 5, 10, 13, 26, 65, 130 11. −29, −1, 1, 29 13. Both $12y$ and 24 can be divided by 1, 2, 3, 4, 6, and 12. The largest of these numbers is 12. 14. 12 15. $10p^2$ (choose the smallest exponent on p) 17. 1 18. 1 19. $6m^2n$ 21. 2 22. 2 23. x 25. $3m^2$ 26. $2p^2$ 27. $2z^4$ 29. xy^2 30. ab 31. xy^2 33. $7x^3y^2$ 34. $-4m^2n$ 35. The greatest common factor is 12; $12x + 24 = 12(x) + 12(2) = 12(x + 2)$ 37. $3(1 + 12d)$ 38. $5(3 + 5r)$ 39. $9a(a - 2)$ 41. $5y^5(13y^4 - 7)$ 42. $4a^2(25a^2 - 4)$ 43. $11p^4(11p - 3)$ 45. No common factor other than 1 46. No common factor other than 1 47. $9m^2(1 + 10m)$ 49. $19y^2p^2(y + 2p)$ 50. $4mn(n - 3m)$ 51. $6x^2y(3y^2 - 4x^2)$ 53. $13y^3(y^3 + 2y^2 - 3)$ 54. $5x^2(x^2 + 5x - 4)$ 55. $8a(2a^2 + a + 3)$ 57. $9qp^3(5q^3p^2 - 4p^3 + 9q)$ 58. $a^5(1 + 2b + 3b^2 - 4b^3)$ 59. $ab^3(a^2b^2 - ab^4 + 1)$

Section 5.2 (page 203)

1. $x + 3$ 2. $p + 6$ 3. $r + 8$ 5. $t - 12$ 6. $x - 8$ 7. $x - 8$ 9. $m + 6$ 10. $x + 11$ 11. $p - 1$ 13. $x + 3$ 14. $k + 4$ 15. $x^2 + 6x + 5$

Try 5, 1. 5 + 1 = 6, yes; 5 · 1 = 5, yes
$x^2 + 6x + 5 = (x + 5)(x + 1)$

17. $a^2 + 9a + 20$; the two numbers that multiply to give 20 and add to give 9 are 4 and 5. $a^2 + 9a + 20 = (a + 4)(a + 5)$ 18. $(b + 5)(b + 3)$ 19. $(x - 1)(x - 7)$ 21. Cannot be factored 22. $(n - 6)(n + 2)$ 23. $(y - 2)(y - 4)$ 25. $(s - 5)(s + 7)$ 26. Cannot be factored 27. Cannot be factored 29. $(b - 3)(b - 8)$ 30. $(x - 5)(x - 4)$ 31. $(y + 3)(y - 7)$ 33. Cannot be factored 34. $(r + 7)(r - 6)$ 35. $(z + 5)(z - 8)$ 37. First, factor out the greatest common factor: $3m^3 + 12m^2 + 9m = 3m(m^2 + 4m + 3)$. Then factor $m^2 + 4m + 3$, ending up with $3m(m + 3)(m + 1)$. 38. $3y^3(y - 5)(y - 1)$ 39. $6(a + 2)(a - 10)$ 41. $3j(j - 4)(j - 6)$ 42. $2x^4(x - 7)(x + 3)$ 43. $3x^2(x + 5)(x - 6)$ 45. We need two terms that multiply to give $3a^2$ and add to give $4a$. The terms are $3a$ and a, since $3a \cdot a = 3a^2$ and $3a + a = 4a$. The answer is $(x + 3a)(x + a)$. 46. $(x - 3m)(x + 2m)$ 47. $(y + 5b)(y - 6b)$ 49. $(x - 5y)(x + 6y)$ 50. $(a - 8y)(a + 7y)$

Section 5.3 (page 209)

1. $x - 1$ 2. $a + 1$ 3. $b - 3$ 5. $4y - 3$ 6. $7z - 4$ 7. $5x + 4$ 9. $m + 10$ 10. $3x - 4$ 11. $3a - 4b$ 13. $k + 3m$ 14. $2x - 5y$ 15. $2x^2 - 5x - 3$; $x - 3$ 17. $6m^2 + 7m - 20$; $2m + 5$ 18. $8y^2 - 2y - 3$; $2y + 1$ 19. Try various combinations of factors of $2x^2$ and 3.

$(2x + 3)(x + 1) = 2x^2 + 5x + 3$: wrong
$(2x + 1)(x + 3) = 2x^2 + 7x + 3$: correct

21. $(3a + 7)(a + 1)$ 22. $(7r + 1)(r + 1)$ 23. $(4r - 3)(r + 1)$ 25. $(3m - 1)(5m + 2)$ 26. $(3x - 1)(2x + 1)$ 27. $(2m - 3)(4m + 1)$ 29. $(5a + 3)(a - 2)$ 30. $(4s + 5)(3s - 1)$ 31. $(3r - 5)(r + 2)$ 33. $(y + 17)(4y + 1)$ 34. $(7m + 2)(3m + 1)$ 35. $(19x + 2)(2x + 1)$ 37. $(2x + 3)(5x - 2)$ 38. $(2b + 1)(3b + 2)$ 39. $(2w + 5)(3w + 2)$ 41. $(2q + 3)(3q + 7)$ 42. $(8x - 1)(x + 6)$ 43. $(5m - 4)(2m - 3)$ 45. $(4k - 5)(2k + 3)$ 46. $(5p + 3)(3p - 2)$ 47. $(5m - 8)(2m + 3)$ 49. $(4x - 1)(2x - 3)$ 50. $(3b - 5)(8b + 1)$ 51. $(8m - 3)(5m + 2)$ 53. Factor out the greatest common factor: $2m(m^2 + m - 20)$. Then factor $m^2 + m - 20$, getting an answer of $2m(m + 5)(m - 4)$. 54. $3n^2(5n - 3)(n - 2)$ 55. $2a^2(4a - 1)(3a + 2)$ 57. $4z^3(z - 1)(8z + 3)$ 58. $y^2(5x - 4)(3x + 1)$ 59. $(4p - 3q)(3p + 4q)$ 61. $(5a + 2b)(5a + 3b)$ 62. $(6x + y)(x - y)$ 63. $(3a - 5b)(2a + b)$

Section 5.4 (page 213)

1. $x^2 - 16 = x^2 - 4^2 = (x + 4)(x - 4)$ 2. $(m + 5)(m - 5)$ 3. $(p + 2)(p - 2)$ 5. $(m + n)(m - n)$ 6. $(p + q)(p - q)$ 7. $(a + b)(a - b)$ 9. $9m^2 - 1 = (3m)^2 - 1^2 = (3m + 1)(3m - 1)$ 10. $(4y + 3)(4y - 3)$ 11. $(5m + 4)(5m - 4)$ 13. $36t^2 - 16 = 4(9t^2 - 4) = 4(3t + 2)(3t - 2)$ 14. $9(1 + 2a)(1 - 2a)$ 15. $(5a + 4r)(5a - 4r)$ 17. Cannot be factored

18. Cannot be factored **19.** $p^4 - 36 = (p^2)^2 - 6^2 = (p^2 + 6)(p^2 - 6)$ **21.** $a^4 - 1 = (a^2 + 1)(a^2 - 1) = (a^2 + 1)(a + 1)(a - 1)$ **22.** $(x^2 + 4)(x + 2)(x - 2)$ **23.** $(m^2 + 9)(m + 3)(m - 3)$ **25.** $(a + 2)^2$ **26.** $(p + 1)^2$ **27.** $(x - 5)^2$ **29.** $(a + 7)^2$ **30.** $(m - 10)^2$ **31.** $(k + 11)^2$ **33.** Not a perfect square **34.** Not a perfect square **35.** Not a perfect square **37.** $16a^2 - 40ab + 25b^2 = (4a)^2 - 2(4a)(5b) + (5b)^2 = (4a - 5b)^2$ **38.** $(6y - 5p)^2$ **39.** $100m^2 + 100m + 25 = 25(4m^2 + 4m + 1) = 25(2m + 1)^2$ **41.** $(7x + 2y)^2$ **42.** $(8y - 3a)^2$ **43.** $(2c + 3d)^2$ **45.** $(5h - 2y)^2$ **46.** $(3x + 4y)^2$

Section 5.5 (page 219)

1. $x - 2 = 0$ or $x + 4 = 0$
$x = 2$ $x = -4$

2. $3, -5$ **3.** $-5/3, 1/2$ **5.** $-1/5, 1/2$ **6.** $8/3, -7$ **7.** $-9/2, 1/3$ **9.** $1, -5/3$ **10.** $3, -5$ **11.** $7/3, -4$

13. $x^2 + 5x + 6 = 0$
Factor: $(x + 2)(x + 3) = 0$
$x + 2 = 0$ or $x + 3 = 0$
$x = -2$ $x = -3$

14. $1, 2$ **15.** $-1, 6$ **17.** $-7, 4$ **18.** $3, -2$

19. $a^2 = 24 - 5a$ Make the equation equal to 0: $a^2 + 5a - 24 = 0$. (Rearrange terms as necessary to put the equation in the regular form, keeping the coefficient of the squared term positive.) Then factor: $(a + 8)(a - 3) = 0$. $a + 8 = 0$ $a - 3 = 0$
$a = -8$ $a = 3$

21. $-1, 3$ **22.** $4, -1$ **23.** $-1, -2$

25. Factor: $(m + 4)(m + 4) = 0$. The factors are the same, so we get only one solution, -4. **26.** 3

27. $3a^2 + 5a - 2 = 0$
Factor: $(3a - 1)(a + 2) = 0$
$3a - 1 = 0$ or $a + 2 = 0$
$3a = 1$ $a = -2$
$a = \frac{1}{3}$

29. $5/2, -2$ **30.** $5/3, -1/2$ **31.** $-4/3, 1/2$ **33.** $1/3, -5/2$ **34.** $-2/3$ **35.** $5/3, -4$ **37.** $2/5, -1/3$ **38.** $-7/3, 4$ **39.** $3/2, -3$ **41.** $5/4, -5/4$ **42.** $3/2, -3/2$ **43.** $2, -2$ **45.** $5/2, 1/3, 5$ **46.** $4/3, -2/3, 1/2$

47. $x^3 - 25x = 0$ Factor out the greatest common factor: $x(x^2 - 25) = 0$. Finish factoring:
$x(x + 5)(x - 5) = 0$
$x = 0$ or $x + 5 = 0$ or $x - 5 = 0$
$x = 0$ $x = -5$ $x = 5$

49. $0, 7/3, -7/3$ **50.** $0, 3/4, -3/4$

Section 5.6 (page 225)

1. Let x = width, $x + 5$ = length
The area is $x(x + 5) = 66$ (length times width)
Solve the equation: $x(x + 5) = 66$
$x^2 + 5x = 66$
$x^2 + 5x - 66 = 0$
$(x - 6)(x + 11) = 0$
$x - 6 = 0$ or $x + 11 = 0$
$x = 6$ $x = -11$

The width of a rectangle cannot be -11, so we reject that solution. The width is 6 and the length is $6 + 5 = 11$. **2.** Length is 8, width is 7 **3.** If x is the length, and $x - 3$ is the width, the equation is $x(x - 3) = 70$. The answer is 7 by 10. **5.** If x is the width, then the length is $2x$. "The width is increased by 2" is $x + 2$. The equation is $2x(x + 2) = 48$; the answer is 4 by 8. **6.** 12, 4 **7.** 2 by 5 **9.** If h is the height, then the base is $2h$. The formula for the area of a triangle is $\frac{1}{2}bh = A$. Using the numbers from our problem, we have $\frac{1}{2} \cdot (2h)h = 25$, or $h^2 = 25$ (cancel the 2's). Solve $h^2 = 25$ to find that the height is 5 and the base is $2 \cdot 5 = 10$. **10.** 9, 6 **11.** Let B be the area of the base. Then $B - 10$ is the height. The formula is $V = \frac{1}{3}Bh$, so that $32 = \frac{1}{3}B(B - 10)$, or $96 = B(B - 10)$, after multiplying both sides by 3. Solve this equation to find that $B = 16$, and $h = 16 - 10 = 6$. **13.** 19, 18 **14.** 4 by 6 by 5 **15.** Let x be the first integer, so that the second of the consecutive integers is $x + 1$. "Their product is 2 more than twice their sum" becomes $x(x + 1) = 2 + 2(x + x + 1)$, or $x^2 - 3x - 4 = 0$, after simplifying. Solve this equation to get $x = 4$ or $x = -1$. If $x = 4$, then $x + 1 = 5$, and if $x = -1$, then $x + 1 = 0$. There are two pairs of solutions here. **17.** 12, 14, 16, or $-2, 0, 2$ **18.** 6, 10, or $-9, -5$ **19.** 5, 6, or $-6, -5$

Section 5.7 (page 233)

1. $(m + 2)(m - 5) < 0$
$(m + 2)(m - 5) = 0$
$m + 2 = 0$ or $m - 5 = 0$
$m = -2$ $m = 5$

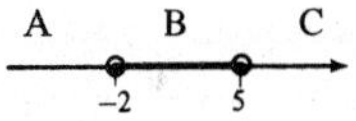

A B C
-2 5

Choose $m = -3$ from A
$(-3 + 2)(-3 - 5) < 0$
$(-1)(-8) < 0$
$8 < 0$
False

$m = 0$ from B $m = 6$ from C
$(0 + 2)(0 - 5) < 0$ $(6 + 2)(6 - 5) < 0$
$(2)(-5) < 0$ $(8)(1) < 0$
$-10 < 0$ $8 < 0$
True   False

The answer is made up of the points from region B, or $-2 < m < 5$.

2. -3 1 **3.** -6 -5

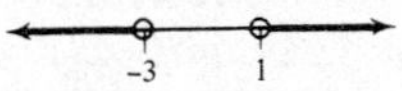

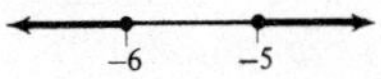

5. -3 3 **6.** -2 2

7. -6  7  **9.** -3 -2

10. 1 2 **11.** -1 5

13. -1 $2/5$ **14.** -4 $1/2$

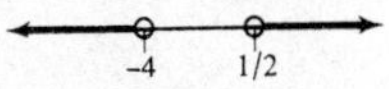

15. $-1/2$ $4/3$ 17. -4 4

18. -2 2 19. -10 10

21. $-3/2$ 5 22. $-1/2$ $1/3$

23. $-1/2$ $2/5$ 25. -4 4

26. -5 5

Chapter 5 Test (page 235)

1. $8m(2m-3)$ **2.** $6y(x+2y)$ **3.** $14p(2q+1+4p)$
4. $3mn(m+3+2n)$ **5.** Cannot be factored
6. $(x+5)(x+6)$ **7.** $(p+7)(p-1)$
8. $(2y+3)(y-5)$ **9.** $(2m-1)(2m+3)$
10. $(3x-2)(x+5)$ **11.** $(2z+1)(5z+1)$
12. $(5a+1)(2a-5)$ **13.** $(4r+5)(3r+1)$
14. Cannot be factored **15.** $(a+5b)(a-2b)$
16. $(3r-2s)(2r+s)$ **17.** $(x+5)(x-5)$
18. $(5m+7)(5m-7)$ **19.** $(2p+3)^2$
20. $(5z-1)^2$ **21.** $4p(p+2)^2$
22. $5m^2(2m+1)(m+5)$ **23.** $-1, -2$ **24.** $1/3, -2$
25. $5/2, -4$ **26.** $2, -2$ **27.** $0, 4, -4$ **28.** Let w be the width. Then $2w-1$ is the length. "The area is 15" leads to the equation $15 = (2w-1)(w)$, or $2w^2 - w - 15 = 0$ after simplifying. This equation factors as $(2w+5)(w-3) = 0$, which can be solved to give $w = -5/2$ (which we reject) or $w = 3$. The width is 3 and the length is $2 \cdot 3 - 1 = 5$. **29.** 10, 3
30. Let x be one number, with $x + 9$ the other one. "Their product is 11 more than 5 times their sum" is $x(x+9) = 11 + 5(x + x + 9)$, which simplifies to $x^2 - x - 56 = 0$, or $(x-8)(x+7) = 0$. The solutions are 8 for x, and $8 + 9 = 17$ for the other number, and $x = -7$, with $-7 + 9 = 2$ for the other number.

31. -6 4 **32.** -3 $1/2$

33. $-2/3$ $1/2$

CHAPTER 6

Section 6.1 (page 241)

1. $x \neq 0$ **2.** $x \neq 0$ **3.** The number 4 makes the denominator equal 0, so the domain is $x \neq 4$
5. $x \neq -5$ **6.** $x \neq 1/2$ **7.** $-3/2$
9. Let $x = -3$: $\dfrac{4(-3)^2 - 2(-3)}{3(-3)} = \dfrac{4(9)+6}{-9} = \dfrac{42}{-9} = -\dfrac{14}{3}$
10. When x is replaced by -3, the denominator becomes $3 + (-3)$, which is 0. This is not a real number.
11. Not a real number **13.** $2/3$ **14.** $m/2$
15. $2k$ **17.** $-4y^3/3$ **18.** $-2x^2$ **19.** $4m/3p$
21. $3/4$ **22.** $9/5$ **23.** $(x-1)/(x+1)$ **25.** $m - n$
26. $a + b$ **27.** $\dfrac{5m^2 - 5m}{10m - 10} = \dfrac{5m(m-1)}{10(m-1)} = \dfrac{5m}{10} = \dfrac{m}{2}$
29. $4r + 2s$ or $2(2r+s)$ **30.** $11s^2/6$
31. $(m-2)/(m+3)$ **33.** $(x+4)/(x+1)$
34. $(2m+3)/(4m+3)$ **35.** 1 **37.** $-(x+1)$ or $-x-1$ **38.** $-(p+q)$ or $-p-q$ **39.** -1

Section 6.2 (page 247)

1. $3m/4$ **2.** $3y^2/2z$ **3.** $3/32$ **5.** $2a^4$ **6.** 2
7. $1/4$ **9.** $1/6$ **10.** $x^3/6$ **11.** $6/(a+b)$ **13.** 2
14. $3/2$ **15.** $\dfrac{2k+8}{6} \div \dfrac{3k+12}{2} = \dfrac{2(k+4)}{6} \cdot \dfrac{2}{3(k+4)} = \dfrac{2}{9}$ **17.** $3/10$ **18.** $4/9$ **19.** $2r/3$
21. $(y+4)(y-3)$ **22.** $6(y-4)/(z+3)$
23. $18/[(m-1)(m+2)]$ **25.** $-7/8$ **26.** -3
27. -1 **29.** $\dfrac{k^2-k-6}{k^2+k-12} \div \dfrac{k^2+2k-3}{k^2+3k-4} = \dfrac{(k-3)(k+2)}{(k+4)(k-3)} \cdot \dfrac{(k+4)(k-1)}{(k+3)(k-1)} = \dfrac{k+2}{k+3}$
30. $(m+6)/(m+3)$ **31.** $(n+4)/(n-4)$ **33.** 1
34. $(y+3)/(y+4)$ **35.** $(m-3)/(2m-3)$
37. $(p-q)/(p+q)$ **38.** $m/(m+5)$
39. $(x+2)(x+1)$ **41.** $10/(x+10)$ **42.** $(m-8)/8$

Section 6.3 (page 253)

1. $7/p$ **2.** $9/r$ **3.** $-3/k$ **5.** $\dfrac{y}{y+1} + \dfrac{1}{y+1} = \dfrac{y+1}{y+1} = 1$ **6.** 3 **7.** $m + n$ **9.** m **10.** y **11.** 1
13. $\dfrac{3}{m} + \dfrac{1}{2} = \dfrac{3 \cdot 2}{m \cdot 2} + \dfrac{1 \cdot m}{2 \cdot m} = \dfrac{6}{2m} + \dfrac{m}{2m} = \dfrac{6+m}{2m}$
14. $(18-2p)/(3p)$ **15.** $(18+3m)/(2m)$
17. $(3y-5)/(5y)$ **18.** $51y/56$ **19.** $m/3$
21. $\dfrac{4+2k}{5} + \dfrac{2+k}{10} = \dfrac{(4+2k) \cdot 2}{5 \cdot 2} + \dfrac{2+k}{10} = \dfrac{8+4k+2+k}{10} = \dfrac{10+5k}{10} = \dfrac{5(2+k)}{10} = \dfrac{2+k}{2}$
22. $7/24$ **23.** $(6-2y)/y^2$ **25.** $(9p+8)/(2p^2)$
26. $(15-12k)/(4k^2)$ **27.** $(7m+4n)/6$
29. $(-y-3x)/(x^2y)$ **30.** $(9x+p^3)/(p^2x)$
31. $(2m^2+4m+4)/[m(m+2)]$
33.
$$\frac{8}{x-2} - \frac{4}{x+2} = \frac{8(x+2)}{(x-2)(x+2)} - \frac{4(x-2)}{(x+2)(x-2)} = \frac{8(x+2)-4(x-2)}{(x-2)(x+2)} = \frac{8x+16-4x+8}{(x-2)(x+2)} = \frac{4x+24}{(x-2)(x+2)}$$
34. $(4m+8n)/[(m-n)(m+n)]$ **35.** $x/[2(x+y)]$
37. $3/[(m+1)(m-1)(m+2)]$
38. $(3y-5)/[2(y+2)(y-2)]$
39. $(m-2)/[(m+3)(m-3)]$ **41.** $3/(m-2)$ or $-3/(2-m)$ **42.** $-3/(y-8)$ or $3/(8-y)$
43. $-3/(y-3)$ or $3/(3-y)$ **45.** $8m/(m+2n)$
46. $10k/(2k+3m)$
47. $(2x^2+6xy+8y^2)/[(x+y)(x+y)(x+3y)]$

Section 6.4 (page 259)

1. $1/2$ **2.** $24/5$ **3.** 12 **5.** $1/4$ **6.** 2 **7.** $2/5$
9. Multiply both sides by 4, getting $4(x/2) - 4(x/4) = 4 \cdot 6$, or $2x - x = 24$, or $x = 24$. **10.** 12 **11.** 2
13. Multiply by 7, getting $7(2t/7) - 7 \cdot 5 = 7t$, or

$2t - 35 = 7t$. Add $-2t$ to both sides, producing $-35 = 5t$, or $t = -7$. **14.** 4 **15.** 1 **17.** 2 **18.** 2 **19.** 5 **21.** Multiply by $2k$, getting $2k\left(\frac{2k+3}{k}\right) = 2k\left(\frac{3}{2}\right)$, or $2(2k + 3) = 3k$. This gives $4k + 6 = 3k$, or $k = -6$. **22.** -2 **23.** -8 **25.** 8 **26.** 12 **27.** 5 **29.** 2 **30.** -1 **31.** Multiply by $y(5y - 12)$, giving $2(5y - 12) = y \cdot y$, or $10y - 24 = y^2$, or $y^2 - 10y + 24 = 0$. Factor to get $(y - 4)(y - 6) = 0$. Write the two equations $y - 4 = 0$ or $y - 6 = 0$, from which $y = 4$ or $y = 6$. Check that both work in the original equation. **33.** 3 **34.** 1 **35.** 1, -24

Section 6.5 (page 265)

1. Let x = the number. Then $\frac{1}{2}x = 3 + \frac{1}{6}x$. Multiply both sides by 6 to get $3x = 18 + x$, or $2x = 18$, or $x = 9$. **2.** 5 **3.** Let x be the numerator of our fraction. The denominator is 5 more than x, or $x + 5$. The fraction is then $x/(x + 5)$. Add 3 to both the numerator and denominator, giving 3/4: $\frac{x+3}{(x+5)+3} = \frac{3}{4}$. Simplify: $\frac{x+3}{x+8} = \frac{3}{4}$. Solve this equation to get $x = 12$. The fraction originally was 12/17. **5.** 6, 9 **6.** 2 or 1/2 **7.** Let x be the number. The reciprocal of x is $1/x$. The problem leads to the equation $x - 2(1/x) = -7/3$. Multiply both sides by $3x$ to get $3x^2 - 6 = -7x$, or $3x^2 + 7x - 6 = 0$. Factor as $(3x - 2)(x + 3) = 0$. Place each factor equal to 0 and solve to get $x = 2/3$ or $x = -3$. **9.** 2/3 or 1 **10.** 1 or $-3/2$ **11.** Let x be the man's wage. Then $2x/5$ is the son's wage. They earned \$336 in four days; in one day they earned \$336/4 = \$84. The equation is $x + 2x/5 = 84$. Solve this to find that the man earned \$60. The son earned 2/5 of this, or \$24. **13.** Let x be the speed of the current. His speed upstream will be $4 - x$ mph, and his speed downstream will be $4 + x$ mph. Since $d = rt$, we have $t = d/r$. The time upstream is $8/(4 - x)$, and the time downstream is $24/(4 + x)$. Since the times are the same, $8/(4 - x) = 24/(4 + x)$. Solve this equation to find that x is 2 mph. **14.** 900 miles **15.** 150 miles **17.** 37 mph **18.** 200 mph **19.** Let t be the time it takes them working together. In one hour, Paul can do 1/2 of the job and Marco can do 1/3 of the job. Working together, in one hour they can do $1/t$ of the job. The equation is $1/2 + 1/3 = 1/t$. Multiply both sides by $6t$, and get the solution of 6/5 hours, or 1 1/5 hours. **21.** 84/19 hours **22.** First, find the time it will take to fill the pool completely. This answer is 18/5 hours. We only want to fill the pool 3/4 full, so multiply: (3/4)(18/5) = 27/10 hours. **23.** Let t be the time it would take Sue working alone. In one day, Dennis can do 1/4 of the job, and Sue can do $1/t$ of the job. Working together, they can do 1/(7/3) = 3/7 of the job. (Here we changed 2 1/3 to 7/3). The equation is $1/4 + 1/t = 3/7$. Solve this equation to get 28/5 or 5 3/5 days. **25.** 100/11 minutes **26.** 33/4 hours

Section 6.6 (page 271)

1. $1/(pq)$ **2.** $2b/a$ **3.** $1/x$ **5.** $x(x + 1)/[y(y + 1)]$ **6.** $n(m + n)/[m(m - n)]$ **7.** $b/[a(b + 1)]$ **9.** $2(2x + 3)/(x - 1)$ **10.** $2k/5$ **11.** First, add $3/y + 1$ to get $(3 + y)/y$. Then invert the denominator and multiply to get an answer of $2/y$. **13.** $(x + y)^2/(xy)$ **14.** $(m^2 + 1)/(3 - m^2)$ **15.** x/y **17.** q **18.** $(r^2 + 1)/(1 - r^2)$ **19.** $(m - n)/4$ **21.** $-(m - 1)/(m + 1)$ **22.** $-(x + 1)/(x - 1)$ **23.** $(y + 1)^2/(y - 1)$

Chapter 6 Test (page 273)

1. $2/(mn^3)$ **2.** $5s(s - 1)/2$ **3.** x **4.** mn^2 **5.** $-1/x$ **6.** $11/[6(a + 1)]$ **7.** $(t - 2)/t$ **8.** $(-k - 6)/[2k(k + 2)]$ **9.** $(3m - 2)/(3m + 2)$ **10.** $(a + 3)/(a + 4)$ **11.** $3(1 + x)/x$ **12.** 3/2 **13.** $-2/3$, 4 **14.** 1/4 or 1/2 **15.** 9/5 **16.** 20/9 hours

CHAPTER 7

Section 7.1 (page 279)

1. 11 **2.** 20 **3.** 29 **5.** -4 **6.** -7 **7.** Let $y = 14$: then $14 = 3x + 5$. Add -5 to both sides, giving $9 = 3x$, or $x = 3$. **9.** 8 **10.** 0 **11.** -2 **13.** 3 **14.** 4 **15.** 7; 1; -1 **17.** 2; 8; 17 **18.** -22; -2; 18 **19.** 9; 3; -15 **21.** 2; 3; -3 **22.** 4; 3; -3 **23.** -3; 5; -5 **25.** 2; 5/2; $-5/4$ **26.** 2; $-4/3$; 13/2 **27.** $-5/4$; 5/6; 7/4 **29.** -4; -4; -4 **30.** The equation $x = 8$ says that x is always 8; the answers are 8; 8; 8. **31.** 3; 3; 3 **33.** -9; -9; -9 **34.** -4; -4; -4 **35.** The equation is $x + y = 9$, and we want to know if (2, 7) is a solution. Replace x with 2 and y with 7, to get $2 + 7 = 9$. This is true, so that (2, 7) is a solution. **37.** Yes **38.** Yes **39.** No **41.** Yes **42.** Yes **43.** Yes **45.** The equation is $x + 4 = 0$, and the ordered pair is $(-5, 1)$. Replace x with -5 (there is no y here) to get $-5 + 4 = 0$, a false statement. The ordered pair is not a solution. **46.** No

Section 7.2 (page 285)

1. (2, 5) **2.** $(-2, 2)$ **3.** $(-5, 5)$ **5.** (7, 3) **6.** $(0, -1)$ **7-22.**

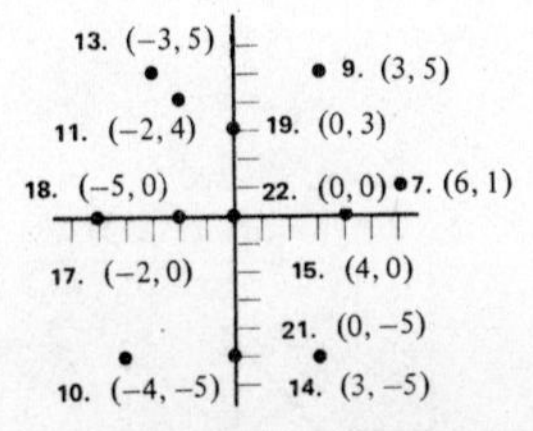

23. I **25.** II **26.** III **27.** III **29.** II **30.** I **31.** IV **33.** The points on the x-axis and y-axis belong to no quadrant at all. **34.** None

35. (0, 0), (−6, 3)
(4, −2), (2, −1)

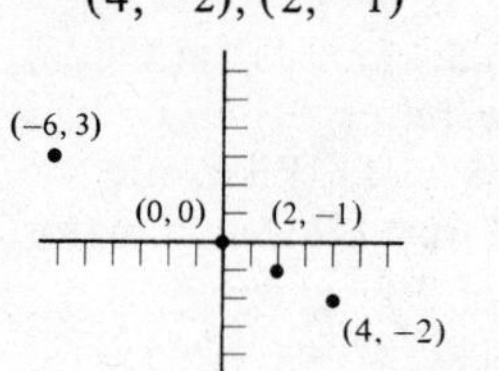

37. (2, −5), (−3, −5),
(0, −5), (−1, −5)

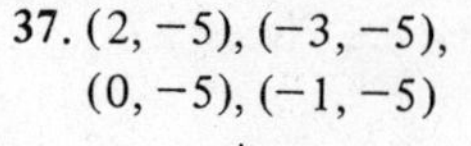
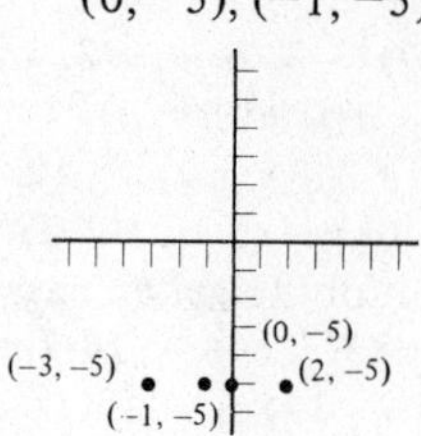
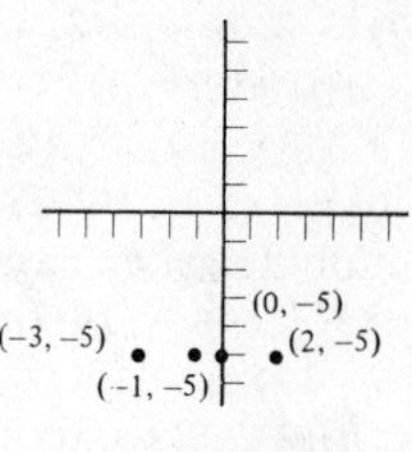

38. (5, −2), (0, −2),
(−3, −2), (−2, −2)

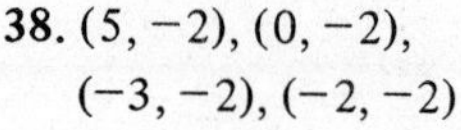
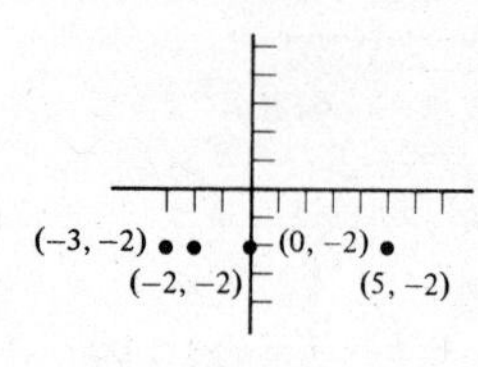

39. (4, 7), (4, 0),
(4, −4), (4, 4)

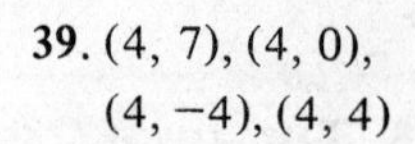
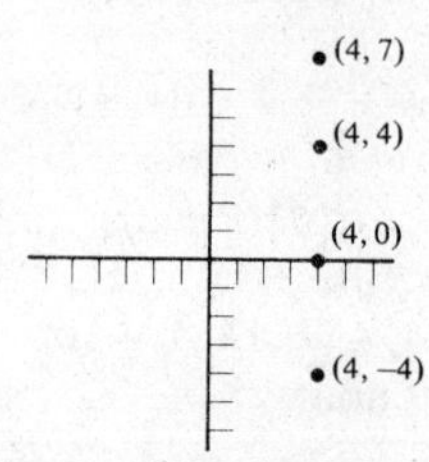

Section 7.3 (page 293)

1. (0, 5), (5, 0), (2, 3)

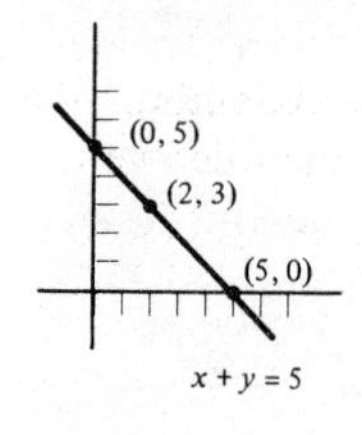

2. (0, −3), (3, 0), (5, 2)

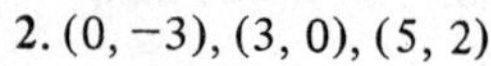
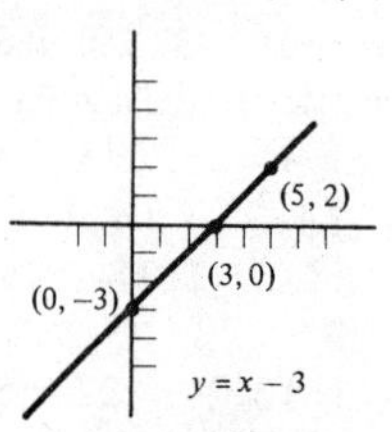

3. (0, 4), (−4, 0), (−2, 2)

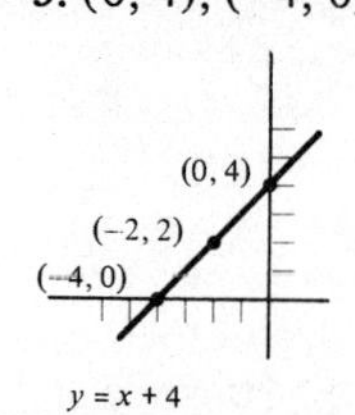
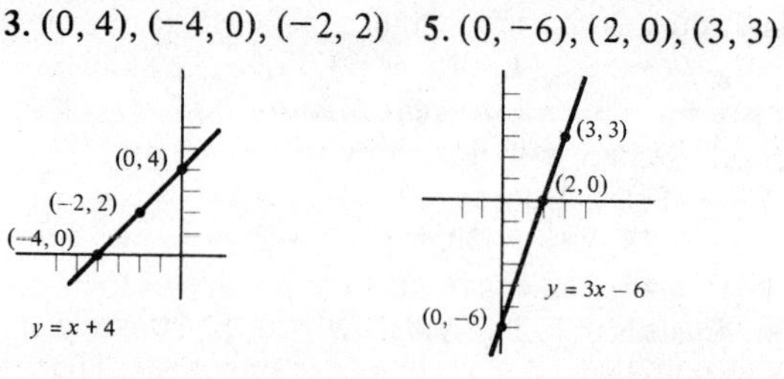

5. (0, −6), (2, 0), (3, 3)

6. (−5, 2), (−5, 0),
(−5, −3)

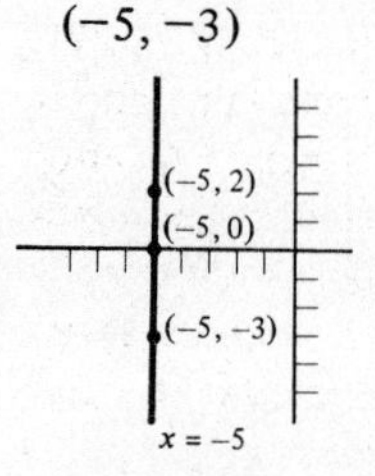

7. (3, 4), (0, 4), (−2, 4)

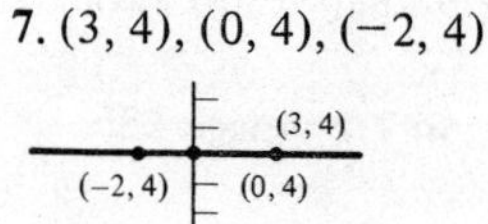
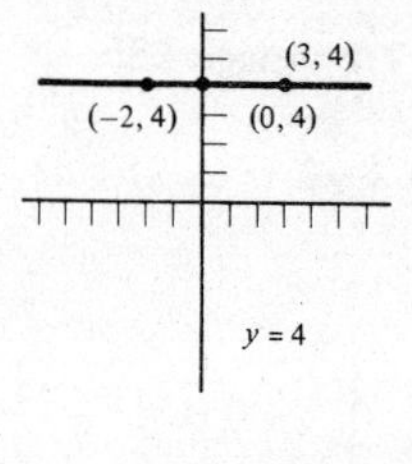

9. (2, 3), (2, 5), (2, −1)

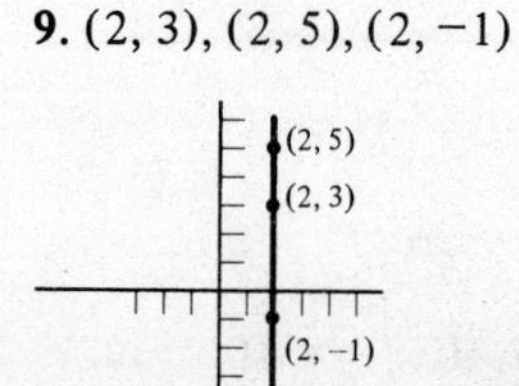

10.

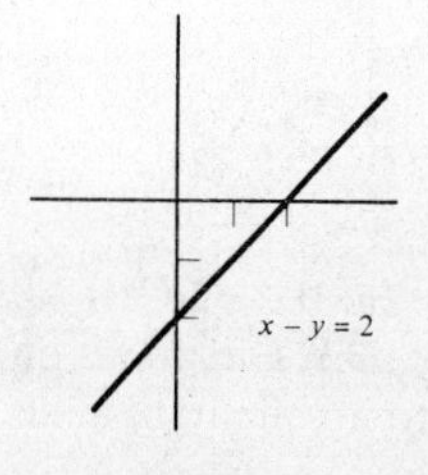

11.

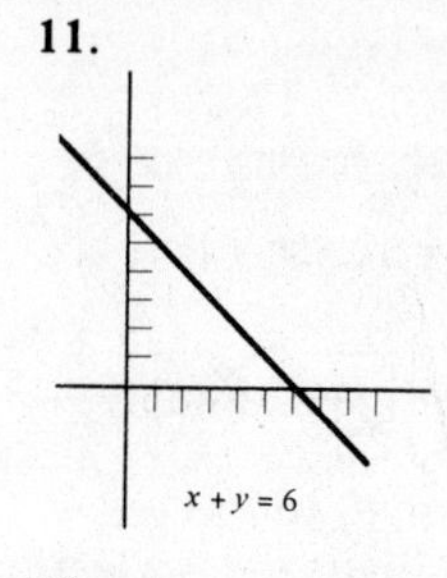

We need three points that satisfy $x + y = 6$. If $x = 0$, then $y = 6$, giving (0, 6). If $y = 0$, then $x = 6$, giving (6, 0). Choose any other number for x, such as 4. If $x = 4$, then $y = 2$, giving (4, 2). Locate these points, and then draw a straight line through them.

13.

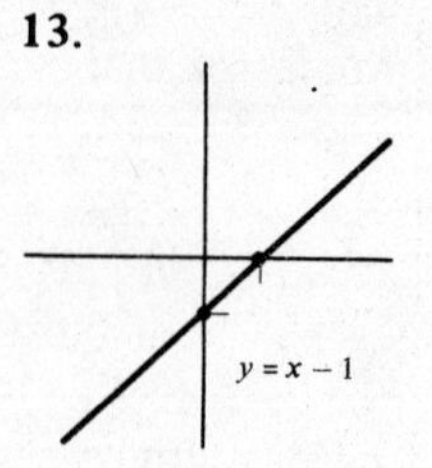

14.

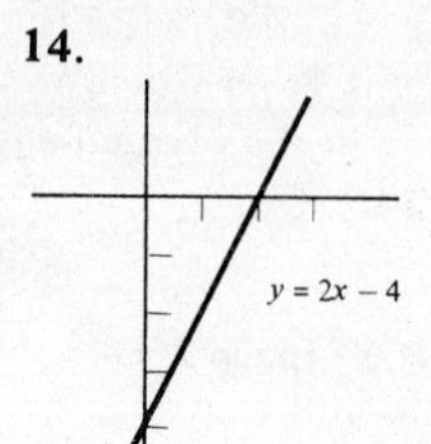

15.

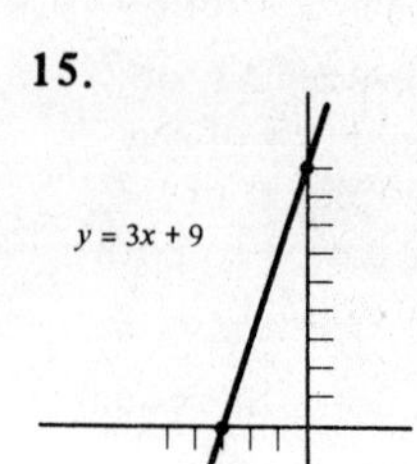

17.

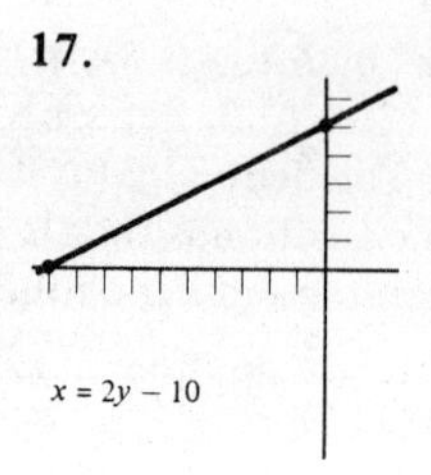

18.

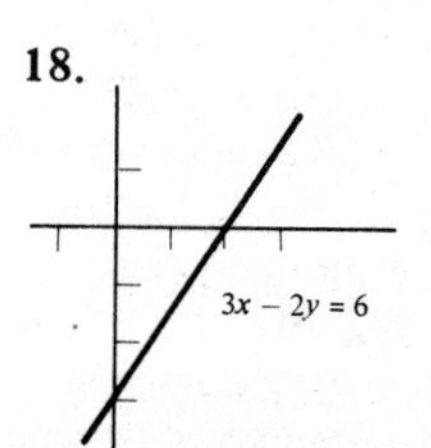

19.

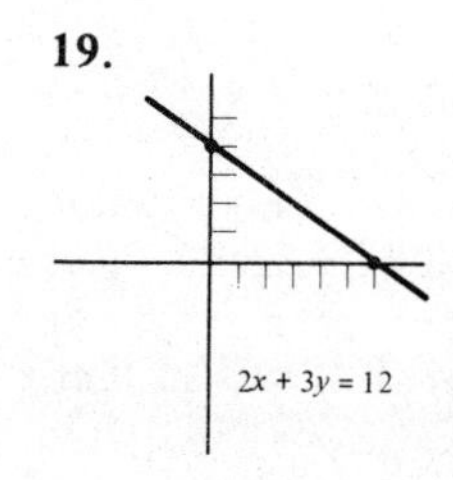

21.

3x + 5y = 15

22.

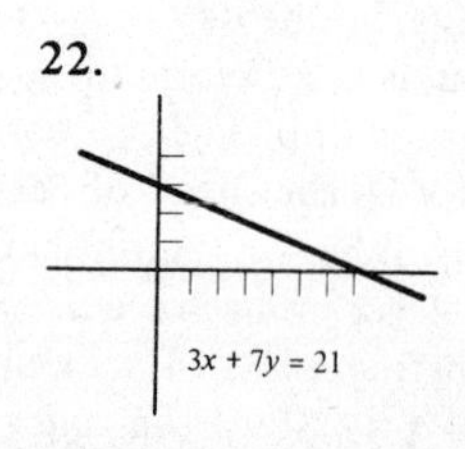

23.

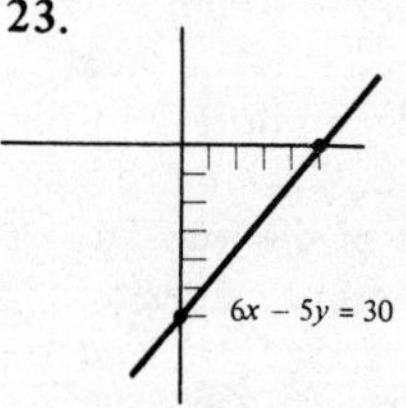
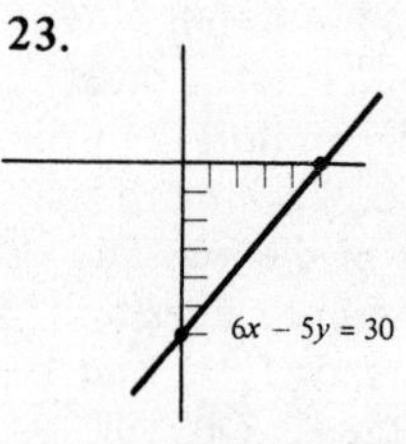

25.

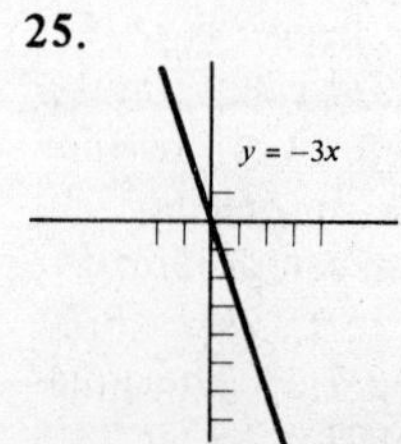

If $x = 0$, then $y = 0$, giving (0, 0). If $y = 0$, then $x = 0$, giving the same point, (0, 0). We need two more points, so choose other values of x (or y). If $x = 2$, then $y = -6$, giving (2, −6). If $x = -1$, then $y = 3$, giving (−1, 3). Locate these points and draw a line through them.

26.

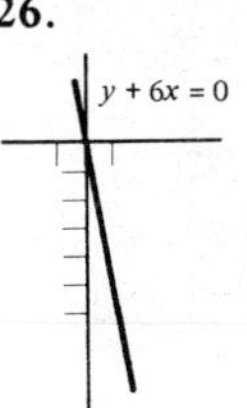

27.

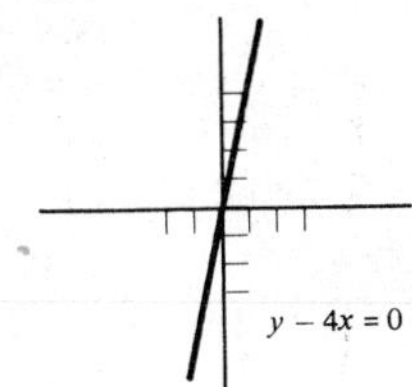

29.

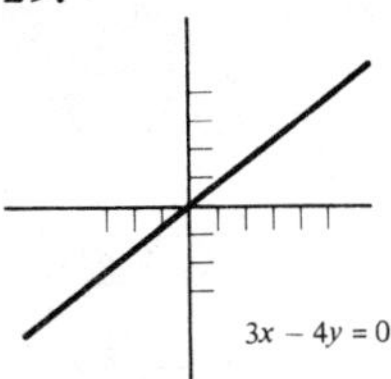

30.

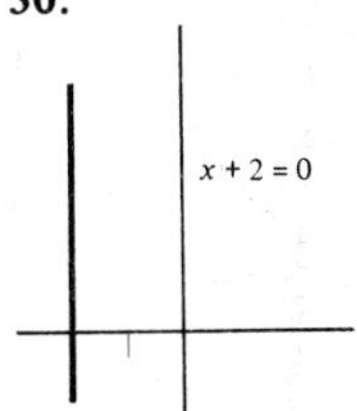

31.

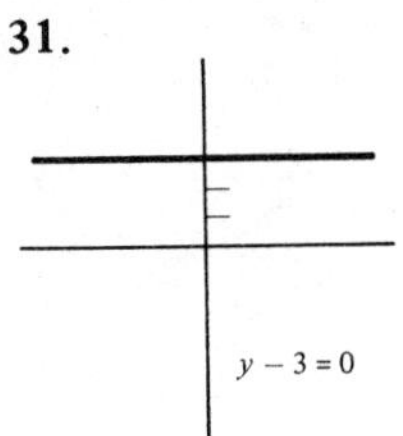

Here y is always 3. Choose any values of x, such as $x = -1$, $x = 2$, and $x = 5$. These produce the ordered pairs $(-1, 3)$, $(2, 3)$, and $(5, 3)$, which can be graphed to get the final straight line.

33.

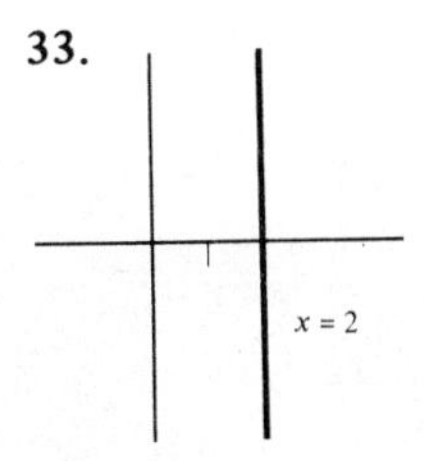

Section 7.4 (page 301)

1.

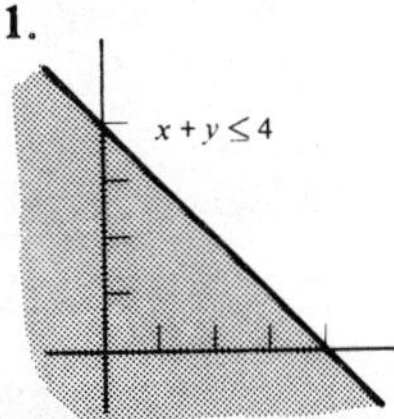

2.

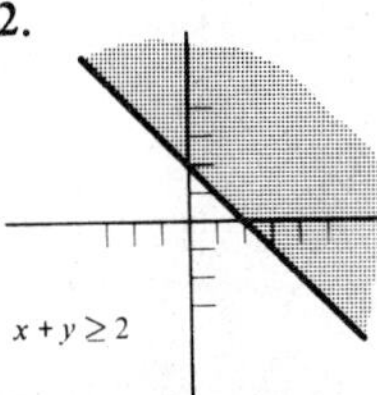

3.

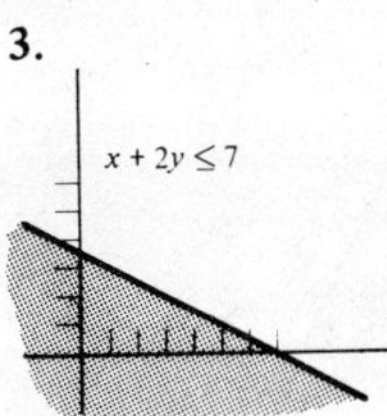

5.

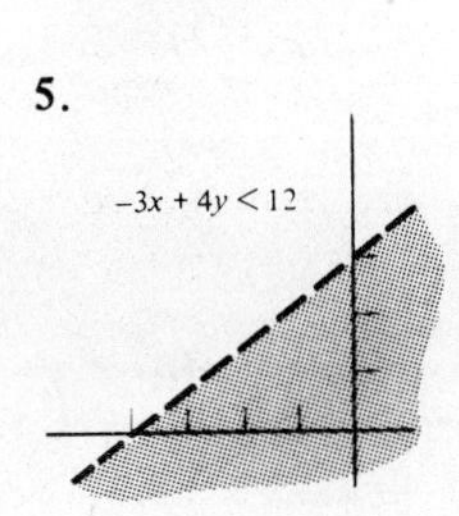

Choose any point not on the line, such as $(0, 0)$. Substitute 0 for x and 0 for y in the original inequality. This gives $-3(0) + 4(0) < 12$, or $0 < 12$, which is true. Since $(0, 0)$ leads to a true statement, shade the side of the line that includes $(0, 0)$.

6.

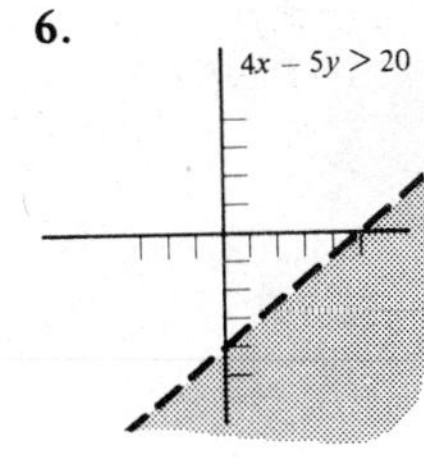

7.

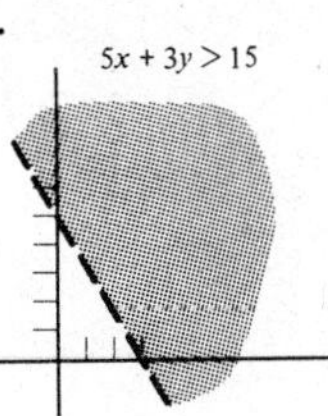

9.

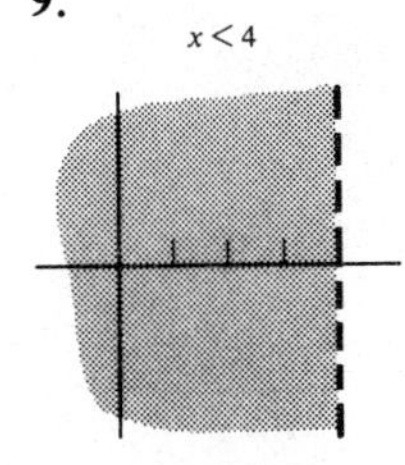

10.

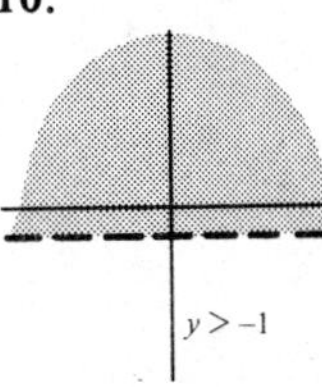

11.

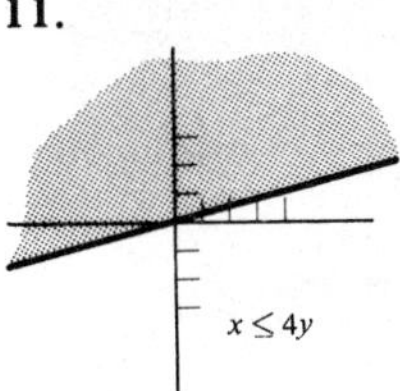

We cannot use $(0, 0)$ as a test point here, since the line goes through this point. We need any point not on the line. Let's choose the point $(5, 0)$. Replace x with 5 and y with 0, to get $5 \leq 4(0)$ or $5 \leq 0$, which is false. Therefore, the side of the line containing $(5, 0)$ is the wrong side; we must shade the other side.

13.

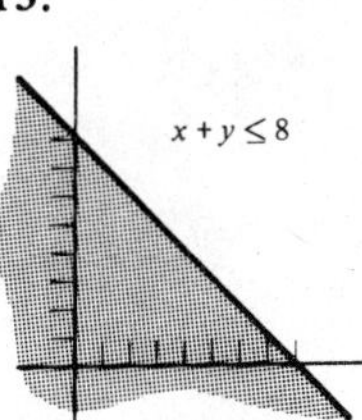

14.

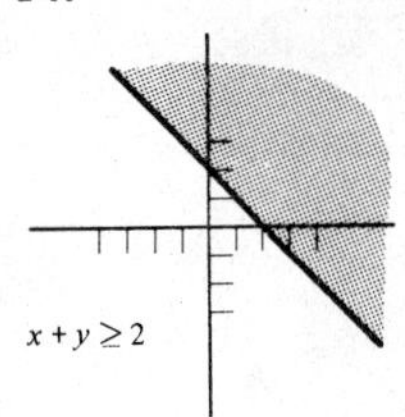

15.

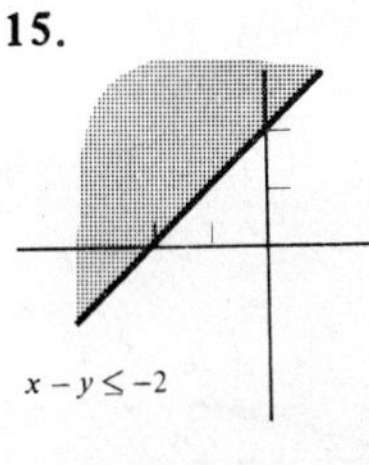

First graph the straight line $x - y = -2$. If $x = 0$, then $y = 2$, giving $(0, 2)$. If $y = 0$, then $x = -2$, giving $(-2, 0)$. Locate these points and draw a solid line through them. The line is solid because we have $\leq$, and not just $<$. Use $(0, 0)$ as a test point. If $x = 0$ and $y = 0$ we get $0 - 0 \leq -2$, which is false. Shade the side of the line not including $(0, 0)$.

17.

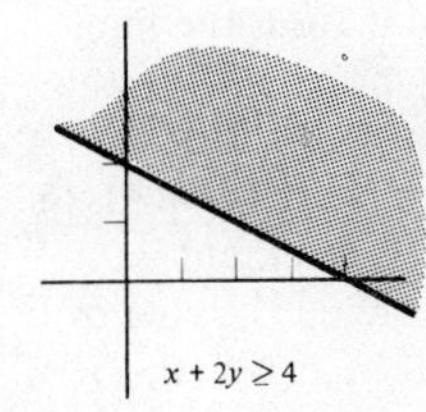

18.

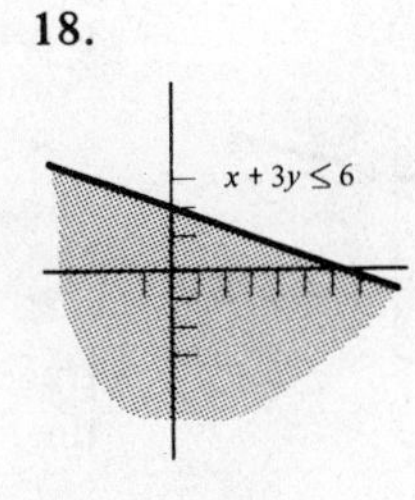

19.

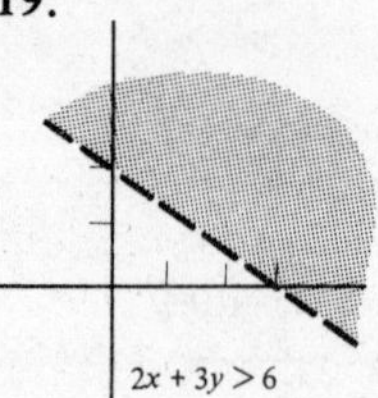

21.

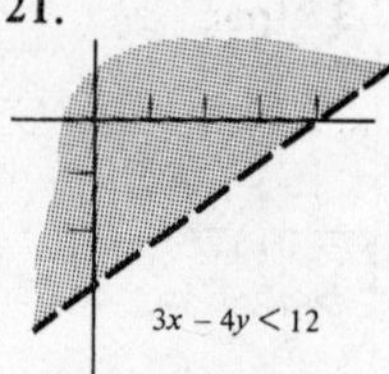

22.

$2x - 3y < -6$

23.

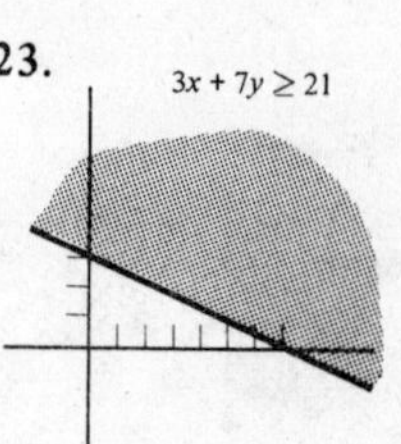

25.

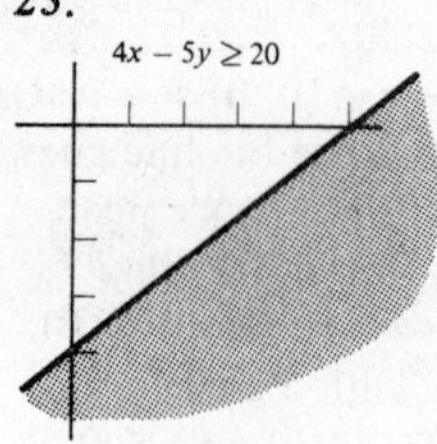

26.

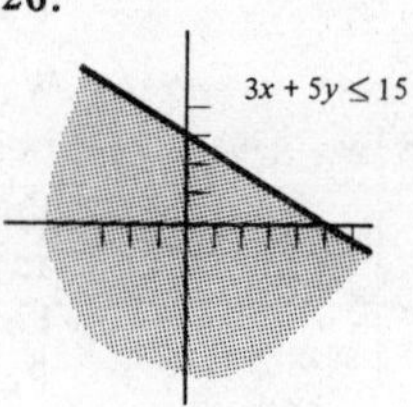

27.

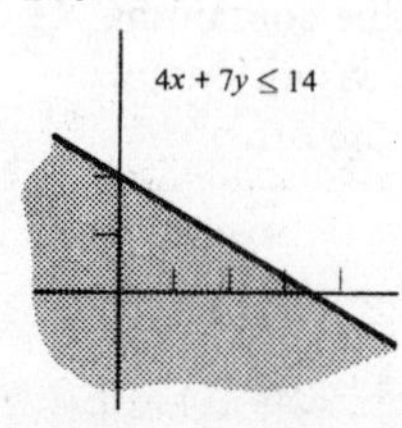

29.

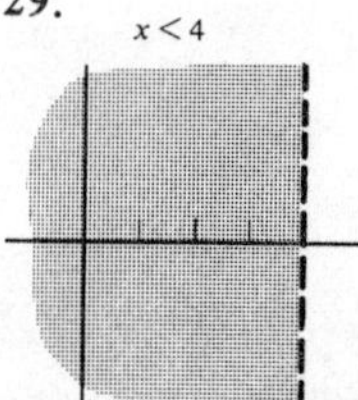

30.

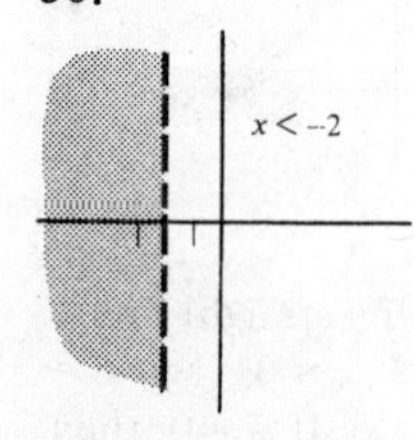

31.

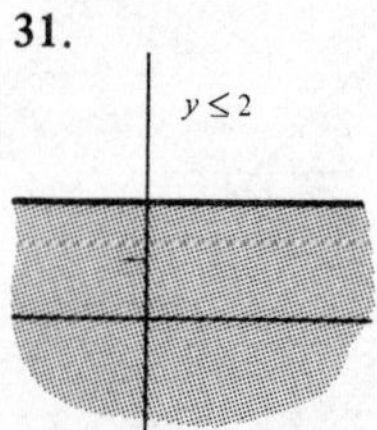

33.

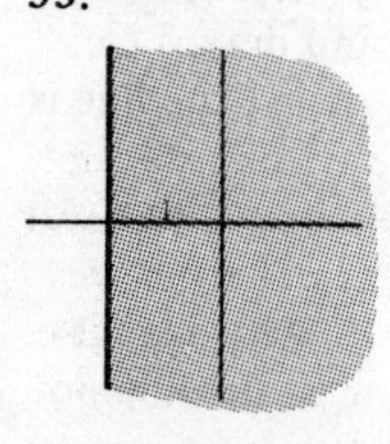

34.

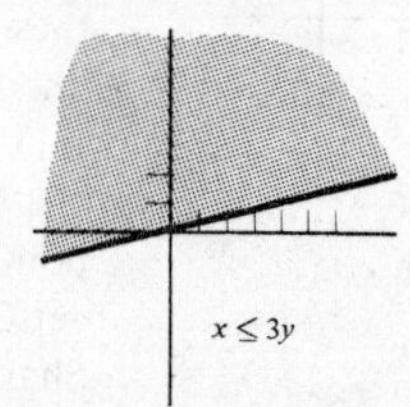

35.

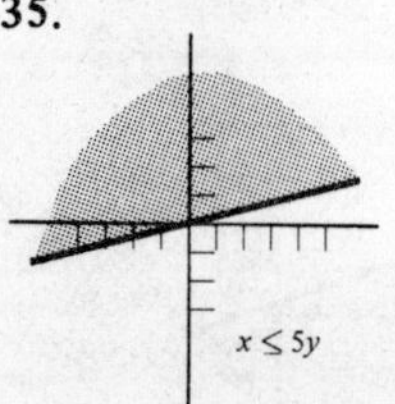

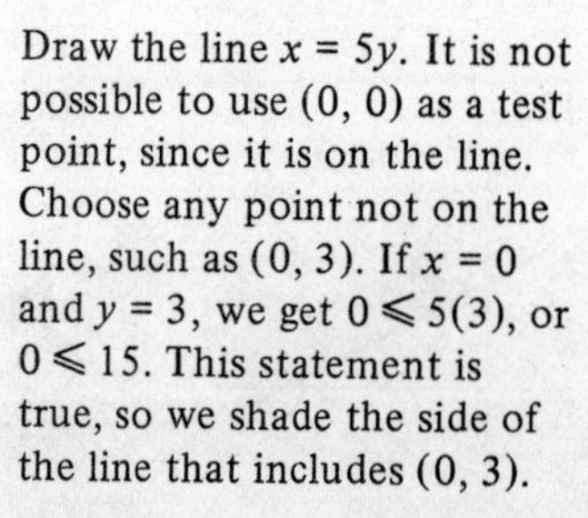

Draw the line $x = 5y$. It is not possible to use (0, 0) as a test point, since it is on the line. Choose any point not on the line, such as (0, 3). If $x = 0$ and $y = 3$, we get $0 \leqslant 5(3)$, or $0 \leqslant 15$. This statement is true, so we shade the side of the line that includes (0, 3).

Chapter 7 Test (page 305)

1. (0, −6), (−2, −16), (4, 14) **2.** (0, −6), (10, 0), (5, −3) **3.** (0, 3), (21/2, 0), (3, 15/7), (7/2, 2) **4.** (0, 0), (6, 2), (8, 8/3), (−12, −4) **5.** (−4, 2), (−4, 0), (−4, −3) **6.** (5, 2), (4, 2), (0, 2), (−3, 2)

7.

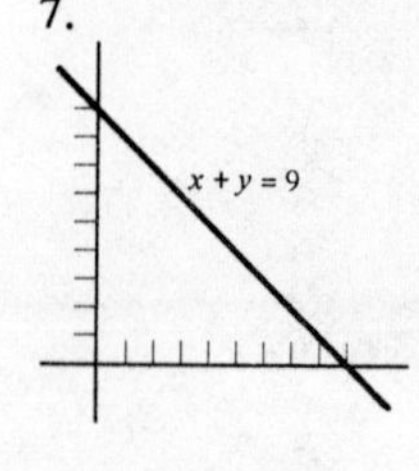

8.

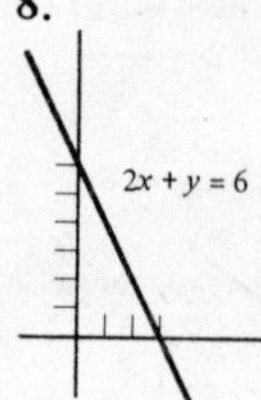

9.

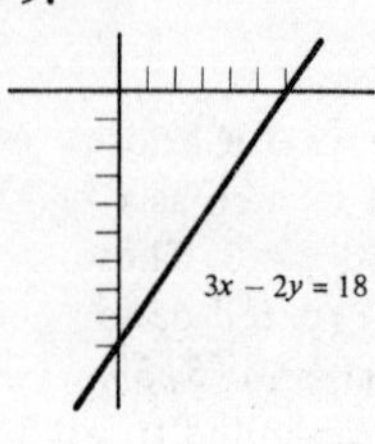

10.

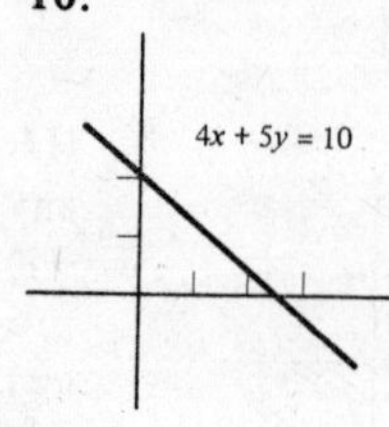

11.

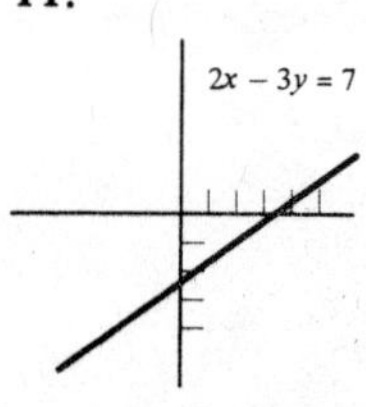

12.

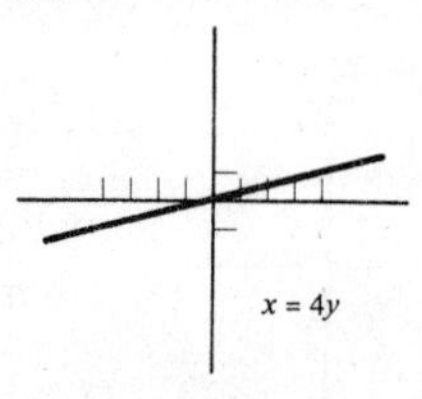

13.

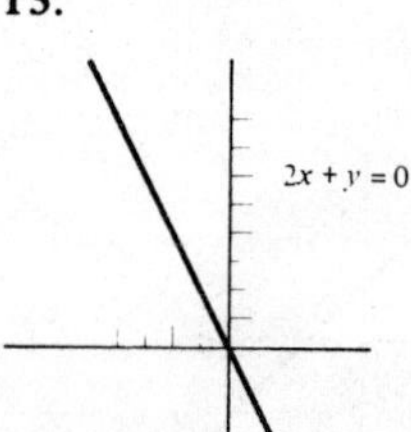

14.

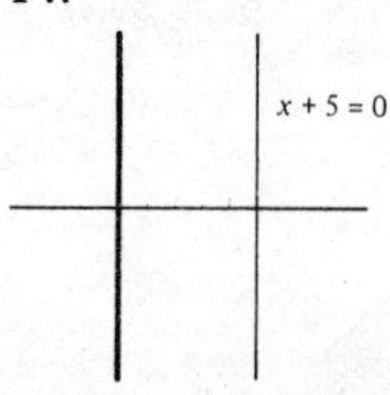

15.

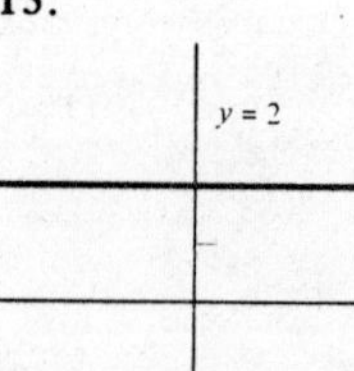

16.

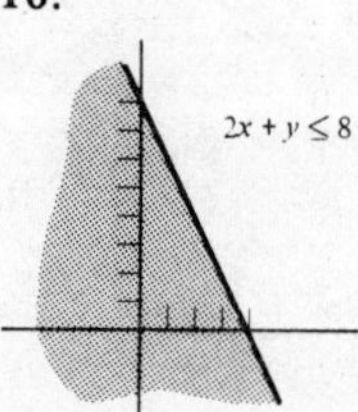

17.

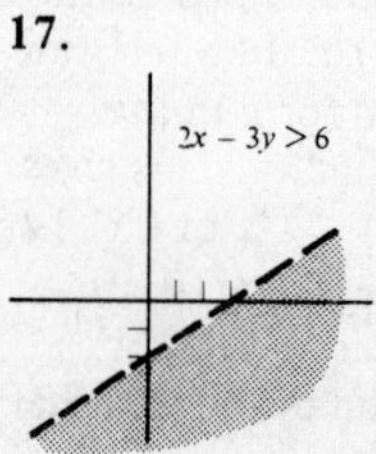

18.

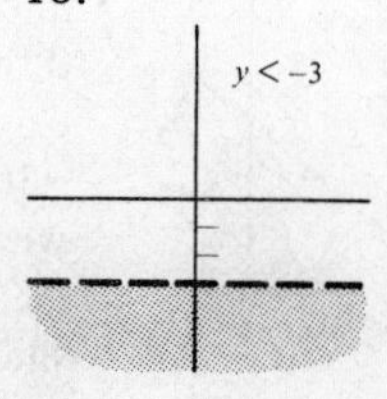

19.

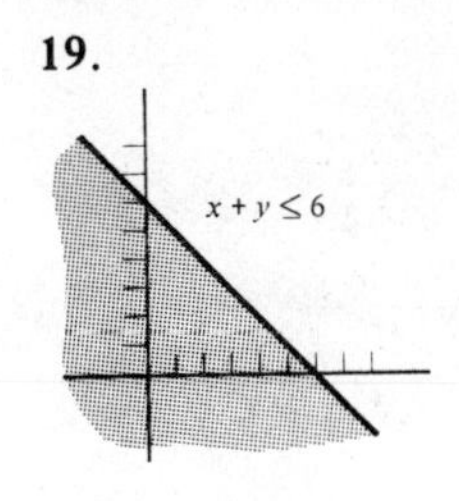

20.

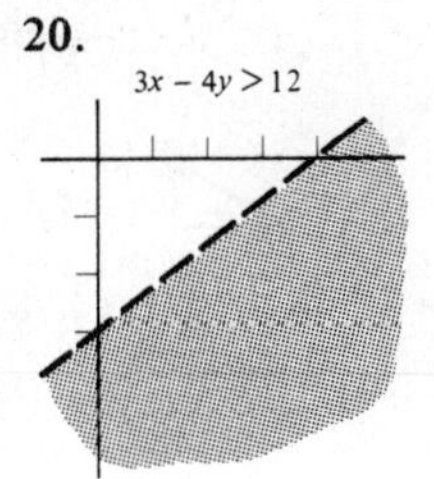

21.

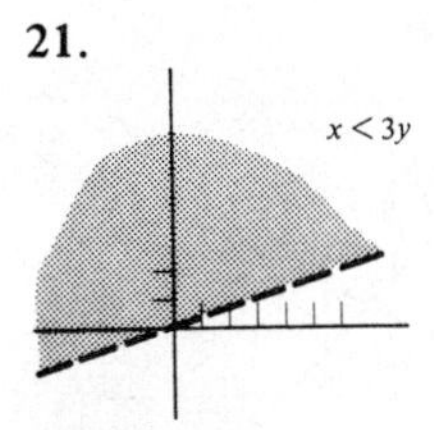

CHAPTER 8

Section 8.1 (page 311)

1. Yes **2.** Yes **3.** Replace x with 4 and y with -2. In the first equation, we get $4 + (-2) = 2$, which is true. In the second equation, we get $2(4) + 5(-2) = 2$, or $8 - 10 = 2$, which is false. This ordered pair is not a solution. **5.** No **6.** No **7.** Yes **9.** No **10.** No **11.** Yes **13.** Graph each line on the same axes. For $x + y = 8$, use the points (0, 8), (8, 0), and (4, 4). For the second line, use the points $(0, -2)$, (4, 2), and (2, 0). These lines cross at (5, 3). **14.** $(1, -2)$ **15.** (4, 8) **17.** $(4, -2)$ **18.** $(3, -2)$ **19.** $(1, -3)$ **21.** $(3, -2)$ **22.** (1, 3) **23.** $(2, -1)$ **25.** $(-8, 6)$ **26.** $(-3, -2)$ **27.** $(-4, -1)$ **29.** (5, 0) **30.** $(-4, 0)$ **31.** $(0, -2)$ **33.** (5, 2) **34.** No solution **35.** Graph each line. The two lines you get are parallel, so that there is no solution for the system. **37.** No solution **38.** Same line **39.** Same line

Section 8.2 (page 319)

1. $(1, -2)$ **2.** (2, 5) **3.** Add the equations, getting $3x = 6$, or $x = 2$. Replace x with 2 in either equation. If we use $x + y = 2$, we get $2 + y = 2$, or $y = 0$. The solution is (2, 0). **5.** (6, 2) **6.** $(3, -4)$ **7.** $(-2, 6)$ **9.** (1/2, 2) **10.** $(1/3, -4)$ **11.** $(3/2, -2)$ **13.** (3, 0) **14.** $(0, -3)$ **15.** $(2, -3)$ **17.** (4, 4) **18.** (5, 4) **19.** One way to work this problem is to multiply the top equation by -3, giving $-3x - 12y = 54$. Then add:

$$\begin{array}{r} -3x - 12y = 54 \\ \underline{3x + 5y = -19} \\ -7y = 35 \end{array}$$

or $y = -5$. Replace y with -5 in either equation to find that $x = 2$, giving $(2, -5)$. **21.** (4, 9) **22.** $(-3, -4)$ **23.** $(-4, 0)$ **25.** $(4, -3)$ **26.** $(5, -9)$ **27.** $(-9, -11)$ **29.** (6, 3) **30.** $(4, -1)$ **31.** $(6, -5)$ **33.** $(-6, 0)$ **34.** (8, 0) **35.** (1/2, 2/3) **37.** (3/8, 5/6) **38.** $(2/3, -5/2)$ **39.** (11, 15) **41.** (22/9, 8/9) **42.** $(-4, 0)$

Section 8.3 (page 323)

1. Multiply the top equation by -1, giving $-x - y = -4$. Then add the two equations, giving $0 = -6$. This false statement shows that the system has no solution. **2.** No solution **3.** No solution **5.** Multiply the top equation by -2, giving $-2x - 6y = -10$. Add the two equations, giving $0 = 0$. This true statement shows that the two equations represent the same line. **6.** Same line **7.** No solution **9.** No solution **10.** No solution **11.** Same line **13.** Multiply the top equation by -2, getting $-4x + 6y = 0$. Then add, to get $11y = 0$, or $y = 0$. Replace y with 0, to see that $x = 0$, giving the solution (0, 0). **14.** (0, 0) **15.** (3, 2) **17.** No solution **18.** No solution **19.** Same line **21.** Same line **22.** Same line

Section 8.4 (page 329)

1. (2, 4) **2.** $(1, -4)$ **3.** (6, 4) **5.** $(8, -1)$ **6.** $(4, -3)$ **7.** Replace x with $2y - 9$ in the first equation, giving $5(2y - 9) + 7y = 40$, or $10y - 45 + 7y = 40$, or $17y = 85$, from which $y = 5$. Replace y with 5 to get $x = 2(5) - 9$, or $x = 1$. The solution is (1, 5). **9.** $(2, -4)$ **10.** $(3, -1)$ **11.** (5, 1) **13.** No solution **14.** No solution **15.** Same line **17.** $(5, -3)$ **18.** $(6, -2)$ **19.** (4, 1) **21.** $(2, -3)$ **22.** $(-5, 3)$ **23.** (2, 8) **25.** $(4, -6)$ **26.** $(36, -35)$ **27.** (3, 2) **29.** (7, 0) **30.** (0, 6) **31.** No solution **33.** Same line **34.** Same line **35.** (6, 5) **37.** (0, 3) **38.** $(2, -3)$ **39.** $(18, -12)$ **41.** Multiply both sides of the top equation by 12, getting $4x - 9y = -6$. Multiply both sides of the bottom equation by 6, getting $4x + 3y = 18$. Multiply both sides of this second equation by -1, to get $-4x - 3y = -18$. Add the two equations to get $-12y = -24$, or $y = 2$. Replace y with 2 in either equation to get $x = 3$. The solution is (3, 2). **42.** $(-2, 1)$ **43.** $(3, -1)$

Section 8.5 (page 335)

1. 43 and 9 **2.** 37 and 19 **3.** 72 and 24 **5.** Let x represent the width, with y for the length. Since the length is twice the width, we have $y = 2x$. The perimeter is 60, so from the formula for perimeter, $60 = 2x + 2y$. Replace y with $2x$, to get $60 = 2x + 2(2x)$, or $60 = 2x + 4x$, or $60 = 6x$, or $x = 10$. Since $y = 2x$, we have $y = 2(10) = 20$. **6.** 6, 6, 9 **7.** Let t be the number of tens, and w the number of twenties. Since there are 85 bills, we have $t + w = 85$. The total value of the tens is $10t$, and the total value of the twenties is $20w$. Since the grand total value is 1480, we have $10t + 20w = 1480$. Solve this sytem to get 22 tens, 63 twenties. **9.** 74 at 8¢, 96 at 10¢ **10.** 249 student, 62 non-student **11.** Let x be the amount at 5% and y the amount at 7%. Then $x + y = 10{,}000$. The interest at 5% is $.05x$, and the interest at 7% is $.07y$. Then $.05x + .07y = 550$. Clear the equation of decimals by multiplying through by 100, to get $5x + 7y = 55{,}000$. Solve the system to get \$7500 at 5% and \$2500 at 7%. **13.** Let x be the number of liters of 90%, and y the number of liters of 75%. Make a chart.

Liters	Percent	Liters of Pure
x	90	$.90x$
y	75	$.75y$
20	78	$20(.78) = 15.6$

Write two equations: $x + y = 20$ and $.90x + .75y = 15.6$. Solve this system, to get 4 liters of 90% and 16 liters of 75%. **14.** 20 pounds of 60¢ and 10 pounds of 90¢ **15.** 30 barrels of \$40 olives and 20 barrels of \$60 olives **17.** plane, 470 mph; wind, 70 mph **18.** Boat is 5 mph, current is 3 mph **19.** John, 3 1/4 mph; Harriet, 2 3/4 mph **21.** 4 girls, 3 boys **22.** 2 girls, 2 boys

Section 8.6 (page 343)

1.

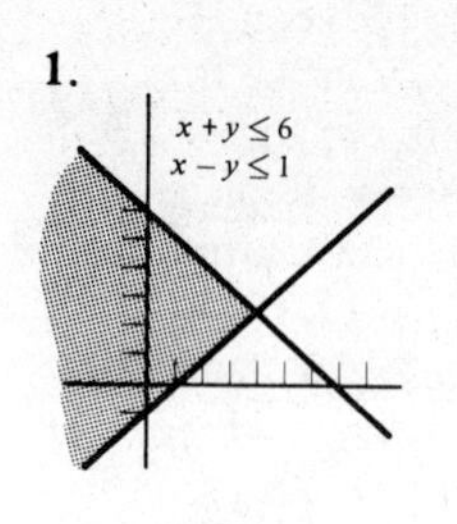

2.

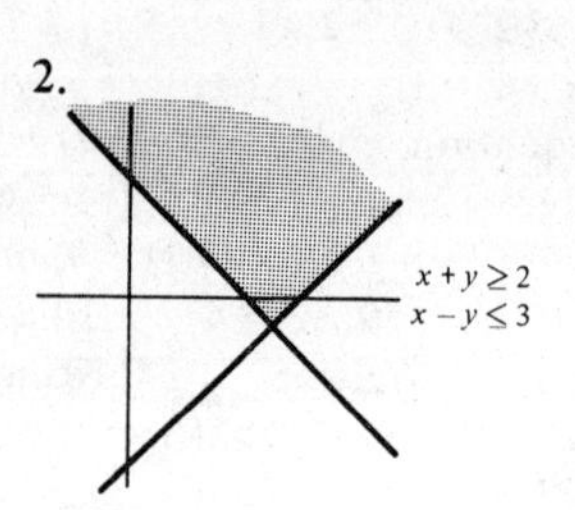

3.

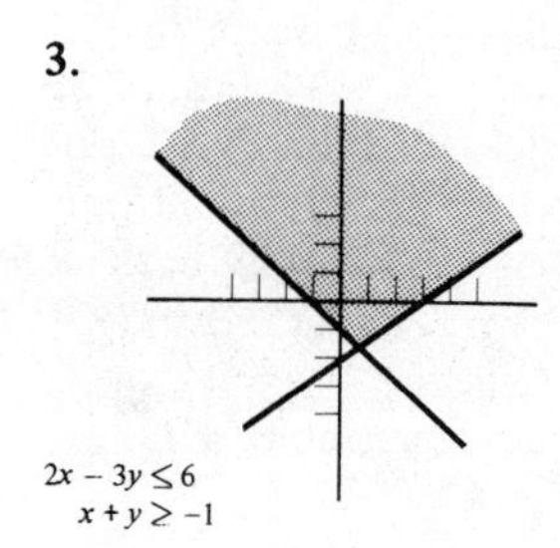

5.

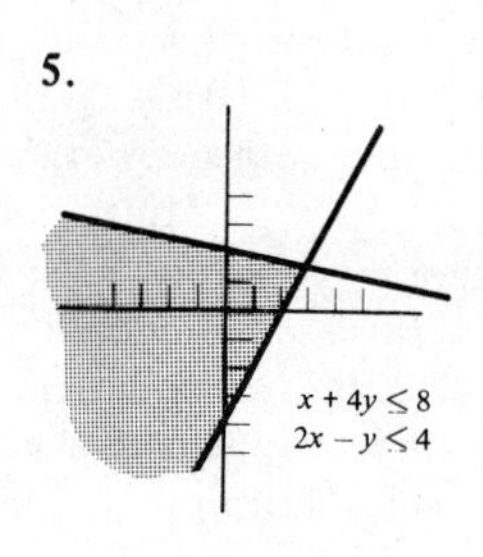

6.

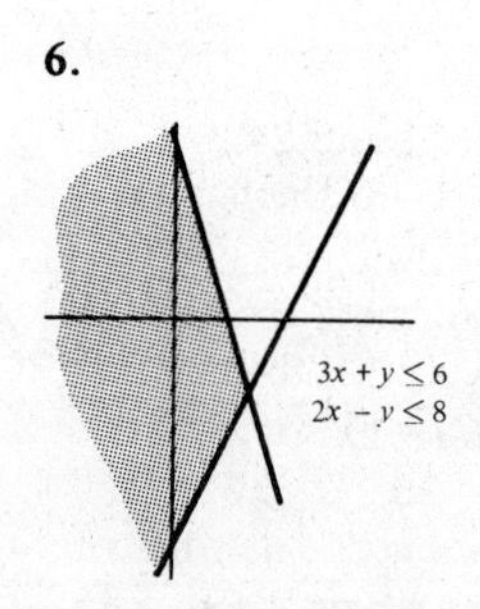

7.

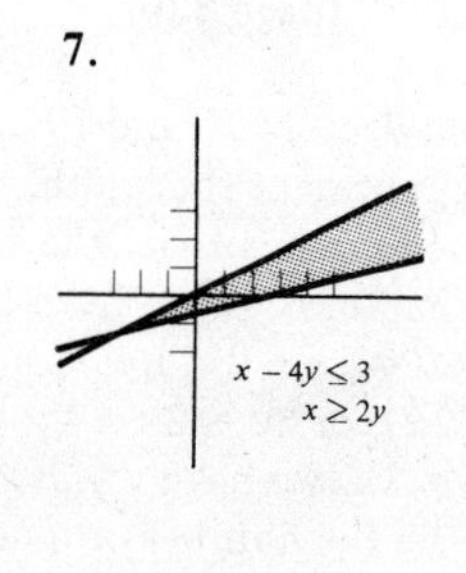

9.

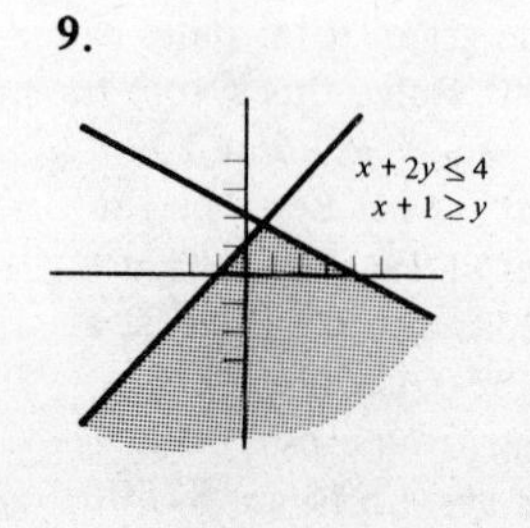

10.

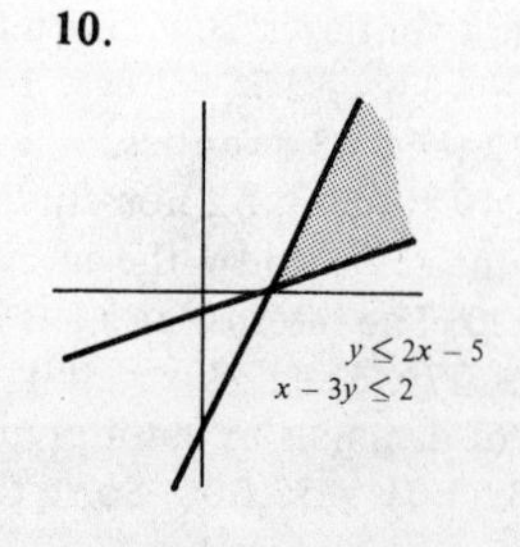

11.

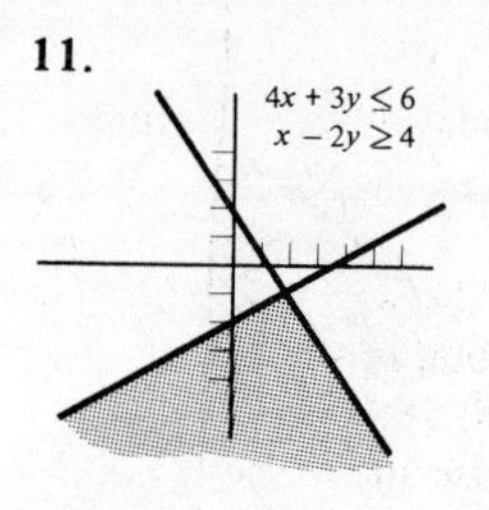

13.

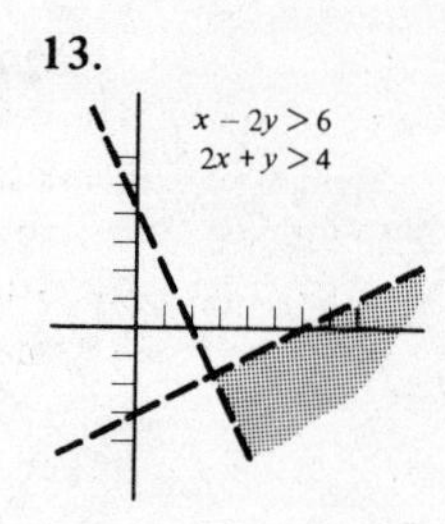

14.

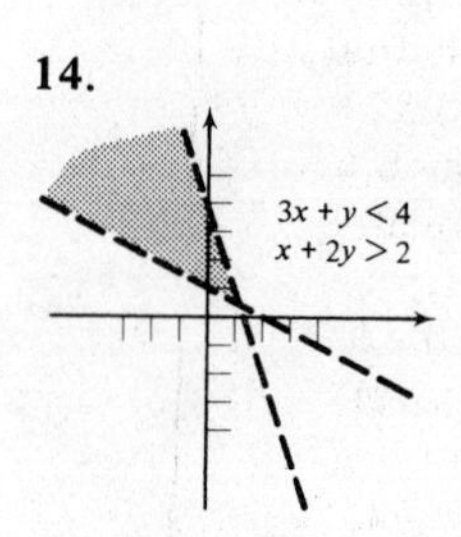

15.

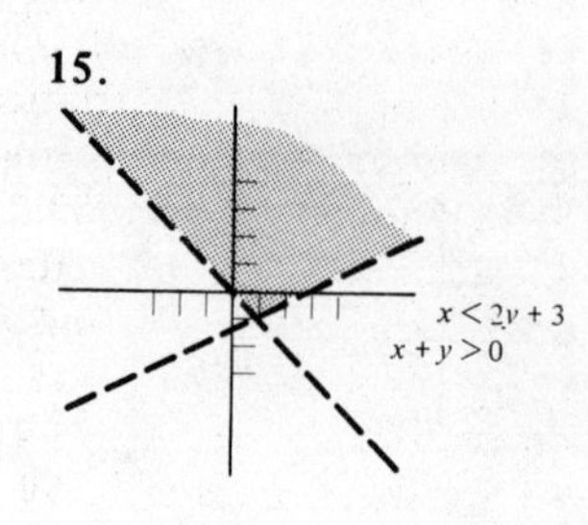

17.

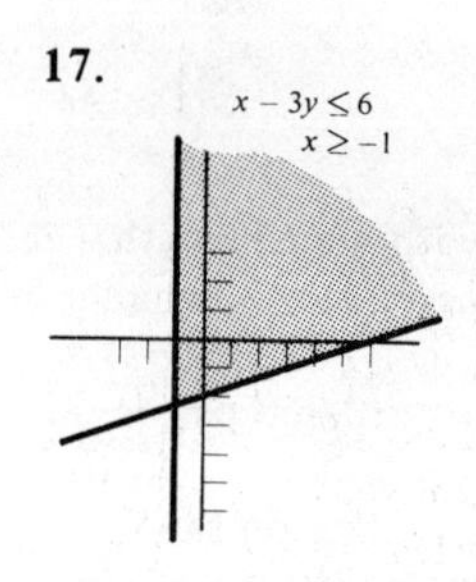

18.

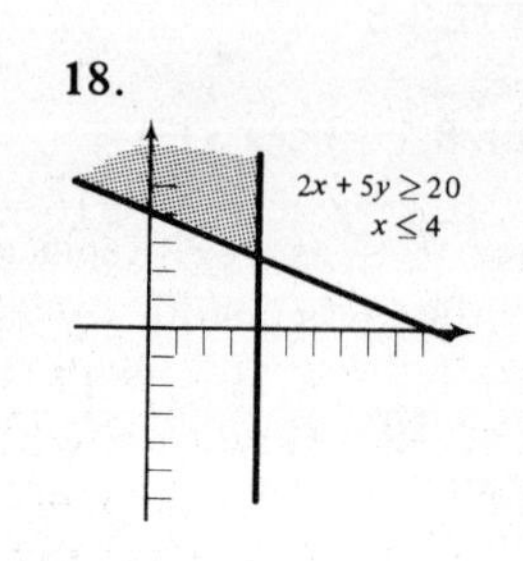

Chapter 8 Test (page 345)

1. (4, −3) **2.** (6, −5) **3.** (4, 1) **4.** (6, 5) **5.** (8, −2) **6.** (2, −5) **7.** No solution **8.** (0, −2) **9.** (2, 2) **10.** (3/2, −2) **11.** (4/3, 6) **12.** (3, −5) **13.** (−6, 8) **14.** 2 at \$2.50, 4 at \$3.75 **15.** 45 mph, and 60 mph

16.

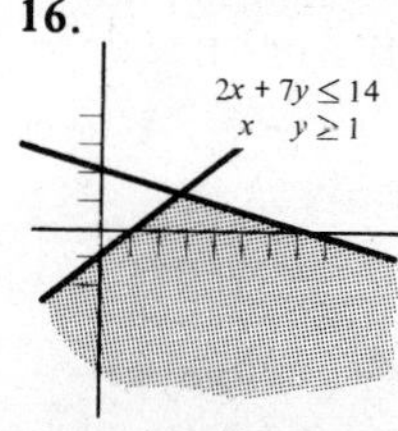

17.

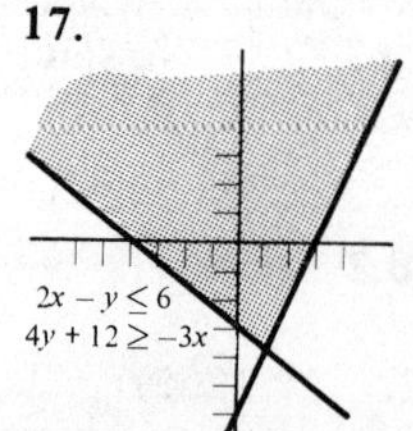

18.

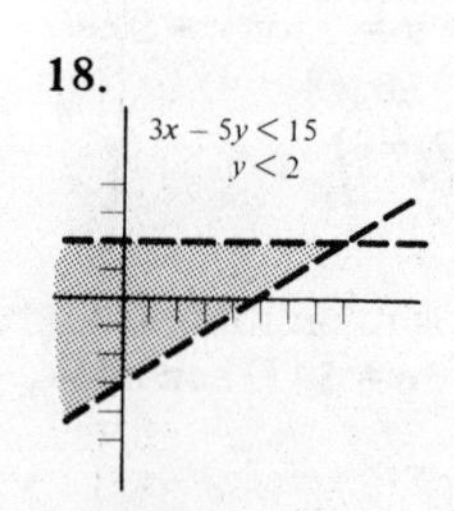

CHAPTER 9

Section 9.1 (page 351)

1. 3, −3 **2.** 4, −4 **3.** Since $11^2 = 121$ and $(-11)^2 = 121$, the square roots are 11 and −11. **5.** 20, −20 **6.** 30, −30 **7.** 25, −25 **9.** 39, −39

10. 47, −47 **11.** 63, −63 **13.** 2 **14.** 3 **15.** 5; the number −5 is not an answer, even though $(-5)^2 = 25$. This is because the radical sign indicates only the nonnegative square root. **17.** −8 **18.** −10 **19.** 13 **21.** 30 **22.** 40 **23.** First find $\sqrt{1681}$, which, from the table, is 41. Then $-\sqrt{1681} = -41$. **25.** 51 **26.** 55 **27.** −70 **29.** There is no real number whose square is −9. This root does not exist. **30.** Does not exist **31.** Does not exist **33.** $\sqrt{16} = 4$, a rational number **34.** Rational; 9 **35.** Irrational; 3.873 (found in the square root table) **37.** Irrational; 6.856 **38.** Irrational; 7.280 **39.** Irrational; 8.246 **41.** Rational; −11 **42.** Rational; −12 **43.** Irrational; 10.488 (find 11 at the left, then look in the $\sqrt{10n}$ column) **45.** Irrational; −14.142 **46.** Irrational; −16.125 **47.** Rational; 20 **49.** Irrational; 23.875 **50.** Irrational; 26.268

Section 9.2 (page 357)

1. $\sqrt{8} \cdot \sqrt{2} = \sqrt{8 \cdot 2} = \sqrt{16} = 4$ **2.** 9 **3.** 6 **5.** 21 **6.** 17 **7.** $\sqrt{21}$ **9.** $\sqrt{27} = \sqrt{9 \cdot 3} = \sqrt{9} \cdot \sqrt{3} = 3\sqrt{3}$ **10.** $3\sqrt{5}$ **11.** $2\sqrt{7}$ **13.** $3\sqrt{2}$ **14.** $5\sqrt{3}$ **15.** $4\sqrt{3}$ **17.** $5\sqrt{5}$ **18.** $5\sqrt{6}$ **19.** $10\sqrt{7}$ **21.** $\sqrt{100} \cdot \sqrt{27} = 10\sqrt{27} = 10(3\sqrt{3}) = 30\sqrt{3}$ **22.** $8\sqrt{2}$ **23.** 40 **25.** $\sqrt{27} \cdot \sqrt{48} = 3\sqrt{3} \cdot 4\sqrt{3} = (3 \cdot 4)(\sqrt{3} \cdot \sqrt{3}) = 12(3) = 36$; alternate solution: $\sqrt{27} \cdot \sqrt{48} = \sqrt{27 \cdot 48} = \sqrt{1296} = 36$ **26.** 45 **27.** 60 **29.** $\sqrt{7} \cdot \sqrt{21} = \sqrt{7 \cdot 21} = \sqrt{147} = \sqrt{49 \cdot 3} = \sqrt{49} \cdot \sqrt{3} = 7\sqrt{3}$ **30.** 24 **31.** $15\sqrt{3}$ **33.** $20\sqrt{3}$ **34.** $12\sqrt{5}$ **35.** $10\sqrt{10}$ **37.** $\sqrt{100/9} = \sqrt{100}/\sqrt{9} = 10/3$ **38.** 15/4 **39.** 6/7 **41.** $\sqrt{5/16} = \sqrt{5}/\sqrt{16} = \sqrt{5}/4$ **42.** $\sqrt{11}/5$ **43.** $\sqrt{30}/7$ **45.** First multiply 1/5 and 4/5, getting 4/25. The square root of 4/25 is 2/5. **46.** 2/9 **47.** 4/25 **49.** Divide 3 into 75, getting 25. The square root of 25 is 5. **50.** 10 **51.** 4 **53.** $3\sqrt{5}$ **54.** $9\sqrt{2}$ **55.** $5\sqrt{10}$ **57.** y **58.** m **59.** $\sqrt{xz}$ **61.** x **62.** y **63.** x^2 **65.** xy^2 **66.** x^2y^4 **67** $x\sqrt{x}$ **69.** $4/x$ **70.** $10/m^2$ **71.** $\sqrt{11}/r^2$

Section 9.3 (page 363)

1. $7\sqrt{3}$ **2.** $14\sqrt{5}$ **3.** $-5\sqrt{7}$ **5.** $\sqrt{6} + \sqrt{6} = 1 \cdot \sqrt{6} + 1 \cdot \sqrt{6} = 2\sqrt{6}$ **6.** $2\sqrt{11}$ **7.** $3\sqrt{17}$ **9.** $4\sqrt{7}$ **10.** $11\sqrt{14}$ **11.** $12\sqrt{2}$ **13.** $7\sqrt{5}$ **14.** $17\sqrt{6}$ **15.** $3\sqrt{18} + \sqrt{8} = 3(\sqrt{9 \cdot 2}) + \sqrt{4 \cdot 2} = 3(\sqrt{9} \cdot \sqrt{2}) + \sqrt{4} \cdot \sqrt{2} = 3(3\sqrt{2}) + 2\sqrt{2} = 9\sqrt{2} + 2\sqrt{2} = 11\sqrt{2}$ **17.** $3\sqrt{3}$ **18.** $-4\sqrt{3}$ **19.** $20\sqrt{2}$ **21.** $-13\sqrt{2}$ **22.** $26\sqrt{3}$ **23.** Simplify $\sqrt{28}$ as $2\sqrt{7}$ and simplify $\sqrt{63}$ as $3\sqrt{7}$. The problem is then $5\sqrt{7} - 2\sqrt{28} + 6\sqrt{63} = 5\sqrt{7} - 2(2\sqrt{7}) + 6(3\sqrt{7}) = 5\sqrt{7} - 4\sqrt{7} + 18\sqrt{7} = 19\sqrt{7}$. **25.** $-12\sqrt{5}$ **26.** $-4\sqrt{3}$ **27.** $6\sqrt{2} + 7\sqrt{3}$ (which cannot be further combined) **29.** $-16\sqrt{2} + 8\sqrt{3}$ **30.** $-2\sqrt{2} - 12\sqrt{3}$ **31.** $20\sqrt{2} + 6\sqrt{3} + 15\sqrt{5}$ **33.** $(1/4)\sqrt{288} - (1/6)\sqrt{72} = (1/4)\sqrt{144 \cdot 2} - (1/6)\sqrt{36 \cdot 2} = (1/4)(12\sqrt{2}) - (1/6)(6\sqrt{2}) = 3\sqrt{2} - \sqrt{2} = 2\sqrt{2}$ **34.** $-\sqrt{3}$ **35.** $3\sqrt{3} - 2\sqrt{5}$ **37.** $\sqrt{6} \cdot \sqrt{2} + 3\sqrt{3} = \sqrt{6 \cdot 2} + 3\sqrt{3} = \sqrt{12} + 3\sqrt{3} = 2\sqrt{3} + 3\sqrt{3} = 5\sqrt{3}$ **38.** $10\sqrt{5}$ **39.** $3\sqrt{21} - \sqrt{7}$ **41.** $\sqrt{9x} + \sqrt{49x} - \sqrt{16x} = 3\sqrt{x} + 7\sqrt{x} - 4\sqrt{x} = 6\sqrt{x}$ **42.** $\sqrt{a}$ **43.** $13\sqrt{a}$ **45.** $15x\sqrt{3}$ **46.** $-y\sqrt{5}$ **47.** $2x\sqrt{2}$

Section 9.4 (page 367)

1. Multiply numerator and denominator by $\sqrt{5}$, getting a final answer of $6\sqrt{5}/5$. **2.** $2\sqrt{2}$ **3.** $\sqrt{5}$ **5.** $3\sqrt{7}/7$ **6.** $6\sqrt{10}/5$ **7.** $8\sqrt{15}/5$ **9.** Multiply numerator and denominator by $\sqrt{3}$ and then reduce to lowest terms, getting $\sqrt{30}/2$. **10.** $3\sqrt{30}/4$ **11.** $8\sqrt{3}/9$ **13.** Here it is only necessary to multiply by $\sqrt{2}$ (and not $\sqrt{50}$), giving an answer of $3\sqrt{2}/10$. **14.** $\sqrt{3}/3$ **15.** $\sqrt{2}$ **17.** $9\sqrt{2}/8$ **18.** $2\sqrt{5}$ **19.** 2 **21.** $\sqrt{2}$ **22.** $\sqrt{2}$ **23.** $2\sqrt{30}/3$ **25.** Multiply numerator and denominator, inside the radical, by 2, giving $\sqrt{2/4}$, or $\sqrt{2}/\sqrt{4}$, or $\sqrt{2}/2$. **26.** $\sqrt{2}/4$ **27.** $\sqrt{70}/7$ **29.** $3\sqrt{5}/5$ **30.** $4\sqrt{7}/7$ **31.** Multiply 7/5 and 10, giving 14. The answer is $\sqrt{14}$ **33.** $\sqrt{15}/10$ **34.** $\sqrt{3}/3$ **35.** $3\sqrt{14}/4$ **37.** $\sqrt{3}/5$ **38.** $3\sqrt{14}/16$ **39.** $4\sqrt{3}/27$ **41.** Multiply numerator and denominator, inside the radical, by p, to get $\sqrt{6p}/p$. **42.** $2r\sqrt{s}$ **43.** $p\sqrt{2pm}/m$ **45.** $x\sqrt{y}/(2y)$ **46.** $m\sqrt{2n}/2$ **47.** $3a\sqrt{5r}/5$

Section 9.5 (page 373)

1. $3\sqrt{5} + 8\sqrt{45} = 3\sqrt{5} + 8(\sqrt{9 \cdot 5}) = 3\sqrt{5} + 8(\sqrt{9} \cdot \sqrt{5}) = 3\sqrt{5} + 8(3\sqrt{5}) = 3\sqrt{5} + 24\sqrt{5} = 27\sqrt{5}$ **2.** $18\sqrt{2}$ **3.** $21\sqrt{2}$ **5.** −4 **6.** 6 **7.** $\sqrt{5}(\sqrt{3} + \sqrt{7}) = \sqrt{5} \cdot \sqrt{3} + \sqrt{5} \cdot \sqrt{7} = \sqrt{15} + \sqrt{35}$ **9.** $2\sqrt{10} + 10$ **10.** $42 - 12\sqrt{35}$ **11.** $-4\sqrt{7}$ **13.** $(2\sqrt{6} + 3)(3\sqrt{6} - 5) = (2\sqrt{6})(3\sqrt{6}) - 2\sqrt{6}(5) + 3(3\sqrt{6}) - 3(5) = (2 \cdot 3)(\sqrt{6} \cdot \sqrt{6}) - 10\sqrt{6} + 9\sqrt{6} - 15 = 6(6) - \sqrt{6} - 15 = 21 - \sqrt{6}$ **14.** $34 + 8\sqrt{5}$ **15.** $87 + 9\sqrt{21}$ **17.** $34 + 24\sqrt{2}$ **18.** $81 - 8\sqrt{5}$ **19.** $(2\sqrt{7} - 3)^2 = (2\sqrt{7} - 3)(2\sqrt{7} - 3) = (2\sqrt{7})(2\sqrt{7}) - 3(2\sqrt{7}) - 3(2\sqrt{7}) + 9 = 4 \cdot 7 - 6\sqrt{7} - 6\sqrt{7} + 9 = 37 - 12\sqrt{7}$ **21.** 7 **22.** 44 **23.** −4 **25.** 1 **26.** 1 **27.** Multiply numerator and denominator by $3 - \sqrt{2}$. In the denominator, we have $(3 + \sqrt{2})(3 - \sqrt{2}) = 3 \cdot 3 + 3\sqrt{2} - 3\sqrt{2} - \sqrt{2} \cdot \sqrt{2} = 9 - 2 = 7$. The answer is $(3 - \sqrt{2})/7$. **29.** $-10 + 5\sqrt{5}$ **30.** $9 - 3\sqrt{7}$ **31.** $-2 - \sqrt{11}$ **33.** Multiply numerator and denominator by $1 - \sqrt{2}$. In the numerator, we have $\sqrt{2}(1 - \sqrt{2}) = \sqrt{2} - 2$. In the denominator, we have $(1 + \sqrt{2})(1 - \sqrt{2}) = 1 - 2 = -1$. The answer is $-\sqrt{2} + 2$. **34.** $(-2\sqrt{7} + 7)/3$ **35.** $-(\sqrt{5} + 5)/4$ **37.** $3 - \sqrt{3}$ **38.** $6 + 3\sqrt{2}$ **39.** $(-3 + 5\sqrt{3})/11$

Section 9.6 (page 379)

1. 4 **2.** 25 **3.** 1 **5.** 7 **6.** 11 **7.** −7 **9.** −5 **10.** 4 **11.** Square both sides, getting $z + 5 = 4$. The solution of this equation is −1. Check −1 in the original equation: $\sqrt{-1 + 5} = -2$ is false, since $\sqrt{-1 + 5} = 2$, and not −2. This equation has no solution. **13.** Add 2 to both sides to get $\sqrt{k} = 7$.

The solution is 49. **14.** 100 **15.** No solution **17.** Square both sides, to get $5t - 9 = 4t$ (be careful on the right side). Solve this equation and check to see that the answer is 9. **18.** 4 **19.** 16 **21.** 6 **22.** 7 **23.** 7 **25.** 4/3 **26.** −3 **27.** 5 **29.** Square both sides of the equation. On the left, you get $2x + 1$ and on the right you get $(x - 7)^2 = x^2 - 14x + 49$. The new equation is $2x + 1 = x^2 - 14x + 49$. Make the equation equal to 0 by adding $-2x$ and -1 to both sides. This gives $x^2 - 16x + 48 = 0$. Factor this equation as $(x - 12)(x - 4) = 0$. Solve each of these equations: $x - 12 = 0$ gives $x = 12$, and $x - 4 = 0$ gives $x = 4$. Now go back to the original equation. $x = 12$ gives a true statement, while $x = 4$ does not. The only solution is 12. **30.** 0, 3 **31.** 5 **33.** 0, −1 **34.** 11 **35.** Add 2 to both sides, to give $\sqrt{4x + 5} = 2x - 5$. Square both sides, to give $4x + 5 = 4x^2 - 20x + 25$. Get this equation equal to 0 as $4x^2 - 24x + 20 = 0$. Factor the equation as $4(x - 5)(x - 1) = 0$. Solve this equation to get 5 and 1. Check in the original equation; the only solution is 5. **37.** 3 **38.** 9 **39.** 9 **41.** 12 **42.** 9 **43.** If x is the number, the equation is $\sqrt{x + 4} = 5$. The solution is 21. **45.** 8 **46.** 1 **47. (a)** Let $a = 900$ and $p = 100$. The formula gives $s = \sqrt{900(900)/100} = \sqrt{810{,}000/100} = \sqrt{8100} = 90$ mph **(b)** 120 mph **(c)** 60 mph **(d)** 60 mph

Chapter 9 Test (page 381)

1. 10 **2.** 31 **3.** 76 **4.** 8.775 **5.** 13.784 **6.** 21.909 **7.** $2\sqrt{2}$ **8.** $3\sqrt{3}$ **9.** $5\sqrt{2}$ **10.** $8\sqrt{2}$ **11.** $5\sqrt{3}$ **12.** $6\sqrt{3}$ **13.** $-\sqrt{5}$ **14.** $8\sqrt{2}$ **15.** Cannot be simplified **16.** Cannot be simplified **17.** $-7\sqrt{3x}$ **18.** $4xy\sqrt{2y}$ **19.** 31 **20.** −1 **21.** $4 - 2\sqrt{3}$ **22.** $11 + 2\sqrt{30}$ **23.** $4\sqrt{3}/3$ **24.** $3\sqrt{6}$ **25.** $\sqrt{3}$ **26.** $3 + \sqrt{7}$ **27.** $-3(4 + \sqrt{3})/13$ **28.** 16 **29.** No solution **30.** 23 **31.** 4 **32.** −5 **33.** −1, −2

CHAPTER 10

Section 10.1 (page 385)

1. 5, −5 **2.** 10, −10 **3.** 8, −8 **5.** $\sqrt{13}, -\sqrt{13}$ **6.** $\sqrt{7}, -\sqrt{7}$ **7.** $\sqrt{2}, -\sqrt{2}$ **9.** Simplify $\sqrt{24}$: $\sqrt{24} = \sqrt{4 \cdot 6} = \sqrt{4} \cdot \sqrt{6} = 2\sqrt{6}$. The two solutions are $2\sqrt{6}$ and $-2\sqrt{6}$. **10.** $3\sqrt{3}, -3\sqrt{3}$ **11.** Take the square root of both sides to get the two equations $x - 2 = 4$ and $x - 2 = -4$. Solve the first equation to get $x = 6$; the second equation gives $x = -2$. **13.** Take the square root of both sides to get $a + 4 = \sqrt{10}$ and $a + 4 = -\sqrt{10}$. Add −4 to both sides of each equation to get the solutions $-4 + \sqrt{10}$ and $-4 - \sqrt{10}$. **14.** $3 + \sqrt{15}, 3 - \sqrt{15}$ **15.** $1 + 4\sqrt{2}, 1 - 4\sqrt{2}$ **17.** 2, −1 **18.** 3, 5/3 **19.** −2/3, −8/3 **21.** 13/6, −3/2 **22.** 22/7, −2/7 **23.** $(5 + \sqrt{30})/2$, $(5 - \sqrt{30})/2$ **25.** Take the square root of both sides to get $3p - 1 = \sqrt{18}$ and $3p - 1 = -\sqrt{18}$. We can simplify $\sqrt{18}$ as $3\sqrt{2}$. Add 1 to both sides of each equation, to give $3p = 1 + 3\sqrt{2}$ and $3p = 1 - 3\sqrt{2}$. Multiply both sides of each equation by 1/3 to end up with the final solutions, $(1 + 3\sqrt{2})/3$ and $(1 - 3\sqrt{2})/3$. **26.** $(6 + 5\sqrt{3})/5, (6 - 5\sqrt{3})/5$ **27.** $(5 + 7\sqrt{2})/2$, $(5 - 7\sqrt{2})/2$ **29.** $(-4 + 2\sqrt{2})/3, (-4 - 2\sqrt{2})/3$ **30.** $(3 + 5\sqrt{2})/5, (3 - 5\sqrt{2})/5$ **31.** About 1/2 second

Section 10.2 (page 391)

1. Take half of 2, which is 1. Square this to get the final answer: $1^2 = 1$. **2.** 4 **3.** 81 **5.** Half of 9 is 9/2; squared, we have $(9/2)^2 = 81/4$. **6.** 121 **7.** 49 **9.** 25/4 **10.** 16 **11.** Half of 4 is 2; squared we get $2^2 = 4$. Add 4 to both sides of the given equation, getting $x^2 + 4x + 4 = -3 + 4$, or $(x + 2)^2 = 1$, after factoring on the left and simplifying on the right. Take the square root of both sides to get the solutions −1 and −3. **13.** $-1 + \sqrt{6}, -1 - \sqrt{6}$ **14.** 2, −6 **15.** −2, −4 **17.** Add −1 to both sides, getting $x^2 - 6x = -1$. Half of −6 is −3, which can be squared to give 9. Add 9 on both sides, getting $x^2 - 6x + 9 = -1 + 9$, or $(x - 3)^2 = 8$. Take the square root of both sides to get $x - 3 = \sqrt{8}$ and $x - 3 = -\sqrt{8}$. Simplify $\sqrt{8}$ as $2\sqrt{2}$; then get the final solutions of $3 + 2\sqrt{2}$ and $3 - 2\sqrt{2}$. **18.** $1 + \sqrt{3}, 1 - \sqrt{3}$ **19.** $(-3 + \sqrt{17})/2, (-3 - \sqrt{17})/2$ **21.** Multiply through by 1/2 to get $m^2 + 2m = -7/2$. Half of 2 is 1, which can be squared to give 1. Add 1 on both sides to get $m^2 + 2m + 1 = -5/2$. Factor to get $(m + 1)^2 = -5/2$. We cannot take the square root of −5/2, so this equation has no solution. **22.** $(9 + \sqrt{21})/6$, $(9 - \sqrt{21})/6$ **23.** No real number solutions **25.** $3 + \sqrt{5}, 3 - \sqrt{5}$ **26.** $(3 + \sqrt{15})/3, (3 - \sqrt{15})/3$ **27.** $(2 + \sqrt{14})/2, (2 - \sqrt{14})/2$ **29.** 1, −1/3 **30.** No real number solutions **31.** $5 + \sqrt{17}, 5 - \sqrt{17}$ **33.** $(3 + 2\sqrt{6})/3, (3 - 2\sqrt{6})/3$ **34.** No real number solutions **35.** $-2 + \sqrt{3}, -2 - \sqrt{3}$

Section 10.3 (page 399)

1. The number in front of the x^2 term is 3, so that $a = 3$. The number in front of the x term is 4, so that $b = 4$. The number left over is −8, so that $c = -8$. (Note: the equation was equal to 0 before we started.) **2.** 9, 2, −3 **3.** −8, −2, −3 **5.** 2, −3, 2 **6.** 9, −4, −2 **7.** 1, 0, −2 **9.** 3, −8, 0 **10.** 5, −2, 0 **11.** Multiply $(x - 3)$ and $(x + 4)$ to get $x^2 + x - 12 = 0$, from which $a = 1$, $b = 1$, $c = -12$. **13.** Multiply out the expression on the left and combine terms to find that $a = 9$, $b = 9$, and $c = -26$. **14.** 5, 14, −5 **15.** Here $a = 1$, $b = 2$, and $c = -2$. Using the quadratic formula, we get $x = (-2 \pm \sqrt{2^2 - 4(1)(-2)})/2(1) = (-2 \pm \sqrt{4 + 8})/2 = (-2 \pm \sqrt{12})/2 = (-2 \pm 2\sqrt{3})/2 = 2(-1 \pm \sqrt{3})/2 = -1 \pm \sqrt{3}$ so that the solutions are $-1 + \sqrt{3}$ and $-1 - \sqrt{3}$. **17.** −2 **18.** $(5 + \sqrt{13})/6$, $(5 - \sqrt{13})/6$ **19.** 1, −13 **21.** $(-6 + \sqrt{26})/2$, $(-6 - \sqrt{26})/2$ **22.** 1, 19 **23.** 1/5, −1 **25.** Add $-3x$ and −5 to both sides, to get $2x^2 - 3x - 5 = 0$. Here $a = 2$, $b = -3$, and $c = -5$. By the quadratic formula, $x = (-(-3) \pm \sqrt{(-3)^2 - 4(2)(-5)})/2(2) = (3 \pm \sqrt{9 + 40})/4 = (3 \pm \sqrt{49})/4 = (3 \pm 7)/4$. Now we can find the two answers. First, use the plus sign; $x = (3 + 7)/4 = 10/4 = 5/2$. Then use the minus sign; $x = (3 - 7)/4 = -4/4 = -1$. **26.** $-5/2, 6$ **27.** -3 **29.** $0, -1$ **30.** $2\sqrt{5}, -2\sqrt{5}$ **31.** $0, -2$ **33.** Here $a = 2$, $b = 1$, and $c = 7$. By the quadratic formula, $x = (-1 \pm \sqrt{1^2 - 4(2)(7)})/2(2) = (-1 \pm \sqrt{1 - 56})/4 = (-1 \pm \sqrt{-55})/4$. Since $\sqrt{-55}$ is not a real number, this equation has no solutions. **34.** No real number solutions **35.** No real number solutions **37.** Multiply both sides by 6 to get $3x^2 = -x + 6$, or $3x^2 + x - 6 = 0$. Solve by the quadratic formula to get the solutions $(-1 + \sqrt{73})/6$, $(-1 - \sqrt{73})/6$. **38.** $4/3, -2/3$ **39.** Multiply both sides by 9 to get $6x^2 - 4x - 3 = 0$. Solve this equation to get $(2 + \sqrt{22})/6$, $(2 - \sqrt{22})/6$.

Section 10.4 (page 405)

1.

$y = 2x^2$

2.

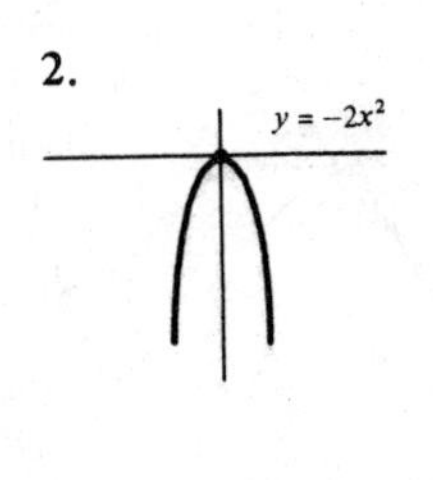

3.

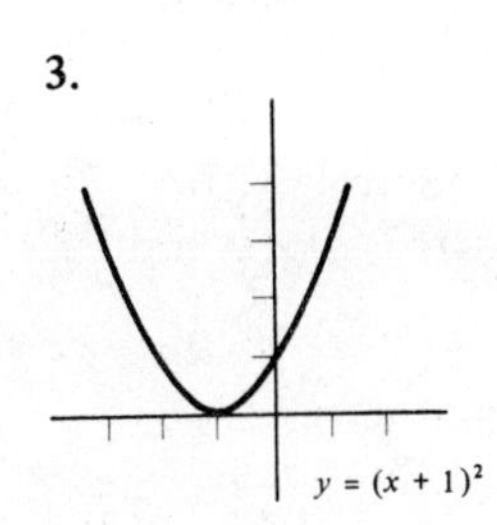

5.

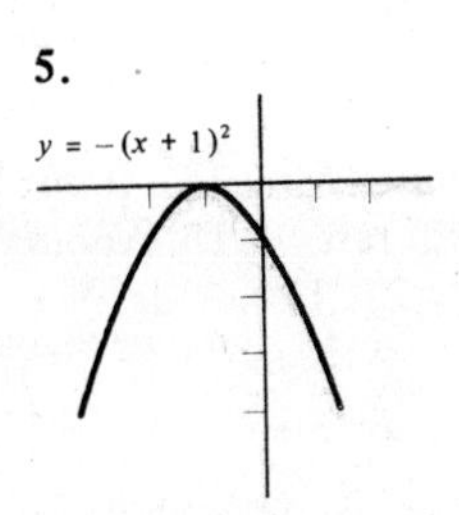

6.

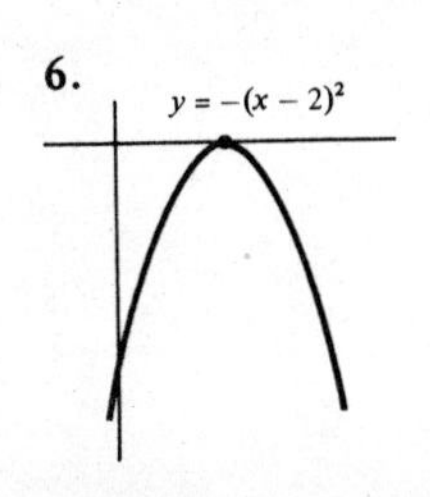

7.

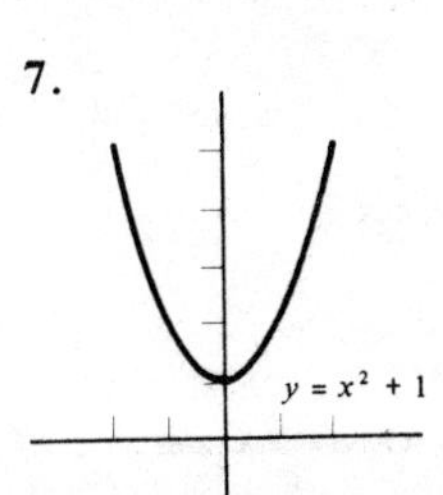

9.

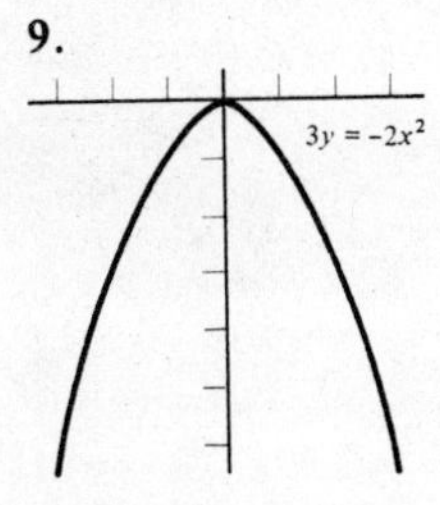

10.

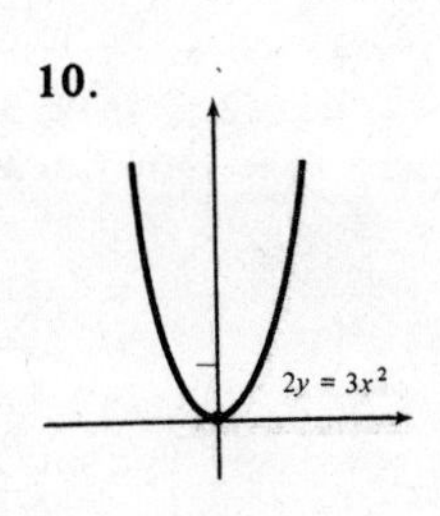

11.

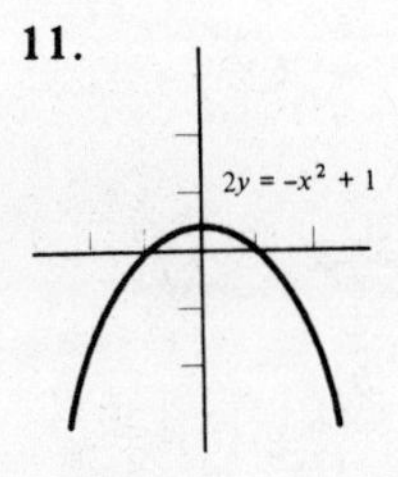

13.

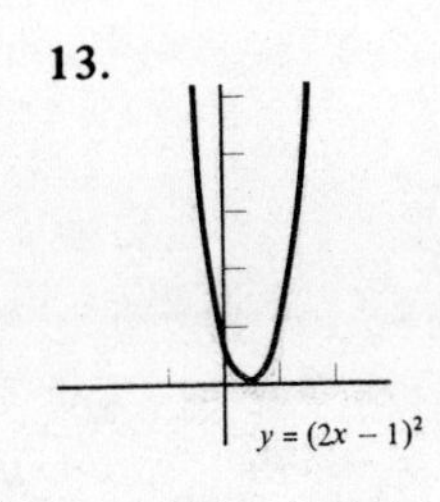

14.

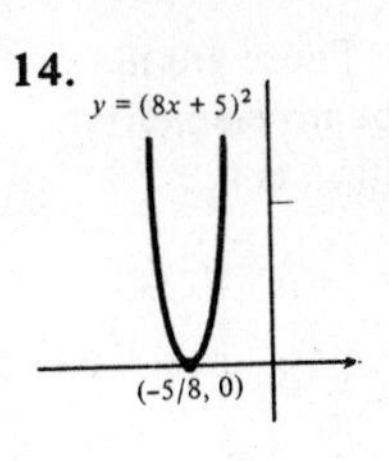

15.

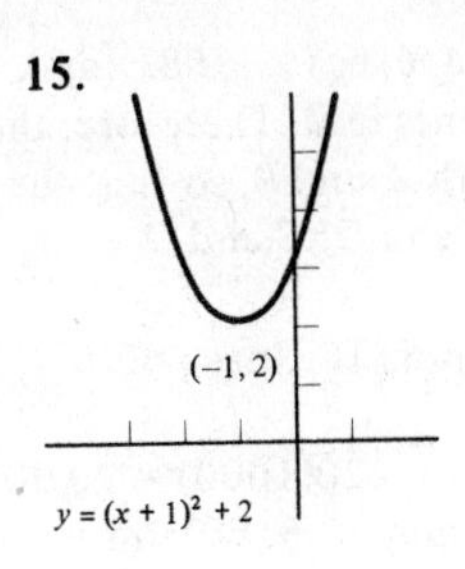

17.

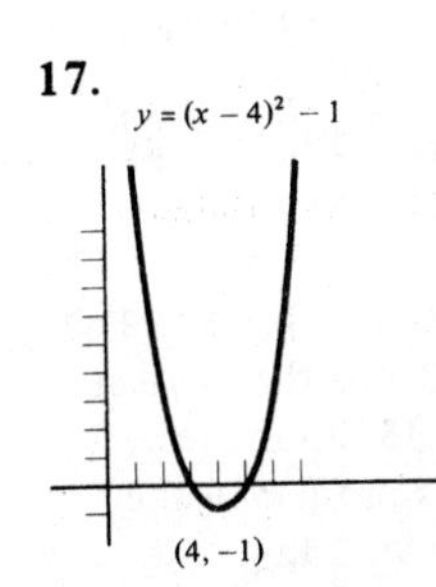

18.

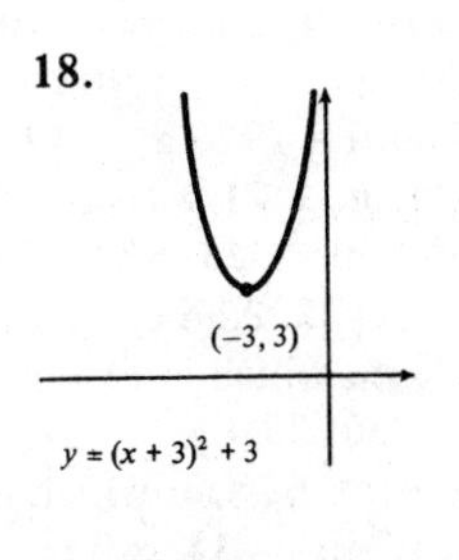

Chapter 10 Test (page 407)

1. $0, 5/2$ **2.** $1/5, -2$ **3.** $\sqrt{5}, -\sqrt{5}$ **4.** $10, -4$ **5.** $(2 + \sqrt{35})/3, (2 - \sqrt{35})/3$ **6.** $5, -2$ **7.** $-3, 1/2$ **8.** $1 + \sqrt{2}, 1 - \sqrt{2}$ **9.** $3 + \sqrt{7}, 3 - \sqrt{7}$ **10.** 4 **11.** $2, 1/3$ **12.** No real number solutions **13.** $(5 + \sqrt{13})/6, (5 - \sqrt{13})/6$

14.

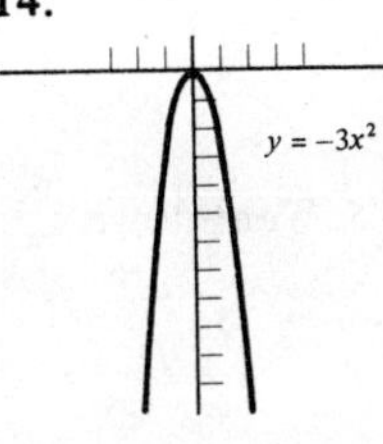

15.

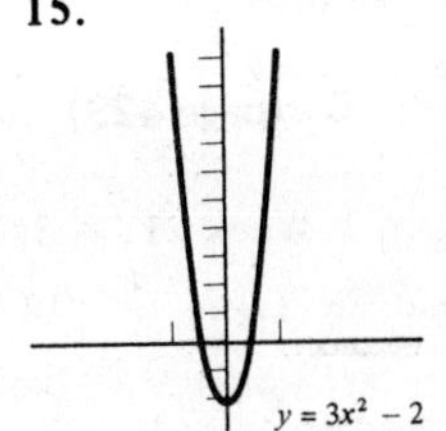

APPENDICES

Appendix A (page 415)

1. {1, 2, 3, 4, 5, 6, 7} **2.** {5, 6, 7, 8, 9} **3.** {winter, spring, summer, fall} **5.** There have been no women presidents to date, so the answer is ∅. **6.** ∅ **7.** {L} **9.** {2, 4, 6, 8, . . .} **10.** {. . . , −15, −10, −5, 0, 5, 10, 15, . . .} **11.** The sets in Exercises 9 and 10, since each contains an unlimited number of elements **13.** True **14.** False **15.** False **17.** True, since 7 is not an element of the given set **18.** False **19.** True **21.** True **22.** False, since not every element of set *D* is an element of set *A* **23.** True, since the empty set is a subset of every set **25.** True **26.** False **27.** False **29.** True **30.** False **31.** True **33.** False; there are 16 **34.** True **35.** True **37.** True **38.** False **39.** False **41.** True **42.** True **43.** False **45.** The set of elements not in set *A* is {g, h}. **46.** {b, d, f, g, h} **47.** {b, c, d, e, g, h} **49.** {a, c, e} **50.** {a, c, e} **51.** The elements common to set *A* and set *D* make up the set {d}. **53.** {a} **54.** {a, b, c, d, e, f} **55.** {a, c, d, e}

57. {a, c, e, f} **58.** {a, d, f} **59.** There are no elements in ∅. Therefore, there can be no elements in both A and ∅, so that the intersection is ∅. **61.** B and D; C and D

Appendix B (page 423)

1. 20 m = 20(1000) = 20,000 mm **2.** 760 cm **3.** 70 mm **5.** 80 mm = 80/10 = 8 cm **6.** 50 cm **7.** .32 m **9.** 5.2 km **10.** 15 km **11.** 7800 m **13.** 6000 g **14.** 15,900 g **15.** 1.92 kg = 1.92(1000) = 1920 g **17.** 8.2 kg **18.** 16.2 kg **19.** 6.94 cg **21.** 810 cg **22.** 42 mg **23.** 9000 ml **25.** 57,000 ml = 57,000/1000 = 57 ℓ **26.** .8 ℓ **27.** 2.96 cl **29.** 12 yards = 12(.9144) = 10.9728 m, which is 11.0 m rounded to the nearest tenth **30.** 29.4 m **31.** 2.0 **33.** 25 inches = 25(.0254) = .635 m, which is .6 m rounded to the nearest tenth **34.** 2.0 m **35.** 196.3 km **37.** 3813.6 g **38.** 590.2 g **39.** 49.9 kg **41.** 7.6 ℓ **42.** 12.3 ℓ **43.** 287.7 ℓ **45.** 36 m = 36(1.094) = 39.384 yards, or 39.4 rounded to the nearest tenth **46.** 87.5 yards **47.** 131.2 feet **49.** 372.8 miles **50.** 528.2 miles **51.** 680 g = 680(.00220) = 1.496 pounds, or 1.5 pounds, rounded to the nearest tenth **53.** 10.8 pounds **54.** 22.2 pounds **55.** 8.5 quarts **57.** 20.3 gallons **58.** 34.3 gallons

Appendix C (page 429)

1. \$11 **2.** Week 21 **3.** \$28 **5.** Went down **6.** Up **7.** 4; 3; 6; 3; 5; 9 **9.** 2; 1; 3; 2; 6; 5; 6; 4; 2
10.

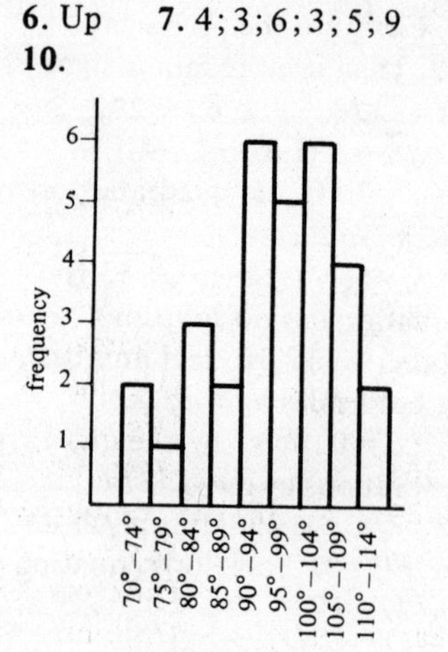

11.

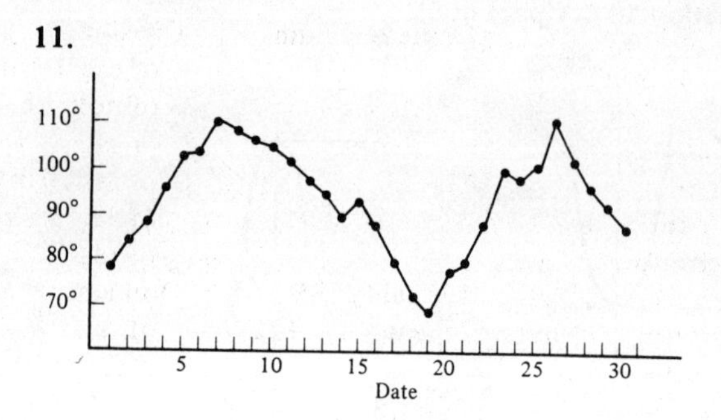

13. About 15 **14.** About 2 **15.** About 10 **17.** Birth rate **18.** About 7/8 acre **19.** Almost 4 acres **21.** 100 inches

Index

What do you think of *Introductory Algebra?*

We would appreciate it if you would take a few minutes to answer these questions. Then tear the page out, fold it, seal it, and mail it. No postage is required.

Which chapters did you cover?

(circle) 1 2 3 4 5 6 7 8 9 10 All

Which helped most?

Explanations _____ Examples _____ Exercises _____ Quick Checks _____ All _____

Does the book have enough worked out examples? Yes _____ No _____

Does the book have enough exercises? Yes _____ No _____

Did you find the Quick Checks helpful? Yes _____ No _____

Were the answers and solutions in the back of the book helpful? Yes _____ No _____

Which topics in the Appendices did you cover? _____________________________

Did you make use of the diagnostic pretest? Yes _____ No _____

Did you make use of the sample final examination? Yes _____ No _____

How was your course taught? Regular class _____ Self-paced _____

For you, was the course elective _____ required by _____________________________

Do you plan to take more mathematics courses? Yes _____ No _____

If so, which ones?

Intermediate algebra _____ Geometry _____ Math for elementary teachers _____

Business math _____ Technical math _____ Nursing (or allied health) math _____

Introduction to math (survey) _____ College algebra _____ Data processing _____

Other _____________________________

How much algebra did you have before this course?

None _____

Terms in high school (circle) 1 2 3 4

Courses in college 1 2 3

If you had algebra before, how long ago?

Last 2 years _____ 3-5 years _____ 5 years or more _____

What is your major or career goal? _____________________________ Your age? _____

We would appreciate knowing of any errors you found in the book.

What did you like most about the book?

FOLD HERE

What did you like least about the book?

College ______________________________ State ______________

FOLD HERE

NO POSTAGE NECESSARY IF MAILED IN THE UNITED STATES

BUSINESS REPLY MAIL

FIRST CLASS PERMIT NO. 282 GLENVIEW, IL.

POSTAGE WILL BE PAID BY ADDRESSEE

Scott, Foresman and Company
College Division

Attn: Lial/Miller
1900 East Lake Avenue
Glenview, Illinois 60025